More Than Curious

A Science Memoir

More Than Curious

A Science Memoir

William H. Press
The University of Texas at Austin, USA

World Scientific

NEW JERSEY • LONDON • SINGAPORE • BEIJING • SHANGHAI • HONG KONG • TAIPEI • CHENNAI

Published by

World Scientific Publishing Co. Pte. Ltd.
5 Toh Tuck Link, Singapore 596224
USA office: 27 Warren Street, Suite 401-402, Hackensack, NJ 07601
UK office: 57 Shelton Street, Covent Garden, London WC2H 9HE

Library of Congress Control Number: 2024948116

British Library Cataloguing-in-Publication Data
A catalogue record for this book is available from the British Library.

http://morethancurious.com
DOI 10.5281/zenodo.10094781 (repository DOI)

First hardcover edition: January 2025

Version 1.1w

Names and events in this work are true insofar as the author's memory and records allow, and except for a small number of pseudonyms adopted. The author welcomes corrections on matters of fact (as distinct from opinion, interpretation, as-remembered, or as-then-thought) and will attempt to include these in future releases.

Cover image credit: NASA/CXC/SAO/Ákos Bogdán; Infrared: NASA/ESA/CSA/STScI; Image Processing: NASA/CXC/SAO/L. Frattare & K. Arcand. The cover image features the galaxy cluster Abell 2744, taken by the James Webb Space Telescope.

MORE THAN CURIOUS
A Science Memoir

ISBN 978-981-98-0176-3 (hardcover)
ISBN 978-981-98-0300-2 (paperback)
ISBN 978-981-98-0177-0 (ebook for institutions)
ISBN 978-981-98-0178-7 (ebook for individuals)

For any available supplementary material, please visit
https://www.worldscientific.com/worldscibooks/10.1142/14068#t=suppl

Praise for *More Than Curious*

"Press has written a riveting book, distinctive in style and theme. He is a top-rank scientist, having attained status and respect as a member of the U.S. academic elite. But unlike most such people, he finds 'ivory towers' constricting and has relished forays into the wider policy and political arena. He addresses important issues, and the narrative is laced and enlivened by intrigues and personality conflicts, recounted with unusual frankness. This enlightening and brilliantly-written book deserves wide readership."

—Martin Rees, University of Cambridge and Astronomer Royal

"In *More Than Curious*, bear witness to the life and times of one of the last of the Citizen Scientists. Bill Press is an astrophysicist like no other, who gifted his brilliance to the work of other scientists, and to multiple U.S. Presidents—through the Cold War and beyond. A natural storyteller, Press's crystalline and lucid accounts of people, places, and things, leave you feeling like a front-row witness to his extraordinary journey."

—Neil deGrasse Tyson, American Museum of Natural History

"This is a superbly interesting book, with personal insights, sometimes funny, often deep, into a half century of science and science's interplay with politics and public affairs. It has been shaped by the author's intense curiosity and creativity and his amazingly broad interests and expertise. Once I started reading, it was near impossible to put down."

—Kip S. Thorne, Caltech, Nobel Laureate

"An epic tale of a scientific career told through unique behind-the-scenes portraits of the people, organizations and issues that defined late 20th century physics and things to which it led."

—Stephen Wolfram, Wolfram Research, Inc.

"As a memoir, *More Than Curious* is a picaresque exploration of high-level bureaucratic and academic politics. Pick it up anywhere, and within ten pages you'll find a sharp (sometimes snarky) insight into some event or public figure that Bill encountered in a long and productive career."

—Stewart Baker, Steptoe & Johnson, former NSA General Counsel

"This book is a gift: William H. Press places his scientific work in the context of the human stories and social and political institutions that swirled around him, writing with elegance and closely-observed detail. The memoir is also an ethnography and history of science in our lifetimes, a rare look not only at a scientific sensibility at work, but of how the scientific establishment relates to the world of politics, economics, and corporate power. I loved it."

—Sherry Turkle, MIT, author most recently of *Alone Together*, *Reclaiming Conversation*, and *The Empathy Diaries*.

"In this well written and vivid memoir, [Press] gives first-hand accounts of hobnobbing with such scientific luminaries as Richard Feynman, Edward Teller, and Kip Thorne, as well as with political leaders including Jimmy Carter and Barack Obama. Throughout, Press conveys the thrill of science and the obligations of its practitioners."

—Alan Lightman, MIT, author of *Einstein's Dreams*

"Press shows himself a perceptive raconteur. His engaging vignettes create a fifty-year mosaic of physics, physicists, scientific institutions, and science policy."

—Steven Koonin, former Caltech provost and U.S. Under Secretary of Energy

"Press, one of the most consequential scientists of his generation, takes the reader on a tour of his life in science through seventy candid, vivid and highly-interesting vignettes… You won't be able to put it down."

—Michael S. Turner, University of Chicago, former Assistant Director, National Science Foundation

"Press represents the best of academic and public service excellence in America. His scientific expertise and contribution to the nation's security underwrite a remarkable life story. The lessons apply to all who aspire for meaning and consequence in their personal and professional lives."

—General Norton A. Schwartz (Ret.), former Chief of Staff, United States Air Force

"[His] book on algorithms is the most dog-eared volume on my bookshelf; this latest one, by the same William H. Press, is yet another wonderful surprise, and I suspect that it too will get more than its share of wear and tear, both in my hands and in those of many others."

—David E. Shaw, D. E. Shaw Research and Columbia University

Foreword by Adam G. Riess

Our understanding of the universe is built on beautiful ideas, supported by often messy and inconclusive data. Ideas are wonderful, but not all are correct; data provides the guardrails that keep science on the path to truth. I was fortunate to meet Bill Press during my first year pursuing a Ph.D. in astrophysics in 1993 at the Harvard Center for Astrophysics. He taught me a potent cocktail for understanding data: computer algorithms, Bayesian inference, and common sense—all the more remarkable since Bill is a theorist. Most scientists encounter these ideas only if they pursue an experimental or observational path.

My first encounter with data as an undergraduate, in Physics Junior Lab at MIT, was embarrassing. My lab partner and I replicated the Stern-Gerlach experiment and plotted the data and error bars against the expected linear trend. We beamed with pride as we presented our results in the oral exam. Our professor, Steven Meyer, whose research focused on measuring the Cosmic Microwave Background, was not impressed.

"Why does the line go through all of the error bars?" he asked. Was this a trick question? Wasn't it obvious that we had done an exceptionally good job? Yes, he explained, but too good, in fact! A healthy third of the data points and their errors should miss the line according to the mathematical definition of uncertainty (for a Gaussian distribution) and some should even miss by two error lengths. It became clear to me that there was more to data reduction and analysis than I had thought. Those error bars were not, as I had thought, mere decorations on the plotted data points.

In my first year of graduate school, I began a project with Robert Kirshner, an expert in supernovae. He proposed that we try to improve the standardization of Type Ia supernovae to make them better cosmic mile markers. Bill Press had developed some theoretical formalism that Bob thought would be useful. The problem was important, since the lack of good distance measurements was holding back our understanding of how the universe was expanding.

Bill Press was known to me not only as a professor in the theory group at Harvard, but also as the lead author of *Numerical Recipes*, a book considered to be the bible of computer data analysis. He was reputed to have a dry sense of humor and was known to not suffer fools gladly—naturally concerning to me as a first-year graduate student. But he also had a reputation for being able to dive usefully into the deep analysis of other people's data without spending the thousands of hours that its observers spent in acquiring it. This made a collaboration with Kirshner natural.

With George Rybicki, Bill had just written a widely cited paper on reconstruction techniques for noisy, irregularly sampled data. This provided an initial direction for our work. I quickly learned how to treat data like a bunch of shady characters at a bar who had just witnessed a crime. Each had a story to tell—some wrong, some right—but with the right questions and cross-examinations, one might hope to get an understanding of what had occurred. Bill had spent time in these figurative watering holes and knew these characters; he knew tricks for spotting liars or hucksters at first glance.

Bill first assigned me the task of analyzing a subset of supernova data in a certain way, and then showing him the results. When I brought these to him, I noticed that his focus was not on the data values themselves, but rather on the relationship between data and quantified uncertainty. It was my Junior Lab experience all over again! You could only understand data if you also understood its errors. You needed to have a theoretical or statistical model of your underlying observation. The observed scatter or jitter in your data had to be "Goldilocks"—just right: If the observed scatter was larger than your model predicted, then your model was missing a source of error; if it was smaller, then you didn't understand your own measurements, or the data had been in some way tortured. Either way, until you got to Goldilocks, you were going to produce a wrong answer or have the wrong level of confidence about what it says. This first task was my initial formative immersion in that figurative crime scene. Did the whole story fit? Who was lying? Whose alibi was too good to be true? Were there more clues available? And, could there be a dog that didn't bark?

Bill was an early proponent of the use of Bayesian analysis in astrophysics. All scientists make use of a general kind of prior understanding of the phenomenon they are studying—if only in choosing where to spend care and time in their observations and analyses. Bayesian methods provide a more transparent and rigorous way to incorporate these prior intuitions, while still always letting the data be the ultimate decider. In his book Numerical Recipes, he put it this way: "Either you become a Bayesian or else you must live in a world with no general calculus of inference." In the memoir you are about to read, he compares Bayes to Constantine, the emperor who converted the Roman

Empire to Christianity, writing that in the course of his career all astronomers either became Bayesians, or else they became suitable only for feeding to lions.

Another learning experience for me during graduate school was Bill's seminar on the Hubble Constant—the rate at which distant galaxies move away from us. The various measurements of it by different observers, each claiming a plausible estimated error, were all over the map. Bill formulated a statistical model in which some of those observations would be considered just plain wrong— and their claimed accuracy would be ignored—while others would be taken as legitimate, even if their errors were found to be overly optimistic. Crucially, he didn't decide arbitrarily which observations were in which category. Rather, using Bayes, he let the collective data itself decide. The value produced by this (what would today be called a) meta-analysis was 74 km/s/Mpc. Amusingly, this is very close to the value my colleagues and I have recently measured with the advantage of the Hubble and James Webb Space Telescopes twenty-five years later.

Indeed, as I now write, the value of the Hubble Constant has again become an exciting topic, a problem that we call the "Hubble Tension," though the range of possibility has narrowed sharply. The most generally accepted standard model of the universe calibrated by the afterglow from the Big Bang predicts a value 67-68 (in units of kilometers per second per megaparsec). Yet, that model harbors a great deal of ignorance. The nature of dark matter and dark energy which make up 96% of the standard model are unknown so we can only assume they are their simplest forms. Intriguingly, the community of observers who directly measure the Hubble constant consistently find a larger value in the range of 70-75. We have spent the better part of a decade making sure we understand our errors (as Bill might have suggested when I was a graduate student) and I think we do. At least no one has found a flaw yet after much intense scrutiny. The more exciting possibility is that we have discovered something new and exciting about the universe. Time (and new data and its analysis) will tell.

I recall now that after hearing about Bill's Hubble meta-analysis, I found myself sitting at a Red Sox game with a fellow graduate student discussing his research problem. He was measuring the movement of water vapor molecular clouds orbiting a black hole in a nearby galaxy. He was trying to measure their mean speed but could not simply measure their average displacement between views because some clouds dissipated and others materialized between observations, confusing the associations. I suggested Bill's Bayesian framework for the Hubble constant, with modifications. It worked amazingly well in this situation. No Bayesian analysis could save the Red Sox in those years, however.

One day, as our work progressed, I proudly showed Bill a set of corrections I had figured out to apply to the data. He asked me how I was sure it was an improvement. It makes the data sharper, I answered. But maybe just by chance, he suggested, the way monkeys banging away at typewriters might occasionally produce a meaningful sentence. He suggested that I do a computer "Monte Carlo" simulation that randomized the order in which I applied the corrections and see if dumb luck could succeed as well as the real thing. This way of thinking was new to me, but I quickly got the hang of it.

Those years were also a great time to learn how to work with large datasets. Astronomy as a field had recently transitioned from the use of photographic plates, which favored a more qualitative approach, to the use of CCDs, which offered digital data and lots of it. Our new supernova data algorithm (later called the Multi-color Light Curve Shape Method, or MLCS) became a key tool for improving the measurement of the expansion rate of the universe at that time. Bill moved on to other problems and to do other things, as he vividly describes in this memoir. I continued to measure supernovae and extended the work to measuring the expansion history of the universe.

As a postdoc at Berkeley in 1997, I was part of the High-Z Supernova Team, and I was in charge of the team's data analysis. We discovered that the expansion rate of the universe was actually *accelerating*, propelled by a mysterious new component now called Dark Energy. I had no way of knowing then that team lead Brian Schmidt and I, along with Saul Perlmutter (as leader of a competing team), would eventually win a Nobel Prize for this discovery. At the time, Brian asked me the crucial question, "How confident are we in this result?" That was the moment when my future as a scientist depended on using some of the lessons I had learned from Bill. One could think of them as merely a useful bag of tricks. But, in combination with common sense and the hands-on experience of our team, these were superpowers.

Bill's life in science and public affairs did not stop with his advising me as a graduate student in those years. As you read what follows, I think you will find his other experiences as interesting, profound, and amusing, as I have.

Adam G. Riess
August, 2024

Author's Preface

A combination of chance family origin, some amount of ability, and unusual luck put me in places and times to observe at close range some significant events and notable personalities in science, and to see and record some noteworthy interactions of science and scientists with politics and public affairs. Along the way, I got to know not just many in my own generation of scientists, but also an older generation whose famous discoveries went back in some cases to the 1930s. Later, I worked with (or for) not just scientists, but also technology capitalists and billionaires, nuclear weapons designers, admirals and generals, and political leaders including two U.S. presidents. Reader, be assured: I was never close to winning any big prize for my own work. But my Ph.D. advisor, Kip, and my Ph.D. student, Adam, each did win Nobel Prizes for theirs—for discoveries that I had in a small way participated in. (Adam has graciously provided a forward to this volume, which you may have just read.)

Over time, I came to think that parts of this story might be interesting to other scientists, or to present or future historians of science, or perhaps even to a general reader or two. The enforced isolation of the COVID pandemic in 2020 left me with the time and motivation to extract a narrative from my files and memory, and turn it into something like a memoir. In fact I did this twice: You are reading the shorter, more circumspect version, most directly the story of the making of a scientist, the body of his and others' scientific work, and some decades of his life in the public realm. I am pleased to make this version available to all who are interested.

A longer, less polished, more personal version, written for my family, is for now unpublishable. It has more stories about sometimes difficult people, and it is freer with my unvarnished judgements about them—not just about their science. But too many of them are alive, and too many of these are still my friends, implying that these are stories not ready to be told. My life apart from science (and apart from my journey to becoming a scientist) is also described more fully in the longer, family version. A time-locked archive will someday make that version available to others. But not soon. In the meantime, I hope the stories in this volume will amuse or instruct.

I decided to publish this memoir openly under a (free) Creative Commons license after discussions with several publishers. Regrettably, there seems today to be no significant commercial market for scientific memoirs of a kind that was once typified by the Sloan Foundation's Series of Scientific Autobiographies. At the time, I found all of the Sloan volumes fascinating—even, or especially, the badly written ones. (A badly written memoir is sometimes more revealing than a slick and polished one.) A market does still exist for streamlined scientist-memoirs with exciting, dramatic arcs ("my path to a Nobel Prize!"), but that isn't *my* life, unfortunately. My anecdotes are picaresque, many, and I hope interesting or funny. Friends in the book business advised me that an academic or specialty press could easily be found to publish this work. But that would yield a tiny readership as compared to open publishing. Physicist Joe Polchinski came to a similar conclusion, posting his 2017 memoir on the ArXiv (albeit under tragic circumstances); Charlie Kennel entrusted his beautifully written 2023 memoir to the University of California's public eScholarship repository; and there are other examples of this trend. I like CERN's Zenodo repository for its promise of maintaining availability for twenty years or the lifetime of the European Union, whichever is *longer*. No commercial publisher offers that deal for keeping a book in print!

Memoir is, by definition, as remembered. From my pre-publication circulation of this work, I have already learned that others remember some events herein quite differently from me. I have corrected, and will try to correct in future updates, things objectively untrue that I wrongly state as fact, as opposed to memory, interpretation, or what I thought or knew at the time. For disputes in these latter categories, I enthusiastically invite participants to record their own versions in blogs, articles, podcasts, books, or social media, and to archive such recollections in one of the several permanent online scientific repositories. Then, future historians (if they actually care!) can try to sort things out.

November, 2023

Table of Contents

1. DNA

In July, 2010, I was part of a group of scientists advising the federal government on what was soon to be called personal genetics, the ability for individual consumers to have their genomes easily sequenced. One of the presentations made to our group was by Anne Wojcicki, the founder of a start-up company called 23andMe. Anne was married to Sergey Brin, the co-founder of Google, so she was one of the richest people in the world.

I have barely any recollection of Anne's talk, but after the talk, she handed out little cards, individually signed by her, which entitled the recipient to have their genomes sequenced for free. At the time, I was a member of President Obama's Presidential Council of Advisors on Science and Technology and subject to various draconian rules regarding bribery and the accepting of gifts, so I didn't take a card. It is possible that, had I taken one and kept it, it might be more valuable today as an autograph with a unique provenance. Sometime later, however, I did order (at retail price) the 23andMe kit. A couple of weeks after spitting into the little plastic vial, I got my results.

Today, DNA ancestry is a well-developed business, and competing companies have big, proprietary databases. In 2010, there existed very little such data. 23andMe could put my saliva through their machine and read my genome. But they didn't, then, have much "ground truth" on verified ancestry. The best they could do with their limited database was to find the fifty customers whose genomes were closest to mine, and then send me the list of those people's self-reported ancestries. I was on my own to draw any conclusions.

My list, when I received it, went like this: 1. Ashkenazi. 2. Ashkenazi. 3. Ashkenazi. And so forth all the way down to: 48. Ashkenazi. 49. Ashkenazi. And then the ringer: "50. Eastern European, perhaps Jewish." I have some news for you, person number 50: Not only are you Jewish, you are doubtless Ashkenazi.

If genetics were truly destiny, my memoir could end here, since I am genetically nearly identical to an inbred population of a few million Ashkenazi Jews concentrated in the United States, Israel, Russia, South Africa, Australia, and (before the Holocaust) Germany and Poland. If genetics were determinative, we Ashkenazim would all have similar lives, like the triplets separated at birth who (if you believe the internet) all became firemen.

That is nonsense, of course. Much more important than genetics is environment, a catch-all word that encompasses our physical and biological surround, the foods we eat, the air we breathe, the microbial ecology of our guts, the influences of our family, our pets, the people we meet, our culture, profession or work, nation, and historical milieu. Those three billion A, C, G, and T nucleotides in our genome might seem to contain a lot of information, but their content is miniscule compared to the quantity individually delivered to my inbred Ashkenazi ancestors and me by lifetimes of environmental happenstance. That reality luckily leaves me with different stories to tell. And I didn't become a fireman.

* * *

Between roughly 1880 and 1924, after which a wave of restrictive immigration laws was passed, something like two million Jews immigrated to the United States from Eastern Europe. My maternal ancestors arrived toward the beginning of this wave; my paternal ancestors near the end.

On the maternal side, I am named for my mother's father, known to me as William Henry Kallick. We grandchildren called him Poppy. Poppy's name was just one of the big and small lies that he told about himself. This was a small one: His name seems actually to have been William *Harry* Kallick. Henry, substituted for Harry, seemed less Jewish. Poppy was born the day after Christmas, December 26, 1890, at 124 Kenilworth Street, Philadelphia. His parents, Lewis and Eva Kallick emigrated from Odessa, Russia (now Ukraine), to Philadelphia a few years before. If not exactly a lie, the name Kallick was not the original name, which may have been Kaliski, a Polish surname and geographical district. According to Poppy, his father was employed as a taster in Seagram's distillery in Philadelphia. He "started swallowing" and became an alcoholic. He died saving people from a burning building when Poppy was a child. Since this report came from Poppy, it is probably not true.

In truth, even the year of William Kallick's birth is not entirely certain. His mother would have been 50 when he was born—very unlikely. In 1936 he went before a judge and swore under penalty of perjury that his birth records had been lost in a fire, and that he was born on December 26, 1894, four years later than we now think. That would have made his mother, Eva, 54 when she gave birth—virtually impossible in those times. My maternal grandmother, Mary Kallick later propagated the story that Poppy lowered his age the better to attract younger women with whom he had affairs.

Poppy was five foot six (not considered short in his generation) and weighed 150 pounds. A photograph of him on the beach at Coney Island in the early 1940s shows a virile man in a skimpy swim suit who looks to be younger than 50 years. He seems proud to be posing for the picture. His is the well-muscled appearance of the man who gets the girl in the old back-of-comic-book advertisements, the one who takes her away from the ninety-seven-pound weaklings—those would be the men on my father's side of the family. In 1955, Poppy tried to change his birthdate back by six years, to December 26, 1888, so as to qualify for social security payments. The attempt failed.

When I was growing up, Poppy was a proofreader at *The New York Times,* and a proud member of the International Typographical Union, Local #6. Proofreading was the highest and most intellectual of the "craft," as opposed to "editorial," positions in the printing trade. Poppy had worked his way up at several newspapers in Philadelphia, St. Louis, and New York, as a young man setting type by hand, and later as a linotype operator. He graduated from the trade-school equivalent of high school in Philadelphia and loved everything about the printed word—albeit grammar and punctuation more than content. In no way was he an intellectual. His daily reading centered on the tabloid *New York Daily News* (not the highbrow *New York Times*) and the *Police Gazette,* a racy tabloid men's magazine.

Poppy set my mother on the path to college and to a professional career as a schoolteacher. She adored him. As an adolescent and adult, it was her father's version of the truth that she believed when, as happened more and more often, her mother would insinuate other interpretations for his behavior. "Your father has the moral fiber of a piss-ant," was one of Mary's favorite aphorisms. After one unexplained absence of several days, William returned home bearing a gift for Mary: several items of expensive lingerie—clean, but used, and monogrammed with another woman's initials. He said that they were a bargain because they fell off the back of a truck.

Mary Kallick, my grandmother whom we called Nana, was three years older than William. She was born Hinda Amdur, probably in what is now Poland, her name changed to Helen Ender when (as a one-year-old baby) she emigrated with her parents, Meyer and Esther, to Philadelphia. In elementary school she became Mary Helen Enders. Names of immigrants could be fluid in this way.

William and Mary were married in Yonkers, New York, on December 30, 1918. Their first child, a son, died in 1919 shortly after his birth.

Their second child, Lewis, was born in 1920. By 1925, when my mother was born, the family had moved to St. Louis. My mother was named Billie after her father. Feminine versions of boys' names were in fashion for baby girls in St. Louis at the time. Lewis died of scarlet fever in 1928 at the age of eight. My mother was three at the time. "Why did it have to be the boy?" Mary is supposed to have said.

Mary, who died in 1958, never seemed to recover from the loss of her two sons. She told my mother that their deaths were God's punishment, but never explained for what. My mother's younger sister Evelyn was born as the result of an accidental pregnancy in 1931. The family had moved to Brooklyn, New York, to an apartment at 705 Prospect Place (later a somewhat better apartment at 1011 Ocean Avenue). I knew Mary (Nana), as a dour woman, the grandparent who always said no. William (Poppy) was the grandparent who could be counted on for yes.

After Nana died, Poppy, nearing seventy and retired from *The Times* with a small union pension, lived alone for a time in his rent-controlled apartment on Ocean Avenue. In a stroke of good luck, he somehow connected with a distant Philadelphia cousin, a widow, Fannie. She soon moved to Brooklyn, and they began living together. Conveniently, as cousins, they shared the last name Kallick. In those times it was disreputable for an unmarried couple to be living together—even when both were septuagenarians!

In April, 1967, when I was a sophomore at Harvard, Poppy took me to Philadelphia to show me the old neighborhood where he grew up. We walked on Front Street between Kenilworth Street and the Delaware River. Poppy said he had worked as a stevedore on the docks. (A couple of years after my visit, this neighborhood and the river were sliced apart by the construction of the Delaware Expressway, Interstate 95.) He located the street corners where his gang had fought and drank beer, and where he and his chums sang close-harmony barbershop quartet songs for nickels from passersby at the corner of Bainbridge and Second, outside of Callahan's Store. As improbable as the latter sounds (now and then), I can attest to the fact that Poppy knew by heart the lyrics—and tenor and baritone parts—for dozens of popular songs from the period roughly 1890 to 1910; and that he carried in his wallet an expired membership card for S.P.E.B.S.Q.S.A., the Society for Preservation and Encouragement of Barbershop Quartet Singing in America, an organization well known to aficionados.

In the 1970s, Fannie's health deteriorated, and she moved back to Philadelphia to be with her two daughters. Poppy gave up his Brooklyn apartment and moved to Belmont, Massachusetts to live with my parents. He died in 1978 after a debilitating series of strokes. Soon afterwards, Fannie's younger daughter Sylvia, then in her mid-60s, wrote to my mother: "Dear Billie, at last we are able to tell you the truth about *our father*, William Kallick."

Fannie, the distant cousin who had moved in with my grandfather after my grandmother died, was no cousin at all. She was my grandfather's first—in fact, only legal—wife. The true story seems to be this: Mary Enders and Fannie Weiss were best friends who worked in a textile factory, sewing. Neither had schooling beyond sixth grade. Around 1910, when both were in their early 20s, they became friendly with a handsome young printer, William Harry Kallick. Before long, Fannie was pregnant. A marriage occurred in 1911, and baby Ada Kallick was born soon after that. A second daughter, Sylvia, followed two years later. One way or another, Mary Enders was still in the picture. Her turn to get pregnant by William came in 1918. Notwithstanding that William was married to Fannie, he and Mary ran off together and were married in Yonkers in December, 1918. Their baby was born some months later and died soon after. Little wonder that, with the death of both sons, Mary might feel that God was punishing her.

There is no record of any divorce of William and Fannie. My grandfather was a bigamist. There is no record in the 1900 or 1910 U.S. census of a family named Kallick (or anything similar) living at 124 Kenilworth Street, the address that Poppy always gave as his birthplace, and to which he led me when we visited Philadelphia together. It may be that he was not born in the United States at all: In an archived U.S. citizenship application dated December 3, 1908, one Harry William Kallick, a printer, age 19, swears that he was born in Odessa, Russia [sic], on October 2, 1889, and that he emigrated to the United States in July, 1892. This could even be the truth, or a version of the truth. Or maybe it was identity theft, someone conveniently using my grandfather's identity, with or without his knowledge. Or maybe he stole someone else's identity. We will never know what parts of Poppy's life story are fabrications. The undeniable fact that I am my grandfather's grandson is one that the reader might want to keep in mind.

2. New Americans

In 1915, Russian army positions collapsed in the face of a massive offensive by the German Kaiser, and the armies of Czar Nicholas retreated in disorder far back into Russia's defensive depth. In Minsk, still barely under Russian control, my paternal grandparents Sholom Press, who was 22, and his fiancée Dvera Steingolz, 21, decided to secretly emigrate to America. They were both escaping and eloping: Dvera's parents disapproved of the religiously non-observant Sholom; and he was in danger of being drafted into the Czar's army.

Sholom and Dvera—in America they became Solomon (or Sol) and Dora—were urban Jews of, in his case, a low commercial stratum. He had completed high school, a rarity for Jews then, and worked in a bank. Dora was a milliner. The two obtained forged papers and traveled on the Trans-Siberian Railway to the large Russian expatriate colony in Harbin, China. Dora found temporary work making hats, while Solomon sought passage to the United States. The plan was for Sol, once in New York, to locate his older brother Barnett and then send for Dora. She accompanied him from Harbin as far as Yokohama, where they were officially married on December 25. Sol sailed from Yokohama on January 23, 1916, on the steamship *Manila Maru*.

As unlikely as it seems, events unfolded exactly according to their plan. Dora, by then well along in pregnancy, sailed on the *Yokohama Maru* on June 19, arriving in Seattle on July 7, from where she traveled by train to New York. Their son Samuel (my Uncle Sam) was born in Brooklyn on October 20. The accepted dates of Dora's pregnancy are just barely biologically possible. But, ninety years later, after the revelations about William Kallick on my mother's side of the family, how could we of the younger generation not wonder whether everything about our grandmother's first pregnancy was, well, kosher? We convinced my cousin Fred (Uncle Sam's only child) to have his DNA sequenced. The results were unequivocally ordinary: Cousin Fred's grandfather was definitely Solomon Press. Sam may have been conceived the very night before Sol's ship sailed.

When I knew him as Grandpa, his business card said, "Sol Press, Cash Buyer." He sometimes took me with him on his business rounds. It went like this: We systematically walked the streets of a chosen, generally run-down, neighborhood in Brooklyn. Grandpa somehow selected a small store, and we went inside. He was a small, nondescript man, who attracted little attention while scouting the place. A fraction of the time, he approached the counter and asked politely, in his thick Russian/Yiddish accent, if the person behind the counter was the owner. If the answer was yes, Sol presented his business card and took out of his pocket a large roll of hundred-dollar bills. His was an immediate offer to buy the entire establishment, including the unexpired lease, the fixtures, and the stock of merchandise—all for cash then and there. Mostly his offers were refused, sometimes angrily, but occasionally the owner welcomed my grandfather as an angel sent from heaven. These were shop owners who knew that their business was failing, but could see no path to a new start.

Grandpa's business depended on his sizing up in a couple of minutes the cash value of the store—and also the actual amount that a distressed owner might be willing to take. The few times I witnessed this, in the early 1960s, the amounts were in the few thousand dollars. The back end of Grandpa's business was to hire an auctioneer, schedule an auction for the fixtures, find other similar stores to whom he could sell the merchandise (at a suitable discount from wholesale), list the property as a sub-let, and so forth. The part I liked was when, in the store, he closed a deal. He would say to me, "Go look around Billy. Take anything you want." Because, now, he owned it all.

According to my father, there were times during the Great Depression when the family was forced to operate a store for weeks or months before he could get an acceptable price for selling it. My father remembers soda-jerking as a kid in a candy store that Sol bought mainly for the fixtures. There was a ladies' lingerie store that proved hard to unload. In better times, Sol would put together somewhat larger deals, collecting money from other investors. Whether one should view his business as capitalism at its purest or as pure economic bottom-feeding is a matter of perspective—even assuming that the two alternatives are actually different.

Sol and Dora had three sons: Sam, already mentioned; Paul, who followed less than two years later; and my father, Frank, who was born in 1924, six and eight years younger than his two brothers. Theirs was very much an Old-World family, with children obedient to their parents. The language of the parents in the home was Yiddish. The newspaper they subscribed to was the *Daily Forward*, printed in Yiddish. Dora read every page. The children understood Yiddish fluently but were not allowed to speak it. They were allowed to speak only English, so that they would be Americans.

Unlike for my maternal grandparents, it would never have occurred to my paternal grandparents to try to pass as non-Jews. They lived their whole lives as Russian-Jewish immigrants in Jewish neighborhoods in Brooklyn. Their determination that their three sons be 100% American was fierce, however. There was no religious observance in the family, except for one anomaly: All three boys went to Hebrew school (summer camp in Frank's case) for just long enough to qualify to be bar mitzvahed at age thirteen. The closest my father came to adolescent rebellion was that he became religiously observant—to the family's disapproval—for some months before his bar mitzvah; but it soon wore off.

Sol's politics, to the extent that they were visible, were somewhere on the socialist left, generally aligned with the *Daily Forward* as pro-union, democratic-socialist, and anti-Bolshevik. The radio at home was always tuned to WEVD—for Eugene Victor Debs—which broadcast a Yiddish hour on Sunday mornings. Years later, when my father was nominated by President Kennedy for a White House advisory committee, the FBI found a photograph of Sol's car in their undercover files, a green Chevy in a Communist-sponsored parade in New York in the 1940s. My father insisted that his father must have loaned the car to a friend.

Family obligation went both ways. Years later, when Frank was a young professor at Caltech, he got a phone call from Sam. "I've decided to run for Congress, Frank, and I need five thousand dollars for my campaign." This was not a small amount—probably three months of Frank's Caltech salary at the time. There was nothing about Sam's solo law practice that evidenced any kind of political connection. My father sent him a check. We never heard another word about his campaign, nor did he ever pay back any of the money. What is likely is that he simply needed the money, and this was a face-saving way to ask his brother for it—a request that, in that family, could not be refused. Many years later it occurred to me that the running-for-Congress story was known to us only via Frank. It is possible that he invented it for my mother, who did not share Frank's absolute sense of family obligation.

Billie first noticed Frank in their high-school honors history class, when she was a junior, he a senior. She had blossomed in high school. She got straight A's. She had the figure of a Betty Grable pin-up girl. Her hair was naturally blonde (or at least a very, very light brown). Thanks to her father, her grammar and diction were excellent, and she often spoke up in class. Frank was the smartest boy in class, she thought. He rarely spoke up, but, when he did, his thinking struck her as intellectual and insightful.

Frank graduated from high school in June, 1941, and started college at the free City College of New York that September. With Pearl Harbor came the draft, but Frank was initially too young to serve—his older brothers both served in the Army—and, as it turned out when he did register, also too nearsighted. He was classified as 4F, "not acceptable for military service." At CCNY, he majored in physics. In his first year, he joined the Physics Club and, by a mailed letter, invited my mother and her friend Elaine to the Physics Club dance—if one can imagine such a thing! Billie herself was dating boys who had been drafted and were about to go to war.

They dated throughout their college years. Billie won a full-support scholarship to NYU, a private university that was considered, intellectually and socially, a cut above CCNY. Dora, Frank's mother, disapproved of Billie not just because future mothers-in-law always disapprove of their future daughters-in-law, but specifically because she didn't believe that Billie was actually Jewish. Blonde hair, NYU, family with a non-Jewish name—Dora was certain that Billie was a *shiksa,* some kind of Delilah intent on robbing Frank of his strength. Dora's facts were wrong, but her intuition was not completely faulty. A part of Frank's attraction to Billie was that she was so *shiksa*-like, yet (theoretically) Jewish. He could have his assimilative cake and eat it, too. Frank was programmed from birth to break out of the ghetto, and Billie, he thought, would help him succeed.

3. Lamont

Everything was accelerated because of the war. City College was pumping undergraduates through its normal four-year curriculum in three years or less, with Frank as a willing beneficiary. He finished his undergraduate studies late in 1944 and was quickly accepted for graduate school in physics at Columbia. The thought of leaving New York City never crossed his or Billie's minds. Promising graduate students in the sciences could earn their keep as teaching assistants in undergraduate courses, meaning that Frank's brothers would not need to continue to support him, an important consideration. After physics graduate students finished their course work (one year, or two at most), they could be promoted to research assistant. But, for this, they had to be accepted for Ph.D. thesis work by a specific professor. When Billie and Frank were married in June, 1946, Frank had finished his master's degree and was looking for a Ph.D. advisor. Billie found work as a nursery school teacher.

In 1947, Columbia University recruited as a professor the prominent oceanographer and geophysicist Maurice "Doc" Ewing. Ewing, a larger-than-life Texan, grew up in the tiny town of Lockney. His ascent to the pantheon of great twentieth century geoscientists was made possible, indirectly, by the sensational murder in 1900 of William Marsh Rice, the Texas real estate, lumber, cotton, and railroad magnate, whose fortune endowed the new tuition-free William Marsh Rice Institute, later known as Rice University. Ewing was in Rice's tenth undergraduate class.

He received his bachelor's degree in 1926 and worked as an instructor while pursuing his Ph.D., awarded in 1931. After that, he moved to Lehigh University, in Pennsylvania, and worked up the academic ladder. In support of the war effort, Doc (as he was now called) moved to the Woods Hole Oceanographic Institution in Cape Cod, Massachusetts, a hub of wartime oceanographic support to the U.S. Navy. Woods Hole was where Doc became known not just to senior oceanographers, but also to Navy admirals and the several New York financiers who were themselves involved in the war effort. Some combination of things in Doc—not least his expansive Texas manner and rough country charm—made him successful at making things happen.

Prior to Doc's arrival at Columbia, Frank showed no particular interest in oceanography, geology, or the newer field of what was called geophysics. As a physics student, he was mathematically advanced and almost entirely theoretical. Columbia was the nexus of a golden age in experimental physics after the war, but Frank doesn't appear to have ever set foot in an experimental laboratory. He was a city boy to the core, thin, pale, scholarly, with thick glasses. The closest he ever came to an ocean was the beach at Coney Island—and he was never a strong swimmer.

Still, for some reason, opportunity or ambition, Frank beat a path to Doc Ewing's door and was accepted as Doc's first graduate student. Whether as a probationary test or as an initiation rite, Doc immediately sent Frank to sea for a six-week oceanographic expedition on the Woods Hole research vessel, RV *Atlantis*. The *Atlantis* was a 460-ton sailing vessel rigged as a Bermuda ketch with triangular sails on her two masts, a jib and spinnaker in front. It had a crew of seventeen, leaving room for just five scientists. Its cruising range was effectively unlimited—as long as the wind blew. On Frank's maiden voyage, the refrigeration unit broke down, and they ate rotting meat for two weeks, with A-1 Steak Sauce to disguise the taste. After that, Frank couldn't stand the taste of A-1 Steak Sauce, but he was a full-blooded oceanographer.

I was born nine months after Frank got back from that cruise. The pregnancy was unplanned. Billie went into labor at a performance of Gilbert and Sullivan's *Mikado* in Manhattan on a Friday night. My parents took a taxi to the hospital. I was born on Sunday morning, May 23, 1948, after a grueling forty-hour labor. By the calendar I was slightly premature, but I weighed nine and a half pounds.

Once at Columbia, Doc Ewing put his unusual powers of persuasion to work in some important ways. He befriended Wall Street banker Thomas W. Lamont and—after Lamont's death in February, 1948—Lamont's widow, Florence. Lamont was a partner at J.P. Morgan & Co. (rising to Chairman of the Board) who had undertaken diplomatic missions for Presidents Woodrow Wilson, Herbert Hoover, and Franklin Roosevelt. Tom and Florence's weekend retreat was a 135-acre estate in Palisades, New York, a hamlet in Rockland County across the Hudson River from the City. Doc persuaded Florence to donate the entire estate, which the Lamonts called "Torrey Cliff," to Columbia University as its new "Lamont Geological Observatory"—stipulating that he, Doc, would be its first director and control all of the donated resources. The gift was made while Frank was still Doc's graduate student. Lamont Geological Observatory opened its doors in 1949. Frank received his Ph.D. just after that. His Ph.D. diploma was signed by Dwight D. Eisenhower, who was briefly President of Columbia between his careers as World War II Allied Commander and President of the United States. Eisenhower was also instrumental in obtaining $200,000 from mining and oil companies as initial funds for Ewing to get the new laboratory going. In a political deal, ten acres of the original estate—the piece on the New Jersey side of the New York border—was deeded to Robert Moses' new Palisades Interstate Parkway highway.

Doc's lesser but still significant piece of persuasion was to convince Columbia to acquire a steel-hulled, three-masted schooner, originally E.F. Hutton's luxury yacht *Hussar*, but now rusting as scrap off of Staten Island. Restored as an oceanographic research vessel and rechristened the RV *Vema*, this 585-ton ship was the jewel in the crown of Ewing's oceanographic enterprise. It made Lamont (the proper noun soon meaning Ewing's institute, not its donor) a peer and competitor to Woods Hole. Frank made several cruises on the *Vema*.

Doc decreed that he and his subordinates would move from the city and live on the Lamont estate. Besides Frank, the original small group consisted of new Ph.D.s Joe (J. Lamar) Worzel and Jack Nafe; and also Angelo Ludas, an expert machinist and first-generation Greek-Italian immigrant, as Lamont's shop foreman. During the war, Ludas had worked at Columbia on the Manhattan Project developing the atomic bomb.

Worzel was an expert on geomagnetism whom Doc had taught at Lehigh and taken with him to Woods Hole and then Columbia. He and Frank were never personally close. Nafe had been in the merchant marine and taught physics at the U.S. Naval Academy during the war (this before getting a Ph.D.). At Columbia, he was a student in I.I. Rabi's experimental physics lab. Rabi was awarded the Nobel Prize in Physics in 1944; his lab was world-famous. Still, Ewing was able to charm Nafe and poach him for the new Lamont. Jack Oliver, a later addition, attended Columbia on a football scholarship, did war service as a Navy Seabee, and soon moved to Lamont as its first resident graduate student.

The tens of rooms in the Lamont mansion (which was known as the Big House) were converted to office and laboratory spaces. Doc, his wife Midge, and his three children lived in the estate guest house while Columbia built for them an imposing new residence. A few hundred yards from the Big House, around a bend in the estate road with shielding woods, there were three small, white clapboard houses where Lamont's servants had lived. Worzel and his wife Dottie (and their three children) were given the largest of these; Nafe, the second largest; Frank and Billie (and me), the smallest. The downstairs in our house consisted of a living room and a kitchen. Up a steep and narrow flight of stairs were the bathroom and three tiny bedrooms under attic dormers. There was a cellar with a coal bin and a coal-fired furnace. When the weather was cold, Frank stoked the furnace with coal in the morning and banked the coals at night. I lived here from age about one to age seven.

What I remember about our family's life in Palisades is an idyllic existence, me running practically wild on the large beautiful estate. I don't recall ever needing to ask my mother whether I could go outside and play, alone or with the other children; I just did it. The back door, off the kitchen porch, opened onto a field of tall grass, with an old hay wagon in it. At some point my father painted the wagon bright red for us kids—my sister, Paula, was born when I was two and a half. There was a rope swing hanging from the limb of a tree. A fenced field of cows was immediately across the dirt road that led to the Ludas house. Our front door led to the gravel road that connected with the Worzels' house. Howie Worzel was a couple of years older than me, Richie Worzel a couple of years younger, so there was no perfect match of playmates. But the Worzels' living room boasted the only accessible television on the estate, so it became the central gathering point. (The Ewings had a television, but we were never allowed to play in their grander house.) Dottie Worzel was in effect the neighborhood Mom. It was she who ran back and fetched the fire extinguisher when Ritchie and I, playing with matches, set fire to the pine needles in the woods behind their house.

Dottie's brother was Albert P. Crary, one of the last heroic polar explorers. The Worzel-Crary connection was that Crary had taken a master's degree at Lehigh while Joe was there with Doc Ewing. When "Uncle Bert" visited, the kids (including me) were willing to turn off the television and listen to his arctic stories. He spent several summers as chief scientist on Ice Island T-3, a several-mile long piece of ice floating in the Arctic Ocean that was inhabited as a U.S. Navy weather station between 1952 and 1954. I still have several letters from him addressed to me ("Billy Press"), with "Ice Island T-3" U.S. Navy postmarks. A few years later, Crary, on his 1960-61 expedition in Antarctica, became the first person to have set foot on both the North and South Poles.

Farther afield than the Worzel house on the Lamont estate, there was a big garden plot of strawberries and rhubarb. Behind that was the Tea Garden, with lilacs and sweet peas and a formal fountain. We kids avoided going there—we thought it was haunted. Behind that, a field of corn. Since in the summer the corn was taller than we were, we made mazes in the cornfield by trampling narrow paths with branching intersections.

Surrounding Lamont, the hamlet of Palisades was not a welcoming community. There were old-money estates with panoramic river views at the top of the palisades. Below the cliffs, in an area called Snedens Landing, a kind of wealthy artists' colony had established itself in the 1930s. *New Yorker* writer E.B. White—he of *Stuart Little* and *Charlotte's Web*—described the place as "steeped in Hudson Valley mists and memories, people make their own wine, stamp out their own copperhead snakes, go picking Dutchman's breeches in the spring. On summer evenings, you can hear the trains across the river, grumbling. There is a good deal of talk about shad." As may be. The urbane residents of his time included Aaron Copland, John Dos Passos, and Noël Coward.

By the early 1950s, when Ewing's boy-scientists invaded and took over the Lamont estate, Snedens Landing's center of gravity had shifted from literature toward "show folk"—successful Broadway producers and actors. Palisade's old wealth and new wealth co-existed in a kind of arm's-length mutual admiration: The super-rich were secretly star-struck; and the stars were secretly jealous of old money. One thing that they agreed on was that the influx of city academics and their families from across the river was a bad thing. Midge Ewing was a Kidder, so she was different and immediately sought after by the Snedens Landing people.

Billie was even more isolated socially than the other wives. She was brash, talkative, inner-city, hyper-correct in schoolteacher diction, and (up to now) self-confident. No one in Palisades would care that she had been valedictorian of Tilden High School, or been one of very few Jewish girls allowed into NYU—and with a full scholarship, no less. She probably didn't know what Dutchman's breeches were, and I'm certain that she never talked about shad. Billie *had* heard of copperheads. I remember her showing me pictures of the snake in books, warning me not to accidentally step on one in the woods next to our house. (That is my closest personal connection to E.B. White.)

Frank, not unlike many career-driven fathers in the 1950s, was for me a largely absent parent. Child-raising was Billie's job. I recall only once Frank's taking me to his lab in the Big House. I remember his handing me an ohmmeter (a device that measures electrical resistance) and telling me to go around the laboratory and figure out which things were conductors of electricity, which insulators, and to record the results on a pad of paper. He probably wanted me to amuse myself so that he could get work done. I made and recorded (in first-grader block letters) many careful measurements.

4. To California

Doc Ewing's boys and the growing scientific staff at Lamont were making discoveries that would prove to be among the greatest in twentieth century geoscience. Lamont's scientists perfected the techniques of underwater seismology. The basic idea was to toss lit sticks of dynamite off the back of a ship like research vessel *Vema*. These would explode underwater and produce acoustic shock waves that traveled through the water into the deep-ocean sediments far below. Geological discontinuities—layers or structures in the sub-ocean rock—generated reflected sound waves that could be measured and interpreted back on the ship. The technique was not unlike using dynamite in boreholes to search for oil, something Doc was familiar with from his Texas oil days—and also a reason for his sending Frank off to get consulting experience with oil company Flying-A when I was two. The family spent the summer in Texas—my earliest retained memories.

Worzel and Ewing are credited with the fundamental discovery that the ocean bottoms have always been ocean bottoms—they are not sunken continents as many had believed (and as hinted at in the name of the Woods Hole oceanographic research vessel the *Atlantis*). Frank, with some assistance from Wencelas Jardetzky, a high-born Russian émigré who washed ashore at Columbia in 1946, developed a body of highly mathematical work on how sound should be expected to propagate, refract, and reflect in various configurations of sub-ocean geology. A few years later, this work was published as a book, *Elastic Waves in Layered Media*, by Ewing, Jardetzky, and Press. Ewing contributed nothing to the book—he understood its mathematics in only the most general way—but he insisted on being a first author on the grounds that his laboratory, indeed his ship, the *Vema*, was responsible for all the data that the book sought to explain. Jardetzky, although he was basically a hired hand, was ten years older than Ewing, so Ewing awarded him seniority as second author, above Frank.

It was unjust. Science is full of stories like this, many worse. Frank somehow managed to get his name before Doc's on the Press-Ewing seismograph, an instrument that he developed with help from Angelo Ludas, whose name appears nowhere. Astronomer Jeremiah Ostriker once explained to me his theory about three-author works: One author does all the work; another takes all the credit; the third is the fall-guy, brought on to take the blame if the work turns out to be wrong. Since Ewing took the credit, and Press did the work, Jardetzky may have been the fall-guy.

The Lamont scientists contributed to, but were appallingly slow to embrace, the greatest discovery of twentieth century geology, that of plate tectonics. In 1912, Alfred Wegener, a German meteorologist and polar researcher, proposed that the Earth's continents, composed primarily of granite, were floating, iceberg-like, on a global lithosphere of denser basalt rock. Even more outlandish, Wegener posited that the continents were slowly moving—drifting around the globe, at times breaking apart, at times colliding. Wegener named his hypothetical original single continent Pangea. This "theory of continental drift" might explain diverse mysteries, such as how seemingly tropical fossils come to be found in what are now arctic regions; or why the east coast of South America looks like it fits so perfectly into the west coast of Africa.

Wegener's ideas were rejected outright by most established geologists and oceanographers, including Ewing and those under his influence. The seabed revealed by Ewing's underwater seismology was made of layer upon layer of undisturbed sediment—miles deep. There was no evidence that continents had ever ploughed through it on their way from one place to another. In this vein, Doc and Frank wrote a paper with the title, "On the Permanence of the Ocean Basins." It proved to be completely wrong.

Ironically, Ewing's team was already responsible for some of the first hard evidence that the continents were drifting—but they failed to interpret that evidence correctly. Ewing's 1947 expedition on the *Atlantis* first discovered what we now call the Mid-Atlantic Ridge, an underwater feature of mountain ranges and canyons that runs continuously down the middle of the north and south Atlantic oceans. Subsequent traverses by the *Vema* allowed the ridge to be mapped in detail by Lamont scientists Bruce Heezen and Marie Tharpe. We now understand mid-ocean ridges—a network of them encircle the Earth—as the "seams" where magma is being forced upward, then spreading outward (in both directions away from the ridges) to form new seabed. The continents don't plough *through* the seabed; they float *on* "plates" of (mostly oceanic) lithosphere, and they move with those plates. When continents collide, mountain ranges are formed—for example, the Himalayas. More generally, when plates collide, one plate is forced underneath the other, subducted down into the Earth's mantle, where it melts into the general convective flow, its material reborn in a distant mid-ocean ridge some hundreds of millions of years later.

These are ideas that could have been developed and verified by the Lamont scientists—but they weren't. I have a clear memory from Palisades of my father using a globe to explain to me the basic idea of continental drift, then explaining why it was surely wrong. (I wish I could reproduce that part!) Doc Ewing's priceless, unintentional gift to the competing groups at Cambridge University (Teddy Bullard and Fred Vine) and Princeton (Harry Hess) was to publish a wealth of oceanographic, seismic, and geomagnetic data—and then to let others make the intellectual leaps that he was too stubborn to make himself, and too autocratic to let those who worked for him make. Frank did eventually embrace plate tectonics—there was no other way. In the 1980s he bought a 28-foot sailboat and named it *Pangea.*

Below the Earth's crustal plates lies the Earth's rocky mantle, which extends down about halfway to the center of the Earth. Below the mantle lies a molten, metallic (mostly iron) core. The exact location of the liquid core was deduced from earthquake recordings ("seismograms") in 1913 by the German seismologist Beno Gutenberg. The connection to Frank's career—and my upbringing in Southern California—is direct.

* * *

Beno Gutenberg was born to a well-off Jewish family in Darmstadt in 1889. He attended Göttingen, the preeminent German university, and received his Ph.D. in 1911. He discovered the Earth's liquid core just two years later. As World War I erupted, Gutenberg entered the German army and was assigned to the infantry. Injured by a grenade, he was reassigned to the meteorological corps, where he assisted in wind forecasts that were essential for the use of poison gas by the Germans. The Kaiser awarded him a medal. Despite his fame as a scientist and his war record, he was unable to secure a permanent university position. The 1920s were difficult times in Germany; and rising anti-Semitism was a factor. Gutenberg supported himself by helping to manage his father's cosmetics factory, working with seismological data in his off hours.

The narrative next shifts halfway around the world. In 1921, the Carnegie Institution of Washington (one of more than twenty charitable organizations established by Andrew Carnegie—not to mention 2,509 public libraries) established a seismological station in Pasadena, California. Although it was legally a separate entity, the Seismological Laboratory (or Seismo Lab) was soon closely associated with the Geology Department of neighboring California Institute of Technology. In 1930, Robert Millikan, Caltech's Nobel-laureate President, and Harry

Wood, the Seismo Lab's director, recruited Beno Gutenberg from Germany to California. The idea was that Caltech would take over the Seismo Lab and that Gutenberg would become its director, joining its two senior scientists Charles Richter and Hugo Benioff.

The transfer took place in 1937. By that time, Richter was famous for his development with Gutenberg of the Richter scale for earthquake magnitudes—to this day the most widely used measure. Unlike Doc Ewing, Beno Gutenberg did not insist on putting his name on everything, and historians of science credit Gutenberg with most of the ideas that went into the Richter scale. Benioff, though less a household name, was famous in the field as the inventor of the Benioff seismograph, a design still in use around the world. Contributing to the discovery of plate tectonics, Benioff would later discover the pattern of deep earthquakes that demarcate the subduction of one plate under another.

In 1955, Gutenberg was, at 66, close to retirement age. Benioff was 60. Richter was 55. I was 7. Benioff expected to be named to succeed Gutenberg as director. Richter didn't want to be director himself—he didn't think of himself as a good administrator—but he was not happy with the idea of the position going to Benioff. Nothing was out in the open, but behind the scenes it was a standoff. Gutenberg and Bob Sharp, the Geology Department chair, hatched the idea of recruiting Frank Press—not immediately to be the new director, but with that succession in mind for the future. Frank was Doc's boy-genius rising star. He was an outstanding theorist, and he also was good at building things in the laboratory—seismometers, amplifiers, sensors. (In his parents' apartment in Brooklyn, Frank had built radios as a teenager.) He published prodigiously. Sharp and Gutenberg made him an offer to come to Caltech and the Seismo Lab as a full professor. Frank says that the discussion of the directorship never came up. I find that hard to believe. He was 31.

I have vivid memories as I was growing up of all of these individuals from social occasions in Pasadena at their houses or at ours. The Gutenbergs, Beno and Hertha, treated Frank like a son, Paula and me like grandchildren. They had almost impenetrable German accents. Beno was a diminutive man, totally bald and practically deaf. He wore a vacuum-tube amplifier hearing aid, about the size of a modern cell phone, except two inches thick and in a silver case. Hertha had been a good friend of Albert Einstein. The Benioffs, Hugo and Mildred, owned a ranch property, and invited our family to visit them there a few times. After Benioff died in 1968, Frank authored his sixteen-page National

Academy of Sciences "Biographical Memoir." Charlie Richter always seemed to me a disagreeable man. His biographer, Susan Hough, relates that he and his wife were committed nudists. When I saw him, he was always fully clothed.

Caltech's offer was a stroke of good fortune for Frank. He was increasingly dissatisfied with life in Doc Ewing's shadow. Billie's life in Palisades was unhappy. Now with her master's degree in education, she wanted a career of her own, and there was no opportunity for one in Palisades. Also, Frank was ready for a shift in his scientific horizons. The Ewing program was ocean seismology, with the emphasis on ocean. Frank felt himself to be more a seismologist and less an oceanographer. The Seismo Lab's program was global seismology which, especially as the discoveries of plate tectonics started coming in, was where the action was.

In the summer of 1955, our family moved to California. We drove across the country in our dark-green Chevy station wagon. This was before the Interstate Highway System. Rural highways were one lane in each direction, with a single white line down the center, solid or dashed. I remember stopping in Tulsa, Oklahoma, to stay at the house of a geologist colleague of Frank's; so we must have driven most of the way across the country on U.S. Route 66, consistent with another memory, of our stopping at Petrified Forest National Monument and buying a souvenir piece of petrified wood in the gift shop.

Southern California was in the middle of its postwar boom. New freeways and new suburbs were sprouting. Between 1950 and 1960, the population of Los Angeles County increased by 25%—but L.A. was already pretty full. Neighboring Orange County's population tripled in that decade, and then doubled again in the next.

In 1955, Frank, Billie, and we two kids moved into the house of a Caltech professor who was away on leave, 885 North Holliston Avenue, just north of Orange Grove Boulevard. The street was lined with tall palm trees. The house belonged to Victor Neher, a Caltech professor. Neher, then in his 50s, was one of Caltech's first Ph.D.s. He had worked with Robert Millikan in the study of cosmic rays. Neher built instruments that were lofted by balloons to high altitudes, detecting particles that would otherwise not penetrate the atmosphere to ground level. By the 1960s, this was a backwater of physics. Particle physics (so called) was being done with synchrotron accelerators, not cosmic rays. Neher was therefore devoting most of his efforts to teaching Caltech undergraduates. One of the physics students whom he remembered later in life was a boy from Mormon Utah named Kip Thorne, about whom we will hear much more, later.

Pasadena is separated from Los Angeles by a range of low hills, the vestigial eastern extension of the Santa Monica Mountains and Hollywood Hills that separate Los Angeles and Hollywood on the south from Burbank and the San Fernando Valley on the north. By the 1950s, Pasadena was a bedroom community for Los Angeles commuters. Earlier in the century, it had been a suburb of stately houses for the well-off, interspersed with orange groves. Altadena, an unincorporated piece of Los Angeles County in the foothills north of Pasadena, was the gateway to the Echo Mountain Resort, a 70-room Victorian hotel at elevation 3,200 feet that was reached by an incline railway. By the 1950s this was long in the past—the hotel burned down in 1905. When I was Boy Scout age, we would hike up into the mountains following the path of the railway to the ruins of the hotel.

The year we lived on Holliston, I was in second grade at Jefferson Elementary School. Billie quickly obtained her provisional California teaching credential and was teaching at the same school—a source of embarrassment to me. What I remember about that year is not school, but the freedom of owning a bicycle in a city with good sidewalks, and also the complete absence of any adult after-school supervision to an extent that seems incredible now. I ranged over about a square mile, to Colorado Boulevard (Pasadena's main street) on the south and Lake Avenue (its second-main street) on the west. I was forbidden to go past Lake Avenue, because that was the bad part of town. Pasadena was the only city in the San Gabriel Valley in which African-Americans (then called Negroes) could live; and this only west of Lake Avenue. This was de facto segregation, enforced by landlords and real-estate agents. The city had two high schools. John Muir High School, west of Lake, was almost all Black. Pasadena High School, at the edge of Eaton Canyon on the east, was all white. Disneyland, in Anaheim, California, opened the same summer that we moved to California. Pasadena's schools were integrated by federally mandated cross-town bussing only the year after I graduated from high school in 1965.

5. Altadena

In 1956 we moved into the house at 1972 Skyview Drive, Altadena. It was the first time that Frank or Billie—or anyone in their families—had ever owned real-estate property. Frank's Caltech salary was $12,000 a year, the equivalent of about $120,000 today. As an elementary school teacher, Billie earned what would be $30,000 today. Compared to any earlier time in their lives, they were well off.

One block north of Skyview Drive, dry mountains jutted up out of the foothills at a rugged slope. Suburbia ended suddenly right there, with only hiking trails and jeep roads penetrating further into the mountains. One block east of our house was a sharp drop into Eaton Canyon, undevelopable land accessible only by a Forest Service road with a locked gate.

Our house was built on a slope, downhill from front to back. Its "improved basement" was a dingy room with yellow pine paneling and a musty smell. We called it the family room, but it was mostly unused except by our pet cats, whose cat litter box added to the odor. Next to the family room was the laundry room and the garage (*two* cars!). On the other side of the garage, directly below my bedroom, was a narrow utility room with its own outside door. The gas furnace and hot water heater were at the back. The space at the front—maybe 12 feet long and 6 feet wide—was intended as a garden shed. It became my "shop," where I did "projects". When I was in fourth or fifth grade, my father built for me a plywood-topped workbench (metal legs, a kit from Sears-Roebuck) and nailed up a Masonite pegboard that I could hang tools on.

Noyes Elementary School was a block away. The school building, a "modern international" design, was built in 1952 with classrooms for about 300 students. I started there in third grade. Billie was now teaching full time, fourth grade at Audubon Elementary across town. Much of her salary went to pay for our housekeeper, Mrs. Erstad, a middle-aged lady of Norwegian descent who came in afternoons. Her duties were light housekeeping, laundry, and cooking dinner for the family every weekday night, the dinner to be left warming on the stove when she departed at five, often before Billie got home. Frank got home at six. In principle, Mrs. Erstad was responsible for watching Paula and me after school. I was supposed to walk directly home. In reality, I just did whatever I felt like. There was an unspoken conspiracy: Mrs. Erstad wouldn't tell my mother that she never had any idea where I was, if I wouldn't.

Transplanted New Yorkers, Frank and Billie embraced Southern California life with enthusiasm. For several years, this found expression as winter weekend car-camping trips to the California desert, cooking on a Coleman stove, the parents sleeping in the back of the new station wagon, the kids in sleeping bags on the ground. The idea of a tent never came up, nor the idea of hiking. It was strictly drive, unpack, cook dinner, sleep, cook breakfast, drive home, repeated at various scenic spots in the Mojave Desert, Death Valley, Borrego Springs, and Joshua Tree National Monument. I somewhat appreciated the beauty of the desert, and the cosmic display of stars in the night sky, but I was much more interested in the camping paraphernalia—the little needle valve and vapor generator that allowed the stove to burn gasoline without exploding; and the mantle on the gas lantern, which was prepared by burning it to an ash that was then somehow capable of surviving the gas flame while glowing white-hot.

The period 1957-58 was the United Nations-designated International Geophysical Year. IGY was the consequence of a slight thaw in the Cold War, enough to allow scientific cooperation between the U.S. and U.S.S.R. Frank was much involved. In the summer of 1958, he and Billie traveled to Russia as part of one of the first scientific delegations. Paula and I were sent to Haskell's Raskells Summer Farm Camp, located in dusty foothills somewhere south of San Luis Obispo. We slept in large semi-permanent tents erected on wooden platforms more than a foot off the ground. That height was important, because the camp was infested with rattlesnakes. Having thrown rocks at rattlesnakes in Eaton Canyon, I was not unduly worried; not so some of the other children. The adult camp counselors carried flat-bladed shovels and were adept at cornering and beheading the snakes. This happened several times in the month that Paula and I were there.

Supporting the camp's flimsy claim to be a farm, every child was assigned a farm animal to feed. I chose the pigs—all five of them—when no other camper volunteered. I could sleep for an extra half hour every morning, because, unlike the other animals, the pigs were fed after breakfast—breakfast slops mixed into the pig feed. I loved the pigs, loved their intelligence. Winston Churchill is supposed to have said, "I like pigs. Dogs look up at us, cats look down at us, pigs treat us as equals." The large sow and boar were each three times my weight, and they delighted in knocking me over in trying to get to my chow bucket before I could empty it into the feed trough. Paula and I went back to the camp a second summer. We learned many leftist union songs, most of which are collected in familiar Woody Guthrie recordings.

In 1958, the United States and the Soviet Union were each considering the ramifications of moving their nuclear weapons tests underground. The more than five hundred above-ground tests, each generating a mushroom cloud of radioactive fallout, had more than doubled the amount of radioactive Carbon-14 in the atmosphere worldwide. Radioactive Strontium-90 from the bombs was detectable in milk and in the bones of children. Above-ground nuclear tests are visible and easily detectable. But, if testing were done underground, how would each side know what the other side was doing? The answer was seismology. An underground nuclear explosion was similar to a small earthquake, but different enough in detail that it might be distinguishable from the constant global background of small earthquakes.

Frank became a consultant to the President's Science Advisory Committee chaired by MIT's president James Killian and Harvard's George Kistiakowsky. Later, under President Kennedy, Frank became a full member of PSAC. One evening in late October, at a very specific time that was known to Frank, and perhaps a classified secret, our family stood at the top of our driveway in Altadena, looking to the northeast. It was a clear night. A sudden flash lit up the low horizon just over the mountains. It was over immediately. You could easily have missed it. It was the scattered atmospheric flash of one of the last open-air nuclear test conducted by the United States, two hundred miles away, at the Nevada Test Site.

Family camping was eventually supplanted by sailing. Frank bought a 24-foot, closed-cabin sailboat, a Lapworth-24, one of the first of that size with a fiberglass resin hull. Billie christened it the *Frilla*, for "Frank-Billie-Bill-Paula." Every Saturday morning for a couple of months, my father packed us two kids (willing or not) into the back seat of the car, and we drove to the factory in San Pedro where the *Frilla* was being built. I always came home itching from the loose fiberglass fibers. Even now, my skin crawls at the characteristic smell of epoxy resin. After the *Frilla* was in the water and Frank had taken some weeks of Coast Guard-sponsored sailing lessons, there were many weekend days when we kids were dragooned into sailing. The *Frilla* was moored at a grubby marina in downscale San Pedro, an hour's drive from home. I had no interest in sailing. Still, I couldn't help picking it up by osmosis. Avalon Harbor on Catalina Island was a six-hour sail. We sometimes stayed there overnight on the boat, and went to the movies in the famous, though by then decrepit, Avalon Casino, sailing back the next morning.

When Frank was away on business trips, family dinners were infrequent. Instead, Billie would drive us to Bob's Big Boy or Taco Tia (Glen Bell's original taco chain, later renamed—wait for it—Taco *Bell.*) Depending on her mood, we might eat there, or she might insist on smuggling takeout food in her purse into a movie theater, the run-down State and Rialto theaters on Colorado Boulevard being her favorites because they ran continuous double features. All movie theaters were single-screen. Multiplex was far in the future. It didn't matter what movies were playing, or what were their starting times, or whether this was a school night. We arrived in the middle, watched the end of one movie, the whole next one, and then the start of the first. At a certain point Billie announced, "This is where we came in," and we would exit and drive home and be sent to bed. I saw a lot of movies this way.

Looking back, it is clear that Billie suffered from a bipolar mood disorder. Double features at the movies on school nights happened in periods of depression. Paula and I learned to recognize our mother's manic periods, although we didn't have a word for them. Suddenly Billie would be *involved* in being our mother, unfortunately not in an emotionally supportive way, but in a mad flurry of educational activities. Billie was raised by a bad mother, and she was a bad mother herself, but she was an excellent teacher. In her manic moods, she substituted teaching—the thing she did best—for mothering.

The fourth-grade social studies curriculum in Pasadena included a unit on farms. I recall Billie's packing us into the back seat of the car and driving east on Foothill Boulevard past Pomona toward San Bernardino. Raymond Chandler's fictitious detective Philip Marlowe drove this route twenty-five years earlier in *The Big Sleep*. It hadn't completely changed: it was still farm country. On secondary roads, we parked on the dirt shoulder alongside fields of (to us unknown) crops, trespassed into the rows, and, with a garden spade, dug up plant specimens for Billie's fourth grade class. We were digging up a large root vegetable (we didn't know what it was) when an angry farmer confronted us. Billie explained that she was a schoolteacher and that we were stealing his whatever-it-was for educational purposes. He told us that it was a sugar beet.

The darker side of my mother's manic phases were when she screamed at us over trivial things. Sometimes, rarely, she hit us. More often, she fined us our allowances. "That's two weeks' allowance," she screamed. I often responded with what I thought was a clever sarcastic remark. "That's four weeks." She marked bold X's on successive Saturdays on the wall calendar by the phone in the kitchen to indicate the weeks I would not be paid. At one point, I was in debt a whole year's allowance; she had X'ed every Saturday on the calendar. I never took this seriously. My allowance was withheld for a week or two at most. Then her mood would change and, for no stated reason, clemency would be granted and I would again start getting paid.

My fourth-grade teacher was Mrs. Plantamura. By now I was wearing glasses for nearsightedness, like both my parents. I was small for my age. At recess, I played exclusively with the girls, who seemed happy enough to include me, even though I could never learn to jump rope. Clarissa Plantamura—she and my mother were on a first name basis—told my parents about this, and they became worried that I would grow up to be what was then called a "sissy". They hired a "catch coach" for me, an athletic, masculine, high-school boy who came to our house two afternoons a week to teach me how to catch a baseball and, by implication, act like a real boy. It was to no avail. Yet, whatever my parents' misguided worries, throughout subsequent life my best friends, buddies, and confidants have been predominantly female; my romantic partners, all female.

The name Plantamura is Portuguese. In the early 20th century, many Portuguese immigrants came to California to work the vineyards. Mrs. Plantamura's husband's father was the leader of a brass band that traveled up and down California in the 1910s and 20s, giving concerts in town squares. In San Diego they played in the Spreckels Pavilion, where public concerts are still held. When our class went on field trips involving long drives in a yellow Pasadena School District bus, Mrs. Plantamura had me sit next to her. On a pad of lined paper, she taught me algebra—how to move the *x* to the other side of the equation and then divide by whatever was multiplying it to get it to stand all by itself. I was captivated.

I was also captivated by the Plantamura daughter, Carol, who at 14 was a dramatic, raven-haired teenager. I was 10. I met Carol when our family visited the Plantamuras at their home. Teenage girls, whom I knew as babysitters, interested me in general: They were so completely different from the girls I played with at school. I couldn't understand how such a transformation could take place. There didn't seem to be a corresponding change for teenage boys, who seemed simply older and cruder extrapolations of the boys in my class with whom I didn't play at recess.

I need to jump out of time to describe some events that happened decades later:

In 1989, I was unexpectedly a candidate for the position of Vice Chancellor for Academic Affairs at the University of California, San Diego, and traveled there for a day of job interviews. The faculty representative on the Chancellor's search committee was the department chair of the Music Department—one Carol Plantamura, now in her mid-40s. We recognized each other immediately. She told me that her mother, my fourth-grade teacher, had died fifteen years earlier. Carol was professionally a soprano who rose to lead roles in second-tier European opera houses before returning to the United States to start a new career as a music professor. She still had the dramatic demeanor that I remembered from thirty years before, and now the imposing physical build and lungs of an operatic soprano. Not surprisingly, she had a beautifully clear speaking voice. Although I quickly established that she was a great raconteur and a fellow cat lover, I didn't get the job.

A few years later, my by-then-ex wife Margaret was working on her second true-crime book. Margaret had already published two detective novels and one true-crime book, and her new subject was a complicated Los Angeles murder case, never solved. The murder victim, D. Scott Rogo, was a journalist who wrote about parapsychology, sometimes in connection with flying saucers and Bigfoot. The police at the time had been assisted by a psychic medium named (as Margaret gleefully reported) *Clarissa Plantamura*.

"It's a very uncommon name," she pointed out. "It can only be your fourth-grade teacher."

I phoned Carol in San Diego. With her characteristically melodious laugh: "Oh yes, of course! Mother had *the gift*." Carol explained that the whole family was gifted in this manner. She recalled a typical conversation at the dinner table: "I had a good talk with Uncle Marco today," her father might say. "And how is he doing?" her mother might ask. "Just fine. He sends his regards." Only later did Carol figure out that Uncle Marco (and many others talked about) were ghosts.

6. Projects

Because of the baby boom, Noyes Elementary School was short of classrooms and teachers, so class sizes were large. Sometimes a teacher was assigned to teach two grades simultaneously in the same classroom, as was the case for my fifth and sixth grade teacher, Mr. Ridout ("Wally" to the other teachers and to my mother). Mrs. Plantamura and my mother hatched the idea that I should be in Wally's combined class, and that I should do fifth and sixth grades in one year. I would then graduate from Noyes and go on to Eliot Junior High School. I don't recall being asked for my approval of this plan.

Mr. Ridout was a fighter pilot in the Pacific in World War II. Like many veterans, he was able to go to college because of the G.I. Bill of Rights, and he then became an elementary school teacher. Our class had a special field trip to an airfield (I forget which) where there were airplanes from the war. Mr. Ridout showed us the kind of plane he flew, a twin-engine P-38 Lightning.

When he wasn't sailing, Frank, my father, worked on weekends. Around this time, he started taking me with him some of the time. Whether this change in habit was to provide additional relief for Billie or represented a new interest by Frank in parenting cannot be known. The Seismo Lab was still at its original location, 220 North San Rafael Avenue, in the Linda Vista hills west of Pasadena. These hills were bedrock poking up through the alluvial basin of the San Gabriel Valley. Since seismographs needed to be on bedrock, the Seismo Lab in those days couldn't be on the Caltech campus. Although purpose-built as a laboratory by the Carnegie Corporation in 1921, the two-story building was a rabbit warren of small offices, shops, hallways, and closets. Gutenberg had retired in 1957, and Frank was by now the Seismo Lab director. He had keys that opened everything. Some locked closets had seemingly not been opened since the 1920s and were filled with electrical and electronic junk from that era. Frank set me up at an unused lab bench and let me take things apart and perhaps succeed in putting them back together. Pretty soon I was taking stuff home to work on (that is, play with) on my own shop bench in the narrow furnace room under the house. Frank bought me a soldering iron, and I learned how to use it.

I have a particularly clear memory of constructing a crystal-set radio under my father's supervision, early in this period. Frank drew the simple circuit—it was the same circuit as his own boyhood crystal set—and he showed me how to wind the inductor coil of thin copper magnet-wire around a toilet-paper tube. Crystal sets have no battery; they are powered by the radio waves themselves. The key component is the "crystal" or rectifier. In Frank's boyhood this was a literal crystal of galena, a shiny lead ore, on which one tried to find a sensitive spot for a "cat's whisker," a small piece of stiff wire pressed against the galena. By 1958 it was possible to use a manufactured germanium diode (poor cousin to the transistor) that was more reliable than the crystal-cat's whisker combination. Frank didn't mind appropriating inexpensive things like resistors or spools of wire for me from the Seismo Lab shop, but a diode was valuable, probably the equivalent of $10 today, so we bought it at Dow Radio Supply on Colorado Boulevard. I remember that it was a 1N34. Over the next few years, I spent many Saturday-morning hours taking the bus and waiting in the single-parts line at Dow. Arriving in due course at the front of the line, I would present the clerk with that week's project's list of components (written in pencil in my childish printing), and he would fetch them from the shelves behind the counter. I pulled crumpled one-dollar bills and change from my jeans pocket to pay.

My electronics hobby peaked in junior high. I subscribed to *Popular Electronics*. I found a junked five-tube AM clock radio and repaired it. After that, I listened to music every night as I was falling asleep. My favorite station was the one that played "slush"—Mantovani and 1001 Strings. Bill Haley, Elvis, Chuck Berry completely passed me by. Not until 1964 did rock-and-roll enter my personal horizon—with the Beatles.

If Dow Radio was the Mecca for electronics hobbyists, their Medina was C&H Surplus, a few blocks farther east on Colorado Boulevard. C&H sold whole pieces of surplus electronics equipment (possibly working, always unlabeled) from World War II and the Korean War. More useful to me, it had whole aisles of self-service bins of cheap components that had been taken out of such equipment. They also sold some non-electronic stuff like camouflage parachute material and survival gear.

Earlier, in sixth grade, my friend Glenn and I had bought a quarter of a parachute—that's how C&H sold it—and sewed camouflage body suits for ourselves using my mother's sewing machine and a McCall's pattern for men's pajamas from Woolworth's. Our parents thought this was cute, and even let us go out in the early evening after dark to "practice survival". What they didn't know was that we were using these occasions to break into the school office at Noyes (plastic shim or coat-hanger methods leaving no visible signs) to read the permanent records of ourselves and our classmates, which were kept in unlocked filing cabinets. I was eventually caught, not by being apprehended in the act, but by mentioning to a teacher some fact that could only have come from those files. My parents were summoned and I was hauled before the principal. It was clear that no one knew that I was breaking and entering at night. They simply assumed that my MO was to sneak behind the counter when the school secretary was away from her desk. I don't recall what my punishment was, but the episode taught me two valuable lessons: "*Don't get caught*," and "*Loose lips sink ships*."

One treasure from the Seismo Lab was a thousand feet of used, yellow, two-conductor blasting wire. This found use when a bin of small, cheap audio loudspeakers appeared at C&H. I bought six. Blasting wire was stiff, single-strand copper. I found that it was easy to snake it through the walls and floor spaces of our house, working only through the openings behind existing switches, outlets, and heating grates. My ostensible purpose was to wire our house for music. The project took several weeks. When it was finished, I conspicuously demonstrated to my parents that I could make music from the record player in my bedroom come out of the heating grate in any room in the house. They told me not to annoy my sister with this, and I promised not to.

It took some months for my parents to discover—I think it was my mother who actually figured out—the true purpose of the project. I had found that the small audio speakers could also work in reverse, as microphones. One of my previous electronics projects was a little two-tube audio amplifier. With that, and headphones, I could listen in on any room in the house from my room. I was surprised at how angry my parents were. However, the only speaker/microphone that they actually made me remove—under watchful eyes—was the one in their bedroom.

Not all my projects were malign. Frank had built the family hi-fi set in the living room from a Heathkit, and he was willing to subsidize my purchase of Heathkit test equipment for my shop. What came in the box of a Heathkit was a pre-drilled chassis, an attractive outer case, hundreds of individual small electronic components, a spool of circuit wire, and a manual with hundreds of numbered mounting or soldering steps. The economic reality that enabled Heathkits was that most of the cost of manufacturing commercial electronics was in the hand labor of the point-to-point wiring and soldering; so the kits could be sold quite cheaply relative to store-bought items. (Only with printed circuit boards and bulk industrial soldering methods, did the Heathkit economic advantage disappear.) I built a vacuum-tube voltmeter (VTVM), an audio oscillator, my own hi-fi amplifier, and an oscilloscope that never completely worked—I must have made some error in wiring it. In truth, the audio oscillator's uses were not completely benign: I used it, along with the amplifier and those same little speakers, to fill the house with an ear-splitting tone at 17,000 cycles per second (not yet known as Hertz). Children's ears extend to a higher range than adults, so my parents couldn't hear any sound at all. They didn't believe Paula when she complained that I was torturing her. And, yes, I had a Gilbert chemistry set, and a subscription to Things of Science, that provided a monthly kit of objects and artifacts for elementary science experiments. I looked forward to the arrival in the mail of its signature small blue boxes.

* * *

Eliot Junior High School was built on North Lake Avenue in 1931. The building was of a boxy, pink stucco, neo-Mission design (complete with bell tower) by the well-regarded architectural firm Marston and Maybury. I don't know who the most famous Eliot alumnus is, but I do know its most infamous: Sirhan Sirhan, who assassinated Robert F. Kennedy in 1968, was two years behind me. His picture is among the seventh graders in my ninth-grade yearbook. In 1959, when I started seventh grade, the junior high school curriculum was mostly unchanged from the 1920s. All students took academic subjects (English, math, etc.) and also industrial arts (shop or home economics) courses. This echoed a time when ninth grade was for many a terminal year. I have clear memories of my three shop courses, but—with a few exceptions—only vaguely remember my academic classes. In seventh grade, I took drafting, also known as mechanical drawing. The work was done on sloping tables, with sharp No. 3 pencils and gum or kneaded erasers on translucent drafting paper. The tools were T-squares, plastic triangles and French curves, a compass, and a little soft brush for clearing away the eraser dust. We drew geometric forms and more complicated objects in plan and elevation orthographic views, in isometric projection, and in two- and three-point perspective. I never learned how to ink my drawings in India ink, the necessary step for printing the translucent drawings by sunlight onto light-sensitive blueprint paper. That would have been in the advanced drafting class, which I didn't take.

I rode my bicycle a mile and a half to school most days. In the second half of seventh grade, I took wood shop and learned to use a drill press, band saw, circular saw, and (to some extent) wood lathe. Everyone made exactly the same projects: a tie rack with beveled edges, countersunk mounting holes, and lathe-fabricated pegs; a wooden box with mitered and pegged joints (and a sliding bottom that revealed a hidden compartment); and a third project that I forget. All the projects were done in soft pine—hardwoods were for the advanced class.

In eighth grade, boys were supposed to take welding and metal shop. I should have done that, and later regretted that I didn't. Girls took two semesters of home economics, mostly sewing. An elite small number of girls (half a dozen per class period) were allowed to take "library practice," that is, work in the school library with the school librarian. Since I still (in fact always) liked the girls in my classes better than boys, I protested to the school administration that it was unfair to restrict library practice to girls, and that I should be allowed into the class as the only boy. My request was granted, and I learned the wonderfully arcane rules of card-catalog alphabetization and memorized large parts of the Dewey Decimal System for cataloging books. I learned to love library paste. Still, in later life knowing how to weld and use metal-cutting tools would have been more useful.

In ninth grade, I took graphic arts, learning to set lead type by hand on a composing stick, picking upper-case letters from (yes!) the upper type case, lower-case letters from the lower case. You had to set the type backwards, right-to-left, and read what you had set upside down. From the composing stick, the type went into a galley tray, where it was tied up in blocks by string. You inked it with a rubber ink roller and printed a proof copy on a small horizontal press. After proofreading and fixing errors, you removed the string and locked-up (i.e., clamped) the type in a chase (iron frame) for printing on the electric-motor-driven platen press.

Printing was the scary part. As the space between the platen and the flatbed opened on each cycle, you stuck your hand into the gap, pulled out the just-printed sheet, and, with your other hand, positioned a new sheet of blank paper. Then you got your hands out of the way before they were crushed by the closing press on its next cycle. No one would allow junior high schools students to do this today! Still, I never heard of a student being injured. Barely less terrifying were the massive paper trimming machines, and the knives and tools used for linoleum block cutting.

My seventh-grade math class was the rare course not out of the 1920s. Quite the opposite: After Sputnik in 1957, a consortium of educators and professional mathematicians, the School Mathematics

Study Group (known as SMSG), undertook a crash effort to modernize the grades K-12 math curriculum. This was the famous—soon infamous—"New Math". Ours was the pilot year for the seventh grade curriculum. Only one teacher, Mr. Benedict, taught the SMSG course; most students took traditional seventh grade math. Our textbook was mimeographed and handed out in sections. The first several weeks were about set theory, an abstract topic with (then) no practical applications, one that had never before been taught in K-12. The SMSG developers weren't crazy—merely misguided. They felt that children were capable of learning the grand concept that mathematics was not just about numbers, but could be about *any* formal system of abstract objects and rules. They may have been right about this—most of us learned set theory without difficulty. The issue was *why*. New Math was accurately seen by the public as more geared to the rarified world-view of pure mathematicians than to teaching something that schoolchildren needed to learn—a point at the time made caustically by the famous Caltech physics professor Richard Feynman (about whom more later). On Fridays, Mr. Benedict devoted half the class hour to telling us WWII war stories. He had been a cryptographer in the Navy, involved in the top-secret exploitation of the U.S.'s breaking the Japanese "Purple" diplomatic code.

Mr. Al Renner was my eighth-grade science teacher. His idea of a field trip was to take our class by school bus to an abandoned, labyrinthine gold mine in the San Bernardino Mountains. In the pitch dark we played sardines, a titillating game for mixed groups of adolescent boys and girls. In the summer, Renner took small groups (of boys only) hiking in the Sierras. I learned to make hot-rod hamburgers, meat wrapped in aluminum foil and cooked on the engine block of the station wagon as we drove, fifty miles for medium-well done.

Renner was also in charge of the after-school club, "the Eliot Electroneers". Obviously, I fit right in. He had relationships with local electronics companies and solicited donations of parts and cash. We had an unlimited supply of then-pricey 2N107 transistors and designed our own circuits, often not very capably. At one point, Al acquired many dozen obsolete telephone switching relays, mounted by tens on 24-inch metal strips (which would, in a telephone switch room, be mounted onto racks). I got twenty for my science-fair project: I built a relay logic machine (a very primitive computer) that could almost play tic-tac-toe. Looking back, it was a pretty good effort: I had written out the whole logic tree in Boolean logic—not unrelated to the "useless" set theory that I learned in seventh grade SMSG math—and then used mathematical rules to simplify it as much as I could, down to the number of relays and relay poles (parallel switches) that I had available.

Then, there was the bar mitzvah thing. Frank and Billie were, as we have seen, atheistic Jews with socialist political leanings, fully in the liberal New York City mold. A few years after we moved to California, our family joined an Ethical Culture congregation. Founded by Jews in the late 19th century, Ethical Culture was one of several so-called "godless congregational" movements that met on Sundays, as if their chapters were churches. The highly regarded Fieldston School in Manhattan was founded by the Ethical Culture movement. Ethical Culture was well-known in New York City, but not well known elsewhere, especially not in California.

Our congregation, which met in La Cañada, was the only one in the northern Los Angeles area. Instead of a religious service, the adults listened to a lecture on some ethical or philosophical topic, followed by time for discussion. Instead of a Sunday school religious education, we children met separately and mostly learned songs or memorized poems with unmistakable moral content. I can still sing about Commodore Grey, who had a pet dog and a pet cat who, though different species, always got along. The lay head of our congregation was Frank Thomas, then already famous as one of Walt Disney's original animators, the "Nine Old Men." Thomas' creations included the stepmother in *Cinderella*, Captain Hook in *Peter Pan*, and the dancing penguins in *Mary Poppins*. He was responsible for the scene in *Lady and the Tramp* where the two dogs fall in love over a plate of spaghetti.

This all notwithstanding, suddenly as I was approaching the age of thirteen, Frank and Billie decided that Ethical Culture was a mistake after all, and that I should be bar mitzvahed, as soon as practicable, at the Pasadena Jewish Temple. A whole-family interview with Rabbi Maurice Galpert was arranged.

The Pasadena Jewish Temple was the only Jewish temple or synagogue for many miles around. It, and Rabbi Galpert, needed to be all things to all (Jewish) people. Officially, the congregation was Conservative, the Jewish denomination that expanded in the United States at the beginning of the 20th century as a backlash against the more liberal Reform movement. However, especially with the influx of post-War immigration to California from "back East," and with Caltech so nearby, there were also many Reform families to be accommodated in Rabbi Galpert's flock. Services, sermons in English, were held on Friday evenings. Traditional bar mitzvahs for boys, and the less-traditional bat mitzvahs for girls, were on Saturdays.

We met in the Rabbi's book-lined study. Frank explained that he had been bar mitzvahed (true) and that he and Billie had always wanted their son to be (less true). The Rabbi explained that bar/bat mitzvah boys and girls had been going to Sunday school for at least two years, and that their families had generally been members of the congregation for years more than that. "Let me speak with you privately, Dr. Press," the Rabbi said to Frank.

When we were ushered back into the study, a deal had been struck. I was to take weekly private Hebrew lessons from the wife of a penurious Israeli Caltech graduate student. I was to attend Sunday school for one year. I would be bar mitzvahed not on my "date" (the Saturday closest to my 13th birthday) but five months later, in mid-October. In Jewish tradition, this was a mark of shame: Only the stupidest children were unable to memorize their required parts of the Bible before their date. Metaphorically, this was God's punishment for my dalliance with Commodore Gray's dog and cat. A part of the deal, not revealed to me, was a financial contribution that Frank would make to the Temple in lieu of years of membership.

And so it came to pass. For a year's worth of Sundays, I sat next to Melanie Michaels, a girl in some of my classes at Eliot on whom I had a severe crush. Then, all the relatives came from back East for the bar mitzvah ceremony. At the reception, my male cousins and I buzzed around Mel like bees around a picnic table. I collected more than a thousand dollars in gift checks, a huge amount of money for me.

Two follow-up stories are in the category of "many years later":

Margaret Geller, Josh Grindlay, and I, all Harvard professors of astronomy, were sitting around at the bar mitzvah reception of another colleague's son. "How much did you give him?" Margaret, who is Jewish, asked me. "Fifty dollars," I said. "Right," she said, "I came up with the same figure." Josh, who isn't Jewish, chimed in that he felt that giving money was too crass. His gift was an FM transistor radio. We just stared at him. There was no use trying to explain five thousand years of Jewish history.

Separately, in the fall of 1987, I flew to Ithaca, New York, for the bat mitzvah of the daughter of my friend and colleague Saul Teukolsky, a Cornell professor, about whom more later. Rachel Teukolsky's assigned *torah* readings, in English, were obscure verses of Deuteronomy that included prohibitions against (1) building parapets with railings less than two cubits high, (2) raping women captured in battle before twenty-eight days had elapsed, (3) punishing the children for the sins of the father, and (4) prostrating oneself on a patterned floor. I wondered silently whether the original author had any inkling that prohibition (3) would become a fundamental tenet of the Judeo-Christian ethic, while number (1) would evolve into a minor zoning ordinance—and numbers (2) and (4) would be seen as barbaric and risible, respectively.

But when it came time for Rachel's *haftorah*, the section of scripture always read that particular calendar week as part of the annual cycle of reading the Old Testament in Hebrew, she recited most clearly, "*Rani akara lo yalada...*" It was *my haftorah* (Isaiah 54:1). Yes, it was October, and she was on her correct date. Perhaps this was part of some Great Plan to finally erase my shame—that particular one, at any rate. Silently, I vowed never to prostrate myself on a patterned floor.

7. Boy Scout

Voluntarily, I attended summer school in junior high. It was something to do instead of being bored at home on dry, hot summer days. The concentrated courses met from 9 a.m. to noon for six weeks. The summer after seventh grade, I took Typing. Sitting in the classroom behind office-style manual typewriters, we did unending speed and accuracy drills. Despite my poor athletic coordination, I turned out to be a good typist. I later came to appreciate this course as the most useful in my entire education. The summer after ninth grade, I took second year French, which was offered only inconveniently at John Muir High School on the "bad" side of town. There was a reason for this, which I will get to.

I never had a regular source of income other than my allowance. I once tried delivering fliers to mailboxes for Mac's Donut Shop (half a cent per flier), but I couldn't compete on volume with his other kid employees—who were delivering their entire stacks into a storm drain. Mac never did any quality control and I was let go. A few times, somewhat later, I did babysitting jobs that my sister Paula turned down. The unruly Kamb twins, Barky and Sasha, were closest to being regulars. Their parents were Caltech professor Barkley Kamb and Linda Pauling Kamb, who was the daughter of Linus Pauling, the two-time Nobel Prize winner (Chemistry and Peace). Barkley Kamb did his Ph.D. work under Pauling and "married the boss's daughter". James Watson's 1968 book *The Double Helix* describes an earlier, wild young Linda Pauling and her role as muse in the discovery of the structure of DNA.

In junior high, most boys became Boy Scouts. Our required shop courses generated merit badges essentially automatically—the shop teacher had only to sign a form—so one could advance through the first few Boy Scout ranks quite easily. I was a member of Troop 1, historically the first Boy Scout troop in the San Gabriel Valley, sponsored by the Pasadena Methodist Church. I quickly went from Tenderfoot to Second Class and then First Class Scout. Glenn Rankin and I were soon known as the troublemakers in the troop. The adult Scoutmaster and his assistants, who included our troop's small number of older Eagle Scouts, did not hide their hostility toward us. Their animus only made us more admired by our peers. We were the rebels—the courageous individualists, the lone wolves. Well, in truth, Glenn and I divided up that responsibility: He was more the lone wolf. I was more the smart-alecky kid who was always sassing the grownups. We worked as a team.

A lot of my authority derived from my generosity in surreptitiously distributing fireworks—firecrackers, cherry bombs, and skyrockets. Our family made trips across the border to Tijuana, irregularly but usually once or twice a year. Billie was still teaching fourth grade, and there was a Latin America unit in the curriculum; she bought souvenirs and craft items for class enrichment. My sister and I were allowed to roam freely on the main tourist shopping street. With all the money that I could save from my allowance, I bought fireworks, by weight as much as five or ten pounds on every trip. Bringing such contraband back across the U.S. border was illegal, but these were simpler times. There was no big drug problem (at least none acknowledged), and Tijuana was viewed as almost a neighborhood of San Diego. Our station wagon, with my cache under the back seat, was routinely waved across the border by the customs officers.

Our troop, consisting of four patrols, each with eight to ten scouts, went on weekend camping trips every month or two. Almost always we camped in the California desert, familiar to me from our family camping trips, with the difference that, in Boy Scouts, you had to carry your own sleeping bag and food in a backpack. I always took some cherry bombs and matches along for amusement on long hikes or in the evening around the campfire. Casually tossing a cherry bomb into the campfire produced generally hilarious results. However, I quickly learned that my stash was more valuable as stock-in-trade than for entertainment. I was able to trade firecrackers for other scouts' supplies at a very favorable exchange rate. My pack became much lighter once I understood this. On camping weekends there were mandatory church services on Sunday morning. We all had to learn and recite the Methodist version of the Lord's Prayer. For Catholics, this was only the change of a few words and phrases, but Jews had to learn the whole thing from scratch.

The Boy Scout camp that I went to in the summer after seventh grade, on Catalina Island, was named Cherry Valley. I was looking forward to helping it live up to its name, explosives-wise, but I was forewarned by a scout who had been there: Right when you got off the boat at the Catalina Isthmus, they inspected your pack for contraband. So, involuntarily, I was a model scout at Cherry Valley. I earned enough merit badges to advance in rank to Star Scout. I had only two ranks to go, Life Scout, and then the coveted Eagle.

Once a year, four Patrol Leaders were elected democratically by the scouts in the troop. Glenn and I both decided to run for office. Our Scoutmaster, Mr. Thomasson, made a speech to the boys, describing the qualities of a good Patrol Leader. They were all things in which I might be viewed as deficient. "Respect for authority," "courtesy to others," and "religious belief" were among them. He was transparently campaigning against me! When the secret ballots were counted, Glenn and I both won. I like to think that this had nothing to do with the fact that the tellers, the two scouts who counted the ballots, were friends of ours; and that we had promised them *many* cherry bombs if we *both* won. I became Patrol Leader of the Ram Patrol. Glenn, leader of the Bears.

It was a great year for Troop 1 before the adults did finally come to their senses and expel me from Scouting. My patrol had off-book meetings, just us kids. We experimented with Drano bombs. We taught ourselves to build match throwers, simple devices made from spring clothespins that, when cocked and triggered, would simultaneously ignite a kitchen match and propel it, flaming, with a range of ten feet or more. We took long hikes in the storm drains that ran beneath the streets of Altadena. Separate from the sanitary sewers, they were relatively clean and almost always dry. We explored, and we drew and distributed maps.

Ironically, it was the storm drains, not the fireworks, that got me kicked out of Scouting: On one of our covert hikes in the underground storm drains, Billy Thomasson, our Scoutmaster's son, got stuck between the iron bars of a massive debris gate that the rest of us were able to shimmy through. He panicked and started crying, shaking off all attempts to free him. It was obvious that, when he calmed down, he would be able to back out of the grate exactly as he got in. We left him with a flashlight. Unfortunately, when he did get home, unharmed, he told his father everything. That was it for my Scouting career. I never got the chance to go for Eagle Scout. Later in life I met and worked with many former Eagle Scouts, often retired senior military officers, and came to understand that their concepts of honor and leadership were somewhat different from mine, and that I might not have fit in well among their number.

In ninth grade, I took an elective journalism class. What stands out about that class had little to do with journalism. It was in Mr. Clement's class that I fell in love with Melanie Michaels, whom I had sat next to in Sunday school for a year. This was a love far beyond anything that I had felt for her at the Pasadena Jewish Temple. This was my first true (and, eventually, first reciprocal) love. It was innocently Platonic—we got no farther than holding hands. School and after-school activities brought us together: We were both elected to the Student Council. Mel was elected Secretary; I was something called Traffic Commissioner, which was about appointing the student hallway monitors who wore bright green triangular sashes as badges of office. And we were co-editors of the school newspaper, *Husky Highlights*. In one issue, we made ourselves Boy and Girl of the Month. Even Mr. Clement was surprised at this forwardness. We went to school dances together, something I had never done before. My parents were suddenly forgiven for having made me take dancing lessons in the Gollatz Cotillion in seventh grade.

Probably the only example of overt anti-Semitism that I experienced growing up had to do with those seventh-grade dancing lessons. Most of the middle-class kids in my school took dancing lessons at the Altadena Cotillion, sponsored by the Altadena Country Club. If your parents were not members of the country club, any other member could sponsor you. Jews, however, were not welcome. Instead, Jewish children schlepped to the ballroom at the Pasadena Civic Auditorium where Miss Gollatz held her competing cotillion—of mostly Jews. I was about as good at dancing as I was at sports.

Poignantly, my crush on Melanie was *doomed* true love: It had been announced that neighborhood school boundaries were being redrawn, and that a small number of Eliot Junior High students would, because of where they lived, graduate not to Pasadena High School with the rest of us, but to John Muir High School on the other side of town. The Michaels family lived at the top of Lake Avenue and were in that small number. At the final school dance, Mel and I clung tightly to each other even as we danced awkwardly. Our special song was "Misty."

So that is why, after ninth grade graduation, I insisted on taking summer-school second year French at John Muir High school. Melanie and I had agreed that she would sign up for the same course, so that we could continue to be together at least for the summer. She never showed up. The class read Saint-Exupery's *The Little Prince*, actually in French. Our teacher, Mr. Swan, claimed to have worked at Brentano's in Paris and to have once sold a book to Ernest Hemingway, perhaps dubious assertions. During my high school years, I did encounter Melanie randomly a couple of times around Pasadena, but—and I really hope this is an original turn of phrase—the magic was gone. A few years after that, I heard that she had dropped out of college and was a topless dancer in the Haight, in San Francisco. She did have the upper-body prerequisite for that line of work; but I never tracked her down to check on what was surely a malicious rumor.

8. High School

My high school locker was on an outside wall of the music building next to the gym. I was getting books from my locker during the class break at 10:55 a.m. on November 22, 1963, when I heard that President Kennedy had been shot. That open locker and its contents, at that exact moment, became a permanent visual memory. The Pasadena High School campus was built for 4,000 students. During my time there, trailer-like temporary classrooms were already being built in the parking lot for the baby boomer overflow enrollment. My high school class had 1,500 students.

Pasadena High School had in its history graduated a few famous alumni: General George S. Patton; Howard Hawks, the film director; Edwin McMillan, the Nobel Laureate in Chemistry. Any ordered list becomes rapidly less famous after that, and they are mostly minor sports figures. There was no sense of history or tradition in the school. We were baby boomer adolescents living in an eternal present. Anne Roosevelt, a granddaughter of Franklin and Eleanor, was in my high-school class and wrote a nice note in my 11th grade yearbook.

Although de-facto racially segregated, P.H.S. otherwise drew on a cross-section of Pasadena. My yearbooks show classmates with Hispanic, Japanese, Italian, Irish, and Jewish names, interspersed among the Protestant majority and Catholics minority. There are a lot of Armenian names. What is striking to me—and would be almost unimaginable today—is how little sense of ethnicity and social class we students had at the time. Families moving to Southern California during the 1950s and 1960s boom years were leaving such labels behind. In place and time, these were peak years for the melting pot. Identity politics was not just a term then unknown; it was un-American and unthinkable.

At high-school reunions in the decades following, it was sobering to watch our self-invented social structure slowly fade. It was still there vestigially: Rick Osborn, our handsome star athlete and valedictorian, held court at reunions decades later, along with his hovering sidekick Mark Swed. Outside of reunions, Rick had dropped out of college, taken up drugs (they said), and became (I heard) a carpenter; while Mark was the chief music critic for the *Los Angeles Times* and a Pulitzer Prize finalist. But we could not escape out ethnic and socioeconomic identities, it turned out. By my 25th reunion in 1990, Suzy Feinstein (whom I had known since eighth grade Library Practice) was more Jewish than I ever recalled; Linda Jacobs had a Ph.D. and turned out to be from a wealthy Lebanese-American family. Amy Uyematsu had become a poet and a voice of the Japanese-American experience. Her parents were interned in the camps during WWII—who knew? In adult life, my classmates had acquired hyphenated identities.

During tenth and eleventh grades, we learned where we fit in this academic hierarchy—learned indirectly by what cohort of students turned up repeatedly assigned with us to the same classes. Behind the scenes, invisibly, we were being sorted into academic tracks: The best students were assigned to the best teachers. The best teachers taught only the best classes. It was a great system—if you were in the top group (and of course white). By the start of my senior year, the big pond of 1,500 had, for most purposes, shrunk to a group of a hundred or so of us who knew each other well, were on our way to good colleges, and (in matters of student influence) ran the school. We were by then the big fish.

In tenth grade, Drivers Ed was the most important class. If you took Drivers Ed in school, you could get a learner's driving permit at age 15 ½. For me, the magic date was November 23, 1963. This turned out to be a Saturday, the day after President Kennedy was shot. Still, my father drove me to the Department of Motor Vehicles office in Pasadena. We filled out the form and waited in a long line. The clerk pointed to a line left blank on the form. "Both parents need to sign." Without missing a beat, Frank said, "His mother is waiting in the car. I'll go get her to sign." We went outside. Frank forged my mother's signature on the form (in a handwriting quite different from his own), and I soon had my permit. In his long career as a scientist, later White House advisor and then President of the National Academy of Sciences, Frank was known as the straightest of straight arrows; but we in the family knew that there were occasions like this when he could be quite bent.

Soon after that, with a loan of infinity months of allowance from my parents—which I knew would eventually be commuted to time served—I was driving my own new Honda S65 motorcycle. Barely deserving the name, the S65 was really a motorbike with only a 65 cubic centimeter engine. The attraction was that its design had motorcycle pretensions.

Apart from my not having to take the school bus any more, one of the biggest changes my Honda effected was providing fast access to the main branch of the Pasadena Public Library. In elementary school and junior high, my cruising range by bicycle included only the Santa Catalina

Branch, whose collection got me through the entire series of Freddy the Pig books—what I liked about them is obscure to me now—and all the Doctor Dolittles. I was able to read Nancy Drew books only at friends' houses, since the public library considered them unworthy children's literature. Ditto for Hardy Boys mysteries (although I didn't like them much anyway). I owned a small collection of the also-deemed-unworthy Tom Swift Junior books. That led a sympathetic librarian to direct me to higher quality juvenile science fiction: Ellen MacGregor's eponymous *Miss Pickerell* series, Carl Biemiller's *Magic Ball from Mars*, Eleanor Cameron's *Mushroom Planet* series, and ultimately to the juvenile science fiction novels of Robert Heinlein.

In the 1940s, Heinlein had written exclusively for adult science fiction magazines, but in 1947, the famous children's books editor at Scribner's, Alice Dalgliesh, convinced him to try his hand at a novel for older children—a market that would later become known as "young adult". His *Rocket Ship Galileo* was published in 1947. After that, new Heinlein juveniles came out about annually. I discovered them in the Santa Catalina Branch about when *Have Spacesuit–Will Travel*, the last in the series, had just been published. I rocketed through the whole series. The plots, in various futuristic but painstakingly scientifically accurate settings, generally involved conflict between a teenage boy and his parents or other authority figures. I leave you to guess who prevailed in every case. Dalgliesh and Heinlein were often at odds over his wanting to introduce more adult themes into the books (read p-o-l-i-t-i-c-s and s-e-x). His penultimate juvenile novel, *Citizen of the Galaxy* (1957) was judged by the Santa Catalina Branch Library as too intense for the juvenile section, because of its themes of slavery and forced marriage—ritual rape being strongly implied. It was classified as adult science fiction, and I had to provide a note from my mother before I could read it.

By the time I was riding my Honda to the Central Library on Walnut Street in downtown Pasadena, where there were no restrictions on what I could check out, Heinlein had parted ways with Dalgliesh and was publishing adult sci-fi novels about yearly. I was on the library's waiting list for each. *Stranger in a Strange Land* (1961), with themes of free love, teleportation, and the eventual emergence of a Nietzschean master race—one of Heinlein's favorite motifs—is considered his most influential work. Charles Manson, whose followers murdered actress Sharon Tate and others in 1969, reportedly modeled his cult on the book.

It is hard to describe, let alone explain, the lifelong effect that Heinlein had on a whole (male) generation of technically minded baby boomers. The stereotypical libertarianism of Silicon Valley entrepreneurs owes as much to Heinlein as it does to the more often cited Ayn Rand. Years later, when I got to know this crowd, I frequently heard echoes of Heinlein in their conversation, and I sometimes would ask, "were you a Heinlein fan?" "Of course." Rand and Heinlein both postured as uncompromising laissez-faire capitalists, but they were different. Rand's protagonists demonstrated heroism by glorifying their individual integrity above the common good. Heinlein's heroes did the integrity thing, yes, but their heroism contrived to solve the future world's problems. Politically, Rand was anti-interventionist. By contrast, Heinlein had been a naval officer and engineer in the 1930s. Several of his works, notably *Starship Troopers* (1959), trot out the right-wing idea that only blooded war-combat veterans should be allowed to vote—because they alone could understand the value of the franchise. But before swinging to the extreme right, Heinlein was a self-described FDR liberal, and there are traces of that here and there in the novels. His Nietzschean supermen and sexually available superwomen often work to benefit the unwitting masses of little people. How could teenage boys in the 1960s possibly be taken in by this right-wing—bordering on fascistic—clap-trap? The question is more, how could we *not* be? We, after all, were going to be the supermen; and we hoped to soon meet the superwomen, especially the ones who (in the adult novels) were constantly in a state of undress.

As avid a fan as I was, I can make the case that I was not completely taken in. I was not politically unaware: I read Walter Lippmann's column every week in my parents' *Newsweek* subscription. I was a member of the Foothill Young Democrats and stuffed envelopes for Lyndon Johnson's campaign against Barry Goldwater in 1964. Heinlein was forbidden fruit.

I started winning math competitions in 11th grade: second place in giving a short math lecture at Cal Poly State College; first place in the problem-solving test at Occidental College. The prizes were either math books or slide rules, sometimes expensive ones. I still have a drawer full of slide rules, now useless except as objects of nostalgia. My chemistry teacher, Mr. Sadoff tutored me after school for the American Chemical Society national chemistry contest. I placed fourth in Southern California, one place below the rank that went on to the national competition. My score did get me invited, along with Mr. Sadoff, to the "National Youth Conference on the Atom," in Chicago. We were sponsored by the Southern California Edison Company, which was uncritically promoting the benefits of its nuclear power stations. Our pictures were in the *Pasadena Star-News*.

Between eleventh and twelfth grades, I participated in a residential summer physics program at Cornell University in Ithaca, one of several science programs around the country sponsored by the National Science Foundation—another post-Sputnik initiative. Many later-distinguished scientists were students in these influential summer programs. In giving an experience in science research to students the summer before they were applying to colleges, NSF had hit a sweet spot. Except for summer camp and staying with grandparents, this was the first time I lived away from home. Far above Cayuga's waters, two dormitories (a boys' and a girls') were given over to high school students in this and other programs at Cornell. There were thirty-six of us in the physics program. My roommate was Ben Orlove from New York City. We ate in Willard Straight Hall with the college-age summer school students. There was a girl that I liked—Judy Lieberman, whom I would meet again later at Harvard (and for fifty years afterwards)—and I worked up the courage to invite her on a movie date. The film was Ingmar Bergman's *The Seventh Seal*, not exactly a feel-good movie. I was clueless.

I won various math awards in 12th grade and accumulated a lot more slide rules. I had the highest score in Southern California on the national math contest exam sponsored by the Mathematical Association of America; but it was a mediocre score compared to the best students in, say, New York City or even Northern California. I commuted on my motorcycle to Pasadena City College for its first-year calculus class. There were too few students at P.H.S. for the class to be offered there. We used the famously abstract textbook by Caltech professor Tom Apostol, a poor choice because it was mostly useless in preparing for the national college Advanced Placement exam. My father gave me his college calculus textbook, written in the 1920s, to study independently. Calculus is unchanging, but the way that it is taught changes appreciably from generation to generation; in this respect it is like Shakespeare. Frank brought home also a copy of the new *Feynman Lectures in Physics,* a book that was generating a buzz not just at Caltech, where Dick Feynman was a professor, but nationally. Feynman's approach was quirky and brilliant. I was enormously influenced by the book (and its two later volumes), first as a student, and later as a professor of physics.

My College Board test scores were all well into the 99th percentile, except for German with a 98. I applied to MIT, Yale, Harvard, and Berkeley. My application essay included the alarming sentence, "Twenty years from now, I would like to be a professor of theoretical physics or of mathematics in a (preferably large) university." All four schools admitted me. Ed Colaianni from John Muir High School, who had been an Eliot Electroneer with me in junior high school, was admitted and went with me to Harvard. We were the only two from the city of Pasadena in our Harvard class. In a freshman class of just over a thousand, Harvard admitted just 29 students from the Greater Los Angeles area, with only about half accepting admission. Only about half of my high-school class went on to college at all, most of those to Pasadena City College, which offered a two-year associates degree.

Margaret Lauritsen was a girl I knew from seventh grade math class. She had disappeared when her family moved to a different junior high school district, and reappeared only in my senior year of high school. The year before, she was in Denmark. Her father, a Caltech physics professor, was on leave at the Niels Bohr Institute. Tommy Lauritsen had been Niels Bohr's postdoc in Copenhagen just before World War II, and he was still a close friend and colleague of Bohr's son Aage (also a physicist). Margaret wore long skirts and loose peasanty blouses, like some lost village girl from a Grimm's fairy tale. Her long, light-brown hair was woven into a thick single braid.

It was logical that Margaret and I should end up paired in the senior-year social hierarchy. Our fathers were both Caltech professors. We were both socially awkward. We were both good at math—from Mr. Benedict's seventh grade New Math class, we both knew some set theory. We fell in love—what else could we do, really? We shared ingenious math puzzles from Martin Gardner's column in *Scientific American*. Margaret taught me to play Go, the game with black and white stones on a wooden grid. (We were both duffers.) Together, we discovered cat's cradle, the string game, and found in the library whole volumes, written by anthropologists, full of instructions for recreating the string figures and associated games of the Polynesians. The Polynesians! You can't get more romantic than the Polynesians! I had read Margaret Mead in social studies class.

9. Getting to Harvard

My parents pushed me to apply for summer jobs—what today would be internships—at several national laboratories, and I was offered a job for ten weeks in the Birge-Powell experimental physics group at the Lawrence Radiation Laboratory, located on the hill overlooking the Berkeley campus. My salary was $75 a week, the equivalent today of about $15 an hour. The official letter that I got, drafted in the Lab's employment office, made it sound like a real job, specifying working hours, benefits (none, basically), and so on. At the bottom of the letter, Birge or Powell (I forget which) had written in pen, "Pay no attention to the above. You can do as much or as little as you like." Rather than drive me to Berkeley, about an eight-hour drive, my father drove me to L.A. airport, and handed me an airplane ticket and cab fare for the ride from Oakland Airport. I struggled to manage a 40 lb. suitcase, a portable typewriter, and a loose pile of books. Described in this way, my parents' attitude sounds unfeeling. I think it was simply a sign of their confidence in me.

The next ten were the loneliest weeks in my life—possibly even in my life since. The "work" was interesting. I taught myself to program in Fortran, keyboarding IBM cards on the IBM 026 punch card machine and then submitting the deck at the computer center window, to be run on the giant IBM 7044 computer, a machine vastly less powerful than today's smart watches—probably less powerful than a high-end toaster. There were two bubble-chamber groups at the Rad Lab. A bubble chamber was a device, about the size of a small truck, for photographing the paths and decays of particles (protons, electrons, and things more exotic) from the (then-) giant particle accelerator, the Bevatron. The group doing outstanding work that soon won Nobel prizes was the Alvarez group. (We will meet Luis Alvarez at length later in my story.) Birge-Powell was the second-rate group. I didn't understand this at the time, of course. Birge was 78 years old, Powell was in his mid-60s. Both were over the hill.

The work of the group was *de facto* supervised by two younger physicists George Kalmus (a Brit) and Bob Ely (an American). Each took a modest interest in what I did during working hours. Kalmus told me that, when he was a child in England during the War, there were no raspberries for jam, but there was a company that made fake berry seeds out of wood chips for inclusion in the artificially flavored jelly.

I made no attempt to meet anyone socially; I had no idea how to. I went to free music recitals and concerts. I spent Saturday and Sunday mornings in the Morrison Library's music listening room, which had turntables, headphones, and an extensive record collection. I went to movies alone. I read books of Shaw plays. I made lists of courses in the Harvard catalog that I hoped to take in the next four years. Margaret and I wrote to each other daily, and we pledged our mutual and undying love. She visited me at Berkeley just once—with her parents—when she was moving into her freshman dorm at the early start of the fall semester. She would be at Berkeley while I was at Harvard.

I had decided to go to Harvard secure in the knowledge that I would be leaving my parents three thousand miles behind in Pasadena. In the middle of that summer, Frank announced that he was leaving Caltech for MIT, where he would become the department head of Earth and Planetary Sciences. He and Billie must have known about this for many months, but Paula and I had been kept in the dark. My precious possessions, "stuff," in Altadena would be packed by anonymous movers and shipped to storage in Boston, where it would be inaccessible to me until my parents bought a house there. In the meantime, they rented a tiny house in Cambridge with a bedroom for Paula (who was entering high school at the Buckingham School) but—aside from a sofa-bed in the living room—no space for me. In the same burst of communication, my father told me what financial support I might expect from now on: He would pay my tuition, room, board, and laundry. I was expected to pay for everything else out of summer earnings or term-time jobs. He did not ask me to pay back the twenty-five weeks that I was in arrears on allowance, commenting only, "That's between you and your mother."

Thus was I cast out into the world at age eighteen, not exactly marooned, not exactly penniless, not exactly orphaned, but with quasi-elements of all three. I was quasi- on my own. My two priorities were to embrace Harvard undergraduate life with the fervor that I had previously shown for learning calculus; and to embrace Margaret with the fervor, etc., etc. These priorities were on opposite coasts, unfortunately.

* * *

Harvard! It is the single institution that most influenced my life. I was an undergraduate from 1965 to 1969, went away for just seven years, and returned as a full professor to stay for another almost twenty-five. It is fair to say that, to the extent institutions can be understood at all, I know the place.

Institutions have a brain, and they also have a gut. The brain of Harvard—the president, deans, and faculty—is in the business of higher education and research. The gut of Harvard—the traditions, the buildings, the creaky processes and procedures, the Admissions Office, the non-academic workforce, the fifty thousand living alumni of the College—is in a compatible but different business, the training of young people for membership in the elite power structure of the United States and (to some extent) world. That function has flourished through three centuries.

I was one of about a thousand incoming freshmen. We were assigned to dorms in and around the Harvard Yard. Upperclassmen lived in the Houses closer to the Charles River. My roommates were Randy Chase, who grew up in New York City and graduated from elite Fieldston, and John Moretti, who grew up in a tough Catholic neighborhood in Cleveland. John was one of Harvard Admission's obvious experiments in upward social mobility. He had what looked like knife scars on his chest and back.

In the week before older students arrived, the thousand of us were supposed to meet each other and learn the rules. Closer to the truth, the preppies, about half the class, met each other; and, separately, we in the other half of the class met each other. This divide diminished in subsequent years, but never disappeared. On my side of the divide, I derived instant status by having been Ben Orlove's roommate at Cornell the previous summer. He was already a legend in our class: In his senior year of high school, he did eight Advance Placement exams and got 5's (the highest grade) on all of them. And that was not something that preppies would admire at all.

I soon learned that most of my classmates were not, by my standards, terrifically smart. This was a shock to me. However, they were all terrifically *something*. That was how Harvard admissions worked: You had to be smart enough to probably not flunk out. (Only fifty or so of us thousand would fail to graduate, albeit some taking an extra year.) Above that threshold, smart didn't count for much. What mattered, according to lore, was that you excelled in some particular thing—whether academic, athletic, political, literary, musical, artistic, or odd-ball other. I now think that this conventional wisdom was retrofitting an explanation to the hidden truth already mentioned, that the admissions process had been tuned, literally over centuries, to populate the power elite. The fact that students already all excelled in something was the symptom, not the disease. The disease was their future success.

President Pusey held a reception for the incoming class, jacket and tie mandatory. There was a formal receiving line, the first I had ever seen. You spoke your name to the first person in the receiving line—the assistant dean of freshmen, say—who would in turn introduce you by name to the next person, and so on. Our dorm proctor drilled us in advance on the protocol. I was fascinated by the fact that errors could propagate and accumulate, so that by the time you reached the end of the line, with the highest-ranking dignitaries, your name might be completely garbled, and it would be bad manners to correct it. It might as well have been a metaphor for the elite that I was being groomed for.

Undergraduate organizations came door-to-door in the freshman dorms recruiting new members. Jim Roosevelt knocked for the Young Democrats. I mentioned that I knew his sister Anne in Pasadena. A political recruiter from farther left queried, as if it was the most ordinary question in the world, "Are you Mao or Trot?" I had no idea what he was talking about. I joined something called *Verein Turmwächter*, the Harvard German Club. They met for dinners and good conversation—for which my high-school German was, alas, insufficient. I did become their volunteer projectionist and watched many German movies (with English subtitles) from the projection booth in Emerson Hall.

Organizations like the Harvard Crimson (the newspaper), Harvard Lampoon (the humor magazine), Harvard Advocate (the literary magazine), or WHRB (the radio station) sought undergraduates who would devote themselves to their respective organizations body and soul. The pitch was subtle: Sign up for our "comp" (competition) and, *perhaps*, we'll *invite* you to be a member. I signed up for WHRB. I had heard that you pretty much had to be a preppy to get into the Crimson, Lampoon, or Advocate.

My first-year physics was taught by a young assistant professor, Daniel Kleppner. Kleppner later became an MIT professor and a founder of the field of ultra-cold atomic physics. He was often talked of as a possible Nobel Prize recipient, but he never won one. I was completely lost for the first month or so of his course, but then, suddenly, everything *clicked* and I could do it all. My math course was second-year calculus, but it was the version for scientists, not the version for would-be math majors—which I had thought myself to be. A student math advisor, a Radcliffe student named Claire Max, warned me that the course for concentrators (Harvard's term for majors) was too hard for any freshman from a public school that wasn't the Bronx High School of Science. Claire, at the time a sophomore, ultimately became a famous astronomer. We became friends decades after this first interaction, of which she has no memory.

I could take pretty much any other two courses I wanted, so I did two history courses: The large lecture course in East Asian History and Culture was normally taught by John Fairbank (the lectures on China) and Edwin Reischauer (the lectures on Japan). However, while on leave as U.S. ambassador to Japan, Reischauer had been stabbed by a Japanese anarchist and contracted hepatitis. So Albert Craig, a lesser light, took over his lectures. In the late 1940s, Fairbank was one of the "old China hands" accused of disloyalty to the United States because they predicted, correctly, the success of Mao and the Communist Party. Fairbank was considered the leading China scholar in the United States in the post-War period.

I found this course almost impossibly difficult. East Asian History was my first introduction to a *real* Harvard course, one with hundreds of pages of reading every week, and the memorization of hundreds of names and dates—then requiring, on exams, the requirement to integrate all these facts into coherent, well-supported arguments. I was used to multiple choice exams. This was very different. The course also gave me a new respect for preppies. They (at least some of them) could actually do this. They knew how to study. I had no idea—made obvious by my first-semester grade, a C+.

My other history course, English Social History, years 400 to 1648, was the opposite in every possible way. The class was small, less than 30 students. It was taught by George Homans, a charming man from a distinguished Old Boston family, himself a great-great-grandson of John Quincy Adams, and a Harvard lifer. He matriculated at the College in 1928 and, other than for service as a Naval Reserve officer in World War II, never left. Although it was an upper-level course, there were no prerequisites. There was hardly any reading. Homans' delightful lectures were not just filled with charming anecdotes—that was all there was to them. I learned that the song "Oats, Peas, Beans, and Barley Grow" referred to the rotation of crops in medieval strip farming; that Hrolf the Walker, a Dane who invaded Normandy in 911, was so called because he was too fat to ride a horse; that Eivar the Boneless—he must have done something, but I now forget what. There was a midterm exam that one could elect not to have counted, and a short term-paper.

I audited two courses taught by distinguished professors—if not quite at the Fairbank/Reischauer level. Politics and Society was taught by Seymour Lipset, an eminent sociologist and theorist of American democracy, and, later, an early Neo-Conservative. The History of 18th and 19th Century Europe was taught by Crane Brinton, a leading historian of France, and a recent past president of the American Historical Association. Lipset's course was difficult. His lectures were dense. I was glad not to be taking it for a grade. Brinton's lectures were much like George Homans, long on charm and short on rigor. In later years, the course, taught at 11 a.m., became known as "Brunch with Brinton". The local term for such courses was "gut". Gut courses were Harvard's way of ensuring that the less intellectually brilliant future leaders of America had a path through the College to a decent grade point average and to a degree not in an unsavory field like physical education or communications (majors that did not exist at Harvard), but, rather, a respectable major of entirely gut courses in government and history. Much later, as a Harvard professor, I taught such a course myself: Astronomy 1.

10. Radio Days

WHRB, which insiders knew to pronounce as "wherb", had captured me in a major way. I spent many hours at the radio station "comping," that is, being trained for full membership. The studios, so-called, were a rabbit-warren of half a dozen rooms in the basement of Claverly Hall, a part of Adams House. In the comp, one had to declare a major and minor field; mine were classical music (called "CM") and controlling. CM meant constructing future broadcasts from the station's large, but not unlimited, access to classical LP records. Controlling (known more formally as studio engineering) meant being on the other side of the glass from the announcer and operating the turntables, tape recorders, and the many switches and knobs on the console. At sign-off, WHRB played not the national anthem (as did all other Boston radio stations), but "Fair Harvard".

A very small number of WHRBies ("wherbies") went on to noteworthy careers in broadcasting. Best known from my Harvard class is Chris Wallace (Fox News)—but he was special from the start as the son of broadcast journalist Mike Wallace. From other epochs, Bruce Morton (CNN) and Scott Horsley (NPR) were in their times nationally familiar names. Martin Bookspan, a couple of decades before me, became the voice of the New York Philharmonic. *New Yorker* music critic Alex Ross was a couple of decades after me. I give this list not because it is impressive, but because it is so thin. A corresponding list of *Harvard Crimson* alumni would go on for pages and include U.S. presidents as lesser lights. There was something willingly proletarian and anti-establishment about WHRB. I could sense it. I loved it. Progressive journalist Sam Smith, who was a whrbie in the late 1950s, once wrote: "WHRB functioned as a counter-fraternity, a *salon des refuses* for all those who because of ethnicity, class or inclination, did not fit the mold of Harvard. Other organizations sought students of the 'right type,' WHRB got what was left over." By the late 1960s, nothing had changed. The radio station became my home, social network, and emotional support system.

Later in life I noticed that my natural roles in larger institutional settings often unconsciously mimicked positions that I had at WHRB. The radio station became—unknowably at the time—a template for my professional life. At WHRB I was regarded as someone who could do almost any job for a short time, but—with the single exception of my quickly becoming Classical Music Director—did not have the necessary attention span to advance through the established hierarchy. I was several times *acting* Chief Engineer, *acting* Program Director, *acting* Station Manager—you get the idea—all of which I did well enough to be invited back the next time an *acting* was needed. I was a talented pinch hitter and a poor team player. Spoiler alert: And so ever since.

Early in December, I was invited to a sort of tea with Mrs. Bunting (as she was called), the President of Radcliffe. Twenty or so of us Harvard freshmen were invited but, to our disappointment, no Radcliffe women. From my summer job in Berkeley, I knew that Dr. Mary Bunting was an accomplished scientist who had recently served as a Presidential appointee on the Atomic Energy Commission. Notwithstanding, she came across to us as a kindly aunt, asking us questions about how we liked Harvard and giving us advice on the best ways to meet Cliffies. She was open about the differences between Harvard and Radcliffe admissions. Harvard liked to gamble, she said, and to take chances on students from diverse backgrounds. Radcliffe preferred to admit a homogeneous class of girls who, often like their mothers, were sure to succeed at Radcliffe. Having made that choice, Radcliffe could be more liberal in its rules because, in the homogeneous population, there would on average be fewer behavioral outliers. I liked Mrs. Bunting's use of a statistical argument about the tails of distributions to make her point. She stepped down as Radcliffe President seven years later, by which time the niceties of her statistics were completely swept away by events of the later 1960s. Soon after that, the Radcliffe admissions office would cease to exist. Harvard admitted both men and women, and it continued to "gamble" on diversity, now for both sexes.

Apart from the radio station, my circle of friends was limited. Randy, John, and I, as roommates, spent endless hours in bullshit discussions of every conceivable subject and made many joint trips to Elsie's Sandwich Shop for roast beef specials, rare roast beef piled high on a Kaiser roll with relish and onion. It cost 50 cents (about $4 today), but our friendship didn't go beyond that. All three Ewing kids from Lamont were at Harvard that year, Peter Ewing in my class; but they were on the far side of the preppy/non-preppy divide. That I mostly hung around with Ben Orlove (from Cornell summer) and Ed Colaianni (from Pasadena) shows how bad I was at making new friends.

I was elected a full member of WHRB early in January. Math and Chinese history, as full year courses, took up from where they left off. In physics, the subject was no longer mechanics, but now electromagnetism, taught by Edward Purcell, who had won the 1952 Nobel Prize in Physics for his discovery of nuclear magnetic resonance—the physical phenomenon behind MRI medical imaging. Ed Purcell and my father had served together on President Kennedy's Science Advisory Committee. Purcell's course used his own textbook, a volume in the Berkeley Physics Course (another post-Sputnik project sponsored by the National Science Foundation). Purcell was a brilliant lecturer, yet soft-spoken and meticulous. He never pulled a rabbit out of the hat, or sprang the surprising answer. Many physics lecturers did do that—physicists always like to prove that they are the smartest people in the room. When Purcell reached a result on the blackboard, you knew *exactly* how he had gotten there. You had *learned* something. He made you feel that *you* were the smartest person in the room.

I needed a course to fill the slot previously occupied by Homans' English Social History. Before final study cards were due, I visited several: a linguistics course, an economics course, a literature course on Dostoyevsky. In the end, I signed up again with George Homans. Spring semesters, he taught a sociology course, Introduction to the Study of Small Groups, an entertaining gut, much in the mold of his history course. I knew I had made the right choice when the student sitting next to me pointed out the whole Harvard football team (by name and starting position). The reading list included a lot of B.F. Skinner, the Harvard professor and pioneer of modern behaviorism, including his utopian novel *Walden Two*. Homans always referred to Skinner in class as "Fred".

And then Gen Ed A, the composition course: What? You thought that I chose to place out of English composition class when offered the chance? A shameful admission for a budding physicist, but I actually enjoyed writing. The honors sections were all different. I had requested the same science-themed section as Ben Orlove, but was mistakenly assigned to another section that met at the same time. My sectionlady was a graduate student of Harvard professor Erik Erikson, the polymath psychologist-psychoanalyst whose own psychoanalysis had been under Anna Freud in Vienna in 1933. We were assigned Erikson's *Young Man Luther*, a psychoanalytic biography, and were then supposed to choose our own historical figure and write our own (much shorter) biography, informed if not by psychoanalysis, then by some other identifiable discipline. I wrote something on the Hungarian-American physicist Leo Szilard, ghostwriter of the famous Einstein letter to Roosevelt that produced the Manhattan Project. My essay wasn't very good, but instead of giving me a bad grade, my sectionlady returned it to me and asked me to submit something completely different. I complied maliciously by writing a reasonably good essay on the manifest flaws of Erik Erikson's psychoanalytic method. She graded it an A- without commenting, and I got an A for the course.

I was learning classical music by scheduling for broadcast pieces I had never heard, by composers that I knew nothing about—and then listening to them for the first time when they were on the air. My other work for the radio station's classical music department was writing "features," one-hour weekly shows on a particular theme, like "Music of the Twentieth Century" or "Choral Masterworks". You picked one or more works, and then wrote four to six minutes of script that was read by the announcer in appropriate segments and supposed to educate the listener. Faking this kind of knowledge was easy: I became facile with the ten-volume Grove's *Dictionary of Music and Musicians*—and there was always the record jacket copy as backup.

Radio work imparts the discipline of working to a deadline. If my feature was scheduled for eight o'clock, then, whether or not I was there with a script, the controller would start the theme music rolling at 8:00:01, right after the Western Union beep (which went over the air automatically at 8:00:00). At 8:00:20, the controller would fade the theme music and cut to the announcer, who *had to have a script to read.* Five minutes late was not good enough. This was a life's lesson: Some commitments are "work to quality". In these, what is important is not primarily *when* they are completed, but *how good* is the final product. Scientific research papers are like this. But other commitments are "work to deadline". You do your best; but if the quality is not up to snuff, then too bad—the show must go on. I soon figured out that almost everything required of me by Harvard was in the work-to-deadline category. I became adept at titrating the quality of my work to A or A- grades in math and physics, B+ or B in my other subjects, and to spend the minimum possible time to achieve just those grades—finishing just at the respective deadlines. I tried never to work harder than necessary.

Still, for the sake of WHRB's listeners, I felt that it would be good for me to actually know something about the music I was forcing on them; so I audited Harvard's one-semester music appreciation course, the famous Music 1: History of Western Music. This course had been given since the 1930s—taught since 1954 by the same lecturer, G. Wallace Woodworth. Except for one year of study in England, Woody Woodworth had been at Harvard since his freshman year in 1920. He was not a serious scholar, but, year after year after year, he was able to muster a remarkable degree of on-stage enthusiasm in his lectures to three hundred students. On the stage in the music department's biggest lecture hall, he raced between the podium, the piano, and a table with a record player. Most of the records from which he played selections were old 78s, not new LPs.

Music 1 was a gut course, but with an asterisk. The exams had a "drop the needle" section, where short selections would be played (dropping the phonograph arm and its needle at a certain place on a record) from works that were on the listening list. The student was supposed to identify the work. There were only a couple of dozen works on the list, and, drawn from 600 years of music, no two pieces on the list were remotely similar. Depending on your auditory memory, drop-the-needle was either completely easy (making the course a gut) or completely impossible, as for any tone-deaf student who made the terrible mistake of taking the course.

Like other Harvard gut courses, Music 1 was full of wonderful, glib factoids: The third movement of Haydn symphonies always has the rhythm of, "Are you the old lady who runs this hotel?" Brahms loved little melodies that chase their own tail, and he then inevitably later inverts them. Vivaldi never met a circle of fifths that he didn't like. Richard Strauss favors the French horn because his father was a hornist. Mozart plays three-card monte: You think you know which of three repetitions is going to be the different one, but you never get it right. The only great composers in the first half of the 20th century were Stravinsky, Prokofiev, and Bartok—Shostakovich was a hack not worth listening to. Copland was the greatest American composer.

It took me years to recognize that Music 1's view of the twentieth century had a parochial Harvard-centric tinge: Copland and Bartok, who both lived in New York City in the post-War period, had Harvard connections. Leonard Bernstein (Harvard class of 1939) also kept a connection to the Harvard music department. The dismissive view of Shostakovich came directly from Bartok, whose dislike of Shostakovich is well-known. For example, the third movement of Bartok's Concerto for Orchestra (1944) incorporates his parody of a main theme from the first movement of Shostakovich's Symphony No. 7, but exaggerated and terminated by a jeering Bronx cheer from the brass. Later that spring, I again tried out for the chorus in the Harvard G&S Players production of *Iolanthe,* and was again rejected. The starring role of the Lord Chancellor was played by an undergraduate, John Lithgow, a name that meant nothing at the time.

Each semester during reading periods, the two weeks between Harvard's end of classes and beginning of final exams, WHRB abandoned its regular programming for so-called Orgy Period, big blocks of broadcast time devoted to a single composer or artist, broadcast twenty-four hours a day. I was appointed Acting Orgy Director, "Acting" only because I was thought to be too junior for the regular title. My job was to coerce people to sign up to program and/or announce these shows. It was easier to do the work myself. On one memorable afternoon I programmed a four-hour Ives Orgy, a four-hour Mendelssohn Orgy, a three-hour Wanda Landowska (the harpsichordist) Orgy, an eight-hour Copland Orgy, and a five-hour Stravinsky Orgy. Jazz orgies I left to people who knew something about the subject. A reporter from *Boston Magazine* came to the station to interview me as Orgy Director, resulting in a talk-of-the-town style piece in their June issue. I came across as an overly precise and pedantic freshman, which was accurate enough.

During Orgy Period, I was on call whenever a member who had signed up turned into a no-show. Over the two weeks in May, I controlled and announced something like sixty hours of air, including my own sixteen hours of Gilbert and Sullivan (four hours on each of four days), and an unexpected jazz DJ shift from midnight to six a.m. The most fun was controlling for "Hillbilly at Harvard," an oddball show that normally ran on Sunday mornings and had been on since the 1950s. Before country music's nationwide revival, this was Boston radio's only country music show. It was put on by a group of station alumni who were by now lawyers and bankers when they were not masquerading as on-the-air hillbillies. During their Orgy programs, they invited listeners, especially female, to come on down to the station and join the party. With thirty people crowding into the small studio room, it almost looked like an orgy. Most of the girls seemed to be high-school students. By contrast, when I was one-manning an overnight Schoenberg Orgy and offered five dollars (then a lot of money) to the first person to call in (male or female, I didn't care), the phone didn't ring.

Margaret and I spent that summer, 1966, together in Berkeley. She enrolled in the summer offering of honors first-year calculus. George Kalmus was happy to hire me back into the Birge-Powell bubble chamber group. Margaret and I couldn't room together overtly. She found me a summer sub-let with a group of older students on LeRoy Avenue, north of campus. She shared an apartment with a group of girls on University Avenue, above a flower shop.

There was a two-week period after Margaret's summer term ended and before I had to go back to Harvard when her parents invited me to stay with them as a houseguest. I slept in the small bedroom at the back of the house that was normally occupied by Margaret's older brother Eric (whom everyone called Butch). Butch was away somewhere, but had left behind his fat, 10-foot-long snake, a boa constrictor who stared at me implacably through the glass of his large box the whole time I was there.

A crisis occurred, sub rosa, when Margaret ran out of the three-month supply of birth-control pills that had been dispensed by the Planned Parenthood clinic in Berkeley. Some backstory is needed: Margaret's grandmother Sigrid emigrated to the United States from Denmark in 1915 with Margaret's paternal grandfather Charles Lauritsen. One of Charlie's projects at Caltech in the early 1930s was to build a large X-ray machine for treating cancer patients. Sigrid accordingly enrolled in medical school at U.S.C. and became the university's first female M.D. graduate. She practiced as a cancer radiologist until she retired in the mid-1950s.

What Margaret knew, in 1966, was that Sigrid was still licensed to write prescriptions. What Margaret also knew was that Sigrid was incapable of calling anyone's bluff. She was in all matters the world's worst poker player, a trait that made her completely manipulable. The transaction went something like this. Margaret: "Sigrid, I need you to write me a prescription for birth control pills." Sigrid: "No you don't. You shouldn't be having sex." Margaret: "If you don't give me a prescription, I'll go get pregnant." Sigrid, immediately folding: "OK, what should I write?" Margaret, triumphant, handed her the empty pill pack. The pharmacist looked at us oddly when we presented the prescription form: Sigrid's printed pad was from the 1940s, and the form gave only a four-digit phone number, something like "SYcamore 7654". Still, he filled it as written.

11. Sophomoric

Sophomore year, Ed Colaianni and I were assigned to Kirkland House. We had a "double": a sitting room, a bedroom with two bunk beds, and a bathroom. Oddly, the bathroom was between the other two rooms; you couldn't get into or out of the bedroom if it was in use. When the original Harvard Houses, including Kirkland, were built in the 1930s, there were odd rooms here and there to house the richer students' live-in servants.

The Kirkland dining room was far superior to the freshman Union, smaller, of course, but also more personal: Within a couple of weeks, the kitchen staff lady who checked us off her list knew us all by name, "Good afternoon, Mr. Press." The staff were Irish women and referred to us as "young gentlemen". The older ladies were in front, serving; the younger and prettier girls were back in the kitchen. Perhaps this segregation was to avoid tempting the young gentlemen with pretty Irish lasses (or vice versa), or perhaps the women with seniority got the easier jobs in front. Either way, the gulf in social class between the students and the staff was glaring and overt. Coming from relatively classless Southern California, I found it hard to get used to. It was Old Harvard's real-life version of *Upstairs, Downstairs.*

Sophomores were not usually hired by the math department as graders, but I filled out the application form anyway and was called in to interview with Dr. Arveson, the head sectionman for Math 21. Arveson was later, and for many years, professor of mathematics at U.C. Berkeley. The new course had expanded from three sections the previous year to ten now, and he was desperate for graders. Maybe just to keep a close eye on my work, he hired me to grade his own section of thirty students.

I enjoyed grading. The students didn't know that they were being graded by a skinny sophomore who had taken the course only the previous year. I quickly learned to recognize fake proofs, where a student would write down a series of true statements, the last of which was what was supposed to be proved—but the statements were not connected by any chain of logical reasoning that I could discern. In an early problem set, students were asked to use the field order axioms to prove that 1 is greater than 0. One student did everything right, but wrote as the last line, "Therefore $1 > 0$, which we know it is." I took off one point for the assertion that his intuition was somehow more meaningful than his correct proof.

Dorm proctors doubled as academic advisors for freshmen, but advisors for sophomores and above were faculty in their major departments. My advisor was an assistant professor in physics, Dr. Snider. Neither of us had the slightest interest in the other. I presented my study card, listing the courses I intended to take. His signature was required. "Is that one too hard?" he said, pointing to Math 213, a graduate course. Five words of advising. "No," I said. One word of rejecting his advice. In later semesters, I dispensed with Dr. Snider entirely. Entering the physics building, I found the first occupied faculty office. I knocked, entered, waved my study card, and said cheerily, "Hi! Remember me?" The professor had never seen me before, but he inevitably responded with something like, "Yes, of course I do! How are things going?" "Great," I said, providing no opening for any further conversation. Embarrassed at actually not remembering me, the professor simply signed and handed back my card.

I can jump far forward in time now for a related story. As a Harvard professor, I signed the study card of a sophomore physics major, call her Ms. C, in no less peremptory a fashion. She never returned. I assumed she either transferred out of physics, or had a different advisor. More than two years later I received a call from the Senior Tutor in one of the Houses, call him Dr. Javert. It seemed that Ms. C had made it all the way to the final term of her senior year with no advisor at all. Every semester, she simply forged my signature on her study card. Now, she was already accepted into law school—not Harvard's. What Javert needed from me was a written statement confirming that the signatures were forgeries, so that he could get her expelled from Harvard before she graduated. I refused. I saw her crime as victimless and, thanks to me, she graduated. I'm sure that she is a fine and resourceful lawyer somewhere. She may even know a little physics.

* * *

Horace Lunt had joined the Harvard faculty in 1949, after finishing his Ph.D. research with the celebrated Russian linguist and literary theorist Roman Jakobson. Lunt believed that Russian was best taught by surfacing its complicated linguistic regularities; that is, by memorizing complex linguistic rules instead of memorizing words and sentences. It is a debatable theory at best. Nevertheless, Slavic A, Harvard's first year

Russian course, was Lunt's legacy. The course was not for the faint of heart. But Ed Colaianni had taken it as a freshman the year before; and Margaret, at Berkeley, had also mastered a first-year intensive Russian course; so I was motivated to give Slavic A a try. I had a romantic vision of Margaret and me in bed together, having sex and—*speaking Russian.* Grammar sections were on Tuesday, Thursday, and Saturday, attendance required. Four hours a week of language labs (listening to tapes in Boylston Hall) were mandatory. Then, because the extremes of Lunt's anti-conversation vision had become blunted over time, we had a two-hour conversation section, once a week in groups of six, with a native Russian speaker.

Genri Yirvrandovich was my conversation sectionman. He had a slim athletic build and spoke no English at all. He was a professional acrobat and mime. He bounced around the classroom, jumped onto desks, grabbed objects and people (especially girls) and tried anything to get us to speak Russian. We eventually discovered that he wasn't a native Russian speaker at all. His mother tongue was Armenian (a completely unrelated language), and he had arrived just three months earlier from Italy, where he had been an extra and stunt double in the Italian film industry. The words "mama mia!" were frequent interpolations in his conversational Russian. Our two hours a week with him was a circus.

My physics course was a disappointment. By contrast, my math course, a graduate level course in complex variable theory taught by David Mumford, was difficult, but rewarding. Mumford was a few years later awarded the Fields Medal, considered the highest honor in mathematics, and he went on to win a MacArthur "genius" fellowship, a Shaw Prize, and the National Medal of Science. I often spent more than five hours on his problem set of five problems, able to solve at most three of them. I audited Richard Brauer's graduate course in real variable theory, a collection of jewels (theorems, actually) that he would take out of some hidden vault, polish, allow us to admire, and then return to the hiding place that would never be accessible to me. Brauer had taught at Königsberg until Hitler came to power in 1933. He was one of the fortunate Jewish scientists brought to the U.S. by the Emergency Committee in Aid of Displaced Foreign Scholars. I briefly considered changing my major to math; but sanity prevailed. I was capable of appreciating higher mathematics, even taking courses in it, but I lacked the unique mathematical clarity of mind necessary to be able to produce it at a professional level.

My Russian language course, Slavic A, was structured to weed out the weak and unfit in the first semester and then, second semester, to increase the challenge for the remaining Darwinian survivors. I was a survivor, but just barely. Every week I studied for many hours just to keep up; and (spoiler alert) I put similar effort into Slavic B, second year Russian, the whole next year. In Berkeley, Margaret's declared major was for a time Russian studies, although she eventually switched to linguistics. Several of our high-school friends did major in Russian studies.

Seen from today, it makes little sense. Why such interest in an arcane, difficult area of study? It is easy to forget today how bipolar the world was during the Cold War, especially so in the 1960s. In space, the Soviet Union was far ahead of the United States. The U.S. Apollo program was only just getting started. The Cuban Missile Crisis was a recent memory. Soviet science appeared equal to or better than American science in key disciplines, especially physics and applied mathematics. Globally, Russia dominated Eastern Europe and Central Asia, and Soviet influence was displacing the West's in the Mid-East and South Asia. In Vietnam, we and the Soviets were fighting what amounted to a proxy land war.

Studying Russian language was, for liberals, a way to extend a hand across the bipolar divide. And, for conservatives, it was a prudent way to better know the enemy. So it was simultaneously a form of protest and an emblem of patriotism. In hindsight, all the effort that I put into studying Russian for two years yielded nothing except, still gathering dust in my self-storage locker, an uncommon Cyrillic-alphabet manual typewriter. It was all a wasted effort. But who could have known?

If I didn't keep myself busy studying, or at the radio station, a kind of depression would set in. Over time, I came to recognize and adjust to a classic bipolar cycle: A few weeks on an even keel, then a few days of manic productivity, followed by a crash to some days of—I thought of it as loneliness, but its right name was clinical depression. Gradually, I got used to this, even holding in reserve challenging projects to do during the manic phases. I could always tell when a manic phase was about to crash. A certain kind of tingling feeling signaled that, typically twelve hours later, I would barely be capable of anything except staying in bed.

Perhaps a contributing factor was my struggle to interpret, or emit, or react gracefully to, social signals. On a bus to visit a high school friend at Bowdoin College, a very pretty freshman from Colby College sat next to me and, after she elicited that I was a "Harvard man" with a sofa in my Cambridge dorm suite, she was crawling all over me. She needed to visit Boston every weekend for an independent research project in primitive art, and she didn't have any friends there to put her up. I couldn't misinterpret this social signal, but I had no idea how to deal with it. I pushed her hands away and told her that I needed the time on the bus to study Russian perfective and imperfective verbs, which I proceeded to do. On another occasion, on an airplane, a girl asked to use my tray table (in addition to her own) so that she could lay out in front of me a selection of the many ID cards in her bulging wallet, including her driver's license. If there was a message in this, I was incapable of figuring out what it was.

A smaller cloud, at this time only barely perceptible in the sky over our lives, was the Vietnam War. When I turned eighteen, I had registered for the draft with the Selective Service Board in Cambridge, because my parents then lived in Cambridge. You didn't change draft boards if you or your parents moved; the initial assignment was permanent. It took my board eight months to "classify" me at all, and then as a 2-S student deferment. Their quota of inductees for the war was still small. On paper, I was a Cambridge resident, a *townie* going to *Harvard*. For a board of upstanding Cambridge citizens, that was something special. I estimated my chances of being drafted as close to zero.

12. Coursework

My physics course in the fall of my junior year was Quantum Mechanics, taught by Professor Norman Ramsey from his own detailed mimeographed notes. He had been teaching the course for decades. The lectures and notes were polished to a satin sheen—any possible question was somewhere already answered in the written materials. Ramsey lectured in the booming voice of an Army drill instructor. In fact, he was raised as an Army brat; during World War II his father was a brigadier general. Ramsey the younger, my professor, was assigned to the Manhattan Project in the War and led the team that did the final assembly of the plutonium bomb on Tinian Island in 1945. When I told my father about my quantum mechanics professor, he laughed—he had taken exactly the same course from Ramsey as a graduate student at Columbia.

Much (if not most) of modern experimental atomic physics in the United States traces to Norm Ramsey and his more than eighty (!) Ph.D. students. For years he was the poster-child example of a scientist whose influence was so finely woven into the fabric of science that no single discovery could be credited to him specifically—meaning that he could never win a Nobel Prize. The Nobel Committee proved this wrong in 1989 when he won the physics prize with a citation that focused on one tiny part of his legacy. They had to find something, since his not getting a Nobel had by then become a scandal.

By this time, I had perfected a minimal-effort approach to doing physics problem sets. If one knew where to look, one could find, somewhere on the shelves of the Physics Library, a worked solution to every problem that any instructor might ever think he was the first to invent. Old physics textbooks contained worked exercises, and there were books in most subfields of physics that collected problems and solutions as study guides for graduate student comprehensive exams, nicely arranged by topic. One particular journal for physics teachers, the *American Journal of Physics* with bound volumes going back to 1933, was full of clever problems and solutions submitted by physics instructors, often from small colleges. An initial investment of my time was required before I became adept at finding things in this literature, but, after that, I did all my problem sets by the "library method," more than twice as fast as before. There was no reason not to be scrupulously honest about it:

My every solution began with a disclaimer like, "This solution is based on *Am. J. Phys*, vol. 19, p. 374," or "Batygin and Toptygin *Problems in Electrodynamics* §4.23". No grader ever rejected my scholarly approach. A grader myself, I guessed that they were probably using my literature-perfect solutions as models.

For math, I took an undergraduate course in abstract modern algebra taught by Chih-Han Sah, an assistant professor who spent his later career at SUNY Stony Brook. This was an easy course, designed for that lower tier of Harvard math majors who would never become real mathematicians—but then, I wouldn't become one either. I learned little, retained over time even less, and got an A+ (not actually a legal grade at Harvard, but Professor Sah's small ornamental salute). My elective course was Economics 1, a huge course supervised by Professor Otto Eckstein, who was later a member of President Johnson's Council of Economic Advisors. Economics was one of the largest majors at Harvard and lectures were in Lowell Lecture Hall, Harvard's largest classroom. Most students in Ec 1 were freshmen who struggled with the math. I found the course easy, and interesting enough that I started reading more advanced economics textbooks on my own.

And then, of course, Slavic B. The Russian Orthodox saints must have smiled on me, because—just that year—the course was revamped and made less rigorous, more conversational. It still required a lot of time studying, but it was not nearly the struggle that Slavic A had been. Marty Chalfie, a whrbie who was rarely around the station because he was also on the swim team, was in my section. Much later, he won a Nobel Prize in chemistry for discovering that a green fluorescent protein isolated from jellyfish could be joined onto many kinds of biomolecules as a visible tracer, the enabling technology for literally thousands of subsequent discoveries in biology.

I still had to earn my spending money. Fall term I again graded Math 21, with the new wrinkle that, instead of grading the individual papers, I was allowed to write out solution sets on purple Ditto masters that were duplicated and distributed to the class. I was never at a loss for good solutions, since I could crib the best solution among the papers submitted—exactly as I imagined my physics grader cribbing from me. I always included an acknowledgment to the student. Spring term, instead of grading, I applied to the student aid office's Faculty Aide program and was assigned to Professor Gerald Holton in the physics department.

In practice, I barely met Holton, and instead worked for his postdoctoral student, Dr. Stephen Hawley (no relation to the similarly named astronaut). The pay was $1.50 an hour, significantly less than I made grading; but the job was more varied. In the 1940s, Holton had been the doctoral student of Percy Bridgman, a professor at Harvard who founded the field of high-pressure physics and was awarded the Nobel Prize in Physics in 1946. Bridgman, suffering from terminal cancer, committed suicide by gunshot in 1961. Holton inherited the remnants of Bridgman's high-pressure work, by this time housed in a single large room in the basement of Lyman Laboratory. That was where I worked. The equipment—hydraulic presses, etc.—was all mechanical and from the 1920s and 30s. After Albert Einstein's death in 1955, Holton had become involved in the curation and publication of the Einstein archive, and he was also editor-in-chief of *Daedalus,* a highbrow intellectual journal. By the time I was a part of his lab, he had very little interest in high-pressure physics.

I got to know Gerry Holton better later, when we were colleagues on the Harvard faculty. Born to a Jewish family in Vienna, he was fortunate to leave Austria in 1938 as part of the Kindertransport organized by British Quakers, and he emigrated to the United States two years after that. In the 1990s, Holton was regarded as a prominent historian and philosopher of science. His and Nina's cocktail soirees attracted a wide spectrum of Cambridge intellectuals.

But, when I worked in Holton's lab, postdoc Hawley's unenviable job was to try to squeeze publishable results out of ancient equipment and an actively uninterested mentor. My job, less unenviable, was computer programming and building electronics for Hawley's futile effort to modernize the lab. The electronics part took me back to my junior high school projects, but now, instead of taking the bus to Dow for parts, I could get (or special order) anything I wanted at the physics stockroom. Also, I didn't have to worry whether anything I built actually worked: Steve would hand me a schematic (circuit diagram). I would build the thing, taking as many hours as I liked at $1.50 per hour, and present him with the result. After that, it was his problem. He spent many hours with a soldering iron fixing my errors—but fewer than it would have taken him to build the things himself. I liked being a very junior experimental physicist, but in truth I had very little talent for it.

Margaret and I wrote to each other much less than before, not because we thought ourselves any less committed, but because we could see the light at the end of the tunnel of our separation from each other. Two lights, in fact: We were planning to spend the summer in Europe—my parents had agreed to let me spend the seven hundred dollars in a savings account that was my inheritance from my grandmother Nana.

And, in Berkeley, Margaret was taking six courses; if she passed them all, she could graduate a year early and move to Cambridge for my senior year. The times had changed enough that we could contemplate overtly living together off-campus—if Harvard would allow it.

The college's strict rule was that all undergraduates should live in the Harvard Houses. House life was a required part of the "undergraduate experience". For those of a certain lofty social class, a year off for foreign travel was also supposed to be part of the Harvard experience. However, fewer students than usual were taking whole years off. The reason was Vietnam. If you were not a full-time student, you were liable to be drafted. The residential Houses were thus overcrowded, and a small number of seniors were therefore being allowed to live off-campus. Kirkland House's quota was twenty, the selection to be made by Master Smithies and his wife.

Mrs. Smithies had certain peculiarities. On Friday afternoons, when girls from Wellesley and Smith were streaming into Kirkland House with their little pastel suitcases for (illegal) weekend stays with their (jock) boyfriends, Mrs. Smithies liked to stand outside on a little balcony of the Master's Residence that overlooked the Kirkland portal and call out to the girls, "Whores! You are all whores!" I wasn't sure that she would be sympathetic to my request to live off-campus with Margaret, whom I carefully described as my fiancée.

My Slavic B experience ended with a peculiar twist. After I took the final exam and was completely done, I received a note from Miss Ketchian asking me to come to her office at a specified time. From the far side of her desk, she said, "I'm having trouble deciding whether to give you a B+ or an A- for your course grade." She hadn't asked me a question, so I didn't say anything. She continued, "Are you thinking of taking third-year Russian in the fall?" I wasn't, but I was not clear about how I should answer. "What's the right answer?" I finally said. She said, "If you were thinking of continuing in Russian, then I would have to give you a B+ to discourage you. But if this is your terminal course, then I can give you an A-." My answer was obvious.

On June 13, at commencement, I watched from the WHRB broadcast area on the steps of Widener Library as Harvard awarded an honorary degree to the Shah of Iran, with the (later, after he was deposed, ironic) citation, "A ruler who has found in power a constructive instrument to advance social and economic revolution." Margaret flew to Boston that same day. On June 15, we flew from New York to Luxembourg on a cheap charter flight. Leasing for three months a tiny Renault, we drove through Switzerland, Austria, Germany, Denmark, Belgium, and the Netherlands, staying in small hotels and pensions, often the ones recommended in Frommer's *Europe on Five Dollars a Day*. London, however, made the biggest impression. We hung out in Trafalgar Square and at Piccadilly Circus. This was the 1968 London of Carnaby Street, *Yellow Submarine*, Twiggy, Mick Jaggar, and David Bowie (none of whom we actually saw). Theater tickets were cheap, and we saw West End musicals with British actors in the canonical American roles: Alfie Bass (not Zero Mostel) in *Fiddler on the Roof*; Keith Michell (not Richard Kiley) in *Man of La Mancha*; and the musical theater debut of a young Shakespearean actress named Judi Dench as Sally Bowles in *Cabaret*. We drove north to Hadrian's Wall where a mistaken note was left on our windshield in the parking lot: "Un bonjour de deux Francais de passage." Our Renault had French plates, in those days still rarely seen in Scotland. We drove south through the Lake District to Stonehenge. There was then no fence around the monument, and it swarmed with tourists climbing on the big stones—homier and different from today's austere distant preservation.

At the end of August, we returned the car in Paris, took the bus to Luxembourg and our charter flight home. There was good news waiting for us. Master (and Mrs.) Smithies had approved my application to live off-campus with Margaret—in sin—for my senior year. Indeed, the times they were a-changin'.

13. Professors

We rented a one-bedroom apartment at 335A Harvard Street, a four-story brick apartment building just five blocks from the Harvard Yard. Margaret was admitted by MIT's linguistics department as a non-degree "special student," allowing her to take graduate courses and (not incidentally) pay full tuition. The MIT department was ground-zero for the world-wide explosion of interest in Noam Chomsky's theory of transformational generative grammar. Margaret took a course taught by Morris Halle, Chomsky's long-time collaborator.

My sole physics course was titled Advanced Quantum Mechanics, but it was really something else entirely. Professor Julian Schwinger was Harvard's most famous theoretical physicist. In 1965 (while I was a freshman) Schwinger was awarded the Nobel Prize in Physics, shared with Richard Feynman and the Japanese physicist Shinichiro Tomonaga. This was the first time in the prize's history that the award was made purely for a theoretical construct, the theory of Quantum Electro-dynamics, "QED". Schwinger was a former child prodigy who had received his Ph.D. at age 21. In 1949, he and Feynman independently proposed what seemed to be two competing theories of QED, but within a year they were shown by Freeman Dyson to be exactly equivalent—just couched in different mathematics. Many thought that Dyson should have shared the prize, but it can be divided at most three ways. (We will see much more of both Richard Feynman and Freeman Dyson later on.)

The two decades following were not kind to Schwinger's approach. Feynman's QED was intuitive, graphical (so-called "Feynman diagrams"), and easy to calculate; Schwinger's approach required long and intricate manipulations of symbols and equations. Feynman, a larger-than-life character, was lionized; Schwinger, marginalized. People said that Schwinger *could* have joined the mainstream; he *could* have applied his universally acknowledged (especially by himself) genius to further important discoveries. What he did instead was to withdraw from communication with colleagues, resign from professional societies, and devote himself to a body of work—essentially a reformulation of all of modern physics—all premised on a counterfactual history where his approach to QED, and not Feynman's, had carried the day. He called this work Source Theory.

Schwinger's lectures were journeys to another planet—his own. Everything the man did was theatrical. He entered the lecture hall in Jefferson Lab on the dot of the hour, walked slowly, slightly hunched over, as if carrying the weight of all physics on his shoulders. He gave no greeting to the class. He went to the blackboard, straightened powerfully as if infused with a sudden jolt of energy, reached with his chalk to the upper left corner of the board as if gesturing to heaven, and, without ever consulting notes of any kind, proceeded to dispense elegant equations across all the blackboards. Every word of every lecture was perfect. It was as if (and probably was) from a memorized script. It was the direct word of God—no, *better* than God—of *Schwinger*. It was a completely useless course, not actually physics at all in any meaningful sense. It was pure performance art. It was mesmerizing. Once started, his lectures paid no attention to the clock: A lecture might go for its assigned hour, or for half an hour into overtime—true art has its own pace.

Later in life I encountered a number of people who, like me, took Schwinger's course in these years. Remembering, we could all only shake our heads in simultaneous admiration and disbelief. My mother told the story of meeting Mrs. Schwinger in the Belmont fish market. Billie would not have recognized Mrs. Schwinger—who was known to be Professor Schwinger's former secretary—except that the well-dressed woman was charging the fish purchase to her account by name. (You could do that then.) My mother introduced herself as the wife of Frank Press the MIT professor. "MIT," Mrs. Schwinger sniffed, ending the conversation. MIT was no Harvard. In his career Schwinger had seventy-three Ph.D. students, four of whom won Nobel Prizes.

Schwinger left Harvard in 1972 at the age of 54 for U.C.L.A. where, it was said, he was given priority to use any tennis court at any time. His office in Lyman Lab, wood-paneled and beautifully furnished, was inherited by Steven Weinberg, a younger field theorist. Weinberg claims that Schwinger left in the office a pair of old shoes with the clear meaning, "so you think you can fill these, sonny boy?" It was made easier for Weinberg to tell this story by the fact that he was himself awarded the Nobel Prize in 1979, and (among acknowledged great physicists) is widely considered the greater of the two.

Professor William Alfred's course in Anglo-Saxon Poetry was even less useful to my career in physics than Schwinger's Source Theory. Still, I jumped at the chance to take it—seniors got favorable positions in the lottery for its limited enrollment, and there were no prerequisites. That year he was more famous than usual: His play *Hogan's Goat*, starring

Faye Dunaway, had just closed on Broadway after running 607 performances. The first half of the course, English 200a, was in effect a foreign language course with grammar drills and vocabulary lists to memorize. Knowing English and high-school German, I found Anglo-Saxon pretty easy to pick up. We didn't have to speak it—no Genri Yirvrandovich here—just be able to read and translate.

When we had the basics of the language down, we started reading the limited Anglo-Saxon literature: an anthology of shorter poems for the rest of the first term; then English 200b, a full semester on *Beowulf*. We were assigned exactly 100 lines to prepare for each class. Alfred quickly learned our names and called on us individually to translate the next few lines as idiomatically as we dared. He must have known by heart all three thousand lines of *Beowulf*. Each few lines would bring to his mind amusing stories to recount, sometimes illuminating of the text, but more often irrelevant. Kittridge, he told us, said of the *Faerie Queene* that every time Spenser took a drink, he changed the allegory. Alfred had a stock of stories about F.P. Magoun, a preeminent Harvard scholar of the previous generation who, Alfred claimed with a straight face, was the model for the cartoon character Mr. Magoo. Magoun taught a famously easy gut course, "Anglo-Saxon Poetry in Translation." Alfred's stories about him were often structured around a dim question by a student-athlete in the front row producing an unlikely eccentric answer by Magoun. "Professor, if the Danes drank kumis [fermented mare's milk] then why did the Saxons drink mead [fermented honey]?" "Because there were more bees than horses." When called on to translate, I was usually prepared, but I learned quickly that I shouldn't waste the class's time if I wasn't. "I'm not prepared," I would say, and with a brief mournful look Alfred would go on to the next student. I got an A both semesters.

I took an applied math course in function theory taught by Bernard Budiansky, one of Harvard's distinguished cohort of applied mathematics professors. This had a lot of overlap with the complex variable course that I had taken as a sophomore, so was easy. But let me jump out of time here for another piece of Harvard history:

There existed at Harvard a rich vein of applied mathematics centered on complex variable and analytic function theory, for which the place was world-famous in its applications to engineering and physics. Professors George Carrier, Bernard Budiansky, Max Krook, and Tai Tsun Wu were in my time the core group of faculty in this field. They rotated among themselves the duties of teaching the advanced courses, two of which I took. This was a field both beautiful intellectually and immediately applicable to practical problems. Harvard graduated many Ph.D.'s and hired outstanding junior faculty in the area. Carrier et al. flourished as, in effect, a small, autonomous department. Carrier was himself awarded the National Medal of Science in 1990.

But by the mid-1990s the field was not the same. Its applications could be better attacked by comparatively brute-force computer calculations—the more so as the sophistication of the computer methods improved. As they successively retired, Carrier et al. pressed for Harvard to replace them with younger scholars like themselves, deeply analytic mathematicians. These appointments were vetoed by one single-minded individual, Paul Martin, who had been made Dean of Applied Science, improbably, because he was a theoretical physicist—a former student of Schwinger's, in fact. Harvard replaced its brilliant old applied mathematicians with young computational scientists, individuals who would have been only B students in Carrier's applied mathematics courses, but whose appointments opened up new fields. Over a generation, Martin's stubbornness became vindicated. Fields of science have an arc: They are born, they grow, they flourish, they wither, and at some point they must die or be killed off by an authoritarian dean.

Back to my senior year as an undergraduate: I took an astronomy course in stellar structure and evolution, taught by a young, high-strung, and very green assistant professor, Richard McCray. There were fewer than a dozen of us in the course—including, as it would turn out, two future Harvard professors of astronomy (myself and Bob Kirshner) and a future Princeton professor of astronomy (Rich Gott). For me, a physics major, McCray's course was my first introduction to theoretical astrophysics—a seed that later sprouted. A less-fertile seed for the companion field of observational astronomy was planted by my fellow student Ken Bechis, later an astronaut, who as an astronomy major was allowed to check out the key for the nine-inch Alvin Clark refractor telescope on the roof of the Observatory (the historical astronomy department building). I spent a few cold nights with him trying to locate known asteroids that, when you found them, were just points of light. From our low hill on Garden Street, surrounded by the lights and turbulence of Boston, Jupiter and Saturn were no more than blurry, dancing images. My observational seed never sprouted. Bob Kirshner's did, and he became a famous observer with many discoveries and prizes. About him more later.

We students beat up on McCray mercilessly. If he handed out a problem set in class, we inspected it and told him which problems we didn't like and wouldn't do. I was often point-man for this "negotiation". He inevitably caved. He was not used to the assertiveness of Harvard upper-level undergraduates. This was exactly how Harvard trained us to be. The last thing McCray's career needed was a public rebellion by the students in his assigned class. We sensed his weakness and exploited it. Of course I got an A. McCray became one of the top theoretical astrophysicists of his generation. He was proud of the fact that so many of us in his first teaching experience did so well later on.

I had two term-time jobs now. I was grading a section of forty-six students in Math 1, the most elementary calculus course. The pay had risen to $6 per student per semester. My other job was continuing to work for an increasingly demoralized Steve Hawley in Professor Holton's basement lab. I spent a few hours a week doing his computer work—for the good of all concerned, my career of building electronics was on hold—but there was by now very little pretense that I was being paid to *do* anything at all. I was there to give the absent Professor Holton the illusion of supporting a worthy student, and to give his postdoc the illusion that his career was not completely in the toilet. I tried hard to fulfil these expectations.

My concurrent career as a computer criminal began in the Holton lab. Computing—not just at Harvard but everywhere—was on the cusp of an epochal change. Harvard had acquired a Scientific Data Systems SDS-940 computer, one of about 60 made. Overnight, the experience of computing was completely different. Before the change, you keypunched your program on IBM cards, brought the deck to the Aiken computer center window, and came back hours later (for a long program, the next day) to pick up your cards and the printed results. There were usually bugs to correct. With luck, you could get in a few runs a day.

After the change, you sat at a teletype terminal in a room full of terminals. You were communicating with the computer in real time and, even more extraordinary, the other people in the room were communicating *with the same computer at the same time*. If your program had bugs, you could correct them in minutes, not hours. You could even have more than one program running at the same time. This was so-called time-sharing. I convinced Hawley that I should move all the lab's data analysis from the IBM to the SDS. Holton had no interest in this, but Hawley had managed to secure a miniscule grant to support his own research, independent of his mentor. My account on the time-sharing system was charged to this grant.

I now spent many hours in the Aiken terminal room instead of at WHRB. Moving the lab's data analysis to the new machine was straightforward. I spent much more terminal time probing the internals of the machine and teaching myself to program in the computer's underlying "machine language". One night the whole machine crashed. When it came back up a few minutes later, I happened to look at our account information, and I saw that I had not been charged for the several hours I had been working. Accounting was being done by the system's *logout* program. The operating system was designed so that no single user could crash the whole machine, but it had never occurred to the designers that users might want to crash their own sessions. I wrote a small program in machine language to do that. From then on, instead of logging out in the normal way, I ran my little program, crashed my session, and avoided Hawley's having to pay anything.

This was a slippery slope. Once computing was, in effect, free, I could in good conscience invite my whrbie friends to be hackers (not yet a recognized word) with me. Initially, we were all using the same Hawley account—taking care to accrue no actual charges. We soon discovered a different system flaw, one that let us read the passwords of anyone who logged into the machine. A breakthrough came when we happened to collect the password of a privileged account—one with the power to create new accounts. We created a slew of (free) accounts for ourselves, thinking that they would be lost among the hundreds of existing user names. This was naïve, as it turned out.

Over Christmas, staying with Margaret's parents in Pasadena, Tommy Lauritsen told me that I was almost certain to be admitted to Caltech's physics Ph.D. program. I had (mostly) A's from Harvard. I had taken my Graduate Record Exams and gotten high scores. But even more important, I had letters of recommendation from Ed Purcell and Norm Ramsey. Purcell actually knew me and likely said good things. For Ramsey, I was just one of a hundred students in his quantum mechanics class. He wrote dozens of upbeat pro-forma recommendations every year. Tommy told me that the letters' routine nature didn't lower their value. No matter how superficial were Ramsey's recommendations, his Caltech colleagues believed that, somehow, he wouldn't send them a bad apple. I hadn't definitely decided to go to Caltech. It depended on where Margaret got into a linguistics Ph.D. program. She had applied to U.C.L.A., which was within commuting distance of Caltech. But if we both got into MIT, we might stay in Cambridge.

Around this time, I learned about a high-paying graduate fellowship awarded nationally to a few students in physics by the little-known Fannie and John Hertz Foundation. I had never heard of this foundation before, and their awards could be used only at a list of ten schools; but both Caltech and MIT were on the list, so I applied. This proved to be a fateful event that, as we will see, set the course of much of my later life. I thus wish I could remember who pointed me to Hertz, or how I learned of their existence. It may have been as random as seeing a poster on a physics department bulletin board.

* * *

In 1657, Edward Hopkins, an early governor of Connecticut, died and bequeathed to Harvard a small sum "to give some Encouragement unto those foreign Plantations for the breeding up of Hopeful youth in the way of Learning." By the time I was at Harvard, his bequest had become a prize, awarded every year to some dozens of sophomores from foreign plantations, the list of which included California. This very minor prize consisted of a single book, chosen by each prizewinner and embossed on its title page with the Harvard seal. The prize was called a "detur" after the Latin "detur digniori" translated as "to the worthier let it be given." My sophomore grades qualified me for a detur. I was a junior by the time Harvard got around to asking me what book I wanted. I had recently read somewhere a favorable review of *The Diaries and Letters of Harold Nicholson, Volume 1*, and asked for that. In quite a different way from my discovery of the Hertz Foundation, this was another fateful event.

It is hard to imagine anyone in modern history with whom I have less in common than with the British diplomat and aristocrat Harold Nicholson (1886-1968). Though he later recanted, he was an early supporter of Oswald Mosely and British fascism. In private he was rabidly anti-Semitic. He was homosexual and married to Vita Sackville-West, herself a lesbian and the lover of author Virginia Woolf. Harold and Vita were devoted to each other in an asexual way—quite different from my excellent relationship with Margaret. Harold was, on and off, an MP in the House of Commons, and he knew everyone in British politics from the 1930s through 1950s. His diary is one of the great English-language diaries, not as good as Pepys surely, but for me in a class with Boswell and (another of my favorites) Anne Morrow Lindbergh. I loved Volume 1. By the time I was a senior, Volumes 2 and 3 had been published, and I devoured them too.

Nicholson died in May, 1968. At that time, I fully intended to have a life as interesting as Nicholson's, even if it did not include hobnobbing with Winston Churchill, Charles de Gaulle, and the Bloomsbury literati.

It was thus inevitable that I should, on January 1, 1969, in the middle of my senior year at Harvard, start my own diary. I typed daily entries (directly into final copy, no editing allowed) for the next twenty years. It became a reflexive habit, producing about a hundred thousand words a year—a good novelistic length. Why I stopped, in the middle of typing a sentence in August, 1990, is a story to be saved for later.

14. Graduating

Three things were important to me in my final undergraduate semester: Computer hacking, radio station WHRB internal politics, and getting into graduate school with a good fellowship. Classwork, a fourth, barely registered on my radar. Events that turned Harvard upside down—climaxing with the forcible student occupation of the main administration building University Hall and a violent early-morning storming of the building by police in riot gear—these registered as important to me only through the lens of WHRB—or through the minor inconveniences of working around them.

WHRB friends Paul Perkovic, Steve Willner, and Ed Belove were my main partners-in-crime as computer hackers. After his graduation, in the early 1970s, Ed worked for Data General Corporation, a time and place described in Tracy Kidder's book *Soul of a New Machine*; he was then an early hire to visionary Mitch Kapor's Lotus Development Corporation. Computer security was in its infancy. As undergraduate criminals, we discovered exploits that much later, we read about in textbooks—independently discovered by others. One was later to become famous as the "buffer overrun vulnerability". We figured out, that if you entered the username "100bbbbbb.SET-1.XXXMDTSWH.OFF", you would be instantly connected as an invisible user with the highest level of security privileges. After some weeks, the system's source code—which existed on a magnetic tape in the computer machine room—was patched to eliminate this vulnerability. But, by then, we had overwritten half a dozen pieces of system software with our own Trojan backdoors (those actual terms then unknown), and we copied our modified versions onto Harvard's master system source-code tape.

We were pretty good, but not by any means as good as the best. In his later 1984 Turing Award lecture, computer scientist Ken Thompson defined four levels of hacker ability: The lowest level of skill involves merely injecting malware into a running computer. The next level (which we achieved) injects it into the source code for the computer's operating system. The level after that (which we missed) injects malware into source code for the *compiler* program of the computer's source code, so that even a fully checked program will compile to an infected binary. Finally, Thompson noted that the compiler itself is used from time-to-time to compile new versions of *itself*. So the loftiest level of exploit (which we never could have imagined), was to inject a Trojan into the compiler's source code, compile the compiler, and then *delete* the malware from source code everywhere. Malware would now be injected into the binary of anything that the compiler compiles, including itself. Yet even a line-by-line security check of the compiler's (and all other) source code will find nothing amiss. If not in *any* source code, then where in cyberspace—a word coined only in 1984 by science-fiction writer William Gibson—does your exploit exist? If a tree falls in a forest…? (This idea of self-reproducing machine code is worth wrapping your head around.)

Computer center staff were actively pursuing us, but they didn't know our real-life identities. Avoiding getting caught became more difficult when we discovered that we were not alone in hacking the Harvard computer's innards. There was someone else, a bumbling amateur, messily discovering for himself the most elementary exploits. We called him the Interloper. We cleaned up after him whenever possible, but he was persistent and unsubtle. Ultimately, we decided to track him down and turn him in to the authorities.

I wish I could say that we caught the Interloper in some clever way. The truth was that a friend overheard a visitor to Kirkland House bragging about his computer hacks. This was obviously our guy. "I know some people who *really* want to talk to you," our friend said to him. His name was Jeff Elman, a senior, a history major, in Eliot House. A few days later we met with him in a midnight meeting.

It turned out that Elman wasn't by nature a computer hacker at all; he was an accomplished phone phreak. Some background: It was legend at WHRB that in the early 1960s, whrbies Charlie Pyne, Tony Lauck, and Ed Ross were the original phone phreaks—hackers who brought the Bell Telephone System to its knees by uncovering systemic engineering weaknesses that allowed just about anyone with a bit of technical knowledge to make free long-distance calls. That group was tracked down by the FBI and arrested—ironically, their phones had been tapped. None of the whrbie phone phreaks were ever actually convicted—unlike their West Coast counterpart John Draper, known as Cap'n Crunch, a friend of Apple co-founder Steve Wozniak, who served time in Lompoc Prison and later worked for Apple. The original whrbies' notes and diagrams were impounded by the FBI. However, copies and reconstructions circulated underground.

How Jeff Elman came into possession of materials and expertise that originated with the original WHRB phone phreaks is a story told in

Phil Lapsley's 2013 book *Exploding the Phone: The Untold Story of the Teenagers and Outlaws Who Hacked Ma Bell* (foreword, not coincidentally, by Steve Wozniak), in which Elman is given the pseudonym Jake Locke. Later in his life, Elman was chair of its Department of Cognitive Sciences, then Dean of Social Sciences, at U.C. San Diego.

Back in 1969, we didn't turn Elman in, but the authorities soon caught him on their own. He was brought up on disciplinary charges and put on academic probation. That caused us to moderate our own hacking exploits considerably. The penalties if caught were now too great.

* * *

I knew that with my Harvard grades, my recommendations from Professors Ramsey and Purcell, and my GRE scores, I would get into any physics graduate school I applied to. The uncertainty lay in where Margaret could get into a linguistics Ph.D. program, and whether we would get fellowships enough to support us in two-student married life. My parents had made clear that their financial support ended with my bachelor's degree. I applied to Caltech and MIT, Margaret to MIT and U.C.L.A. MIT started trying to recruit me some weeks before they were supposed to (by agreement among graduate schools). Larry Rosenson, a physicist whom I had worked for at MIT the summer before my junior year, was chair of their admissions committee. He invited me to visit him at his office and took the occasion to award me—orally and also jumping the deadline—MIT's Karl Compton Fellowship. A couple of weeks later, I had to call Rosenson to say that Margaret had been rejected a second time by MIT's linguistics program, and that Caltech now seemed our more likely destination. Shortly after that, Margaret did receive her acceptance by U.C.L.A., and a small fellowship.

Caltech admitted me, but was cagey about financial support. Tommy Lauritsen unofficially told me that they were certain that I would win an outside fellowship—I had applied for several. Their certainty did little to counter my uncertainty. After applying for the lucrative Hertz Foundation fellowship, I had never heard back from the Hertz people—not a peep. Early one afternoon, Steve Hawley interrupted my paid reverie in the Holton basement lab with the news that the office in Berkeley of Professor Edward Teller—the Hungarian-American physicist universally known as "the father of the H-Bomb"—was trying to reach me. I ran to a phone. Teller's secretary told me to fly to New York City right away—today—and go immediately to the Hertz Foundation's hotel suite at the

American Physical Society annual meeting. They would reimburse me. This was my first introduction to how Hertz operated as the personal extension of Edward Teller and a small circle of his disciples, whom I would soon know well.

Margaret came with me. At the Hertz hotel suite in New York—confusion. They had never heard of me. They accepted my word that Dr. Teller had sent me. They found someone to interview me, a physicist named Hudgins. His first question was, "How would you calculate the temperature inside an exploding hydrogen bomb?" His second question was, "I don't want you to think that this is a loaded question, but what do you think about student activism?" If there was a third question, I don't remember it. It didn't bode well. Margaret, meanwhile, had at the physics meeting located Professor Fay Ajzenberg-Selove, a colleague of her father's and close friend of her family. Fay was born in Berlin in 1926 to a wealthy Jewish family. Her father providentially relocated the family to France in 1930, and, with again exquisite timing, to the U.S. (via Lisbon) in 1940; and he managed to get out with some family money both times. Fay and her husband Wally Selove were now both themselves well off. She took us to dinner at the Russian Tea Room, where we ate a lot of caviar but didn't see any celebrities.

Two days later, back in Cambridge, another call from Dr. Teller's office: I should go immediately to a certain room at MIT and be interviewed by Dr. Lowell Wood and Dr. Harry Sahlin. At least they had names, this time. I rushed to Harvard Square and took the subway two stops to Kendall. This interview was no less peculiar than the first. Wood was a chubby young man with a Mephistophelian beard and matching manner. His laugh was a high-pitched cackle. Hearing that I might want to work with Kip Thorne at Caltech, he asked me some vague questions about gravitational collapse, cackling several times. Sahlin, an older man, by comparison morose, asked me how I might make a hologram movie. I had no idea. They were both friendly enough and encouraged me (with or without a Hertz fellowship) to apply for a summer job at Lawrence Livermore Lab (an offshoot of Lawrence Berkeley Lab founded by Teller in 1952), where they were both staff scientists. I left with the LLL application forms, which I duly completed and mailed.

There followed another radio silence from Hertz for six weeks. Just short of the Caltech's deadline for accepting their inferior fellowship offer, I got a telegram: I was awarded a fellowship, details to follow (I doubted promptly). A week later, I got a thicker and much more businesslike letter from the Livermore employment office offering me a summer job at $700 per month (a huge sum) plus moving expenses. My supervisors would be Dr. Edward Teller and Dr. Lowell Wood.

Arthur Hudgins' question about student activism was prophetic, because a week later, on April 9, the radical anti-war organization Students for a Democratic Society (SDS) invaded Harvard's University Hall, forcibly evicted the deans from their offices, and occupied the building. I rushed to WHRB, whose regular broadcasting was now preempted by full time coverage of these events. An impromptu radical student assembly, meeting in the "liberated," ornately decorated Faculty Room inside University Hall, had just voted to exclude all media from the building—except WHRB and the Harvard Crimson. The time that it took the Crimson to set up and print an edition was six hours, so we were de facto the only media pipeline.

I was sent to University Hall to help secure our lines of communication. Some tough-looking SDSers were guarding the doors, but anyone who looked like a student was allowed to roam freely in and out. In many parts of the building, students had spray-painted slogans on the walls. Locked files were being broken into and rifled, with embarrassing financial records later to be published in *The Mole* underground newspaper. WHRB had commandeered the office of Franklin Ford, the Dean of the Faculty. Via backchannel communication, we explained to Ford that we would keep his office physically safe in return for his continued availability to us as a news source. Of course, he agreed: From the deans' perspective, WHRB could serve perfectly as the administration's unofficial media outlet. So, remarkably, both sides of the conflict thought us useful to them.

I helped string wires out the upper windows of University Hall and into adjacent freshman dorms, where we took over the rooms and telephone lines of students eager to help. Bob Luskin anchored from Ford's office, cutting to reports from Chris Wallace and the others newsies elsewhere in the building and on campus. I felt some vaguely paternal pride, because Luskin had been a candidate in the comp the previous fall, of which I was Comp Director. He went on to become a Rhodes Scholar and, later, a federal prosecutor. Chris Wallace's future success in broadcasting is well known.

At five a.m., while I was asleep in our apartment a few blocks away, riot police with truncheons raided the building, dragged out and arrested the occupying students. Chris Wallace was arrested and used his bail phone call to phone-in a live report from jail. The Harvard faculty met in a closed session and lopsidedly voted "no confidence" in President

Pusey's decision to call in the police. We listened to the supposedly secret debate at the radio station: Unintentionally, the faculty meeting was being held in a lecture room on campus where WHRB had a permanent microphone connection. As a professional courtesy, we invited Crimson reporters to the station to listen with us. Meanwhile, we broadcast the substance of the debate live, with "according to sources" instead of direct audio. It didn't take long for this subterfuge to be discovered. Instead of disconnecting our microphone, the faculty voted to allow us broadcast their meetings live. These Harvard professors could not resist the opportunity of orating to a larger audience. The local student stringer for *The New York Times* was a Crimson reporter named Kerry Gruson, the daughter of Flora Lewis, the famous New York Times foreign correspondent and columnist. I helped her upload audio tape by phone to the *Times* newsroom.

From today's perspective, it is hard to appreciate the significance to Harvard of the events of April, 1969. Student activism against the Vietnam War was, on other campuses, much more violent. Buildings were set on fire, and people were killed. At Harvard, no one was killed, no one even seriously injured. Only a few of the arrested students were subsequently convicted of any crime. In the aftermath of the occupation, there was a student strike, largely (but not completely) over the issue of amnesty for the arrested students with regard to their Harvard "permanent records." Looking back, it seems pretty tame. Still, it is not an exaggeration to point to April, 1969, as the tipping point between Old Harvard and what subsequently became quite a different institution.

Hard now to recall is how very different Old Harvard was from the institution that it subsequently became. Paternalism, in Old Harvard, was an unabashedly good thing. The university was literally and legally *in loco parentis*—acting in place of parents toward its students· That President Pusey could call in riot police to beat with truncheons in effect his own children was, to the Harvard faculty of the time, unforgivable. Chemist George Kistiakowsky and economist Alexander Gerschenkron were passionate about this in speeches before the full faculty.

But they missed the point. What the students were demanding was not a return to safe paternalism, but the end of it. Student activism in the late 1960s was an uprising by a very specific cohort, those older than 18 and less than twenty-something—an ultimately successful campaign whose result was the recognition of this cohort as adult, not just legally, but also practically in matters of university policies and national politics. The impetus for this revolution—first, for the political awareness that was its prerequisite—was the war in Vietnam, and specifically the draft.

Harvard had already changed superficially in my four years. No more coats and ties for dinner. No more parietal hours in dorms. Those changes Old Harvard could easily absorb. What it could not easily accept was the idea that students should be empowered as adults to negotiate the conditions of Harvard student life, and even the educational curriculum itself—and negotiate by a fractious political (not neatly paternalistic) process that, at the time, lacked even basic civility. This loss of *civility* was a mortal wound from which the Old Harvard could not recover. Deans Ford and Watson resigned in 1970, President Pusey in 1971. All were disappointed men—old, white, WASP men, notably. Still, Harvard was a resilient institution. Some version of a New Harvard would replace the old. It had happened several times before in Harvard's 350 years.

Livermore Lab required me to have a physical exam before starting employment. At University Health Services, the physician commented that my sophomore year hernia did not seem to him completely repaired. Really? I left with not just Livermore's form, but also a letter stating that I had a "possible military disability". That week, Margaret and I set our wedding date: September 9, in the Pasadena Jewish Temple, the same as where I had been bar mitzvahed. Our living arrangements for the summer were that we would live close to the campus in Berkeley, and I would reverse-commute (drive in the uncrowded direction) the 40-minute drive to Livermore.

A whirlwind of happenings was associated with the end of the semester and my graduation. I was taking Schwinger's course pass-fail this semester, and I (barely) passed. My other grades were good, and I graduated in Group I with "high," but not "highest," honors. At the congratulatory Phi Beta Kappa luncheon, I got to shake hands with Senator Eugene McCarthy, the guest of honor. McCarthy was a progressive senator from Minnesota and had challenged Lyndon Johnson (and then Johnson's anointed successor Hubert Humphrey) for the 1968 Democratic presidential nomination. His opposition to the Vietnam War won him universal favor among college students. I sat next to a Mr. Shattuck, class of 1901, and across from Judge Wyzanski, class of 1927. Shattuck had served as Harvard's Treasurer during the 1930s Great Depression. Wyzanski was the liberal federal district judge who had recently made headlines by ruling expansively in favor of conscientious objectors to the war. Shattuck, it turned out, had been a minor judge in the Nuremberg trials. Wyzanski: "Mr. Shattuck, did you graduate from the College with distinction?" Shattuck: "I did not, sir, and did you?" Wyzanski: "I am proud to say that I did not. However, let me repeat what Judge Learned Hand once said to me when I was clerking for him…" And so on.

Harvard's 1969 Commencement did not follow the usual script. To head off student demonstrations, an SDS representative was allowed to give a rebuttal to the Harvard President's address. The chosen student was unfortunately so overripe in Maoist slogans that he was booed both by the reunion alumni (for his politics) and by the graduating students (for his hackneyed tropes). President Pusey awarded degrees with the set formulas that, while not Maoist, were equally hackneyed. The script for the dentists was supposed to be, "You are now a part of that ancient and honorable calling that deals with oral disease," but his microphone cut out momentarily, so that only the phrase, "ORAL DISEASE" echoed loudly in the arena between Memorial Church and Widener Library. At the Kirkland House luncheon (I was technically still a House member) a drunken Master Smithies spoke to parents and students: "They say that co-education will eventually come to the Harvard Houses. Let me assure you: As long as I am Master, *there will be no women living in Kirkland House*." He was right about that. A year later, Kirkland House was fully co-ed, and Smithies was no longer Master.

When we drove to my parents' house in Belmont immediately after the ceremonies, a letter from my draft board was waiting for me. My deferment was ended. I was reclassified as 1-A and would be called up for induction into the U.S. Army as soon as the board's monthly quota required it.

15. Livermore

Caltech had an office specifically for helping students avoid being drafted. I contacted them before we left Cambridge. Now in Pasadena, I went to talk to Mrs. Toy. She had already called her Hertz Foundation counterpart, Mrs. Nyholm, and the two had arranged for Dr. Teller, as "the father of the Hydrogen Bomb," to write a letter to my draft board. On the phone, Mrs. Nyholm read me the letter. "William Press's work at Lawrence Livermore Laboratory and the California Institute of Technology is more important to our nation's security than were Washington's boats on the Delaware River, than were Lincoln's railroads and telegraph lines in the Civil War, than were Patton's tanks in World War II…" Those weren't exactly Teller's words, but his were fictions of a similar order.

This was my first glimpse of the contradictory faces of Edward Teller: On the world stage, he truly did great harm. I never doubted this, then or now—the evidence was clear. He was a rabid and irrational anti-Communist, a Dr. Strangelove. Indeed, the character in Stanley Kubrick's 1964 film is transparently based on Teller. Actor Peter Sellers' pseudo-German accent was not so different from Teller's Hungarian. Strangelove's wheelchair and artificial arm echoed Teller's limp and artificial leg—the result of a street car accident in his youth. Teller was a believer that *his* ends justified *any* means. The Oppenheimer security clearance case had demonstrated that Teller had no hesitation in destroying the reputation of someone who stood in his way. But in person, Teller was charming, brilliant, and engaged—if, with his thick accent, sometimes difficult to understand. And he was fiercely loyal to the younger people in his circle, who now included me.

Livermore Lab's high-security fences, guards, guns, and gates were a new experience for me. The Lab was built on a one-mile square, what had been a World War II naval air station. In the 1960s it was still surrounded by Livermore Valley vineyards and pasture land. Chain-link fence and barbed wire defined its perimeter, but there were also many interior fences and manned gates. I was issued a Red picture badge, that color meaning that my security clearance had not yet come through. When it did, I would get a Green badge. The procedure at every gate, perimeter or internal, was to hand the guard your badge—"relinquish control" of it. He would look at it, at you, and then hand it back. I was limited to gates leading into Red areas, only small parts of the Lab. To get from some Red zones to others (without crossing a Green area), I could take one of the communal bicycles that were parked about, exit the outer gate, and bike a half a mile around the perimeter.

Most summer students had already gotten their clearances. For them, there were daily classified lectures on how nuclear weapons worked. We unwashed just sat in our offices. My officemate was a visiting professor from the University of Cincinnati, Lou Witten. He was a man of my father's age, a professor of physics who worked on general relativity theory. Lou jump-started my education in that field, which gave me a leg up when I later started working for Kip Thorne at Caltech. Later, Lou became more famous as the father of mathematical physicist Ed Witten, a pioneer of string theory who was in 1990 the first physicist to receive a Field's Medal, the highest honor in pure mathematics. (More, later, on my encounters with Ed.) Lou's clearance was held up because he was a leftie who in the 1940s had associated with Communist-sponsored organizations. Mine was held up—I found this out only later—because Margaret and I were "living in sin," those the exact words in the confidential FBI field report. The diabolical-seeming Lowell Wood dropped by late afternoon on my first day. He deplored the fact that I wouldn't, for now, be able to work on the classified topic that he had in mind. He promised to leave a stack of books and computer manuals on my desk overnight—they were there the next morning—and to escort me sometime soon to meet Edward, as he called him. Most people said "Dr. Teller".

Early afternoon on a Wednesday, two men with brown beards appeared in my office. One's beard was large and bushy, the stereotype of the absent-minded professor in a 1930s movie. The other's pointed beard and moustache could only have belonged to Captain Nemo. Captain Nemo lifted a chair, turned it toward the blackboard and sat. The Professor went to the blackboard, turned to face us and began: "Of course you start with the field equations." He wrote an equation. Captain Nemo broke in gently: "Perhaps we should tell him why we're here, Jim." "You mean, Jim, he doesn't know?" The Professor, Jim Wilson, was surprised. Captain Nemo was Jim LeBlanc. These were the "two Jims" that Kip Thorne, in a phone call, had suggested that I try to work for. "*Are* you working for us?" they now asked. I told them maybe I was, but I wasn't sure. I'd have to ask Lowell, or "Edward." The Jims were intending to write a hydrodynamics simulation program (a "hydro-code" they called it) capable of calculations in the black-hole Kerr metric. Wilson gave me a paper to study. They told me that they would come back the next Friday.

No one used the word "bomb" at Livermore. Nuclear bombs were referred to as "devices". A hydro-code was what was known in the outside world as a bomb code: a classified, secret computer program capable of simulating the behavior of an imploding (in its trigger phase) and then exploding fission or fusion bomb. The first primitive bomb code was written by the great Princeton mathematician John von Neumann, and ran on the ENIAC computer in the late 1940s. The requirements of bomb codes drove the development of many of the earliest electronic computers, with the U.S. Atomic Energy Commission as sometimes the only customer. In the mid-1960s, Teller and others "on the inside" realized that bomb codes—by then much more advanced than von Neumann's—could also simulate the implosion and then the fusion-powered explosion of astronomical supernovas. Astrophysicists used the term "gravitational collapse" (to a white dwarf star or black hole), instead of the bomb-makers' "implosion". Some of what was in secret computer codes had been declassified, enough to allow Livermore and Los Alamos scientists to publish papers on supernovas and gravitational collapse.

Teller was a main force behind the declassification of bomb code technology, in part out of a genuine scientific interest in astrophysical problems, and in part so that he could recruit more young scientists into Livermore's orbit (including me). Through the 1970s and well into the 1980s, hands-on experience with hydro-codes for astrophysical applications could only effectively be acquired at one of the two nuclear weapons design laboratories, Livermore or Los Alamos. Universities didn't have faculty with the necessary experience to teach the field, and lacked sufficiently powerful computers to do this kind of research. A whole generation of academic computational astrophysicists (and also some other fields of shock-wave physics) learned their craft at the Labs.

Conversely, the new and more powerful bomb codes that powered the redesign of the U.S. nuclear arsenal under President Reagan in the 1980s were in some prominent cases the work of astrophysics graduate students—some of them Hertz Fellows—who after getting their Ph.D.s were "captured" into full-time Lab work. I initially escaped such capture, but was caught by the nuclear weapons establishment in a different way some thirty years later, as we will eventually see.

On their next visit, the Professor and Captain Nemo presented me with five pages of pencil notes—what Wilson had been trying to write on my blackboard—now neatly transcribed by LeBlanc. At the top of every page, it said "Unclassified, but do not copy." There was still some uncertainty about exactly how far the declassification had gone, and there was also a backlash, opposing Teller, urging the Lab to claim as "proprietary trade secret" some things that were legally already declassified.

The Wilson-LeBlanc notes were a revelation for me—a way of looking at the physics of matter and radiation that was different from anything I had learned as an undergraduate at Harvard. The physics itself wasn't new to me, but it was *operationalized* in a way that I had never seen before. It was the difference between tasting a cake (Harvard), and baking a cake (Livermore). The Wilson-LeBlanc pencil notes were not a computer program. They were mathematical equations describing physics, but written in such a way as to make it obvious how to turn them into a computer simulation program.

The Lab allowed Wilson to present papers on his supernova calculations at open international meetings. After his talk, he would always be surrounded by a gaggle of Soviet attendees with questions. Someone later explained to me that when Soviet scientists returned from foreign travel, they underwent a required security de-briefing by the KGB. The smallest snippet of information from Jim Wilson was a golden ticket to future foreign travel. Of course, Wilson never actually revealed anything that was classified.

Subsequently, at Caltech, I never did actually write my own hydrocode. But I referred to my "do not copy" Wilson/Leblanc notes many times—for inspiration on how to operationalize different kinds of physics. My Ph.D. thesis would be, in effect, one particular operationalization of Einstein's field equations of general relativity. Today, the idea of framing physics in a way the enables computer simulations is so taken for granted that it is hard to remember that this was a big paradigm shift. It was happening just when I entered graduate school.

Edward Teller was sixty in 1969 and hadn't personally participated in scientific work of any kind, classified or not, in decades. Instead, he reigned on the ultra-conservative right as a political figure with enormous influence. He used that influence to benefit endeavors that interested him, or to which he was temperamentally loyal, varying from matters with national and global scope (President Reagan's Strategic Defense Initiative in the 1980s, popularly known as Star Wars, to be discussed later), to anything affecting "his" Lawrence Livermore Laboratory, and to matters as small as the Hertz Foundation, or as tiny as shielding Hertz fellowship recipients from the draft. Teller had met retired multimillionaire John Hertz in the late 1950s and had convinced Hertz to redirect his small philanthropic foundation from undergraduate scholarships in mechanical engineering—Hertz' fortune was from founding the Yellow Cab Company in 1907 and the eponymous car rental company in 1918—to much grander graduate fellowships in fields that, in Teller's view, supported the U.S. in the Cold War. Teller also suggested, and personally delivered, an astounding list of right-wing celebrities to serve on the Hertz Foundation Board. In my time they included FBI Director J. Edgar Hoover and Air Force General Curtis B. LeMay. LeMay's statement about North Vietnam, that we should use nuclear weapons to "bomb them back to the Stone Age," led President Johnson to hasten LeMay's retirement from the Air Force in 1965. In 1968, LeMay was Vice-Presidential candidate on George Wallace's third-party segregationist ticket.

Although the Hertz Foundation had a nominal president (Floyd Odlum, a retired financier who had famously guaranteed the funds to develop the Atlas missile when the government was slow to come through) and an administrator, Mrs. Nyholm, all the real power lay with Lowell Wood. In this and many other matters, Wood functioned as Teller's vicar-on-earth, remaining so for three decades until Teller's death in 2003. I was one of the earlier acolytes in the Teller-Wood church. At Livermore, Wood overcame bureaucratic impediments by invoking Teller's name, and Teller, if asked, reliably backed him. It was probably Lowell who had actually written the over-the-top letter to my draft board.

Lowell was a genuine polymath, one of the most creative people I had ever met or would ever meet. At Livermore, already the summer I was there, he was leading a small group of young scientists working on things that had nothing to do with the Lab's mission or sources of funding. The magic words were, "Dr. Teller wants this done." Wood's group, a kind of skunk-works later formalized by the Lab as "O-Group," did accomplish some things. But, equally often, it promised miracles and delivered only viewgraph presentations promising yet more miracles. Lowell's talent as a polymath often did not extend to actually getting a project successfully completed.

But I am not being quite fair: Where O-Group did succeed was in producing alumni and collaborators who went on to achieve great success, becoming distinguished professors, department heads, and deans, and founding significant companies. Computer pioneers John Hennessy (who was later president of Stanford) and John McCarthy (later to win the Turing Prize) were part of O-Group collaborations, as was Nathan Myhrvold, a Hertz Fellow whose company was an early acquisition by Microsoft and who (much later) became known in some circles as the world's most hated patent troll. O-Group members Curt Widdoes and Tom McWilliams invented much of what became the whole industry of computer-aided electronic design. Tom Weaver became, in academia, the world authority on supernovas; and, for his secret nuclear weapons work, won the U.S. government's E.O. Lawrence Medal. My perspective on Lowell Wood is not unusual among those who knew him: He was touched with genius; he was an inspiration to our careers and a brilliant mentor; his scientific reliability was variable; and his ethics echoed those of his mentor, Dr. Teller. Wood, Teller, and O-Group were later the subject of science writer Bill Broad's 1985 book *Star Warriors*. Broad portrays his subjects in generally heroic terms, if occasionally misguided in their fondness for nuclear weapons. I reviewed the book for *Bulletin of the Atomic Scientists*.

Margaret's and my apartment that summer was a couple of blocks from People's Park, a plot of land owned by U.C. that had become a de facto place for counterculture people to hang out. A couple of weeks before we moved in, the university announced plans to build a dormitory on the land and put up construction fences. On what happened to be Bastille Day, the local populace (Berkeley students) stormed the local fortress (People's Park). Police responded in force, and skirmishes continued all day. Margaret and I went out onto Telegraph Ave. to participate. Soon enough someone threw a brick through a police car window, and a swarm of police descended and shot tear-gas grenades. We decided to stay out on the street as long as there weren't gunshots. The tear-gas, while quite unpleasant, left the protesters motivated enough to become even more violent, and the size of the crowd grew. We were mobile enough to avoid the gas for a time, but did finally get trapped in a big cloud. Eyes running and unable to breathe, we stumbled around the block, home. But fifteen minutes later we went out again and rejoined the now-larger crowd.

Harry Sahlin, who was my Hertz interviewer with Lowell Wood, spent a lot of time chatting with me in the Red Badge ghetto. He seemed to have little else to do. Harry was interested in a peculiar set of unrelated things at the fringe of physics, and I was a well-paid, captive audience. We talked about information theory, and some strange old ideas of de Broglie, recently resurrected by the Bayesian statistician Ed Jaynes, that bits and entropy should underlie quantum mechanics. Separately, Harry kept coming back to the idea of a holographic movie. Just like a hologram that could be viewed from all angles, this (hypothetical) movie could be viewed from all times—including seeing the future. Another of Harry's ideas was that, since even irreversible processes are (according to Onsager's Laws, he said) very slightly reversible, it ought to be possible to hear what someone is dreaming by recording tiny motions of their eardrums. When I later summarized all this to Lowell, he told me that Harry had been an ordinary computational physicist until two years before, when he suddenly had gone off the deep end and began talking in this way. Lowell didn't tell me to stop spending time with Harry, however, so I continued to do so. It was like listening to Steve Hawley in Holton's lab: The work was easy and the pay was good.

Margaret and I spent all day Sunday, July 20, 1969, glued to the television, watching the coverage of the moon approach and landing. One of the instruments that the astronauts left on the moon was a long period seismometer that my father had helped design. On Monday, Lowell visited my office with a man whom he introduced as Dr. John Nuckolls. Nuckolls was a small, thin man in his late thirties. His mouth was at all times perfectly horizontal, and he smiled by widening his lips rather than curving them upward. He was the laboratory's best bomb designer, the winner of the previous year's A.E.C. Lawrence Prize for Weapons Development. Nuckolls and Lowell had been arguing about the value of the United States keeping secret the basic concepts involved in designing a hydrogen bomb. Lowell thought that even an uncleared, first-year summer student could design an adequate bomb. Nuckolls said put your money where your mouth is, and they came to visit me.

This was something that I could actually work on that was not Harry's closet of horrors. Wood and Nuckolls did what they absolutely weren't supposed to: They tracked down everything written about H-bomb design in the open literature and brought me copies. Worse from a strict security perspective, they spent hours talking with me about the physics implied by the papers they brought. There is a legal concept, "classification by compilation". That is where they were breaking the rules: No single paper on my desk was classified; but they—who knew the "secrets of the H-bomb"—were selectively choosing what to bring and discuss with me. By the end of the month, I had a sketch, backed by many pages of calculations, of what I thought was a viable H-bomb design.

I was wrong about that, however. Nuckolls won the bet. Later, when my clearance finally did come through (after Margaret and I were married and I was no longer living in sin), I made several trips from Caltech back to Livermore to be tutored by Nuckolls on how things really worked. I had gotten a few tricks right, but missed many of the essential ones. In the decades since, almost everything involved in fusion weapons design has appeared in the open literature. Today, with a selection of papers chosen with insider knowledge, Lowell would almost certainly win the bet. Luckily, there is also much more in the open literature that is wrong or misleading. The rules about "classification by compilation" are thus more relevant today than they were when Lowell and John were breaking them on my behalf.

In my last week, Lowell and Harry escorted me inside the Green Badge Only fence to an appointed meeting with Teller, John Nuckolls, and the current Laboratory Director, Michael May. May, born in 1925 in Marseille, France, emigrated to the United States as a teenager but forever spoke with a pronounced French accent. He earned his B.A. at the age of 19, and a Ph.D. in physics from Berkeley in 1952. He then immediately joined the new, secret Livermore Lab. He liked to relate, "The first guy I met said 'I suppose you know we're actually working on hydrogen bombs?' My jaw dropped."

I was supposed to tell these three what I had accomplished in the course of the summer. All the other summer students had done this as a group. I was alone because, by now, I was the only student still without a security clearance. I decided to talk about three things: My hydrogen bomb, which Lowell assured me would interest Edward even if it was wrong; some thinking about acausality (i.e., time travel) in the Kerr metric in general relativity that was inspired by my Communist officemate Lou Witten; and some of my own thinking about "Schrödinger's Cat" in quantum mechanics—one of the less-crazy things in Harry Sahlin's closet. In the famous thought-experiment of Schrödinger, the cat is quantum-mechanically both alive and dead at the same time. Whether you view this as a bug or a feature tests whether you truly believe in quantum mechanics.

As if it was an oral exam, Teller quizzed me closely on everything I said. When I got to Schrödinger's Cat, his interruptions became more frequent and pointed. Finally, he said, "I am not certain that everything you say is wrong, but I am sure that it is all at least misleading. I should have been *very much misled* if I was exposed to this sort of thinking at your age." I looked at Harry and saw him shaking his head slowly from side to side. As I was leaving, Laboratory Director May said to me, "We are sorry that you were here this summer without clearance." Nuckolls version to me was, "I'm sorry we cannot have a free and frank discussion about your work."

Later, Harry explained how I had stumbled onto the Schrödinger third rail. On precisely the subject of Schrödinger's Cat, Teller had once had an idea of his own. He tried to explain it to Bohr on the train to Los Alamos during the War, but Bohr was dismissive. Only after Bohr was dead did Teller publish the idea—in German—in a collection of essays honoring Heisenberg. Harry located a translation of Teller's essay and (after I read it) arranged for me to have a second interview with Edward, which went much better. "You may be right," he concluded graciously. "I see the way you're *trying* to think about it," he added, slightly less graciously. Teller did like the idea that I would use my Hertz Fellowship to study astrophysics. This was reassuring to me, given that the stated scope of the fellowship was for "applied physical sciences in the national interest," and that I was relying on him to keep me out of Vietnam.

Also in my last week, Harry took it on himself—dropping Teller's name as necessary—to get to the bottom of my withheld clearance. He reached someone named Stinson in A.E.C. security who confirmed that my cohabitation was the sole issue. The result of this investigation by Harry was an agreement with the A.E.C. that, the day after my wedding to Margaret, I would send a telegram to Harry, who would inform Stinson at the A.E.C., who would release my clearance. It happened as agreed. The next time I visited Livermore, I was a newlywed with a brand new Green Badge.

16. Teller

Having described Edward Teller, Lowell Wood, and John Nuckolls as I knew them in 1969, I want to jump forward in time almost fifteen years, to 1983. I was by then chair of the astronomy department at Harvard. On March 23, 1983, President Ronald Reagan declared that the United States was developing a space-based system of nuclear bomb-powered X-ray lasers to shoot down Soviet missiles—his Strategic Defense Initiative popularly known as Star Wars (after the movies). I knew more about the back story of SDI than most. The original concept of an X-ray laser powered by the explosion of a small nuclear bomb was developed by Peter Hagelstein and Tom Weaver (in Lowell Wood's O-Group) and George Chapline, a theoretical physicist at Livermore who had done his Ph.D. at Caltech with Murray Gell-Mann and who alternated between pure theoretical physics and bomb-related work. Lowell brought the unvetted and untested idea to Teller, and Teller took it directly to Reagan. I similarly knew more than was public about how that part happened. My father, as formerly President Jimmy Carter's science advisor, was in close communication with President Reagan's science advisor, Jay Keyworth. I mentioned to my father what my Livermore friends were saying, that Edward Teller had sold the President on a "third generation" of nuclear weapons—at that time the euphemism for the very secret X-ray laser.

It was all a mistake, Frank said. The President had indeed asked to see Teller, but only ceremonially. Keyworth was told to arrange it, but he dragged his feet, because he knew that Edward couldn't be trusted to stick to any agenda. The meeting could be approved only as an innocuous philosophical discussion on arms control. Finally, after reassurance from Teller, it was put on the President's calendar. Edward strode into the Oval Office and, before he could be stopped, laid out for the President the whole crazy Star Wars plan. Outside, in the hall, Reagan's senior aides including Ed Meese were angrily shaking their fists at Jay. Inside, the President told Edward that it all sounded fine and that he would commit to go ahead with it. He may have thought that Teller's proposals had been filtered through advance scrutiny and only brought to him (as most things were) for a ceremonial final nod.

The very first actual test of the X-ray laser concept, with the necessary nuclear bomb, occurred underground in Nevada only well after Reagan's announcement of SDI. Its results were held in a new security compartment, "Sigma 12," that was inaccessible even to people with Top Secret Q clearances, who now included me. It took several years for it to leak out that the reason for all this security was not that the idea worked, but that it *didn't* work. This and other foundational technologies of SDI were fantasies, at best impossibly optimistic, at worst fraudulent—entirely consistent, in other words, with the miracles that Wood's O-Group had previously (not actually) produced.

Already underway before Teller's performance in the Oval Office was a massive upgrading of the U.S. nuclear arsenal, with both new weapons designs and greater numbers. That effort was where most of Livermore Laboratory's effort was—the most money and the most jobs, including the talents of its best weapons designers (of whom John Nuckolls was one). Livermore and Los Alamos, although government labs, functioned as billion dollar-scale corporate pieces of U.S. defense industry. That Teller had hoodwinked the U.S. President into what most of establishment Livermore thought was a scientifically unsound wild-goose chase was not universally appreciated.

Former Los Alamos Lab director Harold Agnew later told me another Star Wars story: He was one of the small audience in the White House when Regan delivered the nationally televised SDI speech. The substance of the speech, roughly a new Fortress America, was clearly at odds with the U.S. treaty obligation under NATO to defend Europe equally. Agnew was sitting between Secretary of State George Schultz and former Secretary of State William Rogers. During the speech, Rogers leaned over and whispered to Schultz, "Were our allies consulted on this?" The incumbent Schultz whispered back, "I have no idea!"

This is all background to a phone call I got in July, 1983, from Mike May. Although no longer Laboratory Director as he was when I was introduced to him as a summer student, Mike was still a power at the Lab. There was a vacancy in the position of Associate Laboratory Director for Physics. Teller, re-energized by SDI at now age seventy-five, was promoting Lowell Wood for the job. May and the establishment leadership favored my other former mentor, the more measured and sensible John Nuckolls. Roger Batzel, the current Laboratory Director, was expected to retire soon, and whoever became the new Associate Director would be the heir apparent. May and others were terrified at the prospect of Lowell's ever becoming Laboratory Director. Their proposed scheme was devious enough to be interesting: *I* would apply for the Associate Director job. Although I had long since flown from the Teller nest, May and the establishment could still represent me as a *second* Teller candidate, an outsider choice to compete with Lowell as the insider one. As a Harvard professor and department chair, and former Hertz Fellow, I had already achieved enough success to be cited as a model of Teller's and Wood's mentorship; so it would be hard for Edward to disown me completely. Then, Nuckolls would emerge as the compromise candidate and get the job.

I agreed to the plan and in late July traveled to Livermore for a day of interviews that went very well—because most of the search committee was in on the plot. I had a scheduled meeting with Lowell, nominally because I had been his student. He kept me waiting half an hour, breezed in with the comment "*Et tu, Brute?*", and quickly left. He may have spent a part of the prior half hour alerting Teller, who now lived across the Bay next to Stanford's Hoover Institution. Word came of a sudden change in my schedule: A car would take me to have dinner at home with Dr. Teller and his wife Mici (pronounced "Mitzie").

At his home, Edward greeted me effusively, put me in a chair in his living room and told me about his theory of gamma ray bursts, then (as later) a problem of interest in astrophysics. He was quick and creative, even in his seventies. In passing, he made reference to some papers of mine, to show me that he had read my CV. We went in to dinner. "We must not speak physics at dinner," Edward said. "The lady [his wife] is not only *not* a physicist, but is *also* Hungarian," This was a Teller joke, since Mici's Hungarian accent was as close to incomprehensible as Edward's. Edward always spoke slowly, as if he was thinking every word in Hungarian, then translating it into German, and only finally into English.

At dinner, just three of us at the table served by a Mexican maid, we talked politics. More precisely what transpired was this: Edward raised one or another of his favorite subjects (importance of ballistic missile defense, futility of negotiated agreements with Russians, the Soviet threat to the U.S., general untrustworthiness of Russians, categorical evil of Russians, general deficiencies of Communism, etc.). On each subject, with great tact and charm, he would draw me out regarding my own views. After I provided some witty and quotable statements revealing of my academic, socialist, Communist, Democratic Party attitudes—in any case enough for him to later use to discredit me as needed—we went on smoothly to the next topic. It was a lot of fun. It was a dream opportunity: Here was *Edward Teller*, hanging on my every word and asking me about *my* political views. I could be the one who succeeded in turning him into a liberal, if I could just find just the right compelling and logical arguments. "You're absolutely right, Bill," he might then say, "I've been completely wrong for the last forty years. I now see the error of my ways. I'm going to announce tomorrow that I support the Nuclear Freeze Movement, and I'm going to start campaigning for Teddy Kennedy."

We moved to the living room for the final act of the drama (or farce). Edward raised the subject of Lowell explicitly for the first time. Lowell was a genius many times over, he told me. Lowell's invention of the key concept in the X-ray laser was the equal of his own invention of the key concept in the fusion bomb. Lowell's invention would make total nuclear defense possible.

"And we need *you* at the laboratory," he said vehemently and unexpectedly. "We need someone with your ability and accomplishment, and your scientific talent. I want to make you a much better offer than this Associate Director position," which he dismissed with a wave of his eyebrows. "Of course, I will have to call the Director to make this offer official, but you can count on it: Come to the Laboratory to do *anything you want*, work in any Division, on any subject that interests you—except administration or policy matters of course—for as long as you like, fully supported with a high salary for the rest of your life!" I thanked him and said I would think about it.

Teller walked me to my waiting car and again thanked me for coming to see him. "Take very good care of him," he said to the driver. I watched out the rear window to be sure that he was not putting a stick of dynamite up the tail pipe.

Mike May's plan succeeded perfectly. Nuckolls got the Associate Director job and then succeeded Roger Batzel as Laboratory Director in 1988. Lowell was never promoted to any position of higher leadership at the Lab.

Since the 1991 collapse of the U.S.S.R., historians have debated—will forever debate—the role, if any, that Reagan's SDI played in the Soviet demise. In nuclear weapons circles the self-congratulatory belief is that, regardless of whether SDI was achievable or was fantasy, the U.S.'s seeming willingness to invest in its astronomical cost was a factor in convincing Soviet president Mikhail Gorbachev that the Cold War was unwinnable, and that the U.S.S.R. must move toward *perestroika* and *glasnost* (restructuring and openness), inevitably leading to the revolutions in Eastern Europe, the rise of Boris Yeltsin, and the dissolution of the U.S.S.R. Most academic historians dispute this version of events. However, in later years, Gorbachev did write that his nuclear advisors told him with certainty that SDI was based on faulty physics and could never succeed; but that, regardless, the Soviet Union would have to start a matching program so as to avoid the appearance of surrendering to the American yoke.

But let us now return to 1969. The beginning of September saw my remarkable transformation. From Berkeley radical-student sympathizer—living in sin with my girlfriend, gassed by the police at People's Park, and (paradoxically?) designing non-working hydrogen bombs at Livermore Laboratory, I became over the course of two weeks a happily married, hardworking Caltech graduate student in physics, living in a compact little house on South Parkwood Avenue that boasted wall-to-wall carpets, a washer, drier, dishwasher (unusual) and central air conditioning (very unusual in Pasadena). Margaret had found our rental house on a trip south to see her parents. Our landlord Mr. Lindeman, a retired small-business owner, lived with his wife in the equally modest house just behind ours on the same lot. Rent was $200 per month. The neighborhood was white, lower-middle-class, Republican: Mr. Lindeman had quizzed Margaret to be sure that we would fit in. She passed the test. That we would both be working on Ph.D.'s—and not in suspect fields like English or a social science—was in our favor.

Our wedding reception was held in the Lauritsen's back yard, beautifully decorated for the occasion. Tommy had strung colored lights over the caterer's tables and starched tablecloths. Margaret and I departed at an appropriate time for our (fictional) honeymoon, to cheers and toasts, riding on her Honda 90 motorcycle, she in back with her arms around me and still in her wedding dress, me in my rented tuxedo.

Caltech was a kind of West Point of physics graduate schools. They didn't call us plebes, but they might as well have. The week before classes, we had a day with ninety-minute "placement exams" in each of four basic subjects (mechanics, electromagnetism, quantum mechanics, and the catch-all of "modern physics"). If you passed, you were qualified to take graduate courses in those same areas. If you failed, you took Caltech's undergraduate courses instead, generally delaying your Ph.D. by a year. These were not easy exams. In electromagnetism my score was 2 points out of 50—and I passed! People with scores of 1 or 0 failed. There were several such. The required first-year graduate courses were much harder than any physics course I had at Harvard. I often spent eight hours or more on a single course's single-week's problem set. Candidacy Exams were given in the spring. If you passed, you were allowed to be accepted for thesis research by a professor; otherwise, you re-took courses and tried again a year later. If you failed a second time, you left.

I missed registration day, which came a few days after the placement exams, because my draft board ordered me to report to the nearest induction center for a pre-induction physical exam. I telephoned their office in Cambridge to understand how this came about. Had they not received a letter from none other than Edward Teller? The helpful woman on the other end of the line found my file. "The letter is here," she told me, "but the meeting notes say that the board has never heard of this 'Edward Teller'."

I thus reported to the induction center at 1023 S. Broadway in Los Angeles at 7:30 a.m. on Monday, September 29, 1970. At 7:50 they started us through. We filled out myriad senseless forms, then started following a red line on the floor from examining station to examining station. The work was mostly done by enlisted medics whose medical skills were rudimentary, especially in trying to find a vein to draw blood. We marched through stations #1 through #14 in shoes and underpants. The examination was undisguisedly Potemkin. At station #15 we lost our underpants and met our first real doctor, who came down the line and checked each person in a couple of seconds: eyes, ears, abdomen, and rectum. Checking my abdomen for hernias and finding none, my doctor was about to proceed to my rectum. I said, loudly, "Doctor, I have two hernias." He stopped long enough for me to show my physician's letter, gave me a more thorough second examination (taking four seconds) and wrote "hern., bil., 1 yr." on my folder. An hour later I arrived at the final station, the summarizing physician. "You fail," he said. "Hernias. One year deferment." He didn't return my smile.

All of the physics faculty knew exactly who I was—Tommy Lauritsen's new son-in-law. There was no use in my pretending otherwise. Caltech was a small place, and my father, Frank, had been gone only four years, so most also knew that I was his son. It did not take long for me to get a reputation as a troublemaker. By long tradition, all exams at Caltech were unproctored take-home exams. However, I refused to sign Caltech's honor code pledge. Ward Whaling, the physics graduate advisor, shrugged and sent me to Graduate Dean Frederic Bohnenblust, to whom I rendered a one-sided, practically Talmudic (some would say Jesuitical), monologue on the ethics of snitching on others—that the only part of the code that I objected to. "I don't have a problem with any of that," he said, yawning. "I'm sure you'll do the right thing." Bohnenblust was later briefly succeeded as acting graduate dean by Max Delbrück, a Nobel Prize winner and one of the great biologists of the 20th century. I had several friendly run-ins with him, too. Decades later, when I became a biologist, I bragged to colleagues, almost truthfully, that Delbrück knew me well as a graduate student.

Harvard had rules, and it had procedures for dealing with students who broke the rules. Harvard's tempering grace came at the end of the process, when (except for the single case of President Pusey's calling in the police to University Hall) a paternalistic mercy would prevail. Caltech also had rules, but no procedures. Its approach to troublesome students was to try to talk them off the ledge or, failing that, to ignore them. A sole exception was the Honor Code Court, which was run by fanatical undergraduates without faculty supervision (this also by long tradition) and which had the power to have students—including graduate students—expelled from Caltech. No sensible physics faculty member ever referred cases involving graduate students to this court, but visiting or new professors sometimes made the mistake of doing so, with consequences always disastrous and disproportionate to the offense for the poor (almost always foreign) student. In my final year of graduate school, the physics department couldn't come up with anyone to be its graduate representative on the court, so I ran unopposed on the announced platform, "I will vote to acquit all students brought before the court, regardless of the facts." I was on the "jury" in just a couple of cases. The facts, such as they were, left me feeling good about honoring my campaign pledge.

In hindsight, I was lucky that my field was physics: The prevailing culture in physics, then and now, was personal and professional arrogance. Caltech (again resembling West Point) saw itself as training for the profession in attitude as well as in competency. The physics faculty likely saw my arrogance as a role model for the other students. I would never have made it in biology, a kinder and gentler field.

I made an appointment to see Kip Thorne soon after the start of the semester, and asked him if I could work for him. He said that if I needed an official thesis advisor to get out of the draft, he would take me on as a first-year student, otherwise not. I told him that I had a temporary deferral that would last at least a year. In that case, he said, I could attend his weekly group meetings informally, and I might similarly come to seminars and hang around with his graduate students in the "relativity interaction room," a kind of a lounge and informal library that was next to his office.

Kip was the youngest full professor at Caltech, then not even thirty. He had been Professor John Wheeler's star Ph.D. student at Princeton, come to Caltech as an associate professor straight out of graduate school, and been promoted to full professor just two years later. Actually, Kip was only Wheeler's second-best student. Richard Feynman, Nobel Laureate and proud fixture on the Caltech faculty, had been Wheeler's best, twenty-five years earlier. Feynman was instrumental in getting the untested Kip appointed at Caltech as successor in a faculty line that went back to Einstein, who had been a regular Caltech visitor in the 1920s—as a fan of Pasadena's sunny Southern California climate and then-resort atmosphere. When Einstein's visits ceased, Richard Chase Tolman became the "relativity professor" in the 1930s and 1940s, then H.P. Robertson in the 1950s and until his retirement. Kip was next in that line. Most professors weren't allocated space for an "interaction room" for their students, but Kip was a star.

A whole series of Kip's weekly group meetings were devoted to discussion, chapter by chapter, of a mimeographed translation of the recent Russian monograph *Relativistic Astrophysics*, by Ya. B. Zel'dovich and Igor Novikov. Zel'dovich, then in his middle 50s, was famous in the West as the father of the Soviet atomic bomb. He had (according to Kip) put aside his classified bomb work to devote himself purely to unclassified astrophysics. Zel'dovich's protégé and younger friend Andre Sakharov was correspondingly known as the father of the Soviet hydrogen bomb, the Russian Edward Teller. Sakharov was just becoming known as a Soviet dissident at this time. Zel'dovich, then and later, remained publicly apolitical. His prestige as a Soviet hero enabled him to run his academic group in Moscow unburdened by interference from the KGB, unlike most Soviet scientific groups.

Kip had somehow become a regular visitor to Zel'dovich's group in Moscow—that very unusual in those Cold War times. He had found an American publisher who would pay for *Relativistic Astrophysics* to be translated into English, provided that Kip would edit the manuscript for scientific accuracy. When the translator turned out to know little physics, this proved to be a major task. That Kip knew some Russian (not much more than I did) helped only slightly. These weekly discussions were chapter-by-chapter tests of whether the edited translation was comprehensible to graduate students. I didn't always understand the physics, but (ever the troublemaker) I complained loudly about the quality of the English prose. The result was that Kip started routing his red-inked chapters to me for copy-editing before they were re-typed and mimeographed for the group. This became part of my routine, the way grading had been at Harvard. It was mostly things like changing the literal "Russian writer of humoristic stories" to the better "Russian humorist". Kip was actually quite a good writer, but he had figured out that he could save time by leaving the routine stylistic editing to me—the ambitious and arrogant first-year student.

The word nerd was not in common use at Caltech at this time, but it certainly described Kip. Born and raised in Logan, Utah, Kip's Mormon-born parents were both professors at Utah State, the state's land-grant college. When Kip came to Caltech as an undergraduate, then went to Princeton for graduate school, it was like shaking the soda bottle before letting it fizz, nerd meets counterculture. In the late 1960s, he was a red-bearded faux hippy, most often dressed in an African dashiki that he had picked up, I think, in Moscow, where many Africans studied during the Cold War at Patrice Lumumba University. Kip and his beautiful red-haired wife Linda, his high-school sweetheart, liked to host hard-rock dancing parties at their sprawling old house in Altadena.

Harry Sahlin called me to say that my clearance had finally come through. A few days later, I had a call from Lowell Wood. The Livermore red-tape department had come up with two rules: Travel expenses could not be paid for non-employees; and part-time employees could not be hired at less than 20% time. Lowell proposed to hire me at 20% time with the understanding that I would bank my salary and use it for travel to visit the Lab. What would I actually do, eight hours a week? "I'm sure that anything you are working on at Caltech is of interest to the Lab." Didn't this arrangement violate the terms of my Hertz Fellowship? "Edward approves, as long as we don't put anything in writing." I visited the Lab for a few days at a time, about once a month. I would drop in unannounced on members of the Teller circle—Lowell, Harry, Wilson and LeBlanc, John Nuckolls, George Chapline, and the younger O-Group people—and shamelessly ask them to unlock their safes for me to browse. (These safes, so-called, were a kind of armored filing cabinet that stored secret and top-secret documents.) My requests were completely against the rules, because I had no work-related "need to know," especially not for every document they might happen to have. Nevertheless, most happily complied.

Nuckolls, who (once I was cleared) was always happy to tutor me in the black arts of nuclear weapons, did one time object on principle. Lowell was there. Nuckolls told Lowell that he didn't think that a mere former, part-time, summer student, me, should be pawing through classified private correspondence between Teller and the Atomic Energy Commission. "I don't see why not," Lowell said argumentatively. The two of them went off to consult Edward, leaving me in Nuckolls office with the locked cabinet. When Nuckolls returned, alone, he without further comment unlocked the repository and left the office. The necessity of maintaining my Livermore connection diminished significantly on December 1, 1969, when the first draft lottery took place. Instead of calling up inductees in an order of their own choosing, draft boards were required to use the order of the lottery results. My birthday, May 23, drew number 319 out of 366 (Leap Day included). My draft worries were over. Others were not so lucky.

Mentioning George Chapline, and having mentioned, in so different a context, Zel'dovich, I am reminded that these two actually had an unlikely relationship. This was years later, when the Cold War was starting to melt. I had by then met Zel'dovich myself a couple of times, once in Warsaw and then in Moscow (to be related later). Chapline, still at Livermore, was by then working mostly on astrophysics, and he was in Moscow for an international conference. Zel'dovich caused an intermediary to make contact with George and make arrangements for the two of them to meet privately. In the U.S.S.R., this usually meant outdoors and in motion. "Walks in the woods," were not just metaphorical. It was clear that each knew a lot about the other's weapons work via classified intelligence sources on their respective sides. Chapline returned from Moscow full of scientific and anecdotal information about the development of Soviet nuclear weapons. It was old, historical stuff—nothing that would then be useful to the United States in any competitive way—but it was surely still technically classified, both in the U.S. and certainly the U.S.S.R. Chapline was willing to talk about the scientific particulars only behind the fence at Livermore. The most likely interpretation was that Zel'dovich wanted intellectual credit—including in the West, including behind *our* fences—for his nuclear weapons discoveries. He was a loyal Soviet, a hero. But apparently he trusted Chapline to share his information only with members of this one peculiar world brotherhood—nuclear weapons designers on the two sides of the Iron Curtain.

Zel'dovich may also have wanted to take Sakharov's reputation down a notch. The latter had by then been awarded the Nobel Peace Prize and was a hero in the West. Zel'dovich told Chapline that he, Zel'dovich, ran the whole hydrogen bomb effort, a reality not appreciated in the West. Sakharov, a young student, played a very limited role, he said. Zel'dovich did credit him with suggesting what in the West was referred to as "the Teller-Ulam trick". Lev Landau's role, in charge of the computational effort, was, Zel'dovich said, much more significant.

In the West, Landau is generally credited as the greatest Soviet theoretical physicist, famous for his celebrated "school" of theoretical physics in Kharkov (later Kharkiv) and then Moscow; and for a series of graduate textbooks on all areas of theoretical physics, written in collaboration with his student Evgeny Lifschitz, that are still in use today. Lifschitz did most of the actual writing, leading to the physicists' joke, "not a word of Landau and not a thought of Lifshitz." (I met Lifschitz a couple of times before his death in 1985.) When the Soviet device worked and prizes were had handed out, Zel'dovich won his third Hero of Socialist Labor medal (the first two had been for the tower fission device and for the airplane-carriable version). Landau refused a medal saying that he wanted only money instead. Nothing happened for a long time; then one day there appeared a huge pile of rubles on his desk. He called all his staff in, said to them "take what you want" and left the room. They left all of it for him. He needed the money for his principal vice—womanizing. Once, in telling off a student, Landau said to him, "You are so lazy that you only have sex with *your wife!*"

When I later came to know many of Zel'dovich's students, they taught me the riddle, "What is the difference between Landau and Zel'dovich?" The answer: "Landau was a great physicist, but *thought* he was a great lover. Zel'dovich *is* a great lover, but he *thinks*...." They used an earthier Russian word than "lover".

17. Gravitational Waves

In addition to spending many hours on course work, going to Kip's group meetings (and copy-editing his manuscript), I was writing a not-very-good science fiction novel, a project that only died a slow death. There was an undergraduate Caltech club that invited famous sci-fi authors to campus; I went to talks by Robert Heinlein, Arthur C. Clarke, Robert Silverberg, Poul Anderson, Jerry Pournelle, and Larry Niven, all of whose works I had read in their entireties. Heinlein came across as stiff and out-of-touch, which, by then, he was. Rumors already abounded that Heinlein's later books were being pieced together from incomplete partial manuscripts by his wife Ginny. Poul Anderson was a good friend of my father's Russian colleague Velodya Keilis-Borok, so I introduced myself to him and we chatted.

On one panel, Pournelle and Niven kept trying to get Sir Fred Hoyle, who was in the audience, to come to the front and join their panel, but he refused. I was sitting next to Fred and Caltech professor Willy Fowler, because I had brought them to the talk. In the 1950s, Fred, Willy, and the Burbidges (a married couple), had first worked out the detailed theory of how the chemical elements are made by nuclear processes in stars. Hoyle's later works were a mixture of good and bad science. He was a proponent of the Steady State Theory of cosmology long after the Big Bang Theory was extensively confirmed by the evidence. Willy was awarded a Nobel Prize in 1983, by which time Fred was in disrepute. But he wrote several science-fiction novels, all of which I enjoyed reading. Willy and Fred will both make continuing appearances in my story.

Clarke was, we now know, prescient. He told us that major changes in life would stem from an exponential spreading of an "information grid," (he called it) to be made possible by communications satellites and computers. He compared this development to the revolutions brought by the piped-water grid, the electric grid, the telephone grid, and so forth. But, he said, this next grid was going to be the Big One. He was prognosticating twenty years before cell phones, the Internet, the first web browsers, and AI. Except for the physical medium being mostly optical fibers instead of satellites, his vision was spot-on.

Statistics suggested that I would pass my Candidacy Exams—most people did—but there were moments of doubt. My cohort of classmates were all closed-mouthed about their weekly problem set scores and first-semester grades, so I had almost no comparative data. On an absolute scale, my scores were abysmal. I rarely got more than half of the assigned problems correct. But, relatively, was that good or bad? I had no way to know. I did know that my classmates were spending more time studying than I was. A high school friend who still lived in Pasadena was going to shoot a self-produced movie in the Caltech steam tunnels with a plot about undergraduates who steal moon rocks brought back by the Apollo astronauts. I wrote the screenplay. (Nothing ever came of this.) The Candidacy Exams were all day on two consecutive Saturdays in April. A week before the first exam, Tommy, Margaret's father, was diagnosed with colon cancer. His major colostomy surgery took place two weeks after my second exam day.

But even after that long second Saturday of exams, I had no idea how I had done. After more than a week of anxious days, I called Kip at home late one evening. He was audibly annoyed. "Oh, you passed with distinction. That's the top twenty-five percent. What did you expect?" The next day, I went to see Leverett Davis, the professor in charge of the exams. I felt pretty cocky. Now knowing that I had done well, I demanded greater transparency for future students, both before and after the exams. Davis listened noncommittally, eventually telling me that my score was number two in the class. "Actually, in effect, number one," he added diffidently, "because a foreign Chinese student always gets the top score, and we discount that."

I still went to classes, but those grades were now irrelevant. At the blackboard, Professor Fred Zachariasen muttered on. I took notes in class (if at all) like this: "In a plasma there are eleven kinds of waves. In a wave there are eleven kinds of modes. In a mode there are eleven kinds of instabilities. Every instability is named after eleven physicists. Many are the names of God."

Kip summoned to his office the six first-year students who wanted to work with him: Me, Jon Melvin, Jay Melosh, Richard Green, a Chinese student named Chen (not the one who beat me in the exam) and a silent bearded fellow whom none of us had ever seen before in classes. He would only take two of us, Kip said. Well-organized as always, he had made a list of six small-bore research topics. We should each take one and work alone on it over the summer. Reviewing what we each accomplished, he would pick two winners. There was only one topic that interested me. It had to do with gravitational waves, a theoretical prediction of Einstein's Theory of General Relativity. None of the others wanted that topic so it was mine.

Why were gravitational waves on Kip's list of projects? In the fifty years since a young Einstein had pointed out their theoretical existence, no one had come up with any process—terrestrial or astronomical—that could produce these waves strongly enough to be detectable. Some relativists doubted their existence at all; the waves might merely be a subtle ambiguity inherent in the four coordinates of spacetime. In that case, a cleverer mathematician might cause them to disappear completely.

But in 1969, about when I was graduating from Harvard, physicist Joseph Weber at the University of Maryland published a paper claiming to have detected gravitational waves of astronomical origin. Weber suspended a three-ton cylindrical bar of aluminum in a shielded vacuum chamber and observed tiny "bell-ringing" events, as if the bar was being struck with the tiniest imaginable hammer at irregular intervals, about three times a day. Weber described how he had eliminated every possible explanation except Einstein's gravitational waves—which, like gravity, could penetrate anywhere, even to the inside of a shielded, isolated vacuum chamber.

The problem was that Joe's waves were far too strong. You could collide known types of stars, or collapse them in supernovas—the waves produced would be far too weak. You could collide so-called neutron stars, which had been predicted in the 1930s and finally observed (as pulsars) just three years earlier in 1967. Still far too weak. You could even collide black holes, at this time purely theoretical constructs that only relativists like Kip believed in. And even that was not enough. Weber's waves were a million times more powerful than any yet-conceived-of source. Kip alternated between believing Weber's results, in which case he wanted (or wanted a student of his) to come up with the brilliant idea about what was producing them; and disbelieving Weber, in which case he or his student should propose where Weber was going wrong—and should help design a new experiment that would not have that fault.

Professor Tommy Tombrello was interested in mounting such an experiment. I spent a lot of time over the summer talking to him and to Jim Mercereau, another professor in experimental physics. I studied the theory of measuring small signals. Vladimir Braginsky, a Soviet experimenter also interested in gravitational waves, was visiting Kip's group over the summer, and I learned a lot from him. Braginsky's English was very good but there were words that he didn't know. Two pieces of his apparatus were separated by a "piece of very thin, perfect non-conducting membrane," he paused, not knowing the right word.

"You can buy cheap at drugstore. For sex." A condom. Russian science was often like that, finding simple solutions that would elude the more sophisticated Western experimentalist.

As the end of summer approached, I wrote up my results on gravitational waves—mostly incomplete ideas, incompletely worked out—for Kip and Tombrello. I had not been up to Livermore for a while, so I scheduled a visit with Lowell and company. Since my previous trip, the Lab had changed my status from part time employee to student visitor. The employment office had sent me all manner of paperwork on this, which Lowell told me to just ignore. One letter was about exchanging my Green Badge, technically now cancelled, for a different one labeled "student visitor".

I decided to test the system by arriving right at the morning rush, when hundreds of people were streaming through the gate. The guard had a rhythm: Take a badge, look at it, hand it back; take the next person's badge, look at it, hand it back; and so on. In perfect rhythm I handed him my badge, which looked exactly like all the others. He took it, looked at it, but didn't hand it back. "Mr. Press," he said with a crocodile smile. "Come with me." It turned out that they had a most-wanted list, and I was on it. In the security booth, he showed me the full-page notice with my large, smiling picture and the words, "DENY ACCESS, CONFISCATE BADGE, DIRECT TO BADGE OFFICE." As punishment, he made me walk to the badge office a mile around the outside fence. I had to spend the morning filling out forms and promising to behave. Purely maliciously, they tried to make me watch all the security training films again. There was just one worth seeing: "The Town without Children," about a supposed Soviet training school for deep penetration agents that was the exact replica of an ordinary American town—but (punch line!) *a town without children.* I liked the meta-ness of it: American stock actors melodramatically over-acting as Russian agents who were themselves acting (spectacularly well, we were to believe) as ordinary Americans—perhaps even employed as actors.

Inside the fence, I wandered into John Nuckolls office and found George Zimmerman also there. George had been a Hertz fellow in my Caltech cohort, until his draft board began closing in. Lowell then hired him at Livermore full time. A letter from Edward (in his case, though not mine) produced a deferment.

Since the cafeteria was outside the fence, after lunch I needed to go back in through a different security gate. The new guard took my new badge and didn't hand it back. "William Press!" he said. The other guard said, “No, they cancelled him this afternoon.” “Thank you, sir,” the first guard said to me, handing my new badge back.

18. Kip

In early September, 1970, Margie Lauritsen was insistent that Margaret and I go on a ten-day vacation with Tommy and herself before our classes started. I felt that I had better things to do than travel with my in-laws. I had joined the American Bookseller's Association and was in process of being certified as a legitimate retail bookstore. This involved having phony letterhead and business cards printed, and then lying on a bunch of forms. The payoff was that once certified, I would be able to order books from publishers as single copies for 40% off the retail prices. At this time, publishing was one of several consumer-facing industries with fixed retail prices. Publishers set the price of books, and it was illegal in most states for a retailer to sell to the public for any less. The retailers were of course willing co-conspirators, since the mandated markups were high. These so-called Fair-Trade Laws were enacted in the 1930s to help small businesses weather the Great Depression. Fair Trade Laws were finally nullified by an act of Congress in 1975. What I wanted to do before classes started was to spend my Hertz Fellowship book allowance. I particularly coveted the ten-volume *Grove's Dictionary of Music and Musicians*, Fifth Edition, which I had relied on at WHRB. Hertz gave me five hundred dollars especially for books (a huge amount then) and they didn't care what books I bought.

Margie's insistence was low key, but unrelenting. A difficult life and husband had taught her how to get her way by persistence. Margaret and I considered it a victory that we were able to change the proposed destination to Hawaii instead of Jackson Hole. In hindsight it is obvious that Margie's intransigence was because of Tommy's cancer prognosis. There might not be many such future opportunities. We were oblivious to this. Tommy's condition was never openly discussed.

We did have a good time in Hawaii. The beaches were full of enlisted military on R&R from Vietnam, met in Hawaii by their wives or girlfriends. "Often met with flowers," I quipped, "because they are badly in need of a good lei." Tommy and I turned out to share a liking for off-highway driving. In our tiny rental car, we circumnavigated the Big Island on mostly dirt roads before making our way to Volcanos National Park. Less accessible then than now, Volcano House was a kind of rugged resort hotel where one spent several days, took all meals in the dining room, and got to know one's fellow travelers—a kind of land-locked volcanic cruise ship. Tommy and Margie both liked this, and they extended our stay there by several days. It may also have been that Tommy was not up to the rigors of daily travel.

When I registered for classes, several professors asked me how the trip had gone. Leverett Davis added, "Are you going to tell us how to run the department again this year?"—his idea of a cute joke. Margaret and I led a dual existence. We were nominally struggling graduate students supported by our fellowships at U.C.L.A. and Caltech; and at the same time, we were that cute young married couple—Tommy Lauritsen's and Frank Press' kids—in demand socially in the hermetic Caltech community, especially the physics department and the Kellogg Lab—which had been founded by Margaret's grandfather Charles Lauritsen. Kellogg Lab was famous for its parties; and the parties were famous for their heavy drinking. Willy Fowler and Tommy Lauritsen were drunken mainstays. Their after-hours alcoholism differed mainly in that Willy was a sweet and jolly drunk, while Tommy tended to become abusive. These two had been friends since their Caltech undergraduate days; both had done their Ph.D.'s under Charlie Lauritsen; both were Caltech lifers. Willy was by now world-famous. Tommy had never fully emerged from his father's shadow—Charlie Lauritsen had died just three years earlier, in 1968.

One large dinner party at Margaret's parents' house was in honor of Aage Bohr, the physicist-son of Niels Bohr and then-director of the Bohr Institute in Copenhagen. Aage and Tommy, contemporaries, had been friends since their pre-War postdoctoral days in Copenhagen. (Aage Bohr did emerge from his famous father's shadow, sufficiently to win the 1975 Nobel Prize in Physics for his work with Ben Mottelson on the structure of atomic nuclei.) Aage had come to Pasadena now to deliver the inaugural Charles C. Lauritsen Memorial Lecture. This was a close-knit community.

Margaret and I went to the party as much for the free meal as for anything else. The occasion brought out the extended Kellogg crowd, from Willy and Ardy, Carl Anderson, and Dick Feynman, down through Stuart Harrison, Alvin Tollestrup (and his post-divorce girlfriend), the Barnses, the Wasserburgs, Ward Whaling, Leverett Davis, Tommy Tombrello, Kip and his wife. And us.

Ardy Fowler was Willy's long-suffering wife. What, specifically, she was long-suffering about had happened so far in the past that Margaret and I had no idea what it was. Carl Anderson had received the 1936 Nobel Prize in Physics for his discovery of the positron, the first known antimatter. Stewart Harrison was the Lauritsen family physician. A footnote in history was that Stewart's first wife, Kitty, met J. Robert Oppenheimer at a garden party in Margaret's grandparents' backyard in 1938, and within a year had become Kitty Oppenheimer. Tollestrup, Barnes, Whaling, Davis, and Tombrello were professors in Kellogg Lab. Gerry Wasserburg was the Caltech geologist famous for his analysis of the first Apollo mission moon rocks; he was a difficult character, often battling Frank on trivial issues while Frank was at Caltech.

In my other life, as a graduate student, I was enrolled in lecture courses given by Fowler (Nuclear Astrophysics), Feynman (Particle Physics) and Thorne (General Relativity). From time immemorial, Willy had scheduled his course at the rude hour of 8:00 a.m. In compensation, the course was ungraded. Feynman was legendary as a lecturer. He made the most complex material so clear that you stopped taking notes—how could you ever forget things that were now so obvious? It was illusion. Half an hour later, you couldn't reconstruct from memory anything he had done. Once I realized this, I took notes whether I needed to or not.

Kip didn't have Dick's flair, but he was a clear and conscientious lecturer. His course was the most important for me, because I wanted to become his thesis student. For a while, I was doing quite badly in it. Not only that: Since my independent project on gravitational wave detectors was winding down—to be resuscitated only if Tombrello decided to do the experiment (which he never did)—Kip had been suggesting to me small, mathematical, research problems for me to cut my teeth on. I was making no progress on any of them. I was depressed, with days that I could barely drag myself to class or to my shared-office desk.

Kip may have sensed my feelings of inadequacy. Perhaps to give my ego a boost, he scheduled me to give a talk to the whole research group on gravitational wave detection, a subject on which he knew I felt comfortable. He picked a date when the eminent Princeton professor John A. Wheeler would be visiting. Wheeler was Kip's own thesis advisor, and, years earlier, had been Feynman's.

More than the usual number of first-year graduate students attended my talk, perhaps because my title was more understandable than most. Along with Kip and John Wheeler, several other faculty were there. Wheeler, a slightly stooped gentleman with a tired expression, borrowed a piece of paper to write down the names of all the students whom Kip introduced him to. I started my talk with a kind of true-false quiz on common misconceptions about Joseph Weber's apparatus. This unmistakably bombed. I jumped forward to my main material. Whenever I needed to write an important formula, there was no room left for it on the blackboard. Whenever I needed to refer to an equation, I had already erased it.

Ten minutes into the talk, Wheeler leaned back and shut his eyes. Thereafter, when he occasionally awakened and looked at the board, there was a look of horror in his eyes, apparent even as his lids sank languorously down again. Polite applause normally followed talks to the group. My final words instead triggered a turbulent stampede for the door. Wheeler remained back from the general melee and gave the blackboard a final long glance, then a final look in my direction. Not horror. It seemed like pity.

Richard Price, one of Kip's advanced students, made a point of telling me, later that afternoon, that it was a good talk. "Really," he added, unconvincingly. "And you finished early," he said, also approvingly. This seemed a variant of the old joke, "The food in that restaurant is *terrible*. Not only that, but the portions are *so small!*"

With the perspective of decades, I think that, apart from poor blackboard technique, mine was actually a good presentation; but, in ways that I didn't then understand, it was poorly matched to its audience. Physicists don't all think with the same style. There are different "dialects". My dialect, although still unformed, already bore the strong imprint of Ed Purcell, plus a bit of Edward Teller and that circle. Purcell's style was, first, to simplify to the simplest cases that preserved the phenomenon of interest; second, to reason with pictures, diagrams, and instinct until you developed some "physical intuition" about the phenomenon; and only then, third, to tease out of that intuition what necessarily had to be the governing equations—and then their solutions. Kip's style was to go as directly as possible from the phenomenon to the equations—however complex they might be—and only then, when the most general case was mathematically well formulated, to look for the necessary simplifications. If Feynman had been at my talk, he might have liked it. He could have "translated" it for Kip and Wheeler. Feynman could speak all the physics dialects.

Kip and Wheeler, along with Charles Misner (another former student of Wheeler's) were at the time writing what became a monumental textbook on general relativity, *Gravitation*, still today in print after almost 50 years. A couple of months later, when I first saw a draft of the chapter on gravitational waves, I recognized its general approach as taken from my talk. I was even credited by name—for "unpublished work". Kip was a stickler for always giving credit, and, on this particular subject, he had absorbed from me a Purcell-style approach. Kip had then convinced Wheeler to throw out their previous less intuitive and more mathematical chapter draft and substitute his new one.

A more effective boost to my ego was being sent by Kip, along with two of his advanced students, to the sixth annual Texas Symposium on Relativistic Astrophysics, held in Austin just before academic Christmas vacation. Relativistic astrophysics, a very new specialty, included the study of some things that actually existed—high-energy cosmic rays, pulsars, quasars—and some things that theorists only hoped existed—black holes and gravitational waves most prominently. The first Texas conference had been organized the year after quasars were discovered in 1964.

Kip insisted that his students be on a strict budget and that we not overspend our cash travel advances. To save on cab fare, Cliff Will, Bernie Schutz and I slipped onto the Terrace Motel's courtesy bus along with the conference participants actually staying there. It was not the best idea: They tried to make us register at the front desk before they would release our luggage. We executed an operation to distract the bus driver and bellman, spirit our suitcases out of the bus, and sneak across the street to our "Austin Motel: Salesmen Welcome." These were lodgings for the lower classes: graduate students; Bill Kaufmann (the sole Astronomer at the largely token Griffith Park Observatory in L.A.); and, surprisingly, Caltech professor Jim Gunn, who had brought along his wife, Rosemary, and was penny-pinching. The rooms were actually pleasant, with even color TV. I hardly minded chasing a two-inch cockroach down the bathtub drain.

The conference was a mostly young crowd, new Ph.D.s and assistant professors. I met for the first time many people whom I would come to know well. Some, like Martin Rees and Stephen Hawking would become famous. Stephen at this time was primarily known for his work with the more senior Roger Penrose, proving that, in some circumstances, the formation of black holes was unavoidable. (Hawking would be dead by the time a Nobel Prize was awarded to Penrose for this work.) Martin Rees soon enough became known as the leading theoretical Astrophysicist of his (my) generation. Although when I first met him, Martin was a radical leftist and scornful of Establishment trappings of authority, he over time became Sir Martin Rees in 1992 and then Baron Rees of Ludlow in 2005. He was President of the Royal Society from 2005 to 2010, also serving as the Queen's Astronomer Royal. At this conference, Stephen's condition was already evident. He could speak, though haltingly, and could walk with difficulty on two crutches. He was just starting to rely on a wheelchair. Others, never famous to the public, were (or became) leading relativists: Kip's co-author Charlie Misner; Peter Bergmann, who had worked with Einstein; Igor Novikov, co-author with Zel'dovich of the book I had copy edited; Yavuz Nutku, a literal young Turk who had just completed his Ph.D. with the famous Chandrasekhar in Chicago; Lou Witten, my helpful, Communist, Livermore officemate; Bill Unruh; Craig Wheeler; Ken Nordvedt, Gary Gibbons. I went out drinking every night with varying combinations of these people. Novikov, a small man, had an astounding capacity for vodka—he was famous for this even in the U.S.S.R. Nutku the Turk discovered in himself, when drunk, a new histrionic command of heavily accented English and became a bottomless fount of anecdote.

Kip introduced me more formally to the senior *eminences grises* who were invited speakers at the conference: Penrose, Tommy Gold (Cornell), Bob Dicke (Princeton), Frank Low (University of Arizona), Iosif Shklovsky (Moscow), Frank Drake (Cornell). Charles Townes (MIT) was someone I already knew, via my father. Dicke was famous for *not* discovering the cosmic microwave background in 1964—he had been beaten out, some thought in an unsportsmanlike way, by Arno Penzias and Bob Wilson at Bell Labs.

Among experimental physicists, Dicke was famous for having invented the lock-in amplifier, also called the Dicke switch. The basic idea was to measure very small effects by switching them on and off at a known frequency, which could then be selectively and precisely amplified. At a later conference in honor of Dicke, Ed Purcell (who had worked closely with Dicke on radar at MIT during World War II) suggested that the way to measure a person's lifetime contribution to the world was to "switch them in and out of history twenty times a second" and then observe what in the world was flickering. With some famous people, the few things they are famous for would flicker a lot, but everything else would remain steady. For others (including Dicke, in Purcell's telling) everywhere you looked many different things would be flickering faintly. Dicke and Purcell died within a week of each other in 1997.

At this 1970 conference, Dicke's obsession was the oblateness of the sun. With Carl Brans, he had developed an alternative theory to Einstein's general relativity. If the core of the sun were rapidly rotating, then the famous observation of the orbit of Mercury that confirmed Einstein's theory would be invalid—and the Brans-Dicke theory might instead be confirmed. A rapidly rotating core would produce a very slight oblateness in the observed figure of the sun. Dicke was attempting to make that measurement. The Brans-Dicke theory was decisively disproved in 1982 by precision observations of the orbit of the so-called binary pulsar—two relativistic neutron stars in close orbit. That and related work won the 1993 Nobel Prize in Physics for Joseph Taylor and Russell Hulse.

Drake and Shklovsky were in 1970 known for trying to make the search for distant extraterrestrial intelligence (using radio telescopes) a respectable field of astronomy, "SETI." Theirs was a mission never completely achieved. I came back from the conference having learned some science but, more significantly, having gotten a sense of the international community that I was in training to join. That was Kip's intent, probably. When I turned in my accounting to the secretary in Kellogg Lab, with ten dollars of my advance unspent, she told me to keep the money.

19. Black Holes

This is what we knew about black holes as purely theoretical constructs in 1971: We knew that there were certain mathematical solutions of the equations of general relativity that behaved like massive bodies, but paradoxically contained no matter. These were the so-called vacuum solutions discovered by Karl Schwarzschild, in 1915, and Roy Kerr, a New Zealander working at the University of Texas at Austin, in 1963. What set Kerr's vacuum solution apart was that it behaved as if it had not just mass, but also rotation. If you find the idea of a rotating vacuum confusing, so did many physicists at the time. Both the Schwarzschild and Kerr solutions exhibited "event horizons," boundary surfaces from within which no material, no light, no signal could escape to the outside. In 1967 John Wheeler coined a name for things with this property: black holes. One way to think about a black hole is that its gravitational pull is so strong that its escape velocity exceeds the velocity of light—so neither light nor anything else can escape.

The term black hole is a virtual Rorschach test for dirty minds. Was John Wheeler, in coining it, blind to all its possible innuendos? No, I don't think so. Wheeler's ability at deadpan humor rivaled that of Buster Keaton. He wore the guise of a perfect gentleman, but his straight-faced absurdities often involved pricking the sensibilities of the establishment. When Werner Israel and others proved that black holes must have the unornamented structure of the Schwarzschild or Kerr solutions, and nothing additional, Wheeler memorably summarized their results as, "Black holes have no hair." Even today, these results are called "no-hair theorems".

As a young man in 1930, while on the boat from India to England, the prodigy Chandrasekhar proved that stars more massive than 1.4 times the mass of the sun that exhausted their nuclear fuel (as all stars must eventually do), had no alternative but to collapse. No stable cold configuration existed. (We will meet an older Chandrasekhar soon.) But collapse to what? The answer was obscure until work by J. Robert Oppenheimer and a student in 1939 connected something real (a star) to something mathematical (the Schwarzschild solution): A gravitationally collapsing star, if it was spherical and not rotating, would in fact become a Schwarzschild black hole. The matter would be lost, and ultimately obliterated, inside the event horizon. After Roy Kerr's discovery, it was presumed—but not known—that the collapse of a *rotating* star would produce a *Kerr* black hole. Black holes, if they formed, were thought to be static objects, frozen, eternal and unchanging. In Russian, the term for black hole was not the literal *chernaya dyra* (which was vetoed by Zel'dovich, because he thought it obscene) but *zamerzshaya zvezda*—frozen star. The idea that a black hole could be perturbed, and that it might react to such a perturbation by emitting gravitational waves, was not thought of. It was a momentary blind spot in science. Blind spots like this allowed graduate students like me to make their first scientific discoveries.

The equations for such a perturbed Schwarzschild solution had been partially worked out by John Wheeler and the Italian mathematical physicist Tullio Regge long before, in 1957. Perhaps because the subject was so abstract, and the term "black hole" had not yet been invented, completing the equations was put aside for thirteen years. It wasn't until 1970 that Wheeler off-handedly suggested that a former student of his at Michigan State University, Frank Zerilli, finish the work. Kip took note of Zerilli's success, and he assigned to his student Richard Price the job of figuring out whether the now-finally-complete Regge-Wheeler-Zerilli equations could be used to demonstrate anything useful about black holes.

In this connection, Kip one afternoon summoned Richard and me to his office. Jim Bardeen, a young professor at the University of Washington, had given a talk somewhere about how gravitational waves might be produced by objects falling into a black hole. Small objects (asteroids, say), might, according to Bardeen, produce high frequency gravitational waves—the kind that Weber claimed to see—when they fell into black hole produced by the collapse of a star. Kip had inherited from Wheeler a liking for making up names for things, and he called this the "Bardeen dimple conjecture". Kip wanted Richard and me, but mostly Richard, to prove the conjecture, using the Regge-Wheeler-Zerilli equations. By implication, he also wanted Richard to take over my tutelage. Kip was, I think, tired of suggesting to me small mathematical research problems that I was unable to do on my own. Perhaps I would do better as Richard's assistant.

Richard Price was more mathematical than I was. He had mastered the currently-in-vogue "Newman-Penrose formalism" (yes, that Penrose), and used it to simplify the Regge-Wheeler-Zerilli equations. The result was still not in a form that could be solved with pencil and paper, the way mathematicians and mathematical physicists then did their work. Richard hoped to make progress by finding limiting cases and approximations, and then piecing together a final result. I said, "Why don't I just solve your equation on the computer and see if Bardeen is right?" "Can you really do that?" "Yes." My optimism was naïve, but not unfounded. Hacking at Harvard had made me unafraid of computers, even as most scientists still regarded them as the esoteric realm of professional programmers. And I had, from Livermore, the Wilson-LeBlanc pencil notes—my own, private, "unclassified, but do not copy" tutorial on how to turn physics into computer code. The program I wrote was very primitive by any modern standard, but the answer that it gave was unequivocal: The Bardeen dimple conjecture was wrong. Gravitational waves were produced, all right, but their frequency and shape had very little to do with what we (on the computer) dropped into the black hole. Instead, they seemed to be properties of the hole itself. The black hole was like a bell, with its own tones and overtones, independent of what object was used to strike it.

This wasn't much of a mathematical discovery—the Regge-Wheeler-Zerilli-Price equations were handed to me on a platter. But it was a paradigm shift, to a view that black holes should be thought of as dynamic objects that responded to perturbations as a bell does to being struck. It took Kip several months to come around to my point of view—he still wanted the black hole to be a static object. But he eventually did (and later wrote a whole book on the subject), and I was invited to give a talk to the whole of Kellogg Lab. This one went much better than my previous talk to the relativity group. My paper, submitted to *The Astrophysical Journal*, was initially rejected by an anonymous reviewer. His many objections basically all came down to, "everyone knows that black holes are static objects". I wrote a rebuttal, and Kip wrote a heated letter to the journal editor. The paper was eventually published as "Long Wave Trains of Gravitational Waves from a Vibrating Black Hole". It should have been my first paper, but by the time it came out, it was my third.

Kip's graduate students never knew in advance how much would be required for their Ph.D. theses. Kip's idea was that we should work *as if* we were already scientists performing research and writing scientific papers; at some point he would decide, seemingly arbitrarily, that you had done enough. You would at that point schedule your final oral Ph.D. defense (a formality) and get your degree. If this happened at the wrong time of the academic year for you to have applied for jobs, then Kip would come up with a short-term postdoctoral appointment in his group. This transition occurred for Richard Price mid-spring. Richard was Kip's fifth Ph.D. student. With his previous four, Kip had established the tradition of an all-night celebration party at his rambling

Altadena home. It was a Friday night; Margaret and I already had tickets to a performance at Beckman Auditorium by The Committee, a San Francisco political comedy troupe, so it was after midnight when we arrived at Kip's.

I pointed out to Margaret the notables: Jim and Rosemary Gunn; Virginia Trimble and (visiting) Martin Rees, who seemed to be together; Richard, of course, who was already quite far gone; Jiří Bičák, Jim Bardeen, Paul Schechter, Phil Callahan, Rolf Sinclair from NSF. Jiří was then a genial Czech postdoctoral fellow. Because of Kip's warm relations with the Zel'dovich group in Moscow, scientists from behind the Iron Curtain were frequently allowed out to visit us. Jiří later became a professor at Charles University in Prague and President of the Czech Academy of Sciences. I had first met fellow graduate student Paul Schechter when we sat next to each other at the placement exams, our first week at Caltech. He looked like a bearded lumberjack. I was impressed at his having been a leader in the radical SDS chapter at Cornell.

Richard passed out at about 2:30. At 3, Kip, Bernie, Jim Gunn, and Jim Bardeen had their Russian-style vodka drinking contest. It was not a knockout since they ran out of vodka after only eight ounces each. To my mind, Gunn, a Texan, won on points. Kip, pretty far gone, approached Margaret and said, leering, "You know something? Your husband is a damn good physicist. But don't tell him I said so." With that, we resolved to stay for the endgame.

At 4:30 a.m. there were us, Gunn and wife, Kip, Linda, and a weird male undergraduate. Richard was asleep in the den, and the undergraduate's girlfriend was out cold on the living room couch. Kip talked with Margaret in Russian. Normally, he got this drunk only in Russia, so this made a kind of sense. He asked (in English) my permission to kiss Margaret, saying that the precedent was well established in the case of all his other married grad students. He assured me this was merely a request, not a requirement for my degree. Gunn was entirely in physical control of himself, but every time someone said something to him he giggled uncontrollably. We left when Linda Thorne was serving coffee to everyone except the undergraduate, who was popping some blue pills. "Try them," he offered to us, "the high lasts about six hours." We declined with thanks.

Jim Bardeen, whom I came to know well and wrote several papers with, was the son of John Bardeen. The father was one of four people to ever win *two* Nobel Prizes, and the only one to win two in physics. Jim, far more than I, bore the burden and advantage of having a famous-scientist father. As a graduate student at Caltech, Jim had worked with Dick Feynman and Willy Fowler. A few days after the party at Kip's, he wandered into my office. He didn't say anything, just, with a pleasantly quizzical expression, started examining all the papers on my desk and looking at the books in my bookcase. After a few minutes of his doing this, I felt obliged to say something. He was actually easy to communicate with—when he talked! Eventually he wandered out without a goodbye.

The episode struck me as more than a little strange. Sometime later, when I met Jim's wife, Nancy, she told me that Jim was the talkative one in the Bardeen family. She described visits to her in-laws. John Bardeen might be sitting alone in the living room. Jim would wander in and sit in a chair facing his father. The two would look at each other. Ten or fifteen minutes would pass with neither saying anything. Finally, Jim would say, "Um, hi, Dad." That made him the talkative one.

Bardeen was a far better mathematician than I was. We worked together on a topic related to Richard Price's work, the so-called Newman-Penrose conserved quantities. Frequently, we talked about an approach in the morning, and I spent the rest of the day struggling to calculate its first steps. At the end of the day Jim wandered back in and—naturally without saying anything—put his completion of the calculation in front of me, let me read it, and then took it away. I wondered whether I should ask him to let me Xerox his notes. I described my situation to Margaret by analogy: "Mr. Hemingway, may I copy your short-story manuscript and submit it, under my name, to my high-school literary magazine, *Nouvel Esprit?*" Margaret only suggested that *Cornucopia* would be a better name for a high-school literary magazine.

However, I began to notice that what Jim worked out were often *my* original suggestions for how to proceed. This was an important self-discovery: I was OK, but not great, at higher applied math. I was much better at *strategy*, charting a mathematical course through an otherwise confusing research problem. Jim was the first of many superior mathematical collaborators with whom I worked, some older than me, some younger, some my own students. It became the pattern of my scientific career.

20. Feynman

Caltech's stars in physics were Dick Feynman and Murray Gell-Mann, both Nobel Laureates. Both were New Yorkers. (Murray's father had added the hyphen to the family name.) Each cultivated a larger-than-life public persona. Each was a prima donna in a different way. Dick presented himself as a salt-of-the-earth Joe Six-Pack, a guy who went to topless bars after work and who liked nothing better than puncturing the hot air balloon of any person the slightest bit pompous. That person, of course, would be Murray, who presented as a hyper-correct pedant, someone who knew everything about everything—and would deign to educate the willing listener at any length—but, even better, the unwilling one. Dick and Murray hated each other, had offices next to each other, and made a show of very publicly consulting each other on deep matters of physics—really only when the instigator thought that he would prove himself (again) smarter than the other. Dick's theater-of-self was the more winning. People (including Margaret and me) loved the Feynman show and we even liked the man himself. Murray was widely disliked.

When Margaret began working on her Ph.D. thesis in linguistics on the Chemehuevi Indian language, Murray (at a Caltech party) impressed her by mentioning that the city Cucamonga, in San Bernardino County, was a Chemehuevi word. (He was mistaken. The word is of Gabrielino origin, meaning that it was in use by the indigenous coastal population around the San Gabriel Mission when the Spanish first settled there in 1771. The Chemehuevis lived in the eastern Mojave Desert and likely came there, by 19th century forced migration, from even farther east.) When Feynman died in 1988, Murray wrote a lengthy and quite nasty obituary for *Physics Today*: Feynman "surrounded himself with a cloud of myth, and he spent a great deal of time and energy generating anecdotes about himself." Both were equally guilty of this. Dick just did it better, and Murray could never forgive him. It was rumored that there was an even nastier version of Murray's obituary that *Physics Today* had declined to publish.

By the summer of 1971, Jim Bardeen had pretty much taken over our Newman-Penrose conserved quantities project and had produced some beautiful results that we were writing up together. I did feel that I had contributed to the project—enough to share the thrill of it, anyway. I related to Kip my amazement at it all falling so neatly into place, after it had made no sense at all for six months. He told me to savor the feeling; it was rare even for good scientists. He then changed the subject slightly. "What do you think about Bardeen horning in on your problem?" I responded that it wouldn't make sense for me to artificially keep Jim from doing calculations that he could do much better than I could. "Well, you better work something out with him," Kip said, "because I don't want you to end up thinking that you've gotten the shaft."

The negotiation with the laconic Bardeen took about thirty seconds. "How about, you be first author," he said. Kip felt that this was an acceptable solution. In the end, I did a lot of the writing. More accurately, Kip did, because he insisted on seeing every draft that I wrote, and returned each completely rewritten, between the lines, in his red ballpoint pen. Drafts in these days before computers were cut-and-paste affairs. Jim wrote his parts out longhand. I typed mine (I had just bought my first, used, IBM Selectric typewriter, the kind with the little removable type ball) and filled in the equations by hand. I noticed that Kip made very few corrections to Jim's sections. Did he really think that Jim was a better writer than I, or was this mere professional courtesy to a colleague? To find out, I began retyping Jim's sections as if I had written them—and they came back from Kip completely rewritten, in red pen.

I learned scientific writing from Kip. A lesson he drummed into all his students was related to the tree that falls in the forest with no one to hear: You haven't actually discovered anything, Kip would say, until you have successfully communicated it to others. Still, when Kip inserted extravagant John-Wheelerisms, I removed them.

Because he was going to be traveling all summer, including to Moscow, Kip made us each write down what we intended to accomplish while he was gone. I wrote that I would calculate with numerical precision the gravitational radiation produced by dropping a small particle into a non-rotating black hole (the R-W-Z equations again), and added orally that I wanted to start thinking about the more interesting features of rotating (Kerr) black holes. Just a few months previously, Roger Penrose had surprised the relativity community by showing theoretically that, even though a Kerr hole was by definition "black," you could contrive to get energy out of it. The "Penrose process" involved putting a particle into orbit very close to (but outside) the hole, then having it decay into two particles, one of which went into the hole, the other recoiling away from it. You could arrange things so that the one particle that escaped had more total mass and energy than the particle that you originally sent into orbit. The extra energy came from the rotation of the hole, which slowed down a bit. I figured that if you could do this with particles, you could surely also do it with waves—maybe with gravitational waves.

I talked my idea around the group using the term "superradiance," which I borrowed from laser physics. Charlie Misner, Kip's senior colleague in Maryland, had a similar idea, but he was lost in the thicket of still trying to explain the faulty Weber experiment. Zel'dovich in Moscow, I later learned, was thinking along similar lines, and maybe not completely independently of me: Zel'dovich's hurriedly written paper proposing black-hole superradiance was submitted to a Russian physics journal in July, just after Kip's visit. Kip surely would have told him about my work. With lasers, superradiance goes hand in hand with "spontaneous emission". That was what I was really interested in finding out. Would a Kerr black hole start spontaneously emitting gravitational waves, losing rotational velocity until it turned itself into a non-rotating Schwarzschild black hole? Would this happen gradually, or in one giant unstable belch of gravitational radiation? In summary: Was the Kerr black hole *stable?* I was to spend much of the next two years working on this question.

Richard Price had accepted an assistant professor offer at the University of Utah, and Jim Bardeen was soon going back to Seattle, so I badly needed a new mathematical partner. Paul Schechter was sharing a graduate student office with a first-year student from South Africa, Saul Teukolsky, who already had a reputation for being mathematical. South Africa—that was very exotic. Saul was obviously not Black, so I assumed that he must be Afrikaans (that is, of Dutch lineage). No, he told me, he was Jewish. Jews in South Africa tell the story of the two brothers from the *shtetl* outside Minsk. The older brother emigrates to New York, and then writes for his younger brother to join him. "Make your way to the harbor at Riga and get on a ship that says U.S.A." The brother complies and ends up in the Union of South Africa.

I started bringing papers to Saul for us to discuss together. Often, he was bothered by the lack of mathematical rigor in a proof which for me was rigorous beyond comprehension. Then, I showed him some computer exercises that I did to check the paper's final result. (From my point of view, I thought I was checking my computer program against a demonstrated solution.) Saul could not imagine anyone doing numerical solutions so readily, and I could not conceive of anyone being comfortable with the horrendous mess of equations in the paper. This was the germ of our lifelong mutual admiration society.

Feynman had announced in class that he was going to give individual oral final exams to every student—ten minutes each. What kind of exam could take just ten minutes? "Mr. Press?" he said leaning out of his office when I arrived at four on the dot. I went in. He shut the door. "Now the spider has caught the fly," his amusement at this joke undiminished on its fourteenth repetition. I stood at the blackboard, and he asked the first question. When I got five words (and one or two chalk symbols) into my answer, he interrupted with the next question. Then similarly for the third. I complained that he was not letting me actually answer the questions. "Oh, you think I don't know exactly what you were going to say?" He got up from his chair with a sarcastic show of effort and went to the blackboard. "You were going to say this...and then this...and then this." He was exactly right. After that, I decided to relax and just enjoy *not* answering the questions. At the end, he said, offhandedly, "You get a B." He must have seen my disappointment, because he added. "It's not an ordinary B. It's a Dick Feynman B. You go out there and be proud of it."

Saul Teukolsky, whose help I needed if I was going to work on Kerr, left for the summer, to South Africa. He was going to marry his college sweetheart, Roselyn, and bring her with him back to Pasadena in the fall. I had no choice but to work my other promise, dropping a particle (on the computer) into a Schwarzschild black hole. It dawned on me only gradually that this was quite a different kind of undertaking from my previous work. My main results on vibrating black holes were conceptual. I was pretty sure I did the calculations correctly, but it was obvious from the structure of the equations that my main conceptual conclusions were right. My work with Bardeen on the Newman-Penrose constants was bulletproof in a different way. It was all equations—no computer modeling was involved. After Jim and I both checked every step enough times, we could be pretty sure that the paper was correct.

"Dropping a particle" was different. There would be a numerical answer—how much energy was produced in gravitational waves. I knew that at least two others were working on the same problem, Jim Ipser, a former student of Kip's in Chicago; and Remo Ruffini, an assistant professor under John Wheeler, in Princeton. I wanted to be *first* with the answer, but I had to get it right. If Ipser or Ruffini got a different numerical answer—and if theirs turned out to be right—my developing reputation as a computational physicist would be destroyed. As a warm-up exercise, I tried to duplicate Zerilli's derivation of his equations—and got a discrepant minus sign on one term. That scared me. If someone else's typographical error led to the wrong answer, I would still get the blame. Physicists are supposed to check everything that goes into their work—that is how the field advances.

Richard Price was still around, luckily. Richard, a direct and opinionated person who often functioned as group leader in Kip's absence, was not (I already understood) a brilliant physicist. But he had an excellent understanding of how equations connected to reality, he was a meticulous and accurate calculator, and, most importantly, he could not be coerced into agreeing with something that he wasn't himself 100% sure of. He agreed to collaborate. We divided the task into pieces, and then the pieces into smaller pieces, and we spent hours arguing about each tiny step—sometimes conceptually, sometimes about whether we had the equations correct, sometimes about my computer program. We both checked and rechecked everything. It took all summer. Eventually, we had a result: The number (in appropriate units) was 0.0104. That value was not large, but it was four to six times larger than previous crude estimates. By the time that Kip was back in town, we had the draft of a paper to submit to *Physical Review Letters*.

Richard took it on himself to call both Jim Ipser (whom he knew well) and Remo Ruffini (whom he had met just once or twice) to tell them that we had a result and were submitting a paper. It was a polite way of saying, "we won!" Jim took the news with good grace. Remo, not so. He had finished his calculation months ago, he said. He had merely put off writing it up while he was vacationing in Italy with his mistress, he said. Surely *that* should not count against his claim for priority over us. Richard told him our answer, 0.0104 and asked what answer he got. Within ten percent of ours, he said. Later in the same phone call it became: within one percent of ours. We weren't sure we believed him.

Kip was visiting John Wheeler at his cottage on High Island in South Bristol, Maine. The handful of houses on the island shared a single phone line, and it took us an hour to get through. Remo may be difficult, Kip said, but he is not dishonest, adding that John Wheeler and he would together decide the question of who got to publish first. What eventually happened, after much *agitazione e emozione* was that a joint paper, almost identical to Richard's and my first draft, was published with four authors: Davis, Ruffini, Press, and Price. This was not a happy experience for me.

Co-author Davis was Marc Davis, a Princeton graduate student who soon quit working with Remo and became an observational astronomer and, later, a colleague and friend of mine at Harvard. Marc and I were never able to compare old timelines down to the day, but a guess is that the truth was this: Marc was making progress on the problem, but with as yet no reliable result. Remo claimed to us that they already had the answer, and he then spun up Marc into high gear. Marc's computer program did then genuinely produce the same answer as ours, but it did so only after he (or at least Remo) had in hand not just our answer, but our whole draft paper. Marc could not have believed his program without our result: He had no Richard Price to check his every step.

Ruffini was a thorn in the side of the relativity community for several more years, leaving in his wake a growing list of people who felt that he had abused them. Kip eventually wondered if his earlier assessment of Remo might have been too charitable. I never heard such an admission from John Wheeler; but by 1975, Wheeler seemed to have cut his ties with Ruffini and, to my knowledge, never again published with him. When I became an assistant professor at Princeton in 1974, I was filling the slot that Ruffini had vacated to go back to Italy, where he spent the rest of a long career.

21. Chandrasekhar

In 1971, Chandrasekhar was by most measures the world's greatest living theoretical astrophysicist. That was no surprise, because it had also been true in 1941, 1951, and 1961. If he had been better known in 1931, the honor would have been his then, too. Chandra had no actual first name: He came from a high Indian caste where one name was enough. When he was a student at Cambridge, England, he adopted the initial "S.", and later the first name Subrahmanyan, but these were only prepended conveniences.

Chandra was the nephew of C.V. Raman, the Indian physicist and Nobel Laureate. In 1971 he had not yet won his own Nobel Prize—that came in 1983. He had been a professor at the University of Chicago since 1936. Princeton had offered him a Chair succeeding Henry Norris Russell, the greatest American astrophysicist to that date, but President Robert Maynard Hutchins sold him on Chicago by asking if he wouldn't rather *make* a Chair famous, than take one made famous by someone else. Chandra once told me that Princeton's evident racism was a factor. President Dodds had assured him that, despite his dark skin, he would be classified as white and thus eligible for full Princeton University Library privileges. Chandra wanted no repeat of the way he had been treated as an Indian in Cambridge by the English; he declined Princeton and accepted the Chicago offer. Chandra and his wife Lelitha became U.S. citizens in 1953. They never had children. Chandra *always* wore a dark suit, white shirt, and tie. His manners were *always* formal.

"Living legend" understates the situation. When Kip, trailed by Chandrasekhar, came into my office and introduced me to him, I felt that I was meeting a marble statue come to life. Specifically, it was the statue of the dead Commander in the Don Juan legend, and I was to be held accountable for my scientific misdeeds—not too many so far in my short career, but I could think of a few. In September, 1971, Chandra, at age 60, had just stepped down as Editor-in-Chief of the prestigious *Astrophysical Journal*, and was unwinding by spending a semester on sabbatical in Kip's group at Caltech. He brought with him a graduate student, Bonnie Miller. It was a matter of pride to Chandra that the average age of his collaborators had never risen above thirty. He thus intended to interact primarily with us, Kip's students.

We went to lunch at the Athenaeum. Chandra ordered a salad, an apple, and coffee, the only vegetarian items. Conversation turned to the question, at what age does a physicist become a has-been. "You are in your 30's still, of course," Chandra said to Kip. "Thirty-one," Kip said, smugly. Chandra advised, "You need not worry until you are 40. Then you must worry. Forty is the critical age."

He continued, "I do not say it, but it is said of me, 'Chandrasekhar at age 40 became tired of the struggle of research and became a journal editor.'" By the time he was done coring the apple, he had arrived at his conclusion: Great minds try to repeat past triumphs. But their triumphs were all too often right but for the wrong reasons, and therefore the attempt to reuse those wrong reasons is doomed.

This was apparently a favorite theme. Chandra carried in his jacket pocket, and now showed us, a place card from a 1955 honorific dinner in Israel. He had been seated next to Niels Bohr. "Because of the alphabetical order of the seating, you understand" On the back of this place card—who knew what else Chandra carried in his pocket!—Bohr had written: $p\,q - q\,p = i\,h/2\,\pi$ (his famous foundational formula of quantum mechanics) and then $(a+b) - (b+a) = i\,\ell/2\,\pi$. Here ℓ was supposed to be a universal quantum of length, an idea that was (in this form at least) utter nonsense.

Chandra summarized diplomatically: "Whether this is right or wrong, you see that his mind was seeking the same success as in his youth." Kip countered politely that astrophysics was so broad a field that no one could become a good astrophysicist before age forty. But Chandra wasn't buying: You cannot become a good astrophysicist before forty, he said, only because the astronomers take that long to accept innovative work. He cited his difficulty in getting his stellar structure papers published in the early days. "And if I did not happen to be the editor of the *Astrophysical Journal*, there would be no relativistic astrophysics published there today," he said pointedly to Kip, who was submitting most of his and his students' work to that journal.

It was Bonnie Miller, Chandra's student, who explained to us that he should be addressed as "Chandrasekhar" by anyone without a Ph.D., but as "Chandra" by anyone with. Bonnie took it upon herself to function as a kind of advance person for Chandra, smoothing his way in interactions with our younger generation, and making sure that we all appreciated his importance (which we did). She and I became good friends, hung around together at work, and socialized after work with our spouses, she often riding home with me on the back of my motorcycle. Her husband,

Ritchie Miller, was a mathematician, a topologist, a visiting assistant professor at U.C.L.A. She was about to get her degree, with employment prospects very uncertain. As it turned out, she was Chandra's last student. Bonnie was a fount of Chandrasekhar anecdotes. In the 1968 U.S. presidential campaign, Chandra so feared George Wallace's segregationist candidacy that he went door-to-door in Chicago slum neighborhoods campaigning for Hubert Humphrey. One can only imagine the reaction of the local residents to the alien appearance of this diminutive Indian man in a dark suit and black tie, a formal manner, and an Indo-Cambridge British accent.

* * *

The unofficial end of summer for relativistic astrophysicists—especially those of us younger than Chandra's pivotal age forty—was the Sierra Conference on Relativistic Astrophysics, a kind of camping trip and tribal gathering that took place just after Labor Day in Tuolumne Meadows, Yosemite National Park. Kip and his wife Linda were going—he was a speaker. Sandor Kovacs, one of Kip's newer students was going, along with his glamorous girlfriend Renee Meil, who was a Pan Am stewardess on that airline's famous Flight 1 and Flight 2, which, with intermediate stops, went around the world in opposite directions. Renee (who pronounced her name *REE-nee*) typically worked the westbound Flight 1 halfway around the world to Tehran or Beirut, had a two-day layover, and then returned on the eastbound Flight 2. That way she could be based in Los Angeles. Flight attendants who did the reverse were based in New York. Margaret couldn't go with me, so for purposes of sharing a tent and meals, I teamed up with Dave Schramm, a postdoc of Willy Fowler and Gerry Wasserburg, previously their graduate student. Schramm was a gentle giant, a champion college wrestler who dominated physically without the slightest intended hint of menace—but you always felt the hint nonetheless. The meals logistics were soon taken over for all of us by Linda Thorne.

Dave and I set off in my aging Buick at 8 a.m. We reached Yosemite around 2, and Tuolumne, on the far side of the park, a couple of hours later. Schramm's tent was huge, about 12' by 12', and he also supplied two tables, two lanterns, a stove, cots and air mattresses. The two of us had it pretty cushy. Sandor was crowded with Renee into a one-person back-packing tent; but when you saw Renee in her shorts and bikini top you didn't feel too badly for him. There was no conference administration to speak of. You just set up your tent and showed up at the open-air lectures the next morning on the grass. By dinnertime there were twenty or thirty tents of other attendees—but no sign of Kip and Linda, or, more importantly, the food that they were bringing. We, plus Jim Wilson from Livermore, pooled what we had for a meager dinner.

Even without any food to cook, Sandor was up just after sunrise to start a cooking fire. He was Hungarian and had escaped as a child with his parents and older sister from Hungary at the time of the 1956 uprising. Renee and I walked a quarter of a mile each way to bring back buckets of water. Returning, we encountered Kip and Linda's battered green VW van—going the wrong way. They had gotten hopelessly lost the night before—in Lancaster, they said, although that was impossible—and were just now arriving. Linda was soon cooking breakfast for us all.

Lectures were in the mornings, and then around a big fire after dinner in the evenings. The afternoons were for hiking. Our first group hike rapidly sorted us by ability: Wilson, Sandor, Schramm, Kip, Linda, me, Renee, from best to worst. At the 9800-foot summit we found, already there, Jay Melosh and his wife, Paul Schechter, four people from Berkeley, and two from Santa Cruz. They had come up the front face, free climbing on the bare rock. The rest of us were by comparison all wimps.

Kip gave one morning's lectures. Among that evening's campfire lectures was a short talk by a geologist on the geology of the Sierras. This (or something like it) was requirement for our waiver of the park's entrance and camping fees. During the talk an argument broke out between the speaker and a physicist who didn't believe that poly-crystalline granite could have formed out of a homogeneous melt. The geologist responded with, "That's just the way it is, brother. Either you believe it, or else you pay your park entrance fee."

* * *

When theoretical physicists can't solve a problem, they simplify it, often multiple times, until they reach a problem that they can solve. That simpler solution sometimes gives insight about how to climb the ladder back toward the original problem. "Assume a spherical cow," is an old physics joke about a physicist who went to work for the dairy industry. My problem with Kerr black holes, mathematically speaking, was that there was no known equivalent of the R-W-Z equation—that one that applied to non-rotating Schwarzschild black holes. Here I was, computer boy-wizard, ready to compute, but lacking an equation to compute with. Gravitation is a "tensor field". That makes its equations horribly complicated. OK, what if I simplified the problem to a "vector field," like the electric and magnetic fields. Maybe I could compute how much superradiance you get when you blast a Kerr black hole with radio waves. No such luck. No one had worked out the necessary equations—and I was not likely to be able to do so. OK, another simplification, to a "scalar field". Scalar fields are the absolutely simplest. A small embarrassment is that there are no known scalar fields in nature. They exist only in theoretical mathematics. Still, the people in Misner's group at Maryland had recently worked out the equations for the Kerr case, the unwieldy Brill-Chrzanowski-Pereira-Fackerell-Ipser equation. (These were all people whom I had by now met. Relativity was a small field.) I would compute the *scalar* superradiance.

Not so fast. The Brill et al. equation was unwieldy not just in name. It had the terrible property that it was "non-Cauchy". Without technical details, this meant that it was hopeless on the computer. OK, yet another simplification: What if I didn't do waves at all, but just imagined a Kerr black hole in a constant, uniform scalar field—a bit like a constant, uniform magnetic field. It wouldn't necessarily have its rotation axis aligned with the field; it could be at an angle. Finally, a problem that I could solve! I could even solve it on my own, before Saul got back from South Africa with his new bride later in September. It took me a couple of weeks to get it right, but I had the fun along the way of talking to Chandra about it. He wandered into my office and let me update him on my progress. Sometimes he asked me mathematical questions related to a problem that he was working on. This was pure flattery of me, since he was himself a vastly better mathematician than I would ever be.

The uniform scalar field problem turned out to have an interesting answer: The black hole *hated* to be misaligned. It swallowed energy and angular momentum from the field, generating a torque to perfectly align itself. This was a result of no practical use, but it was (i) cute, and (ii) demonstrated that *some* physically motivated Kerr calculations were possible. Kip told me to write it up and submit it to *The Astrophysical Journal*. After a long back-and-forth with a picky anonymous reviewer (Stephen Hawking, I later found out), it became my fourth published paper.

Bryce DeWitt was a respected gravitational physicist who did his Ph.D. work under Julian Schwinger and made his career at U.N.C. Chapel Hill and then the University of Texas at Austin. He was a man known for his gentle, softspoken charm. He was in town. The Thornes invited Margaret and me to have dinner with him. The men talked physics over cocktails while Margaret helped Linda get the chicken onto the barbeque grill—that was how things were done, then. Bryce had been Schwinger's graduate student in 1946 to 1950 and reported that, in toto, he spent less than 20 minutes with the great man. He semi-invited me to give a talk at Stanford (where he was visiting for the semester) on "the torque on black holes," evidence that Kip had been telling him about my work. Later, he described first meeting Paul Dirac, the famously eccentric British theoretical physicist who first predicted the existence of antimatter. The great man was out on a stroll, so the young DeWitt tagged along with him, explaining in great detail all the important things he was working on. He finished his monologue and was met by silence, Dirac saying not a word, no question or comment. After a minute or two Dirac said to him, "Do you know of a public toilet nearby?"

Immediately after dinner, Linda urged Kip to go upstairs and work on his book, after which Bryce, Linda, Margaret, and I continued in conversation. Bryce's wife, Cécile DeWitt-Morette, a mathematician and mathematical physicist, was founder and director of the famous physics summer school at Les Houches in the French Alps. I was trolling for an invitation, but Bryce was noncommittal.

The Chandrasekhar lunch group was meeting now at the Greasy, Caltech's downscale cafeteria, rather than the Athenaeum, its upscale faculty club, so that Chandra could get a more substantial vegetarian lunch. Chandra described his regimen for writing books, setting and sticking to an absolute daily schedule. Two years in advance, he commanded his editor at Oxford University Press to meet him at Heathrow airport on a certain date. He would hand him a completed manuscript while changing planes on his way to a conference. But then, he fell a whole day behind the planned schedule. The heavens intervened: He had not included the extra leap-year day of 1960 in his calculation and was thus able to deliver the book on time. One wonders at how close this is to the plot resolution of Jules Verne's *Around the World in Eighty Days*. Surely the world's greatest living theoretical astrophysicist wouldn't spread an apocryphal story!

There were small triumphant moments in the Greasy, one coming when Feynman sat down with us students, and I described to him my intuitive "proof" that gravity must be the only spin-two field in nature. Could he find a flaw? It took him a couple of minutes to understand the proof, then another minute repeating it back to me to be sure he had it straight. Kip and Chandra joined us during this time. There was an interval of about ten minutes during which Feynman was stumped—and everyone got to see. He kept running down various possibilities, then noticing that they were covered by my previous logic. "Yes, that's very clever," he said a couple of times. High praise. Alas, figuring out a loophole, he finally explained to me and the group that I had not excluded the possibility that the field could couple only to vector particles. "Spin-1," in quantum mechanics, or "little balls with snouts" if I insisted that he make the argument classical. "That was clever," he said again, "but I kinda thought that nature [he meant himself] would find a way out of it."

Feynman was famous for his entertaining stories, later collected in books edited by Ralph Leighton. I approach these books skeptically, because some stories that Dick told us students over lunch in the Greasy couldn't be exactly true. His mother, he said, had gone to the Ethical Culture school in New York City. The children were assigned ethical quandaries to think about. When called on, one little girl in the class would always begin with, "Well, I don't really know one way or the other, but *my little brother Felix* says…" The story ends with the big reveal that the girl's surname was Frankfurter. Feynman's mother, Lucille, was born in 1895, as in fact was Supreme Court Justice Felix Frankfurter's sister Estelle. So Lucille and Estelle might in fact have been in class together. The problem is that Felix was actually thirteen years older than Estelle, so the cute part, "my little brother Felix", was an embellishment. OK, it was still a good story. (J. Robert Oppenheimer was another Ethical Culture School graduate.)

A pesky Caltech formality was that I needed to pass an oral examination attesting to my ability to do research. Since I had already submitted four papers for publication, Kip tried to have the requirement waived, but to no avail. My committee was Kip (chair), Jim Gunn, Tombrello, and Feynman. Kip began the exam by announcing that I was going to pass, and that we should waste as little time as possible. Gunn asked me to explain the black hole torque calculation, so that his time wouldn't be completely wasted. Feynman, however, got the idea of that calculation immediately and wouldn't let me waste *his* time by continuing. Instead, he started questioning Kip on matters far above my understanding. By the end of the hour he had disproved Kip's conjecture about the extension of my result to tensor fields and proved his own different conjecture, all by very simple physical arguments.

Kip bet Dick a beer at the Hi-Life Topless Bar that Weber's results would be duplicated, but not necessarily that he was seeing gravitational radiation. Tombrello, looking at his watch, moved that I pass. The vote carried unanimously, in time for me to go with Alan Lightman and David Lee, two of Kip's new students, to visit Sandor in the student infirmary. He had gotten dysentery from Renee, who picked it up in Bangkok on her round-the-world flight. Alan and David worked together on what became known as the Lee-Lightman theory of gravity. Neither stayed in relativity research after getting their Ph.D.s. Lightman worked in astrophysics at Harvard for about a decade, then started writing successful novels. His *Einstein's Dreams* was on *The New York Times* best-seller list for seventeen weeks. He became a professor of humanities at MIT David Lee left science to become a C.P.A. and then vice-president in a large aerospace firm. In 1997 he was a founder of Global Crossing, a global telecommunications startup that rose rapidly to a capitalization of $50 billion. Its 2002 bankruptcy was one of the largest in history. David, who had left the company in 2000, emerged untainted by allegations of accounting fraud. He became a venture capitalist and served as chair of Caltech's board of trustees.

Saul, now back, was working many hours every day on trying to derive an equation that would enable us make progress on Kerr black holes. Unless and until he succeeded, my own work on Kerr was dead in the water. Saul had inherited the local custody of the Newman-Penrose formalism (not to be confused with the N-P conserved quantities) from Richard Price, and was trying to make sense of its thirty-five coupled (and from my perspective, not-computable) Kerr equations. Over what would become eight months, he produced pages and pages of careful handwritten calculations. Chandra often materialized to listen to our discussions. "Perhaps I should learn the Newman-Penrose formalism, so that I can participate in the new excitement," he said to me one day. He was an old racehorse champing at the bit. After he returned to Chicago, over the next several years, Chandra did more than master the formalism. It became the basis for much of his 1983 book on black holes. The old horse did run the race.

A birthday party that Kip insisted on holding for Chandra—against the latter's wishes—was a somber affair. We were in the dark Athenaeum basement sitting around a big table. Kip and Bonnie were trying to keep the conversation with Chandra going for benefit of the rest of us, but it was heavy lifting. Bonnie asked Chandra if he kept up-to-date on affairs in India. "No," he said. "My last contact was when Mrs. Gandhi came to visit me on her last state visit. Before that, it was Nehru who visited me—and I disapproved of him strongly." Then he said, "Let me tell a joke."

We were all immediately apprehensive. If anyone was a joke-telling type, it was not Chandra. "How can one divide sixteen lumps of sugar into three teacups, each having a nonzero odd number of lumps?" Silence. Finally, Saul ventured what we all knew, that it was mathematically impossible, because the sum of three odd numbers was always odd. "Not at all," Chandra said. "Put one lump in the first, one lump in the second, and fourteen in the third." He sat back triumphantly. There was a look of terror on Bonnie's face. She was wondering whether she should pretend to laugh, or just try to change the subject. Saul again took the burden on himself and remarked that fourteen was in fact an even number. Chandra said, grinning, "I would say that fourteen is an exceedingly odd number of lumps to put in a single cup of tea." The color returned slowly to Bonnie's face.

The fall passed without my accomplishing any science of note. I could easily appear to be busy, talking to Saul, or Chandra, or Bonnie. Jim Bardeen was writing a follow-on paper to our previous work and included me as second author, so we exchanged drafts by mail and occasionally talked on the phone. I got a phone call from George Trilling, the physics department chair at Berkeley, dangling the offer of an assistant professorship as soon as I got my degree. But that was years away, I told him. "We like to plan for the future," he said. Kip was pleased when I told him about this, and said that it wouldn't be my only offer and that I shouldn't act hastily.

22. Motivation

Geoffrey Burbidge and his wife Margaret Burbidge were British astrophysicists and longtime friends of Willy Fowler at Caltech and Fred Hoyle at Cambridge. These four had co-authored a 1957 paper, "Synthesis of the Elements in Stars," that became known as one of the most influential scientific papers in the twentieth century. Geoff had the jowls, neck wattles, body fat, and personal manner of an anti-British cartoon depiction of John Bull. Margaret Burbidge was a small, chirpy woman. It was impossible to imagine them in bed together. Jumping a bit out of time, she in 1972 became director of the Royal Greenwich Observatory but was denied the usually accompanying title of Astronomer Royal, the first such case in three-hundred years and quite obviously because she was female. In retaliation, within five years she had emigrated to the University of California at San Diego, took U.S. citizenship, and was elected president of the American Astronomical Association. Geoff, although he became director of the U.S. Kitt Peak National Observatory, remained a British subject. He was known for his support of contrarian astronomical theories. One was his support of Fred Hoyle's steady state cosmology. Another was that quasars were not at great cosmological distances, but were relatively nearby, orbiting within galaxies. Hoyle's cosmology was decisively disproved in 1964 by the discovery by Penzias and Wilson of the cosmic microwave background. Maartin Schmidt's identification in 1963 of spectral lines from cosmological distances in quasars decisively disproved "local" theories of their nature. Geoff never admitted that he was wrong.

But because of his friendship with Willy, Geoff was always welcomed at Caltech. On this visit, he gave an (embarrassing) physics colloquium and met privately with senior professors, including Kip. Geoff was editor of an influential periodical, the *Annual Review of Astronomy and Astrophysics*, which was issued in book form once a year. Kip must have planted the idea: Geoff invited Kip and me, jointly, to write a review article on gravitational waves—the first one of its kind in any journal. This was something that I could do while stalled on other fronts. Review articles are different from research papers. The author is expected to have something to say about all the published papers ever written on the subject—even if it is only to dismiss them with a comment like, "Other authors have also considered the issue, see references 96-122." Kip and I divided up the work and chose the provocative title "Gravitational Wave Astronomy". Gravitational waves were not yet known to exist, much less to be of any use to astronomers! Kip was to be in charge of the section on astrophysical sources, which he parsed in his typically careful way as "A. Sources Known to Exist, B. Sources That Probably Exist, and C. Sources That Might Exist." However, even the first category was highly speculative.

I was in charge of everything else, including the scholarly apparatus of locating and cataloguing all the relevant papers—similar in spirit to stamp or coin collecting. In those days before citation databases and before the web, one proceeded by gathering all the recent relevant papers one could find, then (in the physics library journal stacks), looking up all of their earlier references, then all the references in those, and so on, until the set of relevant references closed on itself. While tedious mechanically—every relevant reference had to be hand-copied onto a 3x5 card—I had a good time tracing ideas backwards in time, and in doing so learning bits of science. I spent long, happy days in the physics stacks. A few weeks later, I was ready to begin writing.

Kip summoned me to his office one morning, saying that he wanted to talk to me about my future. I could tell when I walked in that it was not going to be just happy talk. He was being pressured from several directions to award me my Ph.D. immediately, he said, and he didn't think that I was ready. John Wheeler was soon going to offer me a postdoctoral position at Princeton for next year. Kip wanted me to turn it down. After George Trilling close-to-offered me the assistant professorship at Berkeley, several Caltech faculty (he mentioned Peter Goldreich and Jesse Greenstein, but I surmised also Tombrello, Gunn, and maybe even Feynman) wanted Caltech to match that offer. It would be controversial, because there was an explicit policy that the department not hire its own graduating students. Kip felt that I needed at least a year more "maturing," and that he would be caught in the middle, expected to support his own student, but not, he told me, wanting to. I had the potential, he told me, to become one of the world's top astrophysicists. I was smarter than he was, smarter than the rest of his students. But, by contrast with him and them, I was lazy. I wasn't motivated. He wanted to help launch my brilliant career, but only if I would agree to change and become motivated.

He continued: He had been watching me closely. I was always talking about the TV series that Margaret and I watched regularly, and he had made a point of watching some of them. They were all unworthy. He was particularly appalled by the show *O'Hara: U.S. Treasury Agent*. In my opinion this was a truly great show. Every week O'Hara was assigned to a different sub-agency within U.S. Treasury: Secret Service; Alcohol, Tobacco, and Firearms; Internal Revenue; and so on. The only constant was that, at the end of every episode, when he was about to make the arrest, even the meekest suspect would unexpectedly draw a gun. O'Hara (played by David Janssen) would then draw his own weapon and shout, "O'Hara, U.S. Treasury Agent, stop or I'll shoot," and then shoot the suspect dead, saving the U.S. Treasury the cost of a prosecution and trial. Margaret and I particularly liked it when the now-dead suspect was guilty of a crime like income tax evasion, or failure to purchase required liquor excise stamps.

Kip estimated that I worked—actually worked on my research—about three hours a day. I objected: I thought it was three hours a day for nine months of the year, but 16 hours a day for the three months in total that I might have my teeth sunk into a good problem. That's just it, he said in frustration: The proportions ought to be the reverse. A good scientist could have at most one outside interest. His was sex, he said with a completely straight face. If mine was TV, he could accept that. But no, TV was just one of many other competing interests for me.

That wasn't all: He was appalled at how little I knew about thermodynamics and hydrodynamics. He was right about that. Neither had been part of Harvard's undergraduate physics curriculum. It was left that I agreed to try to work harder, and he assigned me to calculate the adiabatic accretion of gas onto a spherical black hole. It was a straightforward computer calculation that would require me to learn some thermodynamics and hydrodynamics. I didn't think that it would lead to anything interesting. (Spoiler alert: It didn't.)

We submitted our review paper, "Gravitational Wave Astronomy" to *Annual Review* at the beginning of February. It was published in September, 1972. After that time, I never again worked on gravitational waves, their production by astrophysical sources, or their possible detection. Since Kip's and my stories must presently diverge, let me jump out of narrative time on just this subject.

Kip was slow to give up completely on Weber's observations, and he never gave up hope for the detection of gravitational waves. That, along with the properties of black holes, became the defining centerline of his career for the next forty years, with most of his more than fifty Ph.D. students (he counted me as number nine) doing their Ph.D. theses on those subjects. Kip was instrumental in convincing Caltech to hire the Scottish experimental physicist Ron Drever, who began an experimental effort based on laser interferometry. Starting around 1980, this effort joined forces with that of Rainer Weiss at MIT and became the National Science Foundation-funded LIGO project, the "Laser Interferometer Gravitational-Wave Observatory." The idea was to use high-precision lasers to monitor the distance between two mirrors suspended many kilometers apart. A passing gravitational wave would alter that distance by a tiny fraction of the size of one proton. It took twenty years for the two LIGO observing stations, in Washington State and Louisiana, to be built and brought into operation. They observed for eight years, from 2002 to 2010, and saw—nothing.

Most billion-dollar-level experiments would have been terminated at that point, but, due in large part to Kip's continuing evangelism, the NSF funded "Enhanced LIGO" and, after that, "Advanced LIGO". As a Harvard professor, I was often a reviewer of the LIGO proposals to NSF. I always gave enthusiastically positive reviews. This was not due to loyalty to Kip. In truth, I never expected LIGO to find gravitational waves. What I liked and supported was the technology development: The LIGO team, which grew to many hundreds of people, were pushing the frontiers of sub-atomic measurement by many orders of magnitude, far into the quantum domain. This was truly basic research, worth doing for its own sake.

At 09:50:45 UTC on 14 September 2015, LIGO detected the unmistakable gravitational wave signature of two black holes, each of mass thirty times that of the sun, spiraling together and merging. Prior to this detection, no one thought that black holes of this mass even existed. None of Kip's likeliness categories, A, B, or C, in our 1972 review had contemplated thirty-solar-mass black holes—the idea was too outlandish. In 2017 Kip, by then age 77, along with Rai Weiss and Barry Barish, was awarded the Nobel Prize in Physics.

Back to 1972, where I was struggling with the younger Kip's tough love. It was a no-brainer that I had to *appear* to be working harder than before, at least until I got my degree. But did I really want the kind of future that Kip laid out for me—sixteen-hour days with timeouts only for TV or sex—and I could only choose one? I didn't. I didn't have the driving ambition that Kip—or my father, or Saul for that matter—had. Perhaps my lack of drive came from being in Frank's shadow, the way Tommy Lauritsen had been in Charlie's. If so, I thought of it as a blessing. I didn't need to be a slave to ambition. I did what I was doing because it was fun. What I liked was learning, knowing, and discovering—it didn't matter how serious or unserious the subject was. It was curiosity, pure and simple. I had tasted the feeling of being the first to *discover* something, with black-hole vibrations. That was a *feeling* that I wanted more of, but not because of any expectation that it would advance my career.

Margaret and I invited Kip and Linda to dinner at our house. We made a point of showing them my workbench with locksmithing and electronics parts strewn about, her floor loom, the piano, the wind-up phonograph, the box full of tetrahedrons made out of plastic soda straws that had at various times been assembled as a twelve-foot-high replica of a Dyson Sphere. This was supposed to be a gentle pushback against Kip's narrow vision of my future. He didn't appear to get the message.

The next week, Kip was pleased to see me still working at 6 o'clock. He took the occasion to tell me that the department had agreed to hold for me its prestigious postdoctoral title, Richard Chace Tolman Fellow. I could become a postdoc, with that title, whenever I finished my thesis. "When will that be?" I asked. "As early as next autumn, or as late as never," he said a bit testily. "It's up to you." Thus continued his plan to motivate me.

When I complained to Kip that Jim Bardeen had completely stalled on providing a first draft of the follow-up paper to our first one, he told me to fly to Seattle and do whatever it took to extract a paper out of him. I didn't know that co-authorship worked that way; but I was on a plane in thirty-six hours. Jim and Nancy put me up in their modest home. The first day, in his office, Jim was (for him) quite talkative, about every topic in physics except our paper. At dinner, Nancy asked me what we had accomplished, and I said: nothing. Then he'll have to start now, she said. She made him sit at one end of the living room sofa, me at the other end. He wrote one sentence at a time on a pad of yellow paper—it took five or ten minutes—then passed me the pad. I either approved the sentence or sent it back for a rewrite. Nancy sat across the room in a chair grimacing fifty different ways. She had a very mobile face. We continued until well after midnight. I was there for three days, but came back with a draft paper. Even better, I got Jim to agree to sponsor Saul for the Les Houches summer school in France. Kip had used up his quota of influence in getting Cliff Will and me accepted. My flight home was diverted from LAX to Long Beach because of fog.

By the beginning of spring, I was giving "job talks," not because I was actually looking for a job yet, but because I was getting invitations that I didn't want to turn down. My physics colloquium at Caltech went well with Willy, department chair Bob Leighton, Tommy, Feynman, and Carl Anderson in the front row, and with my claque of Kip and the group, Jim Gunn, Virginia Trimble, and Steve Willner (who was now a Caltech graduate student) populating the next several rows. Jim Bardeen had decided, after many months of indecision, to accept a full professorship at Yale, and he had an assistant professor position to fill there.

George Trilling invited me to bring Margaret on my visit to Berkeley. We were fêted and treated like grownups, which seemed strange to us, especially in Berkeley where we had so recently been students being tear-gassed in People's Park. Our sudden, new protocol rank merited attendance at the dinner in our honor by not one, but two Nobel Prize winners, Charlie Townes and Emilio Segrè. Townes had moved to Berkeley from MIT in 1967 and now worked in radio astronomy. His presence thus made some sense. Segrè seemed to be the Nobelist on call for social occasions, a kind of academic Duke of Plaza-Toro. Segrè discovered the antiproton and also two elements, technetium and astatine. He won the 1959 Nobel Prize in Physics. The Duke of Plaza-Toro, in Gilbert and Sullivan's *The Gondoliers*, hired himself out to charity dinners in return for ten percent of the proceeds. I met privately with Townes at the end of the next day. As Provost at MIT, he had been the "closer" who recruited Frank there; now it was Berkeley, and me—two different generations, two different coasts. Townes showed off his pre-production hand-held electronic calculator, the Hewlett-Packard 25—he was on the HP board—which could calculate logarithms, exponentials, sines, cosines, etc. I had never seen anything remotely like it. It was by far the highlight of the trip.

A week later, I gave a talk at the University of Texas at Austin. Bryce DeWitt was my host, and I met his graduate student Larry Smarr, who will figure prominently later in my story. A few weeks after that, I was in Utah visiting Richard Price and giving the same talk. There was still a lot of snow at Alta, and Richard took me skiing for the first time since I was fourteen. He coached me down one real run, lectured me on the physical principles ("the torsional modulus of a ski must be much larger than its longitudinal modulus, but some dissipation is necessary in the longitudinal mode"), but ultimately gave up on account of my athletic ineptitude. I did several runs down the bunny slope—my appropriate level—while he did two serious runs.

Freeman Dyson, the famous British-American mathematical physicist from Princeton, happened also be visiting Salt Lake City. My colloquium dinner was merged with the much bigger dinner reception in his honor at the Fort Douglas Hidden Valley Country Club. I shook Dyson's hand. He had no idea who I was, and I didn't explain.

Kip's plan to motivate me by holding my Ph.D. hostage came to an unexpected end that April. I was working in my office on a Saturday morning. Kip was in his office, unusual for a Saturday. The building was otherwise deserted. Maybe that was what did it. After an hour, Kip sauntered by my office. "I think you should schedule the formalities and get your degree as soon as you get back from Les Houches," Kip said, "so that you can be a postdoc as soon as possible." That was that. I had somehow passed the motivation test. I could go back to watching *O'Hara.*

23. I.O.T.A.

As the summer of 1972 approached, Saul and I had become joined at the hip professionally. This was a partnership that was to last, in one form or another, for more than half a century. We looked to Kip for support and career guidance—he was our academic father figure—but increasingly we looked for scientific advice elsewhere. Saul developed a relationship with the applied mathematicians in GALCIT, Caltech's famous engineering department. We corresponded (in those days by air mail) with Chandra, Bardeen, Misner, and a couple of Kip's former "black hole" students. Alan Lightman and David Lee were now holding up the other pole in Kip's group, working on "other theories of gravity," meaning, alternatives to general relativity, and especially how future experiments could rule them out, hopefully leaving Einstein's theory alone as viable.

One of Kip's last lectures in the relativity class was on gravitational wave detection, much of it presented in exactly my review-article words. Kip was truly a dyed-in-the-wool theorist: He tended to accept uncritically all statements on matters experimental by experimenters, learned these by rote, and was uncritical of their underlying assumptions. Of course, I was hardly without sin in claiming to have experimental expertise when I could barely hold a screwdriver. But still, at the time, I was the more experimental of us two.

Saul and I were turning out serviceable scientific papers, well beyond what was expected of graduate students, but in no way revolutionary. Although Chandra had left, Bonnie Miller was still around. There were natural extensions of her thesis work on gravitating ellipsoidal bodies that, between Saul's ability at math and mine at computing, we could easily do. We tried to include Bonnie in the work, and we would have made her a co-author regardless, but she insisted on withdrawing—even though she had no job offers and badly needed more publications. Even Kip couldn't convince her otherwise. Bonnie's problem was that she was too honest, too nuanced as a person and not greedy enough as a scientist. Separately, Saul and I were cranking out every possible calculation that could be done for scalar fields interacting with Kerr black holes. Unfortunately, with no actual scalar fields existing in nature, these were calculations that impressed only the black-hole cognoscenti.

On a Friday in early June, several of us joined Kip and Feynman at lunch in the Greasy. Kip ate hurriedly and went off to an appointment, but Feynman was enjoying a leisurely conversational pace. Talk was about the lack of progress in particle theory, that most fundamental branch of physics, his own field. "I'll tell you what I notice most," he said. "I'm 54 years old, and I'm still doing better work than the young men [sic] in the field." He drew the contrast to previous times, when the advances in physics were made by the younger (or youngest) people. I said to him, suppose you were the Czar of theoretical physics, with the power to tell anyone what directions to pursue. How long would it take to explain everything, to understand the fundamental nature of the elementary particles? "Two years," he said, without hesitation. And how long will it take with the field's current crop of researchers? "Maybe forever."

Every generation sees the next as not as good as its own, but Feynman's pessimism was not well-founded. The stagnation of particle physics in the early 1970s would soon be replaced by a so-called standard model that didn't "explain everything," but did explain, very elegantly, *most* things. Most of the elements of the standard model were already known at the time of the above conversation: Yang-Mills theory, Glashow's unified electroweak theory, Weinberg and Salam's version of the Higgs mechanism, "asymptotic freedom". I had learned about most of these in Feynman's course the year before.

What needed to change was more about forgetting than discovering. Quarks were not vague manifestations of some Gell-Mannesque symmetry principle or S-matrix breakdown of space and time—forget that. They were real (if unfamiliar) particles that carried conserved quantities (electric charge, "strangeness," and "color"—unfortunate names) and obeyed a version of the Yang-Mills equations soon known as Quantum Chromodynamics (QCD), by analogy with the equations of Quantum Electrodynamics (QED) that had won Feynman his Nobel. Most particles were composites of quarks, just as atoms were composites of protons, neutrons, and electrons. Instead of being bound by electromagnetic forces (the virtual photons that bound the electrons in atoms), quarks were bound together by strongly interacting "gluons". The 1983 experimental discovery of the W and Z bosons with just the masses predicted by QCD convinced almost everyone. Feynman, who died at age 69 in 1988 of a rare liposarcoma, lived to see these changes.

* * *

Three of us in Kip's group had gained admission to Cécile DeWitt's physics summer school in the French Alps: Cliff Will, Saul, and me. That summer's topic would be black holes. Kip was a lecturer. We were all bringing our wives. (Cliff was married to Leslie, Kip's secretary.) The school had dorm accommodations for unaccompanied attendees; and Cécile grudgingly agreed to let Linda stay with Kip in an apartment reserved for lecturers; but, when we telephoned her in Texas, she rejected, with a Gallic gesture that we could almost hear, the idea of accommodating the wives of students. She referred us to the Hotel Peter Pan, a tiny place about halfway down the mountain slope from the school toward the actual town of Les Houches. We wrote to the hotel, in French of course (Cliff was Canadian and Saul had taken lessons at the Alliance Francaise in Johannesburg), and received an answer three weeks later: *L'hôtel est complet*—fully booked. We telephoned Cécile who was, this time, slightly more helpful. There was a local landowner, a Monsieur Laprade, who owned an unoccupied farmhouse a kilometer *up* the mountain from the school. We should write to his adult daughter and ask to rent it. Cécile provided an address. Two weeks later: Success! But with the proviso that, once we arrived, the father would want to interview us and might still reject us as tenants. The daughter provided one hint: Her father, age 89, was a great admirer of Napoleon and might ask us for our views on that subject—apparently still a matter of lively debate in the Haute-Savoie.

The summer school was scheduled for August. Martin Rees invited me to spend July in Cambridge, England, at I.O.T.A., the Institute of Theoretical Astronomy. I was one grain of sand in a sandbox of academic political intrigue. I.O.T.A. was Fred Hoyle's institute, founded by him just five years earlier and housed in a modern one-story building on Madingley Road, adjacent to a sheep pasture. If the building seemed more California than East Anglia, it was because it was a copy of a building at U.C. San Diego that Fred had admired. Recently, Cambridge University had decided to merge I.O.T.A. with its astronomy department and observatory into a single Institute of Astronomy. Fred had agreed, but on condition that the new director be an observer, not a theorist, so that the old I.O.T.A. would be left essentially untouched as a center for theory under his direction. For the new director, Fred favored Caltech professor Wal Sargent (who was a Brit).

Instead, the University named Donald Lynden-Bell, a theorist. Lynden-Bell was a man of charming enthusiasm and right-wing political views. He had been on sabbatical at Caltech my first year there, and I took his astronomy course. In response to Lynden-Bell's appointment, Fred, who had just been knighted and was now Sir Fred, resigned his Plumian Professorship, a chair founded by Isaac Newton, and threatened to resign from the University entirely—a bluff which, when called the next year, did force him to leave Cambridge for Manchester. There was an effort in the summer of 1972 to show the world that I.O.T.A. was not falling apart (although it was). Thus, more than the usual number of summer visitors were invited—including me. (Martin Rees was at the time just five years past his Ph.D. When Sir Fred left Cambridge, Martin was rusticated in Sussex for a year, then returned to Cambridge as the new Plumian Professor and new director of I.O.T.A.)

Margaret and I were already packing for the trip when I went to work on Monday, June 12, 1972. I found Saul in his office and handed him the ellipsoid paper draft to read. Ros was with him. Saul said to me, "I separated the gravitational perturbation equation for Kerr last night."

"Separate" (pronounced sep-a-RATE) is a term of art in partial differential equations. The meaning here was that instead of a set of 45 coupled equations involving all the space coordinates simultaneously, a mess then too hard to solve by computer, Saul had found a way to "separate" out a single equation, easy to compute, from which the full solution could, if desired, be calculated. His equation would soon be known as the Teukolsky Equation.

I could hardly believe it. If true, it was the most important discovery in black-hole theory in several years. He had worked it out twice last night, then slept on it. "He was up at 7:15 this morning," Ros said, in awe of the early hour. It had also worked this morning, and here we were. Saul had called Kip to tell him just before I got there.

Could the same trick be used to separate the electromagnetic equation which Ipser and Fackerell had spent months working on and failed? Saul started calculating. By lunchtime he had the answer: Yes. We spent the afternoon discussing all the different problems that were now amenable to solution. No longer would we have to say that we had solved such-and-such a problem with a fictitious scalar field because the gravitational case was too hard!

Kip met with us to discuss how Saul should publish his result. If it went to a fast journal like *Physical Review Letters*, then Saul would be famous faster, but the equation would be in the public domain. We would have to race everyone else to milk its applications. Kip favored submitting it to a slower journal, like the *Astrophysical Journal*. Saul's priority would be guaranteed by the submission date. We could let the fact of Saul's success—but not the equation itself—leak out gradually. We would have a good six months to be ahead of everyone else. I pointed out that Kip had always before favored the prompt, free exchange of scientific ideas. "This is just too important for that," he said.

Margaret and I flew to England by student charter flight from Oakland. We could not afford her trip at regular airfares. Britain in 1972—before Mrs. Thatcher, before the North Sea oil, before the European Union, before London had become the world financial center, before the arrival of decent cuisine—was a poor country. Average incomes were less than half those of the U.S. I.O.T.A. had arranged housing that they thought suitable for married students. It turned out to be a disconnected tiny sitting room and even smaller bedroom, a hot plate, no refrigerator, and a bathroom two flights down shared with six other people.

We threw ourselves on Martin's mercy. He put in a good word with the Bursar's Office at Kings (his College), and we were soon installed in a tiny (but fully adequate) flat on Benet Street, just a block from the market square in the city center. Rent was £10 a week. Margaret successfully navigated the bureaucratic labyrinth of getting gas and electricity turned on. We both soon learned the secret of impoverished English life: Choices, when offered, were tempting, but illusory. In any situation there was a "done thing". All other seeming options were either unavailable, fiercely expensive, or considered to be in poor taste. You could ask what the done thing was, and you should then immediately choose it. Daily life became much easier after that.

We enjoyed the peculiarities: You couldn't buy milk, except preserved milk, in the (not very) super-market. Instead, the done thing was to take an empty milk bottle from outside your neighbor's door and place it outside your own door. The next morning, an invisible milkman would have replaced it with an (unrefrigerated) full bottle, no questions asked. If you wanted two bottles delivered, you put out two empties. If you wanted a newspaper delivered to your door, you mentioned it to the man selling newspapers at the nearest street newsstand. "Which paper will you be wanting?" "What is the done thing?" *The Guardian* started appearing at our door—we were students of course. We even managed to rent a television for the month, for £5. The done thing, according to the rental store proprietor, was *not* to buy the required, expensive television license; instead, if the TV inspector ever appeared at our door, we should resort to "delayin' tactics". We never found out exactly what these were.

We treated ourselves to one anomaly, a not-done thing. Our first morning in Cambridge, we had wandered into The Blue Boar, Cambridge's fanciest hotel, desperate for breakfast. This was not a done thing. Breakfast was served only to hotel residents. We must have looked so American, so young, and so pitiable, that they made an exception: We could have their elaborate, full English breakfast for 60 pence, a price that they invented on the spot, about $1.50. We went back every day for the whole month we were there. Margaret always wore a nice dress, so that we wouldn't too much lower the tone of the place.

The Cambridge-Caltech connection was strong. Willy Fowler, Jim Gunn, and Wal Sargent were all visiting that summer, a show of solidarity for the beleaguered Sir Fred. Others whom I talked to regularly were Martin, Brandon Carter, Stephen Hawking, Gary Gibbons, Jacqueline Bergeron, and (my officemate) Max Krook, a senior Harvard professor whose applied math course I had once audited. Hawking was based at D.A.M.T.P., Cambridge's legendary Department of Applied Mathematics and Theoretical Physics, and, although mostly confined to a wheelchair, he could drive a little electric car with special controls and often spent mornings at I.O.T.A. He was interested in hearing about Saul's equation and gave me oral messages to convey back to Saul by post, things like, "The Newman-Penrose spin coefficients should vanish on the horizon." The Institute was famous, especially summers, for its atmosphere of relaxed discussion. I spent hours discussing physics and astronomy, sometimes politics, with each of those already mentioned, including the Caltech professors, previously unapproachable on their home turf.

Lunches were often at pubs in the countryside around Cambridge, Madingley and Coton, invariably with a half-pint of room-temperature bitter—the done thing. At one lunch I launched into an impromptu lecture on the implications of the Hawking-Penrose-Gibbons global theorems about black holes. Stephen was sitting next to me. He seemed neither pleased nor displeased, but at least didn't correct me. The British relativists understood, a bit jealously, that, armed with our still-secret equation, Saul and I would be first to prove that Kerr black holes were actually stable—or, even better, find that they weren't—probably in the early fall. Still, it was a race of sorts.

The following Wednesday, at morning coffee at the Institute, a crowd was gathered around the table where Sir Fred was holding forth on his analysis of Bobby Fisher's position in chess. Donald Lynden-Bell came in and stood alone at a corner of the room. It was not clear whether he was being snubbed intentionally, or whether no one dared be the first to desert Fred's camp. But ten minutes later, Donald was still standing by himself in the same spot. My allegiance was to neither side, so I wandered over to chat with this controversial new director. I introduced myself as a student of Kip's, and we talked about current work in black holes. A few minutes later, Martin Rees also joined us. Lynden-Bell was either a brave man, I thought, or else much more arrogant at heart than his manner indicated. His appointment and Fred's impending departure were widely perceived as damaging to the Institute, if not the whole university.

Margaret and I did in fact manage to avoid taking sides. Despite my approach to Lynden-Bell, we were soon invited to tea at the Hoyles, along with Willy and Ardie, and Don Clayton, another visiting American. Clayton was Willy's former student. His was the textbook for the course at Harvard where we students tortured the poor, green Dick McCray. Clayton's later 1975 popular book, *The Dark Night Sky: A Personal Adventure in Cosmology*, was so full of himself that it became known among astrophysicists as *The Universe and Me: Not Necessarily in That Order*. Now, we watched on TV as Stan Smith—my high-school classmate as it improbably happened—defeated Ilie Năstase at Wimbledon. A different evening, Martin insisted that we come to dinner with him, Joe Weber, and Virginia Trimble. Unsaid was the reason: Martin and Virginia had been an item in the past (everyone knew), and he felt uncomfortable with just her and her new, much older, husband. We came prepared to carry the conversational ball, but Virginia easily outdid us.

Yet another evening, we had dinner with Jim and Rosemary Gunn and Lynden-Bell. Donald was the opposite of Fred in every way, including being a Tory of the far right. His perspective was that I.O.T.A. had it too soft, that the University had coddled Fred, and that the new I.O.A. would act in the interests of Cambridge University rather than Fred's international claque (by whom he especially meant Willy Fowler, Wal Sargent, and the rest of the Caltech crowd).

One afternoon, at tea, I was talking to Stephen about some technical details of the Kerr stability problem. He seemed quite interested. He invited me back to his office, and had me shut the door. After about two minutes more of talk he said, "Can you help me to the toilet?" It was exactly Bryce DeWitt's story about Paul Dirac, but now brought up to date with a more modern Cambridge physicist! I wondered how many other stories were in circulation about great physicists who needed to pee.

As the end of the month approached, we loaded up on books from Heffers Bookshop and shipped them by sea mail back to Pasadena. On the last Saturday in July, by taxi, train, and bus, we wrestled our luggage to Heathrow. The flight to Geneva was an hour and a quarter. From there, a taxi to the bus station, and we bought tickets for the bus to Les Houches, with a two-and-a-half hour wait until the next bus. We sat in a well-manicured park, reading and eating Swiss chocolate. Geneva was beautiful, clean, and modern; but it was Francophone, one strike against it for me. The bus departed at 5:30 p.m., crossed into France without any inspection of passports or luggage, and wound its way up into the Alps. At 7:30 p.m. the driver let us off in tiny Les Houches. A middle-aged lady came out of a shop, sized us up as "école d'été de physique" and waved to the lone taxi driver parked up the street. Twenty minutes later we walked into the school's empty dining room. Not quite empty—in one corner Bryce and Cécile, and Saul and Ros, were just starting dinner. We all were the earliest arrivals.

Saul had already seen our rented house: It was very impressive, he said, very peculiar, and very, very far up the mountain, with no road. After dinner we four, with the Press family's luggage, began the trek, scrambling with fifty pound suitcases up steep rocky trails in the dark. Ten rest stops and forty-five minutes later we approached the dark rambling silhouette, all logs and rough-hewn boards, with doors no more than four feet high. It was as if from a museum exhibit.

The house was built in 1777 as a tight, alpine farmhouse-chalet, designed to be snowed-in for the long winters. The main beam of the house was carved with, "Fait le 13 Mai, 1777," followed by the initials of the whole family of original occupants, a cross, and a wheel to ward off the devil. Stooping through the front door, one was in the high-ceilinged great room, like the main hall of some Viking manor, a huge fireplace dominating. Livestock, as well as people, would have been sheltered in this room during the coldest weather. A balcony at second-floor level surrounded the great room on three sides, with bedrooms off the balcony. Back at ground level, another four-foot door led to a dining room and a kitchen with a wood-burning stove. There were electric lights in every room, but the voltage was low, so they glowed only feebly, like candles. The locals believed, according to Cécile, that electricity had weight and had a difficult time making it up the steep mountain. (A better theory was that the gauge of the wires on the utility poles was insufficient, or that they might be iron instead of copper.)

Our original plan had been to share the cost among three couples, including the Wills. However, back in Pasadena, Leslie Will and I had gotten into a tiff over something that seemed important at the time—Kip gave both of us a dressing down—and the Wills had withdrawn from the partnership. As school participants arrived on Sunday, Cécile introduced us to Gideon Rakavy and his wife Ina, both in their forties. He was a professor at Hebrew University in Jerusalem. "Of course, it's none of my responsibility," Cécile said (this remark habitually prefaced her every command), "but you may want to take the Rakavys into your house—they have no other accommodations." So we all helped Gideon and Ina, too, carry their overweight luggage up the mountain.

On one of our first daytime marches up the hill to our chalet we encountered an old man, quintessentially French in appearance, with a white beret, carved walking stick, pantaloon trousers, grey wool socks and walking shoes. "Êtes-vous Monsieur Laprade?" "Oui." We identified ourselves as his tenants. He asked where we were from. Saul's answer made an impression on him. "Sud Afrique," he repeated, nodding knowingly. After some polite conversation we continued our separate ways.

The next day Saul and I hiked up the hill beyond our house to M. Laprade's chalet—Cécile had given us directions. We are here to pay the rent, we announced. He invited us into his study, seated us, and puttered behind his desk filling the inkwell to dip his steel pen. Another fifteen minutes of polite conversation. He asked where we were from. "Sud Afrique," seemed to ring a bell whose sound he couldn't place. He explained that his daughter, who was to arrive in two weeks, handled all the arrangements. Therefore, we could pay him any amount we want; he would write us "un petit reçu" and then his daughter, Mme. Barre, would straighten things out. I paid the full amount, minus the deposit that we had already paid and that the Rakavys had paid, 900 francs in all. He signed the receipt in a spidery hand. As we were leaving, the memory of Sud Afrique finally clicked in his mind. "Did I talk to you yesterday?"

We next saw M. Laprade when he called at our house, took us outside and pointed to the *grenier*. A *grenier*, literally granary, was a separate tiny house, windowless, unheated, tightly constructed, that was built on a raised platform some distance from the main house. It was for storing the winter's supply of grain for eating, plus seed for the spring planting. It was separated from the house so that, if the house caught fire, the family would not then starve. "C'est une petite maison historique," he said. The royal family of Belgium lived in it during the war, hiding from the Germans. Which war, we asked, especially since our house was full of pictures and engravings of Napoleon. His lengthy reply proved to be beyond our language abilities.

More than a week after we were in the house, we discovered that Princeton graduate student Bei-Lok Hu and his wife were living in our grenier, which was fitted out as a tiny windowless bedroom. It seems that Cécile had negotiated a side-arrangement with M. Laprade to rent the grenier separately. There was no toilet. Bei-Lok had been carrying a chamber pot down the mountain to empty at the school. Once we discovered him, we let him use our toilet instead—but preferably when we weren't home.

Later, Ina Rakavy established excellent relations with Laprade. She was educated in France and spoke the language fluently. Also, she happened to be a student of architecture, and she knew of Laprade's famous work: In the French colonization of Morocco after the First World War, he was the leading architectural influence. Indeed, Albert Laprade was a celebrated French architect whose career spanned the decades 1910 to 1960. Who could have guessed?

I'll jump out of time with a much later story about Gideon Rakavy. In the 1990s, I was sitting in my office at Harvard chewing the fat with an Israeli scientist who enjoyed gossip almost as much as I did. On this occasion, she was listing for me all of her colleagues in the Racah Institute at Hebrew University who, in her opinion were unproductive deadwood. She thought they should be retired, if not shot. (She was a person who carried her emotions close to the surface.) "Well, what about Rakavy?" I asked, mentioning that had I shared a house with him at Les Houches in 1972 and had never subsequently seen a single published paper with his name on it. "No, I don't count him," she said. "He and his group work on the atomic bomb project and supposedly they do good work."

"What???" I said. Then, as now, Israel had no publicly acknowledged nuclear weapons program.

Until that moment, I had never actually seen someone's face turn pale white, the exact physiological opposite of a blush. "Oh my God!" After a pause she said, "I thought we were in Israel. I forgot that you are an American."

24. Les Houches

We adjusted to the physics school's daily routine. Margaret and I woke up every morning under several quilts to the sound of oversize alpine cowbells, the kind shown on postcards. A neighbor's herd was driven past our house on its way to daily greener pastures. I threw open the shutters to the brilliant view of the steeply rising green meadow, wildflowers, alpine backdrop, very Heidi-pretty. The Rakavys, whose bedroom was downstairs, were always up much earlier. There was a sink upstairs, accessible from both the Teukolskys' and our bedrooms. Margaret and I had first use, then we knocked on the other door to be sure that Saul and Ros were up. Half an hour later we four emerged into the sunlight and walked down to breakfast. The walk was twelve minutes downhill in the morning, eighteen minutes uphill at night, including a four-minute rest on "Laprade's bench". Breakfast was served in the restaurant (as the school's dining room was called) from eight to nine: baguette with butter (never the croissants and rolls that could be bought in the village), *café*, *thé*, or *chocolat*. We always asked for refills on the *lait chaud*, which we strained through our own precious tea strainer, purchased in Chamonix. Only after dinner was the hot milk properly steamed on the espresso machine instead of being boiled. The restaurant was run by Henri, who was the husband of Nicole, the school's secretary. He was also an alpine guide and a local property owner. The chef's father was the village taxi driver, and the serving girls were related to everybody. We paid for the meals with little tickets; except that when the supply of printed tickets ran out, cash suddenly became acceptable.

The first lecture began at nine—later for lazy lecturers—and ended between ten-thirty and eleven. Margaret and Ros, usually with Leslie Will and Betsie (age eight), took over the lounge at one end of the building; a small scientific library was at the other end. Between the two morning lectures was a half-hour tea break. Lunch was the best meal of the day. The restaurant staff was always very friendly, and one's ability to obtain substitutions—but never extra servings of meat—was limited only by one's ability to communicate in French. There were Gallic idiosyncrasies: the meals served one course at a time, the vegetables after the meat, the salad at the very end, with a superfluity of yogurt, never anything else, for dessert. When there was an afternoon lecture (about half the time), it began at five. The earlier afternoon was reserved for less-formal seminars, conversation, hiking (which we avoided), trips into Chamonix to buy the *International Herald Tribune*, or—as seemed to be our default option—sitting in the lounge passing time. A shifting alliance of Presses, Teukolskys, Wills, and Thornes soon claimed exclusive title to the lounge, in later weeks joined by Lane Hughston, a Rhodes Scholar and student of Roger Penrose, and various of his Anglo-American friends. People accommodated nearer the school hung out in their rooms, but we were not keen to hike up our mountain more than once a day.

Dinner was at 7:30. The kitchen staff worked long hours and were the same at all meals. Above the dining room was a balcony where coffee was taken after the meal, for us according to the formula "*petit café dans une grande tasse avec beaucoup de lait chaud*," soon abbreviated to "*comme toujours*." Typically, we spent a couple of hours after dinner talking or playing cards or reading in "our" lounge before walking up the mountain. The lights in our farmhouse were too dim for comfortable reading. The Rakavys were usually there before us. Gideon was a master fire builder, and they were often sitting in the dark main room staring at a mountain of glowing coals in the massive fireplace (open on all 4 sides). Saul or I lit the wood-burning hot water heater so that we would have hot water in the morning. We all went to bed before midnight.

The lecturers included Kip, Jim Bardeen, and Bryce DeWitt (from the U.S.), Stephen Hawking and Brandon Carter (from the U.K.), Marek Demianski (from Poland), Igor Novikov (from Moscow)—and Remo Ruffini, who gave lectures that were so transparently full of himself that attendance fell to near-zero after the first, to my great satisfaction. I had never forgiven Remo for forcing himself onto my paper with Richard Price. Hawking was exceptionally well-prepared and passed out detailed lecture notes in advance of each lecture. His labored speech was difficult for the English speakers to understand, impossible for the others. Stephen mentioned in passing several results of Saul's and mine, but without ever mentioning our names. It was an early hint of what became obvious later in Stephen's career, that he was exceptionally frugal in giving credit of any kind to others.

Nancy Bardeen sat with us at dinner one Saturday night when Jim was not yet back from mountain climbing. "What is Jim lecturing about?" she asked us. "Haven't you asked him?" I responded. She looked at me as if I should know better. She told us that Jim frequently dreamt that he had told her things, or that he had had long conversations with her. In his dream world, apparently, he was a fluent, talkative person. "Well," she said, perking up a bit, "I'd rather have someone silent who occasionally says important things than someone who just chatters all the time." "Oh, Bill just chatters all the time," Margaret interjected helpfully.

One afternoon Cécile summoned me secretly to her office. She had insider information, she said, that I would be invited to the prestigious "Copernicus" conference in Poland next year, a convocation honoring the 500th anniversary of Copernicus' birth. Saul was not on the invitation list. But now that Demianski had met and heard him here, he hoped to have him invited too—it would depend on how many people didn't accept. But it turned out that this was all merely a preface to the true business at hand: Because of currency restrictions, Demianski was unable to bring from behind the Iron Curtain funds enough for his expenses here this summer. Cécile was making up the difference, but only temporarily. She needed a trusted party—by that she meant someone bent, actually—whose expenses in Poland would be paid by the Poles on condition that he remit to Cécile an equivalent amount in dollars—obtained, presumably, by turning in a false expense account to Caltech. For some reason, she had decided that I was her man. It was another typical Cécile project, especially since it was "none of her responsibility." "Of course, I'll make it slightly worth your while," she said, with the emphasis on slightly.

An after-dinner gathering in the Hawking's flat was an affair of studied normality. Jane Hawking was a pretty hostess, serving hot mulled wine and engaging in conversation with each group of guests in turn. Stephen had labored conversations with a few groups. Margaret and I chatted with Stephen's parents, who were visiting. Mrs. Hawking, senior, was very English in that London, educated sort of way. In the school restaurant, she expected the serving girls to speak English, but she permitted daughter-in-law Jane to translate to French when they could not understand her. Stephen's sixteen-year-old brother was with them. He was a muscular fellow who gulped down his food and then excused himself. His mother told us that he had just failed his O-level examinations and didn't want to do anything in life except become a professional football player and get married. "To anyone in particular?" I asked. "It's always someone in particular, but the someone changes from time to time. He always wants to do the right thing by them." She wore the facade of being oblivious to Stephen's condition, but there was an instant, when she was laboriously helping him to eat a piece of cheese, that the mask seemed to drop; but she quickly recovered.

Towards the end of the month, the simple pleasures of rustic life and black holes were beginning to cloy—particularly with the same company day in and day out. Our inner sphere of eight or ten people, now often including Bryce DeWitt's student Larry Smarr, revolved in a larger constellation of twenty or so, mostly the Americans. Add ten wandering comets with whom we had passing exchange, and finally twenty or so people whom we hardly knew: many of the Europeans and a few camp followers around Remo. Same faces at dinner, in the lectures, passed with nodding hellos in the school buildings, in the village of Les Houches, even on every block in Chamonix, a town of a few thousand people ten kilometers up the valley. Les Houches itself boasted a population of a couple of hundred.

A day's excitement for us might be Margaret being allowed to pay an extra half-franc to secure an unlimited supply of hot, steamed milk—her great sybaritic joy. For others at the school, the excitement was more likely hiking the approaches to Mont Blanc. I was jealous of the people, including Kip, who were busy on their own work, but there was not much work I could do until I got back to my computers. I busied myself writing an outline of my thesis. The thesis would be mostly an unorganized collection of my papers, per Kip's instructions, but I wanted to also include a few dozen pages of explanatory material that would tie it all together.

Then, soon enough, we were back in Pasadena. There, the news was not good. Tommy Lauritsen's cancer had recurred and metastasized. His condition was now thought to be terminal. He was fifty-seven. A succession of family friends began to arrive in town on what were transparently goodbye visits, but always with some other pretext. Margaret and I were often summoned to family dinners of strained bonhomie and always heavy drinking. Willy Fowler was a frequent presence—he was devoted to Tommy. When the out-of-town visitors were Fred and Barbara Hoyle, Willy and Fred drunkenly lamented the sad state of the world. They planned to retire, together, to the family farm. In their minds, there was a single farm, a rose-tinted comingling of Ohio and Yorkshire.

By the time of my thesis exam in October, I had assistant professor offers from Berkeley, Yale, Utah, Harvard, and Princeton—but not the offer that I cared most about, namely one from Caltech. Kip waxed and waned with the moon in his assessment of whether I would get such an offer. I couldn't tell whether the varying factor was the department's enthusiasm or his own. Harvard's offer was the most peculiar. I got a letter from Paul Martin, the physics department chair, offering me an assistant professor position almost off-handedly, and mentioning that the only information that they had about me was my undergraduate transcript—could I please send them something more current. This was, of course, Ed Purcell's doing. His standing at Harvard was such that his merely suggesting my name was sufficient for the department to vote an offer, without any other information or supporting materials.

"You're Bill Press," Feynman said to me, sitting down with us at lunch in the Greasy. "I was reading you last night." He meant my thesis. My exam was two days away. Dick complained amiably about its cut-and-paste nature. It's all right, he said, to write a paper with mistakes, and then write a second paper correcting the first. He was a little dubious about then writing a third paper adding rigor to the second paper's corrections of the first. (He was describing my three papers with Bardeen.) But, he couldn't accept that I should then get to count *all three* as part of my thesis. Also, my earlier papers said that the gravitational equation is not separable, while my later papers say that Teukolsky separated it. "I could tell that Teukolsky, not you, is the *real* master. Who is that guy, anyway?" I silently pointed to Saul, sitting next to me.

My examining committee was Thorne, Feynman, and Gunn. Feynman asked ninety-five percent of the questions, mostly demanding physical—not mathematical—interpretations of things. Then he played devil's advocate and pushed to see if I could really defend my position. The couple of times I wandered off into nonsense, Kip stepped in to ask softball leading questions. It lasted two hours, which was shorter than average. The committee deliberated in private for about two minutes. "Well," Kip said jokingly when they called me back in, "you've squeaked by." Feynman interrupted gravely, "You shouldn't allow any possibility of misunderstanding in this—you should tell him that he did a good job." This was better than the "Dick Feynman B" I got from him when I took his course a little more than a year previously!

For my party at Kip and Linda's house, Margaret and I supplied three gallons of Gallo jug wine, three quarts of cheap brandy (for alcoholic punch), a pound of lox, two pounds of cream cheese, eighteen bagels, a pound each of roast beef and pastrami, two pounds of pickles—and that was just the part of our list that we bought at Fedco. I had posted notices of the party in Kellogg Lab and Astronomy, and the turnout was about seventy-five people. The punch hid the alcohol well, and there was also a keg of beer. Everyone in Kip's group was there, including visiting Charlie Misner, and a good fraction of the Kellogg and Astro faculty. I didn't pass out, and we got home around three. I was now Dr. Press, Caltech's Richard Chace Tolman Research Fellow in Theoretical Physics. I sent a letter to the Hertz Foundation, thanking them and resigning my remaining graduate fellowship.

25. Job, No Job

In late October, 1972, Margaret and I made a trip to the places "back East" where I had job offers. We flew to Boston, visited my parents and sister, and then drove the six hours to Princeton. John Wheeler had arranged for us to stay at Palmer House, the official university guest house. We picked up a key at the university police station. We were alone in the eighteen-room house. The sound of a dozen ticking clocks echoed in empty hallways. Within about three minutes on both sides of every hour, various clocks struck inconsistent numbers of times—only some had been reset from daylight savings time. We went to dinner at the Wheelers' house: elegant, modern, on university-owned land. Mrs. Wheeler—Janette—also impressed us, but not in a good way. She made no bones about her distaste for Margaret's career aspirations. "When I was young, I thought of working, but Johnny pointed out that I wouldn't be able to accompany him on trips, and of course that was that."

Precisely as I finished my dessert, Janette took Margaret to the other side of the room and sat her down for "woman talk," whose purpose was to keep her from participating in my discussion with John. Almost everything Wheeler said to sell me on Princeton had the reverse effect. I should come, he suggested, to bask in the reflected light of Princeton's Great Men past and present: Einstein, Wigner, Schwarzschild, Spitzer. Wheeler's own group had produced Feynman and Thorne (among others). Eventually we did approach some substantive issues. With great delicacy but unmistakable finality he assured me that Remo would not be renewed as an assistant professor when his term expired at the end of next year. Wheeler was fighting to get a tenured position for Karel Kuchař, now also an assistant professor. If Kuchař left whom would he replace him with? "If he leaves, I may go tagging along after him to pick up the crumbs of his genius," he said with Wheelerian hyperbole. Kuchař worked on so-called "canonical" quantum gravity, a field then in vogue, but one that subsequently produced little of significance.

Palmer House was supposed to provide breakfast, but when we got up, we were still all alone in the mansion. I made my way to the physics building, Jadwin Hall. Wheeler was away on a day trip to Washington, but I had appointments with members of his relativity group—almost all of whom were leaving, some leaving with Remo, some because of him. Everyone seemed to know that Remo was being fired. When it came time for my appointment with him, he took me into his office and closed the door conspiratorially. "Certain people have stolen much of my work," he said, "but no one at Caltech." He looked forward to continuing to work with me. I wondered if by that he meant continuing to steal *my* work. It gradually dawned on me that Remo didn't know that he was being let go. Ed Taylor, who co-authored two excellent textbooks with John Wheeler, used to say that John had the ability to fashion a negative message from completely positive comments. That must have happened in this case. Remo heard only the positive. It was not my job to disabuse him.

Jimmy York, in whom mathematical ability and southern redneck charm co-existed in an unusual mixture, was leaving for tenure at Chapel Hill, North Carolina. He took me to lunch across town at the Institute for Advanced Study, where we sat at a table with Freeman Dyson (who remembered me), Tulio Regge, and John Bahcall, who was the newest IAS full professor. Bahcall was formerly on the physics faculty in Kellogg Lab at Caltech; the scuttlebutt was that he didn't get along with Willy, who made his life miserable and forced him out. This was a side of jolly old Willy that I, luckily, had never seen. I found John refreshingly candid. No, I wouldn't be coming to Princeton for the Great Men, he told me. I would be coming because he, Jim Peebles in physics, Jerry Ostriker in astronomy were all talkative guys somewhat senior to me, with a reputation for collaborating. He made exactly the pitch that I wanted to hear.

Over the next twenty years, John was to become my greatest mentor and cheerleader—not always with respect to the content of my work, but often in matters of how science should actually be *done*. I was lucky in that John had one great fault: At first meeting he either loved you or hated you, and he never changed his mind after that. That day, he loved me and, although I didn't at the time know it, this love would in many ways guide the course of my subsequent professional life.

Wheeler, back by train from Washington, took Margaret and me to dinner—again at the IAS cafeteria, a short walk from his home. We drove the three hours to New Haven after dinner. The Bardeens put us up in their new house, which was somewhat nicer than their house in Seattle had been. The problem for me with Yale was that the university's commitment to science was questionable and the physics department was second-rate. Their pitch was: Come here and help Jim Bardeen make us first-rate in relativity. I might have been more receptive if Jim wasn't the way he was—that is, *silent*—and if their new building, which was built on a shoestring budget, wasn't already visibly falling apart.

Also, this far from any mountains to climb (the literal, not figurative, kind), I doubted that Jim would stay at Yale for long. Yale wasn't completely out of the running for me, but nearly so. We drove to Boston the next evening. In fact, after two years at Yale, Jim Bardeen returned to the University of Washington in Seattle and spent the rest of his career there.

The situation at Harvard was somewhat confused. The astronomy department had succeeded in attracting Berkeley theorist George Field as a professor and new director of the Harvard College Observatory (HCO). HCO was not really an observatory, but a set of buildings and substantial historical endowment of money that supported astronomy at Harvard. Being negotiated with the Smithsonian Institution in Washington, D.C., was a proposal that Field also be appointed director of the Smithsonian Astrophysical Observatory (SAO), which had long leased from Harvard a building conveniently next to the HCO—even connected to it by a bridge. The new combined entity would then be known as CfA, the (Harvard-Smithsonian) Center for Astrophysics, with an integrated scientific program that Field would direct. It took a while, and a long meandering interview with physics chair Paul Martin, for me to figure out that my offer from the physics department was not really real. Yes, I would formally be an assistant professor in physics, but it was just a holding action until George took control of the new CfA, of which I would become a part. I had lunch with George at the Faculty Club ("just look for someone who is 6'4" and it will be him") and was favorably impressed; but none of his vague promises could be taken as binding.

The clincher was when I met with Ed Purcell, my scientific hero and mentor since my freshman year. He was very friendly, and we spent forty-five minutes talking about astrophysics in general, my work on black holes, his on the alignment of interstellar dust grains, etc. When there were only ten minutes before I had to go, I tried to get his advice on where I should go next year. "Let me tell you about our 600-foot radio-telescope design," he said firmly, and we spent the rest of the time on that. The message was unmistakable.

Margaret and I flew back to L.A. the next day. In meetings arranged at all three places that we visited, Margaret got no hint of a possible job for herself. Her father was dying of cancer; and she was still a year from getting her Ph.D. So, what it boiled down to was that we really wanted to stay at Caltech. It was thus very welcome news when Kip took me into his office and told me that the physics department had voted me an assistant professor position. There had been a long discussion about the dangers of inbreeding, but, in the end, the vote was 30 to 2, then re-voted to make it unanimous. Tommy, both conflicted and ill, did not attend the meeting. The appointment still had to go through the Caltech administration, but this was a formality.

My future was secure—I thought. Kip started making plans to go on sabbatical leave the next academic year. He would be leaving me, his new assistant professor, in charge of the relativity group. He had just taken on two new students, Don Page and Carl Caves, but he thought that while on sabbatical he could supervise them remotely. I should start thinking about taking on students of my own, he said. Don was something of an oddity at Caltech, a born-again Christian with (as he would earnestly explain if you asked) a personal relationship with Jesus. He had been home-schooled by missionary parents in Alaska in a town "with more feet of snow per year than its population." (That number was around twenty.) In later years, Don was a postdoc with Stephen Hawking in Cambridge, then a professor at Penn State and the University of Alberta in Edmonton. Don's religiosity and charitable nature remained constant throughout his career.

By Thanksgiving week my computer program was working and we pretty much knew that Kerr black holes were stable. Saul and I might have become more famous if they were unstable, but nature didn't cooperate in this way. Still, we were the first people in history to *know* this very fundamental fact about the universe. And we had beaten the competition—there was no sign that anyone else had even been able to derive the Teukolsky equation.

There was, however, a small glitch. Hawking and Penrose had proved very generally—by "global methods" that were way over our heads—that the surface area of a black hole could not ever decrease. This applied in all circumstances. It was a good test of my program. In the computer, I irradiated the Kerr black hole with different frequencies and patterns of gravitational waves, and I calculated the change in the surface area. I should have found an increasing (or constant) surface area in all cases. I usually did. But in one *tiny* corner of parameter space, my program showed a tiny *decrease*. The effect was so small that it couldn't affect our conclusions about stability, but it was not just numerical round-off error. Saul already had ready to submit to *The Astrophysical Journal* his sole-author paper deriving his equation, but we decided to hold off until we could reconcile my code. Most likely it was a small programming bug. But what if it was a tiny slub that, if you pulled on it, unraveled the whole sweater? Would it unravel the Teukolsky equation, which would be terrible for both of us? Or, unlikely but not impossible, would it bring down the mighty Hawking and Penrose, Saul and I felling two giant sequoias with our tiny little axe?

Finding the bug in my program, or the error in the Teukolsky equation, became our obsession. Saul, who marginally had more to lose, checked every line of my code, and thought of other kinds of tests—all of which the program passed. No one in the group knew what we were working on so intensely, or that we were struggling to avoid catastrophe—but might alternatively become famous woodcutters.

November 30, 1972, was the day that the Administrative Committee was supposed to act to finalize my appointment offer. Late in the afternoon, Bob Leighton, the physics chairman, called and asked me to come to his office. He asked me to sit down and said simply, "We were unable to get the appointment through." The Division of Physics and Astronomy had recommended me unanimously, but the other Division chairs, provost Bob Christy, and president Harold Brown (the two of them both physicists) would not go along. "Forty percent of the physics faculty were hired after getting their degrees here," Leighton recited, "and the Committee feels that they are not necessarily the best forty percent." (Tommy Lauritsen was surely in that category.) Since President Brown had concurred in the decision, it was final.

Leighton delivered the bad news to me unemotionally, but Caltech was abuzz. By the time I got back to Kip's office, he knew about it. The Administrative Committee didn't always rubber-stamp actions by the Divisions, but this was the first time in Caltech history that a unanimous recommendation had been rejected. Kip was by turns angry and sad. "I had a dream that you would take over the relativity group, Bill, and I could stay up on my mountain and study astrophysics." Saul was angrier than I was. "That's really for shit," he said several times. The words sounded strange in his South African accent. Tommy's birthday happened to be today, so Margaret and I had dinner at Rose Villa. I waited until after dinner to tell him. He was surprised and disturbed; unspoken were the several ways that this reflected negatively on him.

By the next morning, when events had sunk in, I was angry and disgusted, but not yet depressed. I threw myself into the secret work of understanding our computer anomaly. If the Teukolsky equation was wrong, or my program couldn't be fixed—*then* I would let myself be officially depressed. Tommy located me at the computer terminal in Sloan Laboratory (the only one which always worked, because it was the one that Dean Bohnenblust used). He had been looking all over campus for me. "Willy is in Washington at a meeting along with your father," he said, "and heard the news from him." Apparently, no one at Caltech had dared to call Willy with the news any earlier. "He wants you to hold off doing anything rash until he is back next week." Since I was not anticipating doing anything rash, I agreed.

The morning of Willy's return, he summoned me to his office in Kellogg. His version was that my mother had called my father in Washington to tell him the bad news about me. Frank was pulled out of a National Science Board meeting to take the call. Coming back in, he whispered to Willy "Caltech turned Bill down." "Christ!" said Willy as he left the meeting to himself call Leighton and Christy to find out what had happened. "And when I found out the reason, I just about went through the roof." His voice went up about two octaves in the retelling, as if through the roof.

Willy was Caltech's proudest home-grown product himself. He had enemies on the Administrative Council—chemist Jack Roberts, whose daughter Anne was my sister's best friend in elementary school, was one of them. Willy was convinced that I was a mere pawn—that the action was aimed personally *at him*. This seemed (and still seems) to me unlikely, but I was not about to contradict him. "It wouldn't be right for me to tell you what I plan to do about this," he said. "I don't even want to know whether you would consider an offer from us after all this." His voice started rising again. "This thing has got to be fought out as a matter of *principle*." The last word went into the stratosphere. Ushering me out of his office, he said darkly, in his lowest register, "You always have to know who your friends are."

By mid-December, I was becoming adjusted psychologically to the idea of moving to Princeton. I hadn't officially accepted Princeton's offer, but Wheeler knew from Kip that I was likely to. It would be throwing Margaret's career under the bus. It was not obvious how she would even finish her Ph.D., although finding some path for that was still a priority. By the time of the big relativity conference in New York City (an adjunct to the December annual meeting of the American Physical Society), Saul and I had still not found the problem in our calculations. We had confessed to Kip, who then spent a full Saturday with us going through every detail of our calculations—and not finding an answer.

Coming amidst this negativity, the New York meeting was a validating experience. I was an invited speaker, wedged in the schedule between Stephen Hawking and Kip. The world distant from Caltech seemed to consider me a real scientist. The meeting became a kaleidoscopic montage of everyone I had met professionally in the previous four years. Unusually extraverted, I wandered around pumping people's hands and gossiping. Larry Smarr introduced me to his new girlfriend, previously the unnamed transcendent being for whom he had gone AWOL from Les Houches, traversed the Alps (by bus) to visit in Italy.

Here they all were: Jim Wilson, Lowell Wood, John Wheeler and Mrs. Wheeler, Lane Hughston, Jim Ipser, Bonnie Miller (no Ritchie), Bryce DeWitt, Bob Dicke, Martin Rees (who casually mentioned that he would be returning to Cambridge as the next Plumian Professor), Lou Witten, Remo Ruffini, Steven Hawking and his mother, Bob Wald, Martin Walker, Ken Nordvedt, Victor Ni, Abe Taub, Joe Weber and Virginia Trimble, Bernie Schutz, Franco Pacini, Gary Steigman, Jacqueline Bergeron, Jim Hartle, Jim (and Rosemary) Gunn, Bob Wagoner.

I mentioned casually to Stephen Hawking that Saul and I had a result that was most likely our conceptual or computer bug, but might possibly be the sign of a problem with the "cosmic censorship" assumption in his black hole area theorem. I was not expecting his reaction to be so immediate: He summoned Jim Hartle, I located Saul, and we went to a side room. He pumped us for details, and he grasped them quickly. He then laid out for us a succession of tests and checks that, if we wanted to seriously challenge his theorem, our program would have to pass. I didn't have to like Stephen to recognize that he was very, very smart.

I suppose I have to come out and say it: I didn't like Stephen as a person, then or later. *Of course* I felt anguish and sympathy over his physical condition, but, stripping that away, he was not a nice man. The hints of this in Jane Hawking's post-divorce account, *Music to Move the Stars* (1999), were dismissed by reviewers as the spite of a jilted spouse—this after Stephen left her to marry his nurse—but to me they rang true. I sometimes saw Stephen use his condition to bully others, as insisting that a particular person help him go to the bathroom—a person he disliked, not one of the several who would gladly have assisted.

More substantively, Stephen was slipshod or slow in giving others credit for their work, and fast to claim ideas for himself or favored others when they were already "in the air". The most public example of this occurred in the aftermath of a 1982 conference at Drexel, where Stephen was loathe to give credit to Paul Steinhardt and Andy Albrecht for their important theoretical discovery in cosmology, going so far as to falsely accuse them of plagiarism. The matter came to a head only after Stephen's best-selling book, *A Brief History of Time*, was published in 1988, repeating the slander for the first time in print. With unexpected help from *Newsweek* magazine, Steinhardt was able to amass documentary evidence (including tape recordings of the Drexel conference), proving that Stephen's claims were false. The publisher removed the passage from later printings of the book, but Stephen never apologized or admitted fault. After *Newsweek*'s report, the popular comic Bloom County in July, 1988, ran a strip that had its character Oliver inventing a new physics theory—and receiving a bomb in the mail from Stephen Hawking.

Stephen's good friends Kip and Roger Penrose always defended him, but, I thought, with flimsy excuses. They understood Stephen's defects, I think, but didn't want to see his myth undermined. Yet, myth or no myth, Stephen was, I think, one of the dozen or so greatest theoretical physicists of the 20th century. Number twelve, maybe.

26. Resolution

My two matters of greatest concern rather magically resolved themselves early in 1973. Late one afternoon, Kip told me to go see Bob Leighton. The department secretary was already gone, so I knocked and entered his inner office. "I am now able to offer you an assistant professor position at an annual salary of fifteen thousand dollars," he told me. This was more money than any of my other offers. "How did this come about?" I asked. "You'll have to talk to Willy about that," he said. Willy was at home, recovering from minor urinary surgery. He had promised to piss in every room of Kellogg Lab when he returned, to celebrate his recovery. I conferred with Kip. I badly wanted to stay at Caltech for a year, if only for Margaret's Ph.D., but I had already crossed the psychological Rubicon on ultimately going to Princeton. Could I accept Caltech's offer and then decamp after one year? "You should talk to Willy about that," Kip said. All roads led to Willy, not unusual at Caltech.

I rode my motorcycle to his house on Arden Road. I had decided to float two possibilities, only the first of which was real: (i) I leave Caltech after one year and never come back—that was what I actually intended to do; or, (ii) I leave after one year, go to Princeton for a year or two "for the experience," then return to Caltech. I didn't actually consider the second a viable option. If I stayed at Caltech, the stigma of inbreeding—and of Willy's intervention—would haunt my career forever. Still, I thought that Willy, having won the fight on inbreeding on very personal grounds, would favor the second option, at least as a polite fiction.

Willy told me with some pride that he was peeing well and that he had gotten every single member of the Administrative Committee (whom he referred to as "those sons of bitches") to change their votes—except the chemist Jack Roberts, that biggest son of a bitch of them all. To my surprise, Willy favored only my first possibility. "You stick it to those sons of bitches," he said. "You've earned that." It was a slightly mixed message for me that Willy *didn't* want me to come back to Caltech, but it was the desired outcome overall.

The issue about whether the Teukolsky equation predicted a violation of Hawking's theorem resolved itself quietly. We went through Stephen's checklist. As sometimes in science, there was not a single "aha" moment, but just the elimination of possibilities, with tests and cross-checks at each stage. Hawking was right. We were wrong. It was a subtle coordinate effect. Coordinate effects are the bane of general relativists. Since gravity is able to actually warp space (and time), it can also warp the grid of coordinates (x, y, z, say) that you are using to describe a phenomenon. When you see some effect, it is sometimes hard to decide whether it is real, or just the mistaken application of slightly warped coordinates. Our effect was in the second category. The Teukolsky equation could be written down in several different coordinate systems, all equally valid. We had picked one. It turned out that the only way to be truly self-consistent was to pick one system far from the hole (for outgoing gravitational waves) and a different one near the hole (for waves going down the hole), and then numerically solve a differential equation—a new, auxiliary Teukolsky equation—to relate the two systems. It was a good outcome for Saul's and my partnership. It wasn't a bug in my computer code, and Saul's equations weren't wrong, just slightly incomplete. The answer was something that each of us *could* have figured out, and eventually—when we had to—did.

We finished, polished, and submitted to *The Astrophysical Journal* our two Kerr papers, Saul's with its derivation of the Teukolsky equation, and our joint paper demonstrating computationally that Kerr black holes were stable, and might thus actually be found in nature. When in due course we got back the referee's report, I read the first sentence and guffawed—the anonymous report was so obviously written by Chandra. His 'might-it-not-be-desirable's gave him away. Our result on stability became generally accepted. It was not until 1989, more than fifteen years later, that Bernard Whiting proved completely analytically (that is, without relying on the numerical evidence from a computer) the stability of the Kerr metric. By then both Saul and I had moved on to very different things; but, if I were Saul, I would have felt a twinge of jealousy in reading Whiting's very elegant exposition.

Besides black holes, I was also working on an idea that had come to me while glad-handing my way through the New York relativity meeting. Since gravity warps space, it also bends the path of light rays. If a gravitating object is close to the line between you and a more distant star it is possible for you to see two apparent images of that star, one from rays that were bent "left" around the gravitating object, the other from rays that were bent "right," but both ending up at your eyeball (or telescope). *That* wasn't my idea. Einstein published an article about it in 1936, and Fritz Zwicky had followed in 1937 with a paper that proposed clusters of galaxies as the intermediate gravitating object.

Zwicky was still around at Caltech, retired as a professor and more than a bit crazy. He was a compact man with a thick Swiss accent. We called him the gnome of Robinson Hall (the astronomy building). He used to visit us graduate students to be sure that we knew about his self-described achievements—for example, that he had invented the concept of the black hole in 1934, five years before Oppenheimer's 1939 paper and thirty years before John Wheeler had invented the name (all true, to some degree); and that he had invented the "aeropulse" during World War II (we would call it a pulse-jet), before the Germans used it in the V-1 flying bomb. He also claimed to hold a 1948 patent on the hydropulse and terrapulse, which would "fly" in water and solid rock, respectively. Soon, he said, he would be going back to Switzerland to help the Swiss build a terrapulse for tunneling the Alps. Caltech's astronomy librarian had to razor-blade the Preface out of the printed Zwicky *Catalogue of Galaxies and Clusters of Galaxies*, because it was mostly slanderous personal attacks against his Caltech colleagues, all named.

Up to now (that is, 1973) no actual "gravitational lenses" had ever been seen. My idea was to calculate whether they *should* be seen. At this time, it was a matter of continuing debate whether the universe was "closed" or "open". These were terms of art that described the two possibilities of a Big Bang that either (closed) had mass enough so that gravity's attraction would eventually bring the expansion to a halt and turn it around into a Big Crunch (that term another Wheelerism); or else (open) not enough mass, so that the Big Bang would go on expanding forever. What I had found was a relation between this overall mass density of the universe and the probability that any given line of sight (to any given distant object) should by chance be gravitationally lensed. Roughly speaking, lenses should be very common in a closed universe, less common—but still occasionally there—in an open universe that was compatible with what we already knew. (The universe couldn't be *too* open, or else we wouldn't see all the galaxies that we did see in the sky.) I thought that this was a solid piece of pre-observational theory. That is, it pointed the way to a specific observational program: Observers, go look for lenses, because you'll (probably) find them.

Jim Gunn, to whom I took the idea, liked it and urged me to write a quick paper. I started in. Writing the theory part was easy, but then came the hortatory part where I was supposed to urge the observers to go forth. I didn't how to do this well. I didn't speak "observerese". So I went back to Gunn and pleaded with him to be a co-author and write that section of the paper. He agreed and fairly quickly gave me his pencil manuscript. It was practically illegible, but the prose was very elegant. Kip, who had taught me scientific writing, wrote in clear, school-marmish prose. Jim's writing style was polished and sophisticated. Unrelated to his scientific talents, it increased my respect for him.

I'll jump out of time now to say more about gravitational lenses and the idea of "pre-discovery." "Press and Gunn" was published in 1973 in *The Astrophysical Journal.* It was duly noted by a handful of theorists and observers and then rapidly sank into obscurity. No observers—not even my co-author Gunn—rushed to propose observing programs on major telescopes to look for lenses. Six years later, a team of astronomers accidentally discovered what was apparently two quasars with identical spectra very close together in the sky. News of this spread rapidly among astronomers by phone. "It's one quasar with two images. It's a gravitational lens," I said immediately to the colleague who told me about the discovery. "How do you know that?" he said. "Gunn and I predicted it," I said.

Within days (and not just from me), it became accepted that a first gravitational lens had been discovered. But perhaps because the discovery was accidental rather than purposeful, the formal discovery paper didn't include my paper among the handful of previous papers cited. After that, there was a gravitational lens gold rush, with many lenses discovered in the sky and, over decades, thousands of papers published. Some small fraction of these authors, about 250, were scholarly enough to unearth and cite my paper. Two hundred fifty is considered a large number of citations. But if our paper had *led* to the discovery, its number of citations would be in the thousands. Pre-discovery papers, so-called, are hit or miss in this way. Sometimes they become the foundational papers of whole fields; sometimes they languish in obscurity and are found only by historians of science years too late. Gunn's and my paper was something in-between.

* * *

One week, the physics department colloquium was given by a Princeton professor named Gerry O'Neill. He was an experimental particle physicist, but his topic was the colonization of space. His idea was that the whole Earth would become a giant park, while all humans would live in orbiting Utopias—healthy, happy, and (he said) with unlimited pleasures of all sorts. His wife, a sinuously tall and beautiful woman with waist-length red hair—I expected her to be the high-priestess of this cult—projected his slides for him. The human race would increase its population twenty-thousand-fold, he said. Adding to O'Neill's aura as a crackpot, he looked and dressed exactly like Mr. Spock, the alien on Star-Trek. The audience didn't take his message well.

Bob Kirshner, my Harvard undergraduate friend, by now a Caltech graduate student, asked the most sensible one-word question: "Why?" Fritz Zwicky, on the other hand, chewed O'Neill out for not being visionary enough. "This is all old hat," he announced scornfully. "If you only had looked at my 1948 Halley Lecture, you would have seen that my work went *far* beyond these crude concepts." There was scattered applause in support of our local crackpot over this Princeton imposter.

Paul Schechter, in my entering graduate cohort, was by now well into his thesis work, with Jim Gunn as his advisor. Paul was working on the "mass function of galaxies". That meant, if you took a big volume of space, how many galaxies of each different mass (or mass range) would there be in it? That distribution, plotted as a histogram as a function of galaxy mass, was the "mass function". Paul was analyzing published data in a new way and finding that the mass function could be accurately described by a very simple mathematical formula. Gunn was not very interested in Paul's work; it hadn't crystalized into a clear statement about anything, it was just a curious formula. But it was so beautiful! After Paul showed me the data, I couldn't stop thinking about it.

At this time, little was known about how galaxies formed. In the absence of any real understanding, the community inevitably divided into two camps: "top-down" and "bottom-up". Top-down meant that the biggest structures in the universe—then thought to be clusters of galaxies—should form first and then fragment into galaxies, which might fragment into smaller galaxies, and so on. Bottom-up meant that small galaxies, or maybe the even-smaller globular star clusters, should form first, then aggregate by gravity to make larger galaxies, with clusters of galaxies forming last. At meetings, astronomers nearly came to blows defending one position or the other, rather like the big-endians and little-endians in *Gulliver's Travels*. I didn't have a dog in the fight, but if top-down was correct, it would involve messy radiative hydrodynamics. The physics of each stage of fragmentation would be different and difficult, and I had no idea how to calculate the mass function. If bottom-up were correct, however, the aggregation on each scale would, at least for a while, be driven purely by gravity, which I thought I could calculate. There might be messy hydrodynamics later on for each scale, but that wouldn't affect the mass function. Gravity by itself didn't care about scale, so the same calculation would have to apply on all scales, a property called "self-similarity".

It was surprisingly easy to construct a self-similar solution to the problem of hierarchical condensation under the influence of gravity. I had the answer in a week: The mathematics led to almost exactly Paul

Schechter's beautiful empirical formula! I decided to write a computer program that would simulate the process by following the trajectories of little point gravitating masses, something called an "N-body calculation". This took a few weeks. By later standards the program was very crude, but it showed good agreement with the mathematics and helped validate the assumption of self-similarity, and some other assumptions that I had made. The whole thing was so simple—I couldn't believe that it was not an already-known result. Jim Peebles at Princeton was the dominant figure in this field of cosmological perturbations. His published lecture notes were considered the ultimate authority. I could almost locate the page where he could have done what I did—but he hadn't.

Paul and I published a joint paper on what became known as the "Press-Schechter formalism". It was not a completely happy collaboration, because I wanted to publish and move on, while Paul was a perfectionist who could always think of something additional to tweak (and delay). His solo paper on the empirical data and his mathematical formula came out two years after our joint paper did for just this reason. Our paper, "Formation of Galaxies and Clusters of Galaxies by Self-Similar Gravitational Condensation," became cited by others more than five thousand times.

* * *

Werner Heisenberg invented quantum mechanics in 1925—only Erwin Schrödinger shares credit in the same magnitude. During World War II, Heisenberg worked willingly—some say enthusiastically—on Hitler's atomic bomb project. A fixture of German science in the post-War period, he was elsewhere largely shunned. In 1971 he published an apologist memoir, *Physics and Beyond*, that ought to have been titled *I Wasn't Really a Nazi*. In 1973, thus rehabilitated (in his own eyes, anyway), he toured the United States giving a lecture on the brainchild of his later years, a "Unified Field Theory" that would explain all of physics. At Caltech, his physics colloquium had to be moved to Beckman Auditorium because of the overflow crowd. Feynman introduced Heisenberg with the words, "I never expected that I would get to meet my hero, much less introduce him." Presumably he meant scientific, not humanitarian, hero.

A small private dinner for Heisenberg had been arranged for after the seminar. Kip had something to do with organizing it, so Doug Eardley and I were invited (along with Boehm, Christy, Feynman, Gunn,

Mandula, Ravndal, Thorne, Zachariasen, and Zweig). Heisenberg conversed expansively with Christy, Feynman, and Zweig, while the rest of us around the table listened politely. The discussion was about science administration—how German science was funded, what were the strengths of its organization through the Max-Planck-Institutes, and so on. On all of this Heisenberg was manifestly current: he knew facts, numbers, budgets, and internal politics—the very model of a modern administrator. I thought that perhaps his unified field theory was only an indulgence, not the obsession his talk had made it appear. But, over dessert, he returned to that subject. Zweig was drawing him out, and Heisenberg was responding with a glow of satisfaction. His claims for the theory became more and more inflated as the discussion proceeded.

Feynman had been buttering up his hero all evening, talking about the greatness of his original discovery, etc. Now he suddenly interrupted in quite a different tone, "Professor Heisenberg, I don't believe that your theory makes any of those predictions." Everyone except Heisenberg knew at once what had happened: Feynman's red line had been crossed. Heisenberg may have invented quantum mechanics at age 24, but now, at age 72, he was guilty of intellectual dishonesty. For Feynman this was the sin that could not be tolerated. Feynman continued, "Professor Heisenberg, I'd like to tell you about *my* Unified Field Theory." "Yes?" said Heisenberg, immediately suspicious. "I have a very beautiful equation," Feynman said, "splitzsquirt equals splonck." ("Those are nonsense words," Boehm explained to Heisenberg on his right.) Feynman continued, "This equation makes a lot of important predictions which are all in amazing agreement with experiment. Of course, nobody can understand exactly how I get the predictions, but that's because it's a very complicated equation." Bob Christy came out of shock enough to interject, "Well Richard, I'm sure that either you or Professor Heisenberg will solve the problem before long," and forcibly changed the subject. The dinner ended soon afterward. Heisenberg made a point of shaking hands with Feynman very warmly. Of course, he must have known that Feynman was Jewish.

There were other dinners with famous physicists. Khalatnikov, the Soviet cosmologist and condensed-matter physicist, was so evidently bored at the dinner in his honor that everyone simply stopped talking to him. Instead, at our end of the table, Saul, Doug, Alan, and I discussed how we would convince Sir Isaac Newton that we were really time travelers from the future. My best answer was that Saul would work out a lot of complicated indefinite integrals. Newton had invented calculus and was thus better at it than anyone then on the planet. But, Saul could easily show himself to be even more advanced—his brain held 250+ years of other people's further developments. Doug thought that I should bring along my amateur's 6" telescope mirror—and then break it, to show that it wasn't really so valuable. (My telescope cost $200.) Newton's most ambitious astronomical reflector was less than two inches in diameter.

Caltech's Commencement Day was on a Friday in the first week of June. I was officially being awarded my doctorate, but I had decided not to go through the ordeal of standing out in the sun in a hot black robe. Saul finished his thesis and passed his oral thesis exam in July.

On August 1, 1973, I became an assistant professor of theoretical physics at Caltech. I had my own parking space, with my name stenciled on it. Margaret still had a year to go to get her Ph.D. We decided to move to the beach, partly for just the novelty of it, partly to make Margaret's commute to U.C.L.A. easier. One weekend we explored a piece of geography along the ocean, one block wide and ten miles long, from the Marina, north through Santa Monica, to Pacific Palisades. Venice Beach, officially part of L.A., seemed like our kind of place, a mix of old slum and new gentrification. If I avoided the worst of rush hour, I would have a forty-five-minute commute to Caltech, Margaret a fifteen-minute commute to U.C.L.A. Within a couple of days, Margaret located a rental house, the lower floor of a duplex at 19 Horizon Avenue, four houses from the boardwalk and the sandy beach, a block from the Pavilion, and adjacent Muscle Beach where weightlifters from nearby Gold's Gym perfected their natural tans and were gawked at by admirers. At the time, this was the one and only Gold's Gym. Arnold Schwarzenegger must have been among the specimens that we gawked at on Muscle Beach, but he was only one of many. The house was newly renovated, the interior done in unfinished redwood planking. We moved in at the end of August.

27. Zel'dovich

The Extraordinary General Assembly of the International Astronomical Union was held in Warsaw, People's Republic of Poland, in September, 1973. The IAU was the international professional society to which astronomers belonged, and it held General Assemblies once every three years. The 500th anniversary of the birth of Copernicus was what made this particular meeting "extraordinary". This was my first trip behind the Iron Curtain. There was only one flight a day to Warsaw from Frankfurt, on Lufthansa. My flight was commandeered at the last minute by a German girl's orchestra on a cultural exchange visit. I was about tenth in line when the woman at the counter announced (in German) that there was only one seat left. "I'm alone!" I shouted, and got the seat.

At the hotel, although I had reserved a room with bath, there were none such. Also, the hotel restaurant was not open for dinner. A tourist guide person in the lobby informed me that the official cashier was closed for the day. "How can I get local currency to buy dinner?" He told me to go to my room and wait. A half hour later, he knocked at my door and offered me 60 złotys to the dollar. The official rate was 33, and black-market exchanges were illegal, but I exchanged dollars for a huge wad of złotys. Immediately, I wondered if I was being set up, if the next knock on my door would be the secret police. Better, I should have wondered if I could have bargained for a better rate—almost certainly the case. Glancing constantly over my shoulder to see if I was being followed, I walked to the Palace of Culture and Science, a thirty-story Stalinist monolith, and registered for the meetings.

On the walk back to my hotel, I was still haunted by my black-market transaction. The same guide was hanging around in the lobby. Perhaps *he* was the secret police. Impulsively I gave him 300 złotys. "For your services while I am here," I said. He took the money and nodded knowingly. It turned out to be a very good investment. At breakfast every day, I was the only person who was brought a bottle of mineral water—everyone else got only questionable tap water or tepid tea. And when there were queues (as there always were, for things like dropping off or picking up room keys, which were not allowed out of the hotel), he would materialize and short-circuit the line for me. On the final day, when people were trying to get to the airport, there were no taxis—except one that he got especially for me. I insisted that Claudio and

Sonia Teitelboim ride with me. Both from Chile, he was my co-equal assistant professor in Wheeler's group at Princeton. The two were unlikely to *ever* get a cab on their own, because, with Latin temperaments, they had crossed swords with my lackey on the matter of a six-dollar overpayment to him. "I will allow you to ride only because you are friends of Mr. Press," my lackey said grimly as he put them into my taxi.

There were two days of meetings in Warsaw, then the invitational two-day "Copernicus" meeting on cosmology in Cracow (other subfields had their own sidebar meetings), then a final day in Warsaw. A highlight for me was meeting the Soviets, especially Zel'dovich and those in his group. Because of his bomb work, Zel'dovich was never allowed to travel to the West, but his travel to satellite Eastern Bloc countries was permitted. I heard him make the same joke in English to several audiences: "I am able to achieve Cosmic Velocity One, but not Cosmic Velocity Two." A full understanding of the joke depended on knowing that the literal "Cosmic Velocity One" in Russian was better rendered in English as "orbital velocity," and that "Cosmic Velocity Two" meant "escape velocity". Zel'dovich was a short, crewcut, bulldog of a man. He looked like a wrestler. In his presence, there was no question about who was in charge. I introduced myself to him in a coffee break. "Just 'Zel'dovich', not 'professor'," he said curtly. He knew about me from Kip. "Go talk to Starobinsky," he said.

Alexei Starobinsky was a thin young man in his early 20s with excellent English but a terrible stutter. He was the most mathematical of the Zel'dovich group and worked, in isolation, on Kerr black holes. He had failed to derive the Teukolsky equation before Saul did, but he had been close. Once he knew about Saul's equation (from Kip), he had arrived at exactly the same coordinate-effect difficulty that had caused us so much grief. Saul had found an auxiliary equation that I had to solve on the computer. Alexei had found the exact solution to that equation. This immediately earned him our respect. Saul, also at the Warsaw meeting, spent some hours off with Alexei over the next several days. I wasn't selling black holes anymore; I was selling Press-Schechter self-similar condensation. I again cornered Zel'dovich. He had heard (from Kip) about this work too. "It's wrong," he said. "Go talk to Doroshkevich."

Doroshkevich was a dour man in his mid-thirties. He spoke only very broken English and looked like a KGB agent—although the absence of KGB in the Zel'dovich group was one of few things in the Soviet Union that could be relied on. Doroshkevich put me off with, "I will talk to you about this at 4:00 yesterday." "Tomorrow?" I asked. (I always confused yesterday and tomorrow when I was learning Russian, so maybe he confused them in English.) "No, *yesterday*," he said, annoyed. He meant "today". We met in his hotel room and communicated mostly by equations and drawing graphs. He started by telling me that my work was all wrong, but, after an hour, he was not so convinced and adjourned our meeting peremptorily. He would think about it, he said. The next morning, Zel'dovich, acting more border collie than bulldog, cut me out of the crowd. "Press, I will meet with you at 7:00 tonight in my hotel room." The day's cultural excursion was to the birthplace of Chopin and included a concert. I helped Jane Hawking with the logistical chore of transporting Stephen. We got back to Warsaw just in time for my appointment—no time for dinner.

I made my way through anonymous crowds to the Warsawa Hotel, where all the Russians were staying, and up to Zel'dovich's room. I was carrying my very American-looking attaché case and felt as if I was acting in some film of diplomatic intrigue—here visiting the father of the Soviet atomic bomb. The hotel room was like a doctor's waiting room: Those with appointments sat on the bed waiting for Zel'dovich to finish with previous arrivals. He was talking to Sidney Bludman, who said that he had written the Soviet Academy to invite Zel'dovich to a neutrino conference in Philadelphia. Zel'dovich thanked him and asked for a written invitation, to be sent to him personally. "I collect them," he said with heavy irony, another joke about his inability to travel. An East German arrived and joined me sitting on the bed. Then it was my turn. Zel'dovich first asked how old I was. He then asked me to explain exactly what I had done in my numerical experiments. I explained. When I started to give my interpretation of the experiment, he interrupted. Exactly like Feynman, he wanted only to know the undisputed facts. He would make his own interpretations. He offered a few comments to show that he now understood the problem, and said that he would think about it. He would talk to me in Cracow, he said. He led me out the door to the elevator, shook hands, and turned, speaking German, to the next person in line on the bed.

So, next in Cracow, I cornered him as the meetings were breaking for dinner. "Have you thought about my problem?" I whined. "Come to dinner with us," he commanded. "Us" was Zel'dovich, Novikov, and Doroshkevich. There was a line at the hotel restaurant and Zel'dovich, impatient, led us out of the hotel and a few blocks to a Mlechny Bar, a cheap restaurant serving dairy and light snacks. The Polish waiter pretended not to understand Russian, so the Russians had to tell me what they wanted to order (in Russian), and I then pretended to translate it into English, which the waiter barely spoke—but he knew that I was American, and he had in any case understood their orders in the original Russian. The Poles hated the Russians almost as much as they hated the Germans. Still, older Poles, when they didn't speak any English, often spoke German, but unwillingly. I learned that, when English failed, I could say, "Ich bin Amerikaner. Nun, sprechen Sie Deutsch?" That first part was essential—otherwise they didn't speak any German either.

Zel'dovich disbelieved the results of my numerical experiment, period. He agreed that there might be some nonlinear power law for clustering, but was sure that the exponent was much less than my claimed value of two. He listened to me for fifteen minutes, then said, "We will bet and then no more discussion." The bet, on whether the exponent was greater or less than 1.5, was a bottle of White Horse scotch, his favorite, against a bottle of Stolichnaya vodka. Also, he said he would like the Beatles record "When I'm Sixty Four," and the sequel to the novel *One Hundred Dollar Misunderstanding*. This novel, by Robert Gover, was a satire on American racism, available in the Soviet Union. (The Soviets allowed only American works that showed the United States in a bad light.) The sequel was apparently unavailable.

Out of the restaurant, we walked along quiet residential streets. I asked his view on Watergate. It would all blow over, he said. It was part of the same political system in which the captains of industry planned and paid for the Kennedy assassination. These same plutocrats would, in the end, rescue Nixon, their man. He and Doroshkevich laughed at my assertion that there was no such organized conspiracy. I tried to explain the magnitude of Nixon's crimes, his attempt to use the CIA domestically as a secret police apparatus. But you already have the FBI as your secret police, he said. That's quite different, I said, thinking of the patriotic FBI television series that Margaret and I used to watch.

"It is beyond the intellectual powers of a Soviet citizen," Zel'dovich remarked drily, "to distinguish between a bad CIA and a good FBI."

There was supposed to be a session of short, ten-minute talks, in theory chosen by the conference organizers, but in practice by Zel'dovich. I was hoping to be one of the chosen, but had heard nothing. After the previous night's dinner conversation, I was not optimistic. Martin Rees usually knew what was going on, and I approached him. He told me that Zel'dovich had actually appointed *him* to choose the talks, but during the whole conversation had addressed him as Joe Silk, creating some ambiguity about who was actually appointed. Silk was a British cosmologist who was long a professor at Berkeley. (He returned to the U.K. in 1999 as the Savilian Professor at Oxford.) We went to Novikov, who, as Zel'dovich's deputy, told Martin to pick the speakers. So, in the end, I did get to talk. Zel'dovich then rose and commented to the audience that my result was wrong, but that the whole thing would be settled by a bet and was not worth discussing further.

I'll have to jump out of time to relate how our bet came out. About a year after Nixon resigned, a visiting Russian brought me a bottle of Stolichnaya vodka, the authentic Russian kind, its label autographed with "For Press, from Ya. B. Zel'dovich". "This was your bet with him on whether Nixon would resign," the visitor told me. Zel'dovich had conflated our two discussions and forgotten which one we had bet on. I had won by mistake. The bet on the exponent was never settled. In fact, cosmology advanced in a direction that made the exponent irrelevant. It was not the part of Press-Schechter that turned out to be lasting; probably it was wrong, as Zel'dovich thought. Luckily no one remembered that this was the part that I had thought most important.

* * *

Tommy Lauritsen, Margaret's father, died on Tuesday, October 16, 1973. He was 57. His memorial service, of course non-religious, was in the Caltech Athenaeum. The speakers were Caltech president Harold Brown, former president Lee DuBridge, Bob Leighton (who, years later, became Margie's second husband), one former student of Tommy's (he had very few), and Aage Bohr, who had arrived from Copenhagen the day before with his wife Marietta. Willy sat in the front row and was notable by his absence from the list of speakers. The simple explanation was that he just couldn't do it. The difference between DuBridge and Brown was that the former made deeply held feelings sound trite, while the latter was the other way around. Aage gave by far the best speech, recalling much of Tommy's liveliness as well as his life. The whole thing lasted about an hour.

Aage took me aside and invited me to visit the Bohr Institute in Copenhagen for any length of time and to be paid by them. I thanked him, but it left me uncomfortable. It was nepotism-by-marriage, nothing else. He wanted to prop up my career the way he (and his father Niels Bohr, and Willy, and Fay Selove) had propped up Tommy's. I did subsequently visit the Bohr Institute a couple of times for conferences, but never more than that.

As an assistant professor I had my own parking space, but I had been assigned a space north of Beckman Auditorium, as far away from my office as was geometrically possible given the shape of the campus. I had complained (of course) but to no avail. A couple of weeks after Tommy died, I got a call from a woman in the parking office. I was lucky, she said. A space close to my office had become available, and they were reassigning me. Until they repainted the name, she said, I should look for the space marked "T. Lauritsen."

That last spring and summer in California, I was busy with things that, as far as my career was concerned, were off the main road—maybe even were dead ends. My minimal effort approach as an undergraduate physics major at Harvard had relied on problem books in various areas of physics, so I felt something of a duty to write one on gravitation physics for the benefit of future lazy students. I coerced Saul, Alan Lightman, and (remotely in Utah) Richard Price into becoming co-authors. The activity was not useless—it could be a worthy (lesser) companion book to the magisterial *Gravitation*, by Misner, Thorne, and Wheeler, which had finally come out as a massive volume.

Not useless, but not career-enhancing either; and not something that Saul and Alan should have undertaken before finishing their theses. We did it anyway. For one month, we were each supposed to come up with ten problems a day. We didn't have to solve them at this stage, just make them up, or find them by combing through textbooks and old journal articles. Many came from Kip's course. Sometimes, at lunch, I cajoled Dick Feynman into thinking of problems. By month's end, no author actually had met his daily quota, but we did finish with more than 600 candidate problems. Then, we spent the next year (part time) dividing up the work of producing good solutions. Many problems were lost on the cutting-room floor. The book, as finally published by Princeton University Press in 1975 and still in print today had 475 problems, all with (we thought modestly) excellent solutions.

Kip and others thought that I was similarly nuts to write a paper on how the size of a human (call it two meters in height) is not arbitrary, but, in order of magnitude, is uniquely determined by fundamental physical constants. There was a long history among physicists in thus estimating—purely recreationally—various quantities. The sun has the mass it has (for example) because if it was much less massive, it wouldn't be luminous enough to support life on Earth; much more massive, its lifetime would be too short. One uses physics to convert arguments like this into numerical expressions involving the mass of the proton, the speed of light, Planck's constant, etc. My paper was published in a journal for college physics teachers, earning only scorn from Kip.

But I was unexpectedly vindicated by the arrival at Caltech of Viki Weisskopf, MIT's most famous elder theoretical physicist, to give a physics department colloquium. Weisskopf had worked as a young man with the greats of modern physics: Werner Heisenberg, Erwin Schrödinger, Wolfgang Pauli and Niels Bohr. During World War II, he was a group leader on the Manhattan Project in Los Alamos. He was *almost* a co-discoverer of quantum electrodynamics (along with Feynman and Schwinger): His calculation of the measured Lamb shift gave the correct value; but he was not sufficiently confident in its foundations to publish it.

A special luncheon in the Athenaeum private dining room was given in Viki's honor. "Oh, I have read your paper on the size of man!" he said when introduced to me. Heads turned to listen. "I've been working myself along those lines," he continued, "but I didn't get as far as you did." No one knew what we were talking about, but they were very impressed that I was ahead of Weisskopf in something. Murray Gell-Mann had been Weisskopf's student and made a point of coming up to me and shaking my hand warmly.

Later, there was a luncheon engagement with an associate editor from *Reader's Digest* who was traveling around doing a story on black holes. The Digest's 40 million readers seemed to pull a lot of weight at Caltech; this junior editor was being squired by the Caltech news director, who picked up the tab for our lunch at the Athenaeum. The editor said that he wanted to take the "gee-whiz" out of black holes, and to make them something that the average American could understand, like electric toasters. "Good idea," I said. He felt he must do this soon, because, he said, this bandying about of mystical-sounding scientific terms was promoting a resurgence of interest in the occult, especially witchcraft and psychic power, and this, if not halted, would lead to the rise of new fascism. "A very interesting perspective," I said. Then, he got down to the first electric toaster question about black holes: "Tell me, Dr. Press, do you think that black holes support the Kantian idea of absolute reality?"

28. Princeton

From the very first, we didn't like Princeton, and it didn't like us. University and town were obsessed with the picturesque and quaint. It was a matter of pride that faux-colonial Nassau Street had changed not a bit since the time of Woodrow Wilson and F. Scott Fitzgerald. Jadwin Hall, the relatively new physics building, modern in architecture, was banished half a mile down Washington Road toward Lake Carnegie. Jadwin and neighboring Fine Hall (mathematics) won several architectural awards. I learned later that John Wheeler had been instrumental in pressing the University to depart from the prevailing neo-Gothic in building them, and in including spaces for modern outdoor sculpture in the design. The Wheelers' own home was a striking Mid-Century Modern; he had collaborated with the architect in its design. It was years before, confronted with many such facts, I could admit that Wheeler was not in all matters the conservative defender of the Old Guard that I took him for. He was a radical. Or was he? That was the paradox.

Hibben Apartments, situated another half mile away on the lake, embodied in brutalist concrete the University's stratified academic structure. Hibben (with its neighboring clone, McGee) comprised apartments rented to lecturers and assistant professors, the junior faculty with very little chance of achieving tenure. We were itinerant scholars whose fate was to pass through the university leaving behind no ripple on Lake Carnegie, no academic graffiti that could not easily be removed. We were expected to be grateful for the opportunity to have been at Princeton, adding its prestige to our academic resumes, and departing with fond memories of watching, from our concrete-walled apartments, the geometrical wakes of the undergraduate sculling practice. This just didn't work for Margaret and me.

I made some mistakes right away. On a hot day, I was in shirtsleeves and tennis shoes when I presented myself to the University personnel office to sign the onboarding forms. Logically enough, I was shunted to a line of tie-and-jacketless new employees—the mail-room clerks and computer operators—with then strained apologies when, at the front of the line, they discovered that I was an *assistant professor*. In Jadwin, when my boxes of books started arriving on the loading dock, I made the mistake of borrowing a hand-truck and moving them myself to my office. "We do have help for that kind of thing, Dr. Press," Mrs. Witt,

John Wheeler's kindly, elderly secretary, solemnly rebuked me. That help was all Black, or, as they were still called in Princeton, ten years after the Civil Rights Movement, Negro.

I don't remember who provided me with this useful mental construct: Princeton is two-hundred miles south, and twenty years back in time, from where you think it is. In 1975, no less so than in 1875, Princeton was where the prominent families of the South sent their sons to be educated—a halfway house to the industrial and professional guilds of the Northeast. Harvard and Yale were bridges too far. Woodrow Wilson, Princeton's greatest president before becoming governor of New Jersey and then President of the United States, grew up in Georgia and was at heart an antebellum Southerner. New Jersey was the last of the Northern states to abolish slavery, and it did so, in 1846, by converting all slaves to the status of indentured servants who were "apprenticed for life". Anti-Semitism had its own place in the Princetonian tradition. Murph Goldberger, my department chairman in physics, was Jewish. Extolling everything good about Princeton, he told me that, *already a decade ago*, Jews were admitted to the Springdale Golf Club, home turf of the university's varsity golf team, whose land was owned by the university.

The real business in the physics department was done in senior faculty meetings with junior faculty excluded. Instead of faculty meetings, instructors and assistant professors were expected to attend the weekly Joseph Henry Club (so-called), a forum for lesser issues such as whether the quota for graduate student use of the Xerox machine should be raised from five dollars per student (per year) to ten. Because of its guaranteed captive audience, the Joseph Henry Club was a useful venue for seminars by visiting Great Men for whom a regular Physics Department Colloquium might prove embarrassing—this because many Great Men had nothing, really, to talk about. Once I figured this out, I started sitting at the back of the room, next to a door that led to a small kitchen. From there, another door opened to the hallway. The kitchen door was kept locked, but I had discovered that my building key could quietly unlock it. I escaped from several embarrassing talks—discreetly I thought—through that door.

I was noticed, apparently. Mrs. Witt came to me one day to tell me, apologetically, that a mistake had been made in issuing my building key—which she insisted on exchanging on the spot for a different one. I thought nothing of this until I next tried to escape the Joseph Henry Club, and found that my key no longer opened the secret exit door. Embarrassingly, I was caught out, fiddling with a lock that wouldn't open. Nothing in this event pointed back to John Wheeler—except the style of the exploit. Somewhat akin to Kip's plan to "motivate" me, Wheeler planned to train me to act Princetonian—even if only for the short period that I was expected to be there. He was less open about his plan than Kip had been, but there were times that it came through very clearly. "Janette and I moved many times in my career," he told me. "And every time we arrived at a new place, we *acted like* we were planning to stay forever," the italics, and the message, were clear.

Margaret and I settled into our apartment in Hibben and awaited the birth of our baby daughter. Hibben was an eight-story rectangular box, and each apartment was a duplex with an interior staircase; so the levels with exterior walkways were numbered 1 to 4, with apartments A to T down the length of each straight walkway. The architecture was a scaled-up rabbit hutch where, it was commonly joked, junior faculty were put to breed. We were in Apartment 3M. On the side facing Lake Carnegie, a communal laundry room and a children's play area opened onto a pleasant grassy area that sloped gently to the shoreline. There were benches and swings, and no shortage of advanced degrees among the women doing laundry and minding children. Margaret's Ph.D. in linguistics fit right in. The side of the building away from the lake faced a parking lot and a walkway that crossed Faculty Road and continued up to the campus. At 7:50 every morning, this walkway was a moving line of assistant professors in jackets and ties (with or without overcoats, depending on season) on their way to teach 8:00 classes. One of them—I never found out a name—used to sing an original ditty, "Hippety-hoppety, hippety-hoppety, the Hibben rabbits go to work." His clear baritone carried across the parking lot.

Moving in was a matter of unpacking boxes and trips on Route 1, north to Sears in New Brunswick and south to Mrs. G.'s Appliances in Lawrence Township—the apartments didn't come with refrigerators for some reason—and to a place that sold cheap, unfinished furniture that I could stain and varnish myself. There was a series of largely pro-forma welcoming events, a tea at John and Neta Bahcall's (they, at any rate, a breath of fresh air in Princeton), and a physics department party at the Goldbergers, at which John Wheeler took Margaret and me around to meet the department's Great Men.

That event was educational in itself. Margaret was accustomed to being introduced first, so she held out her hand to each, announcing, "Hello, I'm Margaret Press." After two such, Wheeler gently corrected her with, "This is William Press, and this is Mrs. Press." Over the first months, there were many similar gentle corrections. We took them in stride, attributing them to Princeton's greater formality (re Pasadena) and the more rigid social stratification. It didn't occur to us that there was another effect: At Caltech we were royalty—the princess and prince who united two Caltech royal families. In Princeton, I was just another transitory assistant professor, with Margaret my invisible and pregnant wife.

The rigid pecking order was this: Tenured faculty at the top. Then, undergraduates. The university's primary purpose was to give its undergraduates the full Princetonian experience. Everything else, including any accomplishment as a research university, was secondary. After undergraduates in status came the long-serving administrative staff, people like Mrs. Witt. Then assistant professors and lecturers. At the very bottom: graduate students. And below the very bottom: the *wives* of junior faculty, who were expected to be rarely seen, never heard, and to leave academic house arrest in Hibben or McGee only for driving to the northern edge of town to the supermarket in the Princeton Shopping Center. This did not sit well with Margaret.

Physics faculty at Princeton, as at other first-rate places, had very light teaching loads. My course on general relativity met three times a week. I had twenty or so graduate students, plus six advanced undergraduates. The graduate students were docile and merely took notes. I didn't have to grade them: Princeton's system, more extreme even than Caltech's, was to assess their progress solely on the basis of candidacy exams taken at the end of their first year. My six undergraduates were troublemakers, and I loved them for that. Their mutiny committee (as I called it), Messrs. Grossman and Black, visited me frequently to question my professional ethics in general and the need for some particular assignment in particular. It was only five years since I had led the corresponding mutiny committee in Dick McCray's astrophysics course at Harvard.

To this day, I don't understand John Wheeler's worship of Great Men—a term that he actually used. I respected the historical figures in physics for their famous discoveries. I was awed by some that I had interacted with—Feynman, Chandrasekhar—even by some whom I disliked personally, like Murray Gell-Mann. Stephen Hawking was not yet a Great Man in the Wheelerian universe, but it was already clear that he was going to be. I was with Chandra in believing, as in his story about Bohr, that genius was a transitory thing, and that Great Men should, at some point, be respectfully let out to pasture. Chandra hated Wheeler's apocalyptic style. He told me that the guru who, in Indian fables, goes into the woods to become wise by contemplation, teaches his acolytes to be in harmony with nature. Chandra's interpretation of this for the modern scientist was: Do work that is harmonious with your time, not work that attempts to be disruptive and is most likely wrong. That was his view of Wheeler's work. In truth, I could see merit in both sides of the argument.

Wheeler led a weekly group meeting that included his undergraduates (those being advised on their senior theses), graduate students, postdocs, and his two assistant professors, Claudio Teitelboim and me. One week, Wheeler assigned Thibaut Damour to report to the group on a paper that put forth an alternative to the big bang theory—complete with alternative explanation for the three-degree cosmic microwave background. If viable, this would be important. Thibaut was a French postdoc supported by the French government to spend two years in Wheeler's group. It was clear even from Damour's languorous presentation that the paper was crackpot, "not even wrong" (the ultimate insult among physicists). "Who wrote this garbage?" I asked politely, after airing some more objective criticisms of the paper.

Wheeler stared at me. "Klein," he said, and added, "perhaps you've heard of the Klein-Gordon equation or the Klein-Nishina formula." Yes, I had. With those and many other discoveries to his credit, Swedish physicist Oskar Klein was certainly a Great Man. He had retired in 1962 and was now in his eighties. "If he's wrong, let's write him a letter and tell him so," Wheeler said, adding, "I'll sign it with you." I didn't know whether he was calling my bluff ("if you are critical of a Great Man, you should have the courage to confront him"), or just seizing on a teachable moment for his undergraduates ("this is how science is done").

So, I took the other side: If a person like Klein wanted to pretend to be still making great discoveries, and if it wasn't obstructing the advance of reputable work, then why should we puncture his balloon. "For scientific honesty, of course," Wheeler said. Those were strong words directed against me. Johnny was on a tear. He picked up the phone and dialed Mrs. Witt, then dictated, for all to hear, a pedantic letter—quite insulting to Klein—with each sentence followed by boxes for him to check, true or false, and then return the letter "by way of letting us know your thinking on this matter". I don't think the letter was ever sent; I know that I never signed it.

After Wheeler put down the phone, I related what I had witnessed when Feynman confronted Heisenberg on the latter's crackpot theory. Feynman had reacted only after being goaded in public, and I thought it also pertinent that Heisenberg had been a Nazi. (I didn't know anything about Klein's politics.) Wheeler said that Heisenberg had visited Princeton shortly after Caltech and had given the same talk. "Heisenberg came away from Caltech thinking that Feynman was 'a mere bourgeois physicist' who didn't appreciate the depth of his work," Wheeler said. It is physics lore that Wheeler himself referred to Feynman, his greatest student, as "a mere bourgeois physicist," adept at calculating anything, but unappreciative of Wheeler's grand conceptual vistas. So, it is possible that, in regard to Heisenberg, Wheeler was putting his own words into the latter's mouth.

The invisible elephant in the room was not Klein's work, nor Heisenberg's, but Wheeler's own. His principal scientific interest had now become human consciousness, and whether consciousness had a role in understanding "the collapse of the wave packet" in quantum mechanics and, if so, whether the whole universe might have come into classical existence only when first "consciously" observed by intelligent life. To an earlier group meeting, he had invited two psychologists from Harvard who were supposed to be experts on consciousness. They had become restive, and then visibly eager to escape, when he had surfaced his quantum ideas. This same elephant was in my office later that afternoon when Wheeler appeared there. Shutting the door, he said, "I hope that you don't view my thinking about consciousness as being in the same category as Heisenberg's unified field theory, and I hope that you will tell me if you think it's wrong."

"Oh, no," I said, trying to be tactful, "your work is not in the same class as Heisenberg's."

Once we were fairly settled, we invited the Wheelers to dinner at our apartment, along with Vladimir Braginsky, the Soviet experimental physicist whom I knew from Caltech, and who was visiting Robert Dicke's group in Jadwin Hall. Braginsky was an outstanding physicist, but also a doctrinaire, humorless Communist party member. The evening alternated between us baiting him ("Why is Zel'dovich unable to travel?") and Janette Wheeler baiting us: "Don't you agree that household staff are like children—they love to receive colorfully wrapped little presents?"

I don't think that John Wheeler ever liked me. I don't think he liked me before he hired me, during my two years with him, or afterwards. In his view, I wasn't even up to being a bourgeois physicist. I wasn't even up to being not even wrong.

* * *

Astrophysics, relativity, and cosmology events at Princeton—seminars, colloquia, informal discussions with visitors—rotated among three epicenters, the astrophysical sciences department (Peyton), the physics department (Jadwin), and Bahcall's group across town at the Institute for Advanced Study (IAS). This was a single community—it was the same audience at all the events. The highlight of the week was Astrophysics Lunch, on Tuesdays, in the large private dining room of the IAS cafeteria. Students were excluded. Thirty or so faculty and postdocs sat around the outside of a large horseshoe of tables.

The immortals, Lyman Spitzer and Martin Schwarzschild, sat at the central position, flanked by the demigods John Bahcall and Jerry Ostriker. Spitzer and Schwarzschild together ruled Princeton astrophysics for a generation, passing the torch to Ostriker only in the early 1980s. Spitzer was permanent head of the Peyton department. Born in the Midwest, he was educated at Andover and Yale. In the 1950s, he founded the field of plasma astrophysics, the study of ionized gases and magnetic fields in deep space. He was athletic and a mountain climber. Schwarzschild, a gnome of a man, witty but with a heavy accent, was born into a distinguished German-Jewish academic family—his father, in fact, was Karl Schwarzschild, discoverer of the first black-hole solution to Einstein's equations. Schwarzschild the son emigrated to the United States in 1936, via Norway, and served in U.S. Army intelligence during the war. His famous work on stellar structure and evolution in the 1950s added detailed physics to Chandrasekhar's 1930s mathematical treatment—in Chandra's time, the nuclear reactions that power stars were still unknown.

The Tuesday lunches were conducted like a court martial with those four judges presiding. One never knew *who* would be on trial until one showed up. If a visitor was present, that person was usually the victim. Schwarzschild or Bahcall would lay the trap: "Tell us about something you are working on—not finished work, but just what is on your mind." The visitor would mistake this for a friendly invitation, the chance to pick the brains and get feedback from the famous and up-and-coming scientists in the room. Far from it. Allowing only a brief opening testimony, the presiding judges would interrupt to call on others in the room to criticize, rebut, or belittle whatever the visitor had said. Then, the visitor might or might not be allowed to defend himself. It was brutal, and knowingly malicious, justified only by the fact that everyone in the room always learned a lot—at the visitor's expense.

If there were no visitor, then a similar treatment was given, without warning, to one or more of Princeton's own, usually assistant professors like me. The trap here was, "What do you think about…" some recently announced discovery in any area of astrophysics from any place in the world. If you didn't have an answer, the judges would shame you publicly for your ignorance. After I had gotten this treatment a couple of times, I started getting stomach pains even before lunch while going through the cafeteria line, anticipating that I might be called on. After I mentioned this privately to John Bahcall (in his role as my mentor), he sometimes phoned me an hour before lunch to tell me that I might be called on—and, crucially, what topic I should quickly learn something about.

In later years, when I was established at Harvard, I liked to visit IAS for a week at a time, as a quiet place where I could get work done. Schwarzschild and Spitzer were by then no longer active. If both Jerry and John were to be out of town on the Tuesday, John would designate me to preside over the lunch and call on people. He well remembered how much I had hated being on the receiving end.

Our daughter Sara was born in late November, 1974. In mid-December, Margaret and the baby flew to Pasadena to stay with her mother, and I flew to Dallas for that year's Texas Symposium on Relativistic Astrophysics. All fall, the big excitement was the discovery, by Joe Taylor at Amherst, of "the binary pulsar" (for which they were later awarded a Nobel Prize). This was an object no one ever expected to see: two neutron stars, each more massive than our sun, in a stupendously close orbit—circling each other every eight hours—and with one of the stars rotating on its axis twenty times per second and beaming radio pulses in our direction at that rate. The binary pulsar was a system where the effects of General Theory of Relativity were potentially a million times larger than could be observed in our solar system. The pulsar's rotation furnished a built-in, near-perfect "clock" by which the Einsteinian effects could be calibrated and measured. The binary pulsar was God's gift to Einstein, even if it arrived twenty years too late for him to know about it.

Practically every theorist I knew was, that fall, writing a paper on the binary pulsar, calculating one or another of the subtle effects—tests of Einstein's theory—that were now suddenly accessible to quantitative observation. Saul, with Ed Salpeter at Cornell; and, also at Cornell, Tommy Gold, working alone secretly. In Princeton, Jerry Ostriker, Larry Smarr, and Roger Blandford. And Bob Wagoner and Cliff Will at Stanford. For a while, I was a collaborator on the Ostriker-Smarr-Blandford effort, but by December had removed myself. I wasn't pulling my weight, and I was privately repelled by the feeding frenzy of hurried, and scientifically sloppy, competition. I wanted the practice of science to emulate Chandrasekhar: I wished that everyone would take a year or two, work everything out carefully and correctly, and then publish definitive results. Instead, I was in a ringside seat on the competitive chaos: Bahcall was the associate editor for *Astrophysical Journal Letters*, where most of the quickie papers were being submitted, and I was his favorite anonymous reviewer. I accepted some papers and rejected others.

It got crazy: At the Dallas Sheraton, I introduced Saul and Roger Blandford—they had never met—and they immediately discovered that they were writing competing papers about the same relativistic effect. Still, it was in the interest of both to share their results—which turned out to disagree. Saul took Roger's calculation back to his hotel room and soon found the error. His own calculation was the correct one. Roger, however, when he worked through Saul's calculation, discovered what Saul had missed: Although formally correct, the most important effect was exactly degenerate with uncertainty about the binary's Newtonian orbit—meaning that it could not be measured at all in less than a century. The upshot was that, instead of writing competing papers, they wrote one together, disproving the effect that both initially sought to claim credit for. Saul and Roger explained this result to Joe Taylor at lunch the next day—he was impressed. Then, the middle of that same afternoon, Taylor sought them out to show them a letter from Zel'dovich that had just been hand-carried from Moscow by David Pines: Zel'dovich had also found the same negative result.

This period was formative for me. I learned that, despite the acknowledged thrill of beating one's competitors in the quick sprint—which I had previously tasted in my work on gravitational radiation and Kerr stability—that wasn't the kind of science that I wanted to do. I wanted to have *fun*, and, for me, that meant doing things that were creative, or off-beat, or elegant, or unexpected—and not necessarily fast, nor immediately competitive, nor crowd-following. So (spoiler alert?) right there, I lost any chance of being the kind of scientist who could win a Nobel Prize. I did publish one (sole-author) paper relevant to the binary pulsar, an esoteric discussion of how pulsar data might be analyzed for maximal accuracy. The paper was subsequently cited by others exactly once—by Roger Blandford, who referenced it only to dismiss it as having made unrealistic assumptions.

From Dallas, I was flying to L.A. to spend the holidays with Sara and Margaret at her mother's house. When the flight was diverted to San Diego because of fog at LAX, Kip, Sandor, Willy Fowler, Frank Estabrook, and I grabbed a lucky rental car and crowded in for the two-hour drive to Pasadena. In the back seat, Willy passed around his silver travel flask of bourbon—he always carried one—to Estabrook and me. Sandor was driving, but Kip, in front, uncharacteristically refused to join in.

Then, at the end of the month, I was back in Pasadena for the annual Lauritsen Memorial Lecture. The Lauritsen speaker was John Wheeler. He and I were installed in adjacent visitor offices in Kip's area. I cornered him and asked point-blank, what was my long-term future at Princeton. He made general remarks about the department's desire to promote more *second-term* assistant professors than in the past. I was in my *first* three-year term, and by now I understood that John conveyed negative information at best obliquely. I was already not expecting to remain in Princeton for a second term. Dave Schramm and Dave Arnett, at Chicago, had invited me to give a colloquium and were hinting that it was more than a casual invitation. Evidently, they were getting things lined up to make me a tenure offer. Since Schramm and I had shared his large tent at the 1971 Sierra Conference, when he was Willy Fowler's postdoc, he had become an assistant professor at the University of Texas at Austin, then a tenured associate professor at Chicago. Schramm's meteoric rise was helped by push from Willy (at Caltech) and pull from Chandra (at Chicago), not to mention Dave's own friendly, larger-than-life, physical presence

Kip introduced Wheeler's lecture: "Although John has been sounding crazier and crazier in recent years, it is because he dares to think about problems too difficult for any of the rest of us whose careers are on the line." Wheeler beamed.

29. Metrics

By spring, when I had been at Princeton for a little more than six months, I had acquired all the trappings of a proper Princetonian assistant professor. I was teaching undergraduates and graduate students. I had a graduate student of my own, Paul Wiita, who was struggling to make sense of Chandra's papers on incompressible ellipsoids so that we could apply them to more realistic models of rotating neutron stars. I had an undergraduate senior-thesis student, Adam Burrows, who seemed terrifically lazy, but eventually presented me with a full-blown thesis that neither I, nor Tony Zee, the other member of Adam's committee, could make any sense of. We didn't recommend him for a prize. Adam did his Ph.D. at MIT and eventually returned to Princeton as a full professor, doing pioneering work on exoplanets and brown-dwarf stars.

I was coming up with an abundant stream of research ideas to work on. It was not obvious to me then that these new research ideas were not very good; that realization came only later. Scientists—physicists and mathematicians especially—love to rank each other. "Who is better than me?" "Is Feynman smarter than Gell-Mann?" Alicia Ostriker, when she accompanied Jerry and colleagues to Lahiere's (the de rigueur Princeton restaurant for entertaining distinguished visitors), liked to keep track of how long it took before the conversation turned inevitably to the ranking of colleagues. On average, it was twelve minutes. Ranking colleagues was the astrophysicists' favorite blood sport.

Until the 1970s, such orderings were necessarily subjective. That changed when computer databases made it possible to know not just how many papers a researcher had published, but how many times those papers were *referred to* ("cited") in papers by others. This so-called bibliometrics was—still is—controversial. A wrong paper that everyone in the field likes to take a swipe at can garner a large number of citations, since bibliometrics can't easily distinguish between "good" and "bad" citations, "important" citations or "just-in-passing" ones. One of my papers with Bardeen garnered (over time) 1,726 citations because it happened to be a good compendium of facts about the Kerr metric—but was otherwise quite forgettable.

Nonetheless, citation counts do say *something*. While I was at Caltech, I wrote (or co-authored) papers ultimately cited hundreds and in a couple of cases thousands of times. At Princeton, my not-very-good research ideas—all of which seemed compelling at the time—resulted in papers cited, in some cases, only once or twice. I wasn't doing sloppy work: Re-reading these papers now, I am struck by how pretty they were mathematically—and how irrelevant they turned out to be. Papers don't get cited just for being pretty. Much later, I managed to write a paper that was *never* cited, not even by me.

It was not obvious to my professional boosters (Schramm, Chandra at Chicago; George Field at Harvard; Kip at Caltech) that there was little of importance in my new work. No less in science than elsewhere, you can fool all of the people for a while. John Bahcall wasn't fooled, I think, but it didn't matter. As already mentioned, it was characteristic of John to make snap judgments about people: Either he loved you or he hated you—and, once set, his judgment never changed. He loved me and would write over-the-top glowing letters about me no matter how badly my work went.

I visited the University of Chicago in March, 1975. At the last minute the department Chair, Gene Parker, insisted that Margaret and baby Sara come too, a sure sign that they were going to offer me a job. Schramm met us at the airport and gave us a tour. The Hyde Park neighborhood around the University had only just ticked up toward eventual gentrification. On the one hand, you were safe on the street at night only in the defined area that University police cars patrolled. On the other hand, the real-estate bargains were spectacular. The Schramms paid $35,000 (about $170,000 in 2020 dollars) for a 3000 square-foot high-ceilinged apartment in a beautiful old building. The university put us up overnight in the Quadrangle Club, on campus. The next morning, Parker handed me the offer letter: associate professor with tenure (promotion to full professor a formality after just a couple of years) starting at twice my Princeton salary.

Chandra was by now working on his book on black holes, and he was effusive in complementing me on my black-hole work—his contribution to the recruiting pitch. Yet, interestingly, when the book finally came out in 1983, its references to my work were disappointingly sparse, often replaced by references to later work by Steve Detweiler, Chandra's own student. Saul fared much better than I did. At this meeting, however, it was all praise. Chandra then brought up another matter that I had planned to leave alone: My student, Paul Wiita, had found an actual mathematical error—not just a typo—in one of Chandra's papers on incompressible ellipsoids. That *never* happened, and there had been an exchange of correspondence before Chandra had acknowledged that it was, in fact, an error.

Now, he made oblique amends with the animated telling of an Indian fable: There was a prince who never told a lie, and his chariot rode a foot above the ground. Then, in a battle, his advisors induced him to call from a tower that the enemy leader was dead. The statement was not a lie because there happened to be a deceased elephant of the same name. The prince's forces beat the demoralized enemy, and the prince's record for truth was in principle unspoiled. But his chariot was not fooled and now clunked along at ground-level.

I had this image still in mind when I met with postdocs Bob Wald and John Friedman. They seemed glum. They were my contemporaries, but their professional chariots were just clunking along, while mine was floating. They clearly didn't think I deserved to be the prince of relativistic astrophysics. I had started with all the advantages—undeserved. Plus, what I was actually better at was (from their perspective) merely self-promotion. Notably, I was being merchandized by Schramm, himself the emperor of self-promotion. It didn't seem fair. I privately thought they were right. It wasn't fair. But it was fact, and I was going to take advantage of it.

There was to be, at the end of June in beautiful Varenna, Italy, an international conference on black holes and neutron stars. Attendance was by invitation only—it was one of a prestigious annual series labeled "International School of Physics *Enrico Fermi*". I wasn't invited. Neither was Larry Smarr, nor Saul, nor, for that matter, any former members of Kip's group, nor anyone associated with Chandra's group in Chicago. It was not hard to understand why: The conference was organized by Remo Ruffini. We, on Remo's enemies list, were cut out. When Claudio Teitelboim happened to mention that one of Wheeler's grant accounts held $1,200 in unallocated funds, he and I hatched a plan: We would hold a competing conference in Princeton, "International School of Physics *Joseph Henry*". Ours was low-budget, but it was also invitational: You had to be under the age of 30—and friends of Remo need not apply. You also had to be willing to sleep on the floor in a sleeping bag, many to a room. We attracted 20 participants, a number of whom later remembered the event fondly when, years afterward, they had become famous professors. The talks were good, and the conference banquet was a hot-dog roast on the lawn behind Hibben Apartments.

Apart from Varenna, Ruffini was not advancing much professionally. Martin Rees and I discussed this late one night in Martin's visitor office at IAS—he seemed to always be there, working day and night. Martin felt that Remo did not deserve to be—alone among all active relativists—unemployed. "After all, his only crime is selling his own work too aggressively, and I shouldn't think that any of you Americans would hold *that* against him." Martin was always (then and now) anti-American in this way, but he never let it get in the way of friendships with individual Americans.

An exclusive summer conference that I *was* invited to, thanks to Kip and Zel'dovich, was to be held at a resort on the outskirts of Moscow. It was cancelled at the last minute. There were supposed to be just seven Americans (Kip, David Pines, Jerry Ostriker, John Bahcall, Fred Lamb, Jim Bardeen, and me) and, very unusually, we were encouraged to bring our families. Kip's Soviet friends told him that there were two theories about why it was canceled. Either it had been supported by the losing side in the internal political struggle over who would succeed Academician Mstislav Keldysh as president of the Soviet Academy of Science. Keldysh was three times a Hero of Socialist Labor and a key figure in the Soviet space program. He was a member of the party Central Committee and a delegate to the Supreme Soviet. In the event, he was succeeded as president of the Soviet Academy by Anatoly Alexandrov, who is now remembered principally as the designer of the ill-fated reactor at Chernobyl. Or, the other theory, the conference was cancelled by the KGB because of the large number of Jews on the Soviet organizing committee (not to mention among the invited Americans).

So, as in earlier summers, Margaret and I went to Cambridge, England instead, this time with Sara, now eight months old. King's College rented us a flat at No. 8 Whichcote House, Springfield Road, that was by British standards palatial— full large kitchen, two bedrooms, "lounge", study, all electric, and wall to wall carpets, for £20 a week. Sara slept in the pulled-out bottom drawer of the bedroom dresser.

It was an exceptionally good summer at the Institute of Astronomy; or maybe it just seemed so because—outside of Princeton—I was now known as a real person, on my way, it was thought, to a professorship at Chicago. Martin Rees was by now out from his Sussex exile and ensconced as the Plumian Professor (Fred Hoyle's old position). Willy Fowler no longer visited, but in his place was a newer crowd: Director Donald Lynden-Bell and locals Gary Gibbons, Bill Saslaw, Michael Rowan-Robinson, Bernard Jones, Peter Meszaros; visitors Jerry Ostriker, Jim Gunn, Wal Sargent, Joe Silk, Virginia Trimble, Bohdan Paczynski, Beatrice Tinsley, Richard Gott, Vincent Icke, Gary Steigman, Margaret Geller, and many others. Rumors had swirled that Stephen Hawking was going to leave Cambridge for a professorship at Caltech, where he had just spent a year on sabbatical in Kip's group. But by summer, 1975, he was back in Cambridge, though only as a Reader. He was belatedly awarded a professorship at Cambridge in 1977, and became Lucasian Professor (the chair held by Newton, Airy, Charles Babbage, and Paul Dirac, among others) in 1979.

Beatrice Tinsley, I knew, was one of the two candidates with whom I was competing for the professorship in astronomy at Harvard. There was well-founded pressure there to hire a woman, and she was excellent. With Gott, Gunn, and Schramm, she had recently published (their names in alphabetical order) a paper with the then-provocative title, "An Unbound Universe?" In a cheeky tone, but with irrefutable scientific rigor, the paper brought together and analyzed the evidence that there was, in fact, not enough matter in the expanding universe to bring about its eventually turn-around and collapse. There was not even enough matter to significantly ever slow down the expansion. The universe was manifestly "open", not "closed". This conclusion was counter to what most observers (outside of Caltech) thought was true, and counter to what most theorists *wanted* to be true. The paper was instantly famous and, over succeeding years, it proved to be prescient. The next big piece of that puzzle waited for a quarter-century, until two groups found independently that, far from slowing down, the universe was actually accelerating. (More on that later.)

Rich Gott, Beatrice's co-author, was my former classmate at Harvard. He had gone on to do his Ph.D. at Princeton with Spitzer, then to Caltech for a postdoc with Gunn, where we overlapped for a year. From undergraduate days on, Rich was one of the most eccentric people I ever knew. No, he was the *most* eccentric—excluding those belonging in the psychiatrists' Diagnostic and Statistical Manual. But I say this with affection: I loved talking to Rich. It was prerequisite to have time on one's hands. He was a Southerner with a slow drawl who liked to examine any issue at length from all possible sides and angles—and, in his case, from more than three dimensions: Time-travel was one of his favorite topics. Another was the famously unsolved four-color map theorem. As an undergraduate, Rich had discovered a proof that turned out to be the same as one discovered by Kempe in 1879. That proof had stood for ten years before it was shown to be incorrect. "Ten years!" Rich would say. "Ten years! If I had lived then, I could have been a famous mathematician for ten years!" Over time, Rich's eccentricities took a toll on his scientific reputation, but at the time, in 1975, he was a rising star. Gunn was his strong supporter.

If Tinsley was the principal obstacle between me and a Harvard tenure offer, Gott was the principal obstacle to my being promoted to tenure at Princeton. I knew from John Bahcall's continuing indiscretions to me that the physics department (Murph Goldberger, chair) had come up with half a position for me, conditional on the astrophysics department (Lyman Spitzer, chair) coming up with the other half of a joint appointment. Astrophysics did have a full position available, but Spitzer wanted to hold it intact for Gott's eventual return. While the matter was being settled, Gott was rusticating in Cambridge as a Fellow of Trinity College. Jerry Ostriker was said to be balancing in the middle. He was Spitzer's impatient heir-apparent and thought—I was told this only many years later—that letting Lyman appoint a protégé would accelerate the old man's transition to emeritus.

That summer, a glimpse of this intrigue came through when Jerry took me aside to give me his take on the Gott-Gunn-Schramm-Tinsley paper—not the science, but the personalities. He wondered who actually *wrote* the paper. "Schramm has the animal energy, of course," he said, "but I doubt that he knows how to write." I mentioned that Gunn could write elegantly. "No, I think that it was written by Gott," he said. Beatrice, Jerry thought, was the fall guy, the one to take the blame if—as he thought—the paper was wrong.

But the paper was not wrong! Tragically, Tinsley's life ended before she could get the credit she deserved. Unable to find a job with her husband in Texas, she had divorced him and moved to Yale in 1974, in the process losing custody of her two children. In 1978 she was diagnosed with cancer. She died in the Yale infirmary in 1981 at the age of 40. That the universe was indeed "unbound" became established only in the late 1990s.

I didn't know, that summer, that my candidacy at Princeton had already imploded. I learned the details later from, naturally, John Bahcall. Murph and Lyman had a meeting with the Dean, who told them that it was up to them to divide the baby (me) one way or another. John met with Lyman and, sensing some flexibility in the latter's negative position, urged him to make calls about me to Thorne, Goldreich, and Chandra. But, at a meeting with the physics principals the next day, Lyman (per John) "sat there like a bump on a log and didn't say one thing in your favor, or give one word of encouragement to the physics department for them to hire you." Afterwards, John asked Lyman if he had talked to any of my references. "I called Jim Gunn." And that was that. Gunn recommended Gott, who arrived in Princeton as an assistant professor in January, 1976, and was awarded tenure two years later. We were friends then, and are still today. If Rich gravitated toward unproductive and eccentric areas of research during his long career at Princeton, it may in part have been because Jerry Ostriker was active in all the productive areas, and Rich may have wanted not to be seen as Jerry's understudy.

That I had a valid reason for disliking Spitzer made it even harder for me to understand why everyone considered him a great theoretical astrophysicist. He did careful, obvious things, both in the pioneer years of plasma astrophysics and afterwards. He could never have commanded such respect, I thought, as a theoretical *physicist*, a community where cleverness, not just diligence, was coin of the realm. Nor could I understand why Lyman seemed generally well-liked by colleagues. A couple of years later, I sat next to Alan Lightman at a conference talk by Lyman, who credited no one in the field for their contributions except his own graduate students. "But I won't try to remember their names, since there have been so many of them," he said dismissively. Alan, who with Stu Shapiro had done important work in exactly the same subject, came to a slow boil.

This was also the summer that I got to know Margaret Geller well, another lifelong (though at times rocky) friendship. She was a Princeton Ph.D., supervised by Peebles, who had gone on to be a postdoc at IOA. She was only the second woman to get a Princeton Ph.D. in physics and had horror stories about Princeton that she enjoyed telling in an animated fashion. So, we bonded immediately. She had become a post-doc at Harvard—that job offer more due to a rave recommendation from Charles Kittel, her Berkeley undergraduate advisor, than from Peebles, who wrote only faint praise. Geller's analysis of Peebles was not too different from mine: He was jealous of his own narrow field (in which he was the acknowledged world expert) and wouldn't stick his neck out for anybody, not even his own students. It was useless to go to him for advice on a scientific problem: If the problem was within his jealous perimeter, he would say nothing; if it was outside, he would have nothing to say.

Geller later became a full professor and senior scientist at the Harvard-Smithsonian Center for Astrophysics. She co-discovered the fact that galaxies and clusters of galaxies are organized into "large-scale structures", something that surely would have won a Nobel Prize had the assignment of credit been less murky. Peebles was awarded the 2019 Nobel Prize in Physics for his work in exactly the part of cosmology that he for decades guarded so jealously. At the time of that final recognition, he was 84 years old.

Going to Jerry Ostriker with a scientific problem was an entirely different experience. Earlier in the spring, I had gone to him with a question that came from my student Paul Wiita's work. "You know," he said, "I've been thinking about something quite similar. Why don't we all write a paper together?" Jerry was famous for these Faustian bargains. His name on a paper guaranteed that it would be taken seriously. He would contribute to any joint effort an hour or two of creative discussion, and he would read and usually improve the paper in final draft. What he wouldn't generally do was much real work. In the case of Wiita, I turned down Jerry's offer—with trepidation. Not long before, I had been at an Institute astrophysics lunch where Jerry lectured on his own theory of galaxy formation in a talk that bore considerable resemblance to a recent preprint by Gott and Thuan. Daring where my colleagues feared to tread, I spoke up, "Jerry, what is the relation of this work, if any, to the recent work by Gott and Thuan?" "It's not *completely* independent," he said breezily. But, he added to me after lunch, caustically, "You know, Gott and Thuan were *my* students, not the other way around. They *could* have written their paper *with* me."

Martin Rees was Jerry's opposite, a constant source of research ideas freely given away, and an outstanding communicator. Martin was so polished as a speaker that one could have listened to him read used-car classified ads from the Sunday paper, and they would have seemed not just interesting, but like unified pieces of some grand intellectual edifice. From summer, 1975, in Cambridge, I took away several ideas that, germinating over the next couple of years, helped me out of my Princeton work slump.

Trinh Thuan, incidentally, was in Cambridge for two months that summer, after which he was supposed to return to his postdoc position at Caltech. When he went to the U.S. Embassy in London to get his return visa, he learned that he was now a stateless person. He had been in the United States on an exchange visa but, now that there was suddenly no longer any South Vietnam, there could obviously be (they told him) no further "exchange." Could he then come to the United States as a refugee? Obviously not, because he had already been in the U.S. and left voluntarily for the summer in England. His job, belongings, furniture, were all back in Pasadena, where he was still paying rent on his apartment. The only good news was that, as an officially stateless person, he could not be deported anywhere by the U.K. The eventual "solution" was that the embassy in London gave him not a visa, but an official letter to the airline stating that the normal penalty for transporting a visa-less foreigner would be waived in his case. Then, when he arrived in Los Angeles, he would be at the mercy of the immigration officials; except that in the U.S., as in the U.K., a stateless person could not be deported.

The summer passed pleasantly with fried duck and pancakes at the Peking Restaurant on Burleigh Street; visits to the Fort St. George pub on Midsummer's Green; cream teas (tea, sugar, milk, hot water, bread and butter, jam, cucumber sandwiches, scones, cakes, and whipped double cream) at The Orchard in Grantchester or the Garden in Coton, where we often ran into the Gunns or Lynden-Bells; and dinners at the Red Lion Inn in Grantchester.

Rich Gott took us to high table at Trinity: smoked salmon, roast pork, dressing, asparagus, potatoes, ice cream and cookies, sardine on toast, and wine of course; then up to the port room for the candlelit port ceremony: the vote on whether to have claret in addition (it always passed), then three rounds of port, madeira, claret and (since this was the summer) sauterne; then also fruit, cheese, cigars and snuff, the last from a large silver snuffbox. Margaret finished off the claret bottle on her turn, a so-called buzz, which entitled her to a free glass when they brought out the next bottle. Rich told us that he thought of Trinity as an interstellar generation starship, self-contained and surviving through the ages—generation upon generation—by means of long-held, closely maintained traditions. Its crew had by now long forgotten exactly what was their destination, and what their mission was supposed to be when they arrived. Robert Heinlein's 1941 novella *Orphans of the Sky* had a similar premise.

30. Ticket Out

Our way home from England involved a detour to a relativity workshop conference in Wales. Bernie Schutz, Kip's former graduate student, was by now a professor at Cardiff University. He and Roger Penrose had organized the conference. Gregynog was the university's conference center, a huge Tudor manor house set in 750 acres of parkland. We were arriving a day early. A girl who could not have been even eighteen escorted us through the long hallway, the dining room, the front hall, up two flights of carpeted marble stairs, to our bedroom (which included a crib for Sara). We explored the empty house: sitting rooms, meeting rooms, common room, senior common room, a library with ten thousand old books, a music room with pipe organ and two pianos, and about *fifty* bedrooms. The refectory, when we made our way there at dinnertime, was empty and dark, but two places had been set by unseen hands for us, dinner platters of cold cuts and trimmings. We sensed the presence of ghosts—centuries of Tudor Princes of Wales. But we learned the next day that the manor was actually a re-creation built in the mid-nineteenth century by a Welsh cement magnate. A few years previous, it had been featured in *The Concrete Quarterly*: Its walls were made of cement. In the 1920s it was a musical center and hosted concerts by Elgar and Holst, who played the pianos that we now plinked.

The next day, at coffee in a paneled room after dinner, Penrose told me about a mathematical game that he had invented, a kind of jigsaw puzzle with just two kinds of pieces, differently proportioned diamonds, that could cover the whole plane—but not in any regular pattern. The game was to get as far as you could in the covering before you ran into trouble—produced a gap that neither diamond would fit. I asked about some details, and he ran back to his room, returning with a handmade set of wooden pieces. Although he could prove that the whole plane could be covered (it was a simple proof involving the Fibonacci numbers that even I could understand), he had never actually seen an example with more than a couple of dozen puzzle pieces. This was catnip to me. Implicit in his proof was an algorithm that could be programmed on a computer. I promised to send Roger a computer-drawn "blueprint" that would show him a solution with many thousands of puzzle pieces, and eventually made good on this promise. In the 1980s and 1990s, aperiodic lattices including Penrose's were found to occur in nature, at the atomic level, in so-called quasicrystals, now a pretty subfield of condensed-matter physics. A Nobel Prize in chemistry was awarded for their experimental identification.

Back in Princeton, that fall, I was teaching an undergraduate class in mathematical methods of physics to a class of exactly five enrolled undergraduates. Five was a magic number. If any one of them dropped out, the class would be cancelled and I would be assigned to teach some unwelcome number of freshman lab sections. John and Janette Wheeler held a welcoming "Einstein-Maxwell Picnic". My curiosity on how (and why!) John insisted on naming events after Great Men was mostly played out, but I did wonder why Bohr had not been included.

Even after so many decades, I am ambivalent about the Princeton that I experienced in the 1970s. It launched me professionally in a way that Caltech could never have done. It was the active center of the universe in theoretical astrophysics, one of my fields, and (via Einstein) the historical center of the universe in my other field, relativity. Spitzer and Schwarzschild were the great astrophysicists of their generation. Ostriker (along with Martin Rees on the other side of the Atlantic) was the rising star of his. Everyone who was anyone in astronomy visited Princeton frequently. Caltech, which owned Palomar Observatory, ruled the observational roost, but the strong Caltech-Princeton-Cambridge (U.K.) axis benefitted all three institutions. After I had my Chicago job offer, I asked Martin for advice. He told me that it would be a mistake to leave any job at Princeton, even a nontenured one. I should hold on there as long as I could. "After all, it is the best place in the world."

On the other hand, I hated the place. I hated its petty academic jealousies, its small-town small-mindedness. I hated the genteel Southern veneer, where you were never supposed to even hint to a person's face what you really thought of them, but made sure behind their back that everyone else in town knew. It didn't help that I was, myself, an arrogant young scientist with a strong streak of paranoia, an infant daughter, and an unemployed, unhappy wife. To give it its due, Princeton was changing: It had been announced to the junior faculty with some fanfare that we were now allowed to use the senior faculty shower on the third floor of Jadwin Hall—the one used by senior faculty after their tennis games. We were told to bring our own soap.

I had a figurative ticket out of Princeton, a ticket to Chicago, but it was going to expire within weeks. There was only so long that I could keep Schramm and Chandra waiting for an answer. Harvard was still in play—I spoke regularly by phone with George Field, and he was trying to get me an offer. But the situation was complicated in ways unique to Harvard. These bear mentioning.

Outside of the professional schools (Law, Medicine, etc.), all Harvard departments were under the purview of the Dean of the Faculty of Arts and Sciences. There were other lesser deans, but he—it was then Henry Rosovsky—was "the Dean". Rosovsky later famously turned down an offer to become president of Yale (its first Jewish president), leaving Yale's actual first Jewish president to be Rick Levin, appointed in 1993, whom I came to know well. (But that is much later in the story.)

As at other universities, each department at Harvard had its faculty "lines" to fill. At Harvard, these were literally lines drawn in a bound ledger book jealously guarded by Miss Johnson, the Dean's septuagenarian secretary. Each department had a few pages with blue-ruled rows that were labeled down the page by calendar year. The pages were divided into columns by vertical red rule-lines. The names of tenured professors were inscribed by Miss Johnson (in pencil) on lines that corresponded to their future dates of retirement—at the time, a rigidly enforced age sixty-seven. When a retirement came, that professor's (younger) successor would be entered, one column to the right, on a future retirement year's line farther down the page; and a *line* would be drawn, diagonally, connecting the old and the new in an unbroken chain. Those were the faculty lines. The whole thing was called a "Graustein Chart" after the long-dead dean who invented it.

When George Field was recruited from Berkeley as the first joint Harvard-Smithsonian director of the new Center for Astrophysics, Miss Johnson's ledger connected him to the retiring Fred Whipple. However, the next retirements were years distant, so George had negotiated with the Dean for three new lines in Astronomy. One can be sure that Miss Johnson disapproved of the anomaly. Al Cameron had been recruited to the first of these new lines. My offer, if I got one, would be the second. The remaining line was to be filled with an optical observational astronomer.

Departments voted on tenure offers, but a numerical plurality was meaningless, because every voting professor was required to write a confidential letter to the Dean explaining his vote. One might vote for one candidate and then, concocting an explanation, recommend a different one to the Dean. It was a given that one's letter to the Dean should try to undercut the positions of professors with contrary views.

From the letters received (or otherwise, by some mysterious process), the Dean would come up with three to five names of candidates, and himself write to experts at other universities. Those letters, typed by Miss Johnson, were formulaic: Which of these "men" (only later, "people") is *the best* in the world in his field. The locution wasn't *the best that we can attract* or *among the best*. It was *the best*. That formulation had worked for Harvard since time immemorial; but by the 1970s it was becoming frayed around the edges. Recipients of the Dean's letter would generally telephone someone they knew in the department to find out who the department really wanted, and then (if they concurred) warp their letters to make that person the best in the world—in some carefully defined subfield. Of course, the Dean knew this was happening and would correct for it in interpreting the responses.

The appointment process was not then finished—not nearly. For each appointment, the Dean would ask the President of Harvard, at the time Derek Bok, to schedule a so-called Ad Hoc Committee. Two or three outside experts were invited to travel to Cambridge. They, plus one or two Harvard faculty from outside the requesting department, would meet with the President and Dean, generally for a full morning. Testimony would be heard from the department chair and other professors, pro and con. The Ad Hoc members didn't vote; they only "advised" the President, who, when the meeting was adjourned, generally took them to lunch in a private room at the Faculty Club and rarely indicated what his decision—the final one—might be. President Bok generally delivered his verdict, by phone to the department chair, the day after the Ad Hoc meeting.

Are we done? Not necessarily. While the President might simply, after all this fuss, approve the candidate originally voted by the department, he might instead—this happened about a third of the time—send back to the department a completely different name—or even more than one. It was then up to the department whether to accept one of these, or make no appointment at all.

This was all at the time opaque to me, except insofar as John Bahcall's sources provided not-always-reliable glimpses. I was elated when George Field called me in mid-September to say that the astronomy department had voted to offer me a full professorship. (It was another quirk of Harvard that full professor was the only tenured rank. At Chicago I would have started as a tenured associate professor.) George knew, though I didn't, that several faculty were planning to write opposing letters to the Dean. I was devastated when, sometime later, Bahcall told me that Dean Rosovsky's letter to him solicited opinions on (alphabetically listed) Arnett, Press, Silk, Tinsley, and Wagoner. Except for Beatrice, these were people a decade senior to me. Their published corpus vastly outweighed mine. Dave Arnett, at Chicago, was someone I admired, and he was a part of the effort to recruit me there; but he had been Al Cameron's Ph.D. student, and Cameron (now at Harvard just one year) was supporting him.

The Ad Hoc was scheduled for mid-October. As the date approached, the motivational phone calls from Field became fewer and less upbeat, and Bahcall's sources all went radio silent. John's interpretation, like mine, was that the astronomers in Cambridge were now expecting a decision in favor of Dave Arnett—or possibly Beatrice. "If for any reason you don't want to go to Chicago, I can get you a five-year appointment here at the Institute," John told me. He knew how unhappy I was in Princeton's physics department. He didn't understand that I hated the entire Borough and Township of Princeton and possibly even the adjoining townships of Lawrence and Hopewell.

Chandra happened to be visiting Princeton during the final two weeks of my waiting. He and Lelitha were longtime friends of the Schwarzschilds and often stayed at their house. He once told me that these were "unofficial" visits and that his last "official" visit to Princeton had been in 1967. I was impressed by his making the distinction, as royalty might. We had substantive discussions about who we might recruit to Chicago as assistant professors—Doug Eardley was a top candidate. Chandra told me that Chicago wanted me for my best qualities—scholarship and depth—while Harvard, if they came through at all, wanted me for my least attractive—marketability, charisma, and salesmanship. He was exactly right about this. My Chicago ticket was looking better and better.

George finally called me the day after the Ad Hoc. He was downbeat. It was unusual that there was as yet no verdict. "I may have failed to present your case in the best light," he said, cryptically and ominously. A glimmer of hope was that two of the outside members (names kept secret, even from him, until the meeting) were old relativity friends of mine: Charlie Misner and Stan Deser. Deser had been in my office at Princeton just the previous spring, where he had noticed the page proofs of *Problem Book in Relativity* (now finally about to be published) and had been wildly enthusiastic.

I later learned the source of Field's pessimism: One of the outside letters had made a whole series of unfavorable points about me in the form of rhetorical questions. Called to testify, Field was blindsided by

Bok's and Rosovsky's going down the list and demanding answers to each, rather than letting him present the positive case. The letter might have been Gunn's—then or later, I never asked either Field or Gunn.

After more than a week of silence, I got a phone call from Alex Dalgarno, the astronomy department chair (a rotating position subordinate to George Field's). "The recommendation of our department that you be made an appointment has, at last, been acted on by the Dean and President." I waited for the bad news. "And they have *approved* it."

I knew immediately that I would pick Harvard. It was irrational, but the very real merits of Chicago—even Chandra's final pitch—simply evaporated in my mind. Was this a baby-duck-patterning thing, because I had been a Harvard undergraduate? There must be a parallel universe in which I never left California, in which I was a Berkeley undergraduate with Margaret, then went to graduate school at Caltech and worked with Kip, just as in our universe's timeline. In most ways I might be, in that universe, not very different from in this one. But in that universe, I surely would have gone to Chicago instead of Harvard.

31. Tenures

I couldn't formally accept Harvard's offer until I received the "Sir I Beg To Inform You Letter", an engraved document that read, "Sir: On behalf of the Harvard Corporation, I beg to inform you that you have been appointed to the position of ____ commencing ____." The word "Professor" and the date were always filled in by hand. And I couldn't get the letter until I had an interview with Dean Henry Rosovsky in which he would tell me my salary and, if necessary, erase from my mind any doubts about accepting Harvard's offer. Rosovsky was an economic historian who specialized in Japan, and had been chair of the economics department. His corner office in University Hall overlooked the New Yard, with a view of Memorial Church and Widener Library, to which, when I entered, he gestured grandly, as if to say, "this says it all."

"Have you ever been in this office before?" he asked. I allowed as how, actually, I had spent a lot of time in it in 1969 when it was occupied by student radicals. He frowned. "But I was just an engineer for WHRB," I added, and the frown disappeared. Leading up to the discussion of money, he told me about the "Harvard community" and the "Harvard system". We were all a family. We took care of our own. He had a favorite anecdote about an economics professor who as the result of a stroke became "really a vegetable," but they kept him on full-time salary for three years. I volunteered that the story seemed of limited relevance to me at my age and career stage. He again frowned.

"We professors are all the best," he told me, "so it isn't right for us to try and judge each other. Some scholarship takes years to bear fruit." Thus, he said, my salary would be based on my calendar age. At twenty-seven, I would be the youngest current full professor by five years, and my salary would be $23,000.

"So, I'll be paid the least of *all* Harvard professors?" I probed.

"Let me check," he said, opening his desk drawer and taking out a loose-leaf notebook. He turned to the last page and scanned it, counting. "No," he said. "There are twenty full professors who are being paid less than you will be." He looked back at the list, frowned again, and said as if to himself, "I'll have to do something about that."

Secretly, I was elated. The offer was only $2,000 less than Chicago's. I had been expecting much worse. There was one further thing to be negotiated. Because I had never been an actual astronomer before—I considered myself a physicist—and because the whole Center for Astrophysics thing was so new and unproven, I wanted a joint appointment in the physics department as a kind of safety net. Thanks to Ed Purcell (whose acquaintance I had recently renewed when he visited Princeton for several weeks), Physics had already voted me a courtesy title. But what did that title mean, if my "line" was 100% in Astronomy?

Rosovsky was sympathetic. He harbored his own doubts, I think, about the Center for Astrophysics. He would make sure, he explained, that my Harvard title was not "Professor of Astronomy and Physics," but instead "Professor of Astronomy and *of* Physics". That second *of*, so easy to miss, was important at Harvard. It signified that I was a full voting member of both departments and, more importantly, that if I ever decided to make my primary home in Physics, my line would move with me. This was exactly what I wanted. Later, I paged through the Harvard Catalog, noting that some joint titles did have the second *of*, others did not—those poor bastards!

Fifteen years after these events, Henry published a wry and witty book, *The University: An Owner's Manual*. His generic description of recruitment interviews like mine gives a sense of his ennui about it.

> The next part of the interview is also depressingly familiar. I have to listen to a recitation of everything that is wrong with Harvard, Boston, Cambridge, our departments, salaries, and so on. Housing is too expensive; public schooling is bad and private education too expensive; the spouse sees few job opportunities; the department is much too small; good graduate students are going to Princeton; there is little collegiality at Harvard; *und so weiter*. There is some truth in all of these assertions, and their detailed and loving presentation is part of the bargaining game.

He must have thought me an easy mark: I was too young and uninformed to raise any of these issues! I do make a brief, anonymous appearance in Henry's book as "the youngest person awarded a tenure post, a twenty-six-year-old astrophysicist from Princeton" (he was exaggerating by one year). That mention hints at why my approval had been delayed for a week: George Field was recruited from Berkeley to clean the Augean stables of astronomy at Harvard. He had been given the precious gift of two new faculty lines. Al Cameron, in the first of these, must have seemed to Bok and Rosovsky as an unconventional, risky appointment. Al was a Canadian nuclear physicist, most of whose career had been at the Canadian government Chalk River Laboratory in Ontario. Emigrating to the United States and switching fields to nuclear astrophysics, he had kicked around in various academic appointments, then moved to NASA's Goddard Institute in New York City, with an academic affiliation at—of all places—Yeshiva University. Yeshiva's medical school and law school were highly regarded, but for Harvard to hire, from Yeshiva...an astrophysicist? Bok and Rosovsky had swallowed hard and approved that appointment. In fact, their reluctance was misguided: Cameron was one of the great nuclear astrophysicists of his generation, in a class with Willy Fowler, the Burbidges, and Fred Hoyle. Then, changing directions, Al became one of the great planetary scientists of his time, responsible for much of the now-accepted explanation for the formation of the solar system and origin of the Moon.

But now, to fill his second new line, George had brought them a twenty-six-year-old [sic] boy-wonder with, already, the reputation of a spoiled brat. It was a testament to their confidence in George that they approved me at all—but that confidence was not without limits. Stan Deser later confirmed to me that Bok and Rosovsky had given George a tough grilling at the Ad Hoc meeting. Henry, he said, was the more negative of the two. Derek seemed the more adventuresome.

My new astronomy colleagues (about whom the reader will presently learn much more) held a lunch for me in the Faculty Club. That night, we all came down with food poisoning. My symptoms were mild, but one colleague ended up in the hospital emergency room. I hoped it wasn't an omen. This was in October, and (according to the Sir I Beg To Inform You letter, which I soon received) I would not become a Harvard professor until July 1, 1976. I determined to enjoy my remaining time in Princeton no matter whom I offended.

One day, when I gave John Wheeler a ride to the Institute, we saw Lyman Spitzer drive into the parking lot, followed by another car. Wheeler said to me, "I see that the new Director and his wife are taking separate cars, now." "What?" I said, picking up on this. "Have they announced the new IAS Director?" Small-town Princeton was on tenterhooks over would succeed Carl Kaysen. "I spoke out of turn," Wheeler said, shutting down. In Princeton, this was gossip of a high order. I spread the news to all of my colleagues. Except that a couple of weeks later, historian Harry Woolf, not Spitzer, was announced as the new Director. I learned later that Spitzer had been offered the job, but turned it down when he discovered how much of his effort would have to be in fundraising.

Was it something in the air, or was it just my overheated imagination—because John Wheeler was himself acting strangely. This became obvious when department chair Murph Goldberger summoned to a meeting in his office the five physics faculty who worked in relativity and cosmology: The senior professors Wheeler, Dicke, and Peebles, and (unusually) the junior faculty Claudio Teitelboim and me. Claudio was recently back from Moscow, where he was treated like royalty, with his own car and driver, and scheduled meetings with various Soviet Academicians. His father, whom he was visiting, was now head of the Chilean government-in-exile, the surviving remnant from Chile of Allende's Communist administration.

Our meeting's purpose, it seemed, was to grill Wheeler, in our presence, on what progress he was making in new recruiting for the relativity group. I would be leaving at the end of the term, Claudio a year after that. Many entering graduate students wanted to work in relativity, but Wheeler was willing to take only students who would work on human consciousness—and of these there were none. He had allowed his postdocs' terms to end without replacing them. To my knowledge there was no new recruiting going on. Wheeler said that he had in mind a few people as one-year visitors, and gave several names, all in my opinion unsuitable. Then, pressed to name possible new junior faculty, he rattled off a whole list of names, all unsuitable or unavailable. And it was obvious that he knew this. The meeting ended inconclusively.

Not long after, Wheeler appeared in my office. "Eardley," he said, and repeated, "Eardley. Eardley. Can he pop down here for a *consultation?*" Doug had just been promoted to tenure at Yale and, in any case, he and Diane viewed New Haven a temporary resting place while they searched for positions in California, the only place they really wanted to live.

We wondered if John's inaction stemmed from a terminal illness; but he seemed fit and healthy. It was Larry Smarr who cracked the case. When Larry visited Texas (his Ph.D. alma mater), Bryce DeWitt swore him to secrecy and told him that Wheeler was being courted there and—against all odds—seemed to be interested in leaving Princeton that very summer. Wheeler had visited Texas twice, keeping even Mrs. Witt in the dark as to his whereabouts. Larry being sworn to silence—his telling his wife Janet, me, Margaret, and our friends Nick and Mary Jane Woodhouse didn't count—we had to sit on this information and watch matters unfold.

We did tell the Eardleys when Doug was formally invited to visit Princeton. He gave a great talk and told everyone who asked, as politely as he could, that there were no possible circumstances, none, that could move him to Princeton. At Doug's private meeting with the senior professors, Bob Dicke pointedly asked Wheeler, directly, what were his eventual retirement plans. "I would *normally* expect to be here for several more years," Wheeler said, "unless I get run over by a bus *next summer*." Doug, reading him perfectly, could not help laughing, and the others wondered what he found so funny about Wheeler's being hit by a bus.

When Wheeler visited Texas again, and Mrs. Witt *did* know where he had gone, we knew that an announcement must be close. On a pleasant spring day, Wheeler popped into my office for a consultation (as he would say). "I have this clock inside me that I can't ignore. I've decided to move to the University of Texas at Austin." He wondered whether Kip Thorne could be attracted to Princeton, a complete red herring.

The Woodhouses were already back in their native England, Nick at Oxford and Mary Jane now a barrister in the Inns of Court (and later a judge). Larry and I sent them an overseas cable referring to the previous summer's hit movie and also John Archibald Wheeler's initials: "JAWS OPENING AUTUMN TEXAS FOR LONG RUN BANG" One paid by the word, and eight words was the minimum. "BANG," was the international cable abbreviation for exclamation point—at least I thought so. The story all eventually came out as this: Wheeler had asked for a (rare) exception to Princeton's mandatory retirement age, and the university had refused. He then wanted the physics department to appoint Karel Kuchař to succeed him, and the department had refused. He was fed up. Texas offered a star salary and had no mandatory retirement age.

At the group's lunch the next week, Wheeler blew up his paper lunch bag and smashed it with, indeed, a loud bang. "I have some cherry bombs at home," I said. "Shall I bring one in?" His eyes lit up. But I decided not to follow through after Margaret observed that he would undoubtedly set it off in his office, and that everyone's first thought would be, "Oh my God, Johnnie has *shot himself* rather than go to Texas." Famously, Wheeler liked to tell people, always in a soft voice, "When I was a small boy, I blew off the end of my finger with a cherry bomb; when I was older, I learned to play with dynamite; then, at Los Alamos, I worked on the A-bomb; then, the H-bomb; and now, I'm working on...the Big Bang."

Given the circumstances, it was not in Wheeler's nature to go gracefully. When the department proposed a gala retirement party to celebrate his thirty-eight momentous years at Princeton, an event to be held as a dinner dance at the Prospect faculty club, his answer was that Bill Press was leaving too, and that he would only attend such a party if the two of us were equally honored. Murph called me into his office to tell me, not without irony, that I was required to attend the event, had to sit at the head table with Johnny, and would have a very, very brief speech made by someone (they weren't sure whom) in my honor. And so it unfolded, late in May. The Wheelers pointedly arrived more than an hour late, and they left early. I danced alternately with Margaret and with the beautiful Sonia Teitelboim. "But I am not a Marxist," she called out across the room as she and Claudio were leaving, wagging her finger at me. (Claudio's father, a prominent Chilean politician, was still in exile in Moscow.)

Jerry Ostriker gave me a different kind of send-off. He appeared in my office, closed the door, and took me to task for (he said) my describing his work to others as sometimes sloppy and occasionally incorrect. The charge wasn't exactly true, but it wasn't exactly false either. Jerry was truly a great theorist, but great theorists in astronomy didn't need to have high batting averages. Ostriker's reputation was based on his being brilliantly insightful in perhaps one paper in ten. Still, he was probably right to criticize me. I could have been more diplomatic in my assessments. But he was too late. His blows barely landed: I was already feeling like a Harvard professor.

32. New Colleagues

Historically, astronomy at Harvard went through several Ptolemaic epicycles of alternating eminence and mediocrity. In 1839, the Harvard College Observatory was established on Observatory Hill, about three-quarters of a mile from the Harvard Yard. In 1847, a telescope of world-class, the Great Refractor, was commissioned. At fifteen inches it was the largest in the United States and remained so for twenty years. Astronomy then consisted of the visual determination of stellar positions, and the anecdotal description (with sketches) of planets, comets and nebulae. An eighth moon of Saturn, Hyperion, was discovered at HCO in 1848. The observatory's separate allocation of the Harvard endowment supported one or two astronomers at a time. No formal courses were offered. By the 1880s, the enterprise had sunk into obscurity.

A rejuvenation occurred with the advent of photographic observation. In the 1890s, director Edward C. Pickering (of course a Harvard graduate) established a continuous nightly survey of the heavens, one photographic plate at a time, amassing over many decades an eventual collection of several hundreds of thousands of the glass plates. One of Pickering's innovations was to hire a small army of women for the repetitive work of measuring, on the plates, the positions of millions of stars. These "Harvard computers" (also known as "Pickering's harem") collectively became internationally famous. Several individuals, Annie Jump Cannon and Henrietta Swan Leavitt in particular, were known as the equal of any male astronomer. But, during Pickering's long tenure of forty years as director, the work became routine, another long period of mediocrity.

Harlow Shapley, recruited as the new HCO director in 1920 after Pickering's death, was a fresh breeze. He had studied under Henry Norris Russell at Princeton and was not a Harvard alum. Shapley knew how to command media attention—he had once worked as a newspaper reporter covering crime stories—and used Harvard's bully pulpit to his (and Harvard's) advantage. Clarence Darrow invited Shapley to testify in the 1925 Scopes trial as an expert witness in favor of Darwin's theory of evolution. Still, Shapley did bet on some lame astronomical horses. It was at that time a matter of debate whether the Milky Way constituted the whole cosmos—an "island universe" of stars surrounded by infinite void—or whether the nebulae (observed only as small blurry patches of light) were themselves galaxies like the Milky Way, but much more distant. Shapley defended the island universe hypothesis in a famous debate at the National Academy of Sciences and, in the opinion of most present, won the debate. It was some years before the evidence from Edwin Hubble's observations in Pasadena convinced him to reluctantly change his mind.

Under Shapley, Harvard first established, in the 1930s, an Astronomy Department that could offer courses. Shapley's protégé Donald Menzel was its first chair. Shapley arranged for Radcliffe College to award a Ph.D. degree to Cecilia Payne (later Payne-Gaposchkin), so that she could teach courses to Harvard undergraduates. When I came to Harvard, Cecilia, in her mid-70s, was a professor *emerita*, still teaching the occasional seminar. Frances Wright, her friend and contemporary, still offered a course in celestial navigation, the fossil of a World War II instructional curriculum for naval officer cadets. By the 1970s, the students taking the course were mostly from families with ocean-going yachts—a rare stratum even for Harvard.

The two women had traveled around the country in 1930, visiting observatories in the West and camping on the way. To me, these were two old ladies who sometimes showed up at faculty meetings. No one ever told me that it was Cecilia who first discovered that the stars (and therefore the whole material universe) were composed mostly of hydrogen and helium, and not silicon and iron as had been believed.

Over a period of two decades, Don Menzel and Bart Bok, both Shapley protégés, managed, mostly by academic inbreeding, to bring the department to another nadir of mediocrity. I never met Menzel, and met Bok only once or twice. But when I arrived at Harvard, the ten voting tenured professors in astronomy included Menzel's former students David Layzer, Chuck Whitney, and Owen Gingerich; his "grand-student" Bill Liller, and Bok's former student Ed Lilley. The collective lifetime contribution of these individuals to astronomy was very, very small—in the case of three of them, nonexistent—and they comprised half of my new department. In department meetings, Lilley spoke up mainly to complain about a Massachusetts law that forbade his carrying a gun on campus.

I could perhaps be more charitable: Owen Gingerich did change fields to become a distinguished historian of science, a world expert on Copernicus. David Layzer did little himself, but, of his seven Ph.D. students, three became prominent astrophysicists: John Bahcall, Joe Silk, and Bob Rosner. I once asked Bahcall what it was like to work under Layzer. "The better I knew him, the more strongly I was motivated to get my degree *quickly*," he said.

Harvard astronomy was not alone on Observatory Hill. In 1955, Harvard astronomer (but not alum) Fred Whipple had negotiated to move the Smithsonian Astrophysical Observatory, a small branch of its well-known Washington, D.C., parent, to Cambridge. Whipple became its new director. SAO was administratively separate from HCO and was housed in its own building, but connected to the Harvard building by walkways on each floor. After, and because of, Sputnik, SAO rapidly expanded. James Baker, a famous optical physicist, designed a telescope ("Baker-Nunn") that could be used to track Soviet satellite orbits. A dozen of these, each weighing 3.5 tons, were operated by SAO around the United States through the 1960s and 70s.

SAO also ran Operation Moonwatch, a network of more than one hundred amateur astronomers with small, government-provided tracking telescopes. These small telescopes, a foot or so in length, were also used by the CIA to locate prospective ICBM targets in the Soviet Union: Agents parachuted into Soviet territory would make their way to designated sites and photograph the sky at specified times, then somehow make their way out. From the streaks of artificial satellites in the photos relative to fixed stars, an exact location could be deduced. Lowell Wood—not always a reliable narrator—told me that a friend of his was one such agent who never came back.

Later, SAO became the home of the Central Bureau for Astronomical Telegrams, an international agency for promulgating astronomical discoveries that was founded in 1882 to coordinate comet observations. These "telegrams" continue today, but emailed and posted on the internet.

Compared to HCO, SAO had all the money. Compared to SAO, HCO had all of Harvard's prestige. There should have been a natural synergy, but Whipple and HCO director Leo Goldberg (a student of Menzel's) instead jealously non-cooperated. SAO grew in size but not stature. Now there were *two* mediocre institutions on Observatory Hill. That was the situation when the two parent institutions, one in the Smithsonian Castle, the other in the Harvard Yard, selected George Field to direct both. George, a Berkeley professor of astronomy, was an unlikely choice. A theorist, he had never run anything. That he got the job was due to Jesse Greenstein and Ed Purcell. Greenstein was a student of Menzel. But after escaping Harvard (and escaping Shapley, whom he detested), he kept an abiding interest in Harvard matters. He was elected to the Harvard Board of Overseers, and was known to Harvard president Derek Bok (no relation to Bart). Purcell was, as we already know, at least as influential. Each of these two happened to have worked with Field previously, and they recommended him.

Before I arrived, Field had (as we saw) engineered Al Cameron's Harvard appointment. On the Smithsonian side, a first great coup was to bring Riccardo Giacconi and his entire group of more than a dozen people. Giacconi was known as the father of space-based X-ray astronomy. His group's *Uhuru* satellite, launched by NASA in 1970, increased the total number of known X-ray sources in the sky many-fold. Later, long after Uhuru, he led development of the Einstein Observatory (the first fully-imaging X-ray satellite, launched in 1978) and the Chandra X-ray Observatory, named for Chandrasekhar (launched in 1999). He shared the 2002 Nobel Prize in Physics.

In the 1970s, Giacconi's group was already based in Cambridge as a part of American Science and Engineering, a for-profit MIT spinoff with contracts from government agencies and others. Giacconi was made a Harvard professor, but did not use up a "line," because his salary came from Smithsonian and his NASA contracts. At Giacconi's insistence, a second professorship was awarded to Herb Gursky, his sidekick. Herb was never an independent intellectual force but, in effect, gave Riccardo two votes in the department. AS&E's stock had made Riccardo quite well off. The company developed many of the original X-ray scanners used at airports and was briefly notorious for developing an X-ray scanner that could be used by De Beers in South Africa to scan exiting miners—it could find diamonds that they swallowed to smuggle out. Riccardo lived in a beautiful old Cambridge house on Craigie Street. The house was so large that it had been subdivided into a duplex. David Rockefeller, Jr., lived in the other half.

Alex Dalgarno, an Anglo-Irish atomic physicist, had come to Harvard astronomy in 1967. He was low key, and ran a group with more connection to the chemistry and physics departments than to HCO. He seemed to be no great fan of the changes implied by George Field's arrival.

So, accompanying the five deadwoods, there were five of us who were thought to be alive: Field, Cameron, Dalgarno, Giacconi (with two votes), and me.

Field controlled the purse strings for all junior appointments. On the Smithsonian side, he could appoint, by fiat, any number of three-year postdoctoral fellows—positions that would become the prestigious Center for Astrophysics (CfA) Postdoctoral Fellowships that exist today. On the Harvard side, he could tell the Astronomy Department how many assistant professorships he would pay for—with the implied threat of refusing to fund an offer to anyone he disapproved of. With this kind of autocratic power, it was a testament to George's personality that he was well liked—at least at first.

Before I arrived at Harvard officially, I had flown to Boston to participate in the department meeting that decided on three new assistant professors. Brian Flannery, Bahcall's postdoc at IAS, had been appointed a year earlier. Before the meeting I huddled with George, so that there would be no light showing between us. George asked me whether I favored Cameron or Gursky to succeed Dalgarno as department chair. The latter would give Giacconi not just two votes, but control of the whole department. I told George that if Gursky became Chair, I would accept Bahcall's standing offer to stay at Princeton for five years. In the event, the chairmanship went to Cameron.

George already knew (and clearly had already invisibly influenced) the ordering that the search committee, chaired by Whitney, presented: Alan Lightman, Ed Turner, Josh Grindlay (Giacconi's candidate), Margaret Geller, Tranh Thuan (Jerry Ostriker's favorite candidate), Ned Wright, down through about fifteen people. At the meeting, the deadwood made a futile attempt to elevate Wright on the list. He had been at Harvard since freshman year (in my Harvard undergraduate class, as it happened) and would continue the noble tradition of inbreeding. In truth, Ned was a much better astrophysicist than his sponsors and went on to a distinguished career at UCLA. The offers went to the top three names, and all, within a few weeks, accepted. I looked forward to having Alan as a colleague again.

Margaret Geller should have gotten one of the positions. Her contributions to astronomy proved to be much greater than those selected. But for Giacconi's two votes, and perhaps the fact that she and Lightman were both theorists, she might have. Field subsequently underbid the Harvard position with an inferior three-year Smithsonian postdoc offer that, despite Geller's faculty offers from Dartmouth, Haverford, and Yale, and feelers from UCLA and Bell Labs, she accepted. I thought that this was a career mistake on her part.

Larry Smarr was also coming to Harvard. Field had offered him a CfA Fellowship, but Larry instead secured an appointment as a Harvard Junior Fellow. This, despite the mundane name, was a uniquely prestigious Harvard position, roughly equivalent to assistant professor, but without any obligation to teach. The Society of Fellows had its own dining room, kitchen, and chef in Eliot House, where it held weekly banquets, well lubricated by the contents of its private wine cellar. During his time as a Fellow, Larry was elected steward of the Fellows' wine cellar, where he laid down vintages for future years. Ned Wright, sponsored by the astronomy deadwoods, also became a Junior Fellow.

My student Phil Marcus came with me to Harvard, although he remained formally a Princeton graduate student. That can go either way: It is understood, when a professor moves, that his Ph.D. students can be automatically admitted, if need be, to the new department's Ph.D. program. When I moved to Harvard, I had an unexpected additional student, a kind of stowaway. Clifford Taubes was a first-year student in the astrophysics department who had come to realize that Princeton was a big mistake. Taubes was a mathematician at heart and had little in common with Jerry Ostriker, who was assigned as his first-year advisor. I agreed to rescue him. As a new Harvard graduate student, he worked (notionally) on relativity with Smarr and me for a year, then found the right Ph.D. advisor in Arthur Jaffe. He was very successful after that, becoming a full professor of mathematics at Harvard and winning the Shaw Prize in Mathematics in 2009. (His acceptance speech thanked Larry and me for the rescue as described.)

I had to learn to be a Harvard full professor. Roughly, that meant that I should get my way not in the manner that I was used to—by being a juvenile spoiled brat and having a temper tantrum—but rather by conducting myself as if it were my *due*, that any outcome other than what I wanted was *unthinkable*. George Field should have been my mentor on this, but I found him remote and inaccessible. One had to make an appointment several days in advance with the Director's Office just to get five minutes with George. Invariably, in these brief moments of mentoring, he would say, "you must learn to take the *high road*, Bill."

I tried, and it sometimes worked. I had been assigned to a corner office in the newest of the Observatory buildings, but it was smaller than David Layzer's corner office on the other side of the hall. His had been enlarged by the incorporation of the small office next to it. There was another such small office next to mine, fortuitously empty. I explained to the Observatory's Associate Director for Administration, a practical Boston native named Bob Reed, that the situation was *unthinkable*. It was not my *due*. One morning, a burly facilities man arrived at my office. I thought he was there merely to scope the job of removing the partition wall. He knocked on the wallboard in several places, then, with steel-toed boots, simply kicked the wall in. The few metal framing verticals were removed a couple of days later.

I definitely wasn't on the high road when the National Science Foundation sent me for review a proposal by one of the greatest then-living physicists, Lars Onsager. Onsager, originally Norwegian, had single-handedly invented the field of non-equilibrium statistical mechanics. His Nobel Prize was awarded in 1968. He had retired from Yale at age seventy and moved (with a star salary) to the University of Miami in Coral Gables. My problem with his proposal was twofold. First, it was crackpot, nothing at all to do with the work that he was famous for. Second, it had clearly been written by its co-Principal Investigator, a man whose salary was a big chunk of the proposed budget. I had met this person at a relativity meeting and come away thinking that he was a con job. Now, I wrote an immoderate review, very nearly accusing Onsager of scientific fraud for putting his name on the proposal. Definitely the low road. My review went into the mail in early September.

It would have taken about a month for NSF to collect the reviews (there would be several), anonymize them, and send them to Onsager, along with their decision on his grant. Onsager died at his desk, of an aneurysm, on October 5, 1976. The obituary didn't say that he died while reading his mail, but that is what I imagined. "You killed him," Phil Marcus told me. "He read your review, his blood pressure spiked, and he popped an artery. You've killed your first Nobel Prize winner." Phil seemed pleased. His thesis advisor was now someone important.

As a new Harvard professor, I was entitled to sit at the "Long Table" for lunch in the Harvard Faculty Club, where professors were seated, elbow to elbow, simply in the order that they arrived. I took Larry to lunch, and we were seated opposite one another. President Bok came in, spotted the men sitting next to us, and strode across the room. "How are you...very good to see you..." etc., etc. Wow, we thought, the President of Harvard standing right next to us! Then Daniel Moynihan came in. He had just been elected Senator from New York. He also came over and exchanged pleasantries with our neighbors. Wow, we thought, Daniel Moynihan standing right next to us! We were so star-struck that it didn't occur to us to wonder whom we were sitting next to.

33. Jimmy Carter

My father was summoned to Washington for a meeting with new president Jimmy Carter, who asked him to become his Science Advisor. In Carter's administration, the position was one of seven top advisors to the president. The science advisor was also director of the Office of Science and Technology Policy in the White House, with then about thirty professional staff. This meant my parents moving to Washington. In D.C., they rented a dingy apartment near the National Cathedral and put the Belmont house up for rent. Frank got a phone call from Joe Califano, the Secretary of Health, Education, and Welfare, whom he had never met. "Frank? I think we've got a job for Billie." Carter had instructed the Cabinet that his administration was going to be a big, happy family, all pitching in.

Jack Ruina, then MIT Vice President, gave Frank a going-away party attended by four previous presidential science advisors, all of whom were in Cambridge: Killian (Eisenhower), Kistiakowsky (also Eisenhower), Wiesner (Kennedy), and Hornig (Johnson). Also, Lord Solly Zuckerman (the former British science advisor) and assorted MIT brass, Viki Weisskopf and John Deutch among them. Deutch fell in love with Sara and kept telling me what a lucky man I was—he had four sons. The party was, in effect, Frank's induction ceremony into the club of big shots.

The effect of my father's elevation on us was modest. He let me rifle his MIT office for physics and math books to borrow, long-term, while he was based in Washington. Books were vastly more important before email and the internet. For theorists, a well-curated collection of texts and monographs, close at hand, was often the difference between success and failure in a project. My father and I both knew that the books I borrowed would likely never be returned. Margaret, Sara, and I got to attend Frank's swearing-in in the White House Rose Garden. The weather was nice. Dick Atkinson was being sworn in as National Science Foundation Director at the same time. The two families stood on the platform near President Carter who, in person, had more of a commanding presence than came through on television. He spoke off-the-cuff for ten minutes. I was holding Sara, who kept pointing to the president and repeating forcefully, "That's not Jimmy Carter." The president shook hands with each of us. Sara would not give him her hand, so he shook her bare foot. To my sister Paula, he said, "You are very beautiful." To Margaret, "Pleased to meet you." It was unclear whether either, or both, should be offended. After the ceremony there was a reception, minus the president, in the Cabinet Room. Sara sat in the Secretary of Defense's chair and scribbled on his White House note pad.

After that first occasion, when I was in Washington for meetings, mostly at NSF, I visited Frank in his grand, high-ceilinged corner office in the Old Executive Office Building, inside the White House fence. With family, belying his reputation for absolute discretion, he let himself tell tales out of school: The senior staff requested use of the White House swimming pool, and Carter refused. Senator Frank Moss wanted to be the next head of NASA and Frank was finding a way to block it without leaving fingerprints. Bob Frosch became NASA Administrator. Once, Frank called me with a request: Jimmy, Rosalynn, and Amy wanted a little telescope to look at the planets. What kind should Frank get for them? I recommended an eight-inch Celestron. He ended up borrowing a five-inch Questar from NASA, also a fine instrument.

At the same time, my father's becoming a big shot further distanced him from me. I had gotten used to telling him about projects or scientific directions that interested me. One weekend, I was sailing with him on his little sailboat (still the Frilla) when he interrupted me: "Who put you up to telling me about that?" I was dumbfounded, but I understood instantly that everyone approaching Frank now had an agenda. I could avoid such suspicions only by communicating less with him.

Frank took me to lunch in the White House senior mess, a tiny wood-paneled basement room with about a dozen tables. Very unimposing, the menu ran to club sandwiches and hamburger plates. The waiters were Filipino U.S. Navy stewards, a tradition from the epoch when the Philippines were a U.S. colony. In the cramped space, one could listen surreptitiously to the conversations of other senior White House staff. Thirty years later, when I had a White House badge myself, security had become tighter, and only the elite blue-badges (whose wearers had unlimited West Wing access) were allowed into the senior mess.

When Frank had been in office less than a year, Si Ramo, Caltech's board chair, visited him and offered him the job of Caltech president, to succeed Harold Brown. This was one of the two jobs to which Frank most aspired—the other being president of MIT. Aware that he would never get a second chance, he nevertheless told Ramo that he could not accept, not even delaying the start by three years, because doing so would make him a lame duck within the White House before Carter's expected second term. There was no such second term, of course, Carter losing to Ronald Reagan in 1980. Frank was a candidate for his other ambition, MIT president, in 1980, but failed in part because of the clamor of the institute's engineers to have one of their own appointed. There had traditionally been an alternation between scientist and engineer in the position, though not always rigidly. Paul Gray, an electrical engineer, was appointed.

At work, I had gotten myself on both the graduate student admissions committee and also the Center Postdoc selection committee, so I was up to my ears in reading hundreds of folders. This was generally boring, but it was a back-door way to influence the intellectual direction of the whole place. A guilty reward was the occasional laugh at the expense of a misguided applicant. One such was Mr. Prasad [name here changed], an Indian who was somehow an undergraduate in Newfoundland. Instead of the usual three letters, he had only one which began, "Mr. Prasad is not up to the usual quality of our students here in Newfoundland." Then the color of the ink changed and it went on. "He has a much inflated view of his own talents." The writer must have looked at this and thought, "No, they might still admit him," because he then added in pencil, "Mr. Prasad has no interest whatsoever in astronomy."

Not just relativity, but many other areas of physics (and other sciences) were being transformed by the increasing availability and power of computers. Astrophysics, although its realm spanned stars, galaxies, and the cosmological universe, was at its base, *applied* physics. I had been in the lucky vanguard—at Caltech in the right place and right time—to gather relativity's low-hanging computational fruit. Other young scientists were doing the same in different subfields of science. I realized that there were areas of computational physics that could be harvested for application to astrophysics. My conscious decision was to continue moving away from relativity and into these other areas.

Fluid dynamics, the study of fluids and gases in motion, was one such area. Paul Wiita, Larry Smarr, and I wrote a paper suggesting how turbulence might explain the synchronization of rotation of close binary stars, whose same sides always faced the each other (like the Moon to Earth). Saul Teukolsky and I calculated how close-binaries might form in the first place by the dissipation of orbital energy into fluid modes. Saul, Phil Marcus, and I found a new way to calculate the circumstances under which a rapidly rotating disk might instead choose to become a ring.

This was an idealized Chandrasekhar-like calculation, but we thought that it could apply both to the formation of the solar system, and to the formation of galaxies.

I taught a one-semester graduate course in relativistic astrophysics, designed to complement the course in general relativity (proper) taught by Steven Weinberg, the star theoretical particle physicist whom Harvard had captured from MIT four years earlier. Weinberg had first showed in 1967 how electromagnetic forces and nuclear forces might be simply two aspects of the same fundamental force, something known as electroweak unification. For this he shared the 1979 Nobel Prize in physics. Relativity was something of a lark for Steve, only barely related to his already-famous work. He had taught himself the field and written a somewhat quirky textbook—an amazing investment for a lark. Weinberg, who was fifteen years older than me, became a friend and career mentor to me, but it was an odd sort of mentorship: He gave good advice freely, and I almost never took it. At lunch one day, Steve told me that, at a conference the week before, Jerry Ostriker had given him a pointed message to carry home: "Tell Bill that he should stop having so much fun in his work and do something important." It was clear that Steve agreed with this. He, like Kip earlier, thought that I could work harder.

Steve Weinberg was a younger member, although no longer very active, in the mysterious JASON Advisory Group. At the time of Sputnik's wakeup call in 1957, scientific advice to the United States military came predominantly from the same men who had been senior scientists in World War II—and they were now in their sixties or older. Postwar, much of defense science was given over to the civilian sector, and younger scientists (except for very few directly employed by the government) saw no reason to learn about defense problems. Sputnik changed that. A group including Eisenhower science advisor George Kistiakowsky, Pentagon director of research and engineering Herb York, and Princeton professors Eugene Wigner and John Wheeler, determined to create two institutions that would recruit bright scientists into defense work. Their concept was to have both full-time professionals and part-time amateurs. The Institute for Defense Analyses was founded as a not-for-profit defense "think tank". After sputnik, its mission and funding were expanded. It was the home of the professionals.

York (who was later the founding Chancellor of U.C. San Diego), Wheeler, and the others selected a group of star young physicists who were professors at universities. These were the amateurs. They were invited to organize themselves as a group that would meet during the summers (administratively supported by IDA) to work on classified defense problems—especially any that were too difficult for the professionals. The JASONs were paid a daily fee that was generous by academic standards, but within industry norms. Originally, the group was called Project Sunrise (likely a Wheelerism), but by 1960 they were known as JASON, an acronym that stood for exactly nothing. From the beginning, the group was self-governing. Murph Goldberger (my Princeton department head) was the first JASON chairman. Twenty years later, in 1977, most of the original crew were still members, plus a few later additions, including Curt Callan, another Princeton professor. Curt called me in early May and invited me to join the group. I had plans already for most of the summer, but found a way to come for two weeks in June. I had no idea how momentous a decision this would turn out to be, resulting by now in more than forty summers' involvement with JASON work.

The summer of 1977 became one of continuous travel. The JASON study was in La Jolla, an upscale district of San Diego surrounded by ocean on three sides. The actual venue was, improbably, Cummins Hall, the girl's dormitory on the grassy campus of the Bishop's School, a private boarding school. Because the school was empty in the summer, it was available to rent. Beds were taken out and replaced by metal desks, blackboards, and document safes. I was used to Top Secret facilities like Livermore, with gates, guards, and guns; but for JASON, the government was satisfied, apparently, with the fact that the dormitory had only one entrance, where a sleepy private guard was stationed.

I knew some of the JASONs already: Roger Dashen and Freeman Dyson from IAS, Fred Zachariasen from Caltech. Jonathan Katz, Bahcall's prodigy postdoc, had joined a year previously. I knew Dick Garwin of IBM slightly, from his gravitational wave experiment. Others, I knew by reputation: Walter Munk, the famous oceanographer; Ken Case, from Rockefeller; Henry Foley, from Columbia. In 1948, Polykarp Kusch and Foley had co-authored the discovery paper of the anomalous magnetic dipole moment of the electron. This was the experiment that confirmed Feynman's and Schwinger's independently developed theories of Quantum Electrodynamics (QED). But when the Nobel Prize was awarded in 1955, it was shared between the more senior Kusch and Willis Lamb, who did a related experiment on the hydrogen atom. Foley did little of note after that. I liked him immediately. He had a clear-eyed view of JASON. They had all been whiz kids, but had grown old together, while they let no one else into their exclusive club. Some JASONs resigned or became inactive during the Vietnam War (which had ended just two years previously). What remained of JASON was the change-resistant hard core. Three years previously, JASON's government sponsors gave them an ultimatum: Bring in a new generation of JASONs or face extinction—funding all cut off. Callan was the first new JASON; I was about the fifth.

There were two principal topics being studied that summer. The first was sub-ocean "internal waves" that, some thought, might be produced by our nuclear missile submarines and make them locatable and thus vulnerable to preemptive attack—negating their value as a strategic nuclear deterrent. The second was climate change—possibly the first systematic study of this very new subject by the U.S. government. In my usual groove, I did some computer simulations of the relatively simple, though nonlinear, differential equations that each of the two studies generated. Margaret and Sara were back in Boston so, on weekends, I patrolled the beaches looking for pretty girls—just looking.

I learned from Freeman Dyson that he had once undergone the Pasteur anti-rabies treatment. Despite its reputation, he said, it was not particularly painful. But one in six hundred patients died from encephalitis carried in the rabbit brains used to make the vaccine. What he most noticed during the treatment, he said with a completely straight face, was his own insatiable appetite for lettuce.

I was home for only a few days before Margaret, Sara, and I set off for Cambridge, England. By now, this was familiar ground. Churchill College, a "new" college built of brick in 1958 on the outskirts of the city, rented us a semi-detached modern flat, one of several laid out amid flower gardens on the far side of the playing fields from the College proper. The view from our front door was a grand vista, manicured grass stretching seemingly to infinity in all directions and, on the horizon, the skyline of the town with the King's Chapel. I had a short walk to the Institute of Astronomy.

The usual gang of Americans were there. Martin Rees held a party for all of us in his rooms at King's. These had once belonged to Lord Keynes. Some worn wood paneling had only recently been removed, exposing a wall of murals painted by Bloomsbury-circle artists Vanessa Bell and Duncan Grant. Naked men and clothed women cavorted—in the paintings, not at the party. Of the Bloomsbury crowd, Dorothy Parker famously quipped that they painted in circles, lived in squares, and loved in triangles. The other historical attraction was the ancient W.C., the very one on which Keynes had formulated modern economic theory.

One weekend we drove to Oxford to visit the Woodhouses, our friends from Princeton. Nick was now a proper Oxford don in Waddham College. He and Mary Jane had a tiny row house on Observatory Street. She also rented an apartment in London near her law chambers and commuted to Oxford on weekends. After putting Sara to bed, we drank several bottles of wine while sitting around an ancient miniature fireplace that burned peat bricks—for us a novelty.

I had just a couple of days back in Boston, then traveled to Waterloo, Ontario, for GR8, the triennial general relativity meeting. GR7, three years earlier in Tel-Aviv, was my first introduction to the international GR community. Now I seemed to be part of its establishment, a member of the organizing committee, no less—Kip's and John Wheeler's doing. John was himself on the committee. Waiting at Newark Airport for our flight to Toronto for the first committee meeting, he said to me, "Let's make a list of who should be the invited speakers." I reeled off names. He added a few and rejected none of mine. The meeting, like so many international committees, devolved into endless nitpicking. John went silently to the blackboard and transcribed our list in his hyper-legible hand. I was only surprised that he had not brought his favorite-colored chalk. He then sat down, still saying nothing. After more fruitless discussion, someone said, "Well, what about John's list?" It was quickly adopted. One of the GR8 sessions was "Computer Techniques," chaired by Larry Smarr. In later years, this session was remembered as the meeting that founded a new field, numerical general relativity.

34. Protvino

My U.S.S.R. visa came through just a day before I was scheduled to fly from JFK to Moscow on Pan Am's direct Moscow flight—misnamed, because one actually had to change planes, to a stripped-down Boeing 707, in Copenhagen. The flight, both legs, was almost empty. I was participating in a formal U.S.-U.S.S.R. scientific exchange, a three-week workshop on astrophysics at a country location outside of Moscow. Our American delegation consisted of David Pines (unofficial delegation head), Peter and Susan Goldreich from Caltech, Ed and Mikka Salpeter from Cornell, Fred Lamb and Don Lamb (who were identical twins and both astrophysicists), Jon Arons, Stu Shapiro, Doug Eardley, Brian Flannery, and two token experimentalists, Walter Lewin and George Clark (both from MIT). Exchanges like ours, where American and Soviet scientists could meet in surroundings that allowed private interactions, were still rare. Ours was the first in astrophysics, and was possible only because of Zel'dovich's clout.

We were met at airport customs by Rashid Sunyaev, one of Zel'dovich's two lieutenants. Rashid's work was known to us, but, as with many of the Soviets, he had never traveled to the West. Both scientifically and in person, he was a man with hustle, known with respect as the Soviet Ostriker but (I thought) with a bit of Soviet Ruffini.

A side note: the reader may find peculiar my consistent use of "Soviet," never "Russian". That was the political correctness of the time. Although dominated by ethnic Russians—the Georgian Stalin one great exception—the U.S.S.R. maintained the pretense of being multi-ethnic and multi-national. If you referred to someone as Russian, you would be corrected by the other Russians: "Soviet". This was especially true of Jews, who were sometimes made to list "Jew" instead of "Russian" as their nationality on their internal passport. Ironically, Sunyaev was the rare case of a non-Russian who had risen to high professional visibility. He was a Tatar, from Tashkent.

The U.S.S.R. Academy of Sciences (ANSSSR) had its own hotel in Moscow, an inefficiency typical of the Soviet command economy. The Academy was allocated resources according to its needs and political clout. It was easier for it to get a whole hotel than to get operating funds to pay some other part of the vast bureaucracy for hotel rooms. The Academy hotel was dingy and not very clean. The bar of soap was the smallest I had ever seen, the towel threadbare. We were staying in Moscow just one night. Rashid instructed us to go to dinner at the Hotel Warsaw and then left. At the restaurant, it took hours for our food to come, and it had no resemblance to what we had ordered. But, it was an upscale banquet, Soviet style—with a special exorbitant tourist price to match. There was entertainment, a slush-rock band (minus heavy beat) with here and there a bit of tango thrown in. The crowd was well-dressed and, in appearance, ethnically diverse—more like the United States than like Poland or Germany, say.

When the bus picked us up the next morning, the Soviets were already on board. The conference was in Protvino, a new town two hours bus ride to the south. Protvino was completely owned by the ANSSSR. It housed the twenty thousand scientists, technicians, and support staff for the Serpukhov high-energy particle accelerator. The main street had twenty stores—better stocked than those in Moscow—one large restaurant, our hotel, and a "Palace of Culture" with meeting rooms and a movie theater. The rest of the town was apartment blocks of uniform design. We couldn't at first understand why it took so many people to keep their accelerator running. Fermilab, the equivalent American facility, in the Chicago suburbs, employed fewer than two thousand. The answer became clear on the tour that afternoon, when they showed us row upon row of machine shops, one of which, incredibly, was making machine screws from rod-stock. Thousands of parts that, in the United States, would simply be ordered from a catalog were here made in workshops. In its peaks—Serpukhov was one, Sputnik had been another—Soviet science and technology were the intellectual equal of American. But the underlying industrial economy was vastly more primitive.

The Soviet delegation was led by Zel'dovich (of course); Isaac Markovich Khalatnikov, the director of Moscow's Landau Institute of Physics; and Roald Sagdeev, the director of the Moscow Space Research Institute, known as IKI, roughly equivalent to one of the U.S.'s large NASA labs. Sagdeev traveled frequently to the West and was a Communist Party member, but he managed to be also something of a free spirit. His teenage son, along for the conference, wore a Haifa University sweatshirt that Roald had brought back from a trip. Free spirit or not, Sagdeev later won his government's Lenin Prize in 1984 and Hero of Socialist Labor in 1986. He was elected to the Supreme Soviet in 1987. Soon after the fall of the Soviet Union, however, he emigrated permanently to the United States, settling at the University of

Maryland. He was briefly visible to the public when he in 1990 married Susan Eisenhower, granddaughter of the U.S. president.

Iosif Shklovsky, the prominent Soviet radio astronomer (whom I had met five years earlier at the Texas Conference) was another big-wig. The little-wigs were mostly younger members of Zel'dovich's Moscow group. They stayed the whole three weeks. A rotating assemblage of researchers from other groups and other institutes, few staying more than a couple of days, filled out the Soviet side. Zel'dovich probably arranged it this way to keep out the riff-raff. He was unable, however, to exclude two dour "scientists" from IKI who were—word of this spread within the first ten minutes—the conference's KGB minders. Behind their backs, we called them 007 and 008. Mostly we simply avoided the minders, but, after 008 came down with kidney stones and had to be evacuated back to Moscow, Stu Shapiro spent a night out drinking with Rashid and 007. When Rashid became drunk, he referred to 007, apparently without irony, as "my director".

Velodya Keilis-Borok, my parents' close Russian friend, later explained to me why. IKI was not a purely academic institute. It designed not just scientific satellites, but also spy satellites, and it was secretly administered jointly by ANSSSR and the KGB. Our minder 007 was a KGB colonel with the title Vice-Director of the institute. He made all decisions that were not purely scientific. It was remarkable that all the Soviet scientists except Sunyev treated him so dismissively; but they were under the protection of Zel'dovich or Sagdeev.

For the Soviet participants, the three weeks were a lavish junket, paid for out of someone else's budget—I never understood whose. We had nightly banquets with caviar, smoked sturgeon, endless zakuskis (hors d'oeuvres), vodka, and Georgian wine. There was always at least an hour of formal toasts. Sober, David Pines, our delegation head, was an unctuous toady. Inebriated, his toasts were argumentative, with allusions to Soviet misdeeds. Zel'dovich—careful man—preferred to make no toasts himself, but to call on others. Thus called upon, the attractive young woman who was a technical translator offered, "To extra-terrestrial astronomy and appropriate cooperation between the U.S. and U.S.S.R." Zel'dovich responded with: "So very brief, and so very *on the line* politically!" to much laughter. Sunyaev toasted with a long story about his Muslim ancestry, camels in Tashkent, his right—preserved, he claimed, under Soviet law—to have four wives. I lost the thread at that point. Zel'dovich had learned that my father was in the White House and he called on me to offer a toast related to that. I made up a lame joke about sons who ask about their biological parentage (it fell flat), then offered a toast to "the only thing able to come out of a black hole," namely scientific cooperation between the U.S. and U.S.S.R. I had already learned always to say "cooperation" and never "collaboration," because the latter, translated into Russian, was tainted by the meaning "collaborationist"—with the Nazis in World War II.

The scientific sessions were rough-and-tumble, very much in the style of Landau, whose ghost then still presided over all of Soviet theoretical physics. The speaker could be, and usually was, heckled by the audience on any point lacking clarity. We saw this in the extreme when our nominal senior host in Protvino, the Director of the Institute of High Energy Physics, insisted on giving a talk on his own theory of gravity. He was a party member, a vice-president of ANSSSR, and a deputy member of the Supreme Soviet. His talk was such utter nonsense that it started a big row. Leonid Grishchuk, a young member of the Zel'dovich group, ambled to the board like a huge Russian bear on the hunt. He was soon joined up front by his (safer) elders Zel'dovich, Lifschitz, and Khalatnikov, all shouting among themselves and the speaker in emotional Russian. Lifschitz occasionally recovered his English enough to explain to the Americans that the debate was ridiculous—it was all in his book with Landau; then he returned to the fray. It was eventually decided that the senior party member would continue through to the end of his presentation, which prompted us younger Americans all to walk out and go to dinner. The braver young Russians (Grishchuk, Dmitri Kompaneets, Sasha Polnarov, and Andrei Illiaranov) soon joined us.

Soccer practice in the afternoon was mandatory. So that the contests would not be completely one-sided, a couple of the better Soviet players were assigned to the American side—and I was assigned to the Soviet team. After a few first attempts, my teammates never knowingly sent the ball my way. However, with foresight, I had brought on the trip a glow-in-the-dark frisbee, and we introduced our Soviet colleagues to that nighttime game, which I was better at.

I also brought with me a bottle of White Horse scotch, to pay off my bet with Zel'dovich—the actual bet, not the mis-remembered one on Nixon's resignation that he had mistakenly paid off to me with an autographed bottle of vodka. I explained the situation, and he accepted the gift. A couple of days later, Peter Goldreich told me that Zel'dovich had come to him and hinted that more than anything in the world he would like to be made a present of a glow-in-the-dark frisbee. "But Press has the only frisbee," Peter said. "This I know," said Zel'dovich, "but Press has already given me the bottle of White Horse." So, doing the right thing, I had to seek out the father of the Soviet atomic bomb and deliver, with a straight face, the memorable line, "Yakov Borisovich, please allow me to present you with this glow-in-the-dark frisbee." He accepted it graciously.

About one day out of three was given over to all-day excursions. We spent half a day cruising the Oka River in a modern planing riverboat driven by water jets. It had forward-facing benches that seated two on each side of the aisle. The engine was very loud. That was the whole point: Americans and Soviets paired off on the benches for frank conversation, while 007 sat glumly on a bench reserved for him, right over the engine. The afternoon was a mushroom hunt in a dark forest where—you guessed it—Americans and Soviets paired off to gather mushrooms. The Soviets were supposed to know which were the poisonous ones, but this subterfuge was so thin that, ultimately, all the gathered mushrooms had to be thrown out.

One could only hunt mushrooms for so long. In pairs, we wandered back and joined a group of sunbathers on an array of haystacks in the middle of a sunny field. The men took off their shirts, while the translator girls stripped down to their underwear, in which they looked like chubby Soviet track stars. For some reason, we had no double-oh characters along with us this day, but I wanted to be careful. I took Dmitri Kompaneets aside and asked if there was anyone on the excursion who might be "unreliable" for us to talk to. "In the Zel'dovich group, everyone is OK," he said, "but I don't know about the others." "Are you sure about Misha?" I pressed. I had gotten bad vibes from Misha, who had traveled alone to the West—usually a bad sign—and whose father was a high Soviet official. "Misha is a good boy," Dmitri said. "Don't worry about Misha."

Peter made a point of having long, individual frank talks with many of the younger Soviets. Some, especially the Jews, were quite bitter about the situation in the U.S.S.R. Dmitri was exceptionally outspoken. His father, a famous physicist, had died the previous year. Dmitri was convinced that as a Jew his father was not given the best medical care—drugs from the West, for example. Almost no Jews were now getting into the universities, he said. His case had required his father's influence. Materially, Dmitri was quite well off. He had his own car, something almost unheard of among his peers. The car's radio had a shortwave converter with five settable buttons—all tuned to Voice of America's various frequencies.

Yet another excursion was to Buffalo National Forest, a preserve where they were trying to cross-breed the nearly extinct European bison with the American variety, and trying to keep alive a few precious beavers, all but wiped out by 19th century trapping. We should consider ourselves privileged to see the actual beaver dam, they said. I kept to myself that in New England, beavers were considered a pest, their dams dynamited.

A different day, we rode for two-and-a-half-hours to Yasnaya Polyana, the Tolstoy family estate. Our guide there was a young Slavic-featured blonde, a French-language student working on a summer job. Since we didn't speak French, she spoke in Russian, in turn translated by our chubby IKI interpreter girls, usually only after much giggling consultation among themselves. Shklovsky sometimes offered his own translations, which were more amusing. We kept hearing about the large and beautiful house in which Tolstoy lived. Finally, we arrived at a stand of trees. "This is the house," says the guide. There was no house. It was the emperor's new clothes. We requested another attempt at translation. "This is the house!" insisted the guide, and our chubby translators all nodded in agreement. Pointing at trees, the guide proceeded to describe the nonexistent house room by room as we walked through them. (One of the Soviet astrophysicists told us later that Tolstoy sold the house during his lifetime, and it was taken away in pieces.)

When all possible local excursions had been exhausted, excursion days became "picnics". A truck would be sent out early in the morning. When we arrived at some forested location, the table was already set with white linen, caviar, vodka, wine, and even, rarest of delicacies, Pepsi-Cola (bottled in Moscow). Then, when everyone was drunk, they stoked the coals and cooked shashlik, long skewers of lamb, at least two to a person. You dipped it into a peppery sauce, and ate it with ripe, whole tomatoes and cucumbers.

The American neutron bomb was the subject of much discussion between the Soviet and American scientists. Our Soviet colleagues unabashedly admired the freedom and economic prosperity of the West, but they kept a small quantum of belief in Soviet idealism—the longing for a classless society with collective values favoring culture and science over money and greed. In June, two months before our conference, the *Washington Post* had published an article about the neutron bomb. The concept was a nuclear weapon that produced very little blast, but a copious flux of neutrons. These could penetrate buildings and bunkers without damaging them, but would kill people. Secretary of Defense Harold Brown, a former Livermore Lab director (and Caltech president) quipped that the neutron bomb was the perfect capitalist weapon—it killed people without damaging property. The unfortunate quote was picked up and amplified by the Soviet media. We Americans were challenged to defend our country's policy—if we dared.

Coincidentally, I knew a lot about the neutron bomb. I had visited Livermore Lab two weeks before the *Post* article came out. Larry Smarr was there consulting with the Jims, Wilson and LeBlanc, on his general relativity computer program. Larry avoided all classified work, but he held a security clearance, because you couldn't use the Livermore computers otherwise. The two of us were in LeBlanc's office. "Well, what weapons are you working on?" I asked LeBlanc. Larry cringed. He hadn't learned that Livermore people loved to talk about their work and ignored need-to-know rules if you expressed the slightest interest. Soon, LeBlanc was drawing diagrams on the board and explaining all manner of clever weapons designs, including, as it happened, the neutron bomb.

Obviously, I couldn't repeat any of his lecture outside the Livermore fence. Zel'dovich knew about my Livermore connection, I think, but he didn't tell the others. His own continuing work on weapons was a secret from the Moscow astrophysics group. I knew from my Livermore friends that it existed. Still, I felt an obligation, here in the heart of Mother Russia, to defend the American bomb—not least because it was a Livermore design. The defense went like this:

U.S. and NATO defense posture in Cold War Europe was predicated on the scenario of a massive Soviet armored attack through the Fulda Gap, overrunning Frankfort and continuing west to the Atlantic, much like Hitler's blitzkrieg in World War II. Personally, I never thought that there was the slightest possibility of this. It was the classic proverb, "Generals are always prepared to fight the last war." A thousand years of history taught that Russia always absorbed attacks into its own depth, only then initiating massive counter-attacks on weakened forces. It never attacked first. This was as true for World War II's eastern front as it had been with Napoleon. Soviet designs on Western Europe, which were real, were to be achieved by subversion from within. Nobody in Washington asked me what I thought, however. The Fulda Gap scenario was U.S. dogma.

NATO's forces were incapable of stopping this imagined attack. The U.S. looked to nuclear deterrence, threatening to nuke Moscow if an attack happened. Since the U.S.S.R. would then launch a nuclear attack against American cities, no one on either side thought that this deterrence was credible. What we would actually do, it was thought, was to use nuclear weapons on the massed Soviet forces—and not on Russian soil. The problem with this was that it would reduce to radioactive rubble a large number of European cities and additionally kill hundreds of thousands by fallout. No one believed we would do this. So, really, we had no deterrence at all.

The actual utility of the neutron bomb was not to kill civilians in bunkers. It was to stop the Red Army, to kill troops in tanks and armored personnel carriers. The underlying scenario was ridiculous, but the military response to it was, in the topsy-turvy Cold-War world, rational. I explained all this to my Soviet friends. They listened with interest, but were unconvinced. Evidently, after all the negative press, President Carter himself become unconvinced, because he cancelled the program, triggering new outrage, now from the national security community on the other side of the issue. When I asked my father about the whole mess, he began, "Here's the most charitable case that one can make for the president's actions..."

The last week of our three-week workshop, everything fell apart. Zel'dovich, it turned out, was still battling behind the scenes to keep Vitali Ginzburg's whole group from making the trip to Protvino and taking over the conference. We temporarily captive Americans were *that* unusual—or maybe the caviar and vodka were just that good. Ginzburg and Zel'dovich, both early protégés of Landau, were enemies on a melodramatic scale. Ginzburg was famous for the Ginzburg-Landau theory of superconductivity. That work, done in 1950, waited until 2003 for its much-deserved Nobel Prize.

One morning, when we woke up, we were told that Zel'dovich and his whole group had departed in the middle of the night, and that Ginzburg and a group of twenty would be arriving soon. They never arrived. We Americans sat around for a day without even a picnic—much less an excursion. We had to make do with an evening of drinking and dancing with the pretty waitresses from the restaurant, one of whom brought in a record player. At midnight, the restaurant doors were locked from the outside. We were trapped. I seemed to be the only one who was concerned. I broke through a sealed delivery door at the back of the restaurant kitchen and went to bed. Zel'dovich, Sunyaev, and a few others came back the next day, embarrassed. We had a couple of desultory scientific sessions, then we all returned to Moscow (and the dirty Academy hotel) several days sooner than planned. When our bus reached the outer Moscow Ring Road, Zel'dovich asked the bus driver to stop at the first Metro station. The last I saw him, this three-time winner of both the Order of Lenin and Hero of Socialist Labor, he was disappearing down the subway station stairs, not visibly special among the other Soviet citizens in the crowd.

35. Academics

Committee meetings in the astronomy department could be cutthroat. We were doing a search for two new assistant professors. One position was earmarked for a solar astronomer. The other was for any field, on the basis of the candidate's merit. Bob Noyes, a Smithsonian professor, was the solar astronomer on the search committee; I was the theorist; Ed Turner, the optical observer. Turner was only an assistant professor himself, but he had accepted a tenure offer from Princeton, so we treated him like a grown-up.

There were three obvious top candidates: Bob Rosner, John Huchra, and Paul Schechter. Not entirely coincidentally, they were in exactly our three fields: Rosner studied the Sun; Huchra was an optical observer who did his Ph.D. at Caltech with Wal Sargent; Schechter was a theorist who, after Bahcall had seemingly soured on him, took a second postdoc at the University of Arizona, a great place for observers, not so great for theorists. The two others were already at the Center for Astrophysics as postdocs. I preferred Huchra to Schechter, given what I knew about both. I could ally with Turner on our committee and make this happen. Except that a funny thing happened at the last committee meeting before our recommendation to the full department. Noyes announced that he could not possibly support Huchra and would oppose him in the full department meeting. We asked why. "It's just something that I feel," he said.

I understood the play immediately. Noyes knew of my aversion to solar astronomy. So, I would be against Rosner, he thought. He assumed that Paul, as a theorist, was my actual candidate of choice. He reasoned that by giving me Schechter, he would ensure that I would refrain from vetoing Rosner. It was a well-executed maneuver, and the department voted offers for Rosner and Schechter. Despite any misgivings, I couldn't bring myself to speak out against Paul at the department meeting. George Field luckily rectified matters by promoting Huchra to a permanent Smithsonian position, which he happily accepted.

Alan Lightman and I wrote a paper together on the development of correlations among galaxies, then a hot topic. I mention it not as an important piece of work, but rather as an example of a topic, once hot, that advances the field and then creates its own irrelevancy. Peebles first noticed that the correlation function of galaxy positions—roughly, how many galaxies should you expect to find, on average, at a given distance from a given galaxy—was a universal and unexplained function. The study of galaxy correlation functions led directly to the discovery of structure in the universe on much larger scales (first by Margaret Geller and John Huchra, among others). That led in turn to the development of computer codes that could simulate in a "standard cosmological model" how large-scale structure evolved to produce galaxies.

These models agreed beautifully with observation—including reproducing Peebles correlation function. The latter hadn't been explained in any deep sense, but it was no longer very interesting—galaxy correlations became just another well-established consequence of the standard model. As an assistant professor, Alan was doing good science, but his heart was already in writing—poems, essays, and eventually novels. I was one of few who knew this—we were sworn to secrecy. Alan's eventual move to MIT's humanities division freed him to do what he really loved.

Alan and I submitted our paper to *The Astrophysical Journal*, and got a favorable, innocuous, report from the anonymous referee. We made the referee's small suggested changes and resubmitted the paper. In the second round the same referee suddenly discovered a slew of new things to object to. He had been flying under false colors in the first report. Now, he seemed eager to delay the paper forever. He slyly offered to read the paper carefully a third time, should we decide to resubmit.

I was angry. Against all protocol, I was going to find that referee and give him a piece of my mind. I called Jim Peebles first. "Are you the referee?" "No, I'm clean," he said, and then added with some relish, "You mean there is *another* referee who takes issue with your work? Well, keep trying, the phone book isn't that long." "I'm up to P already," I pointed out.

Next, I called Paul Schechter, whom I had not talked to for six months or more. "I'll ask you bluntly," I said, "are you the referee?" "Yes," he said, "and let me ask you bluntly, am I going to get the assistant professorship at Harvard?" He was quite surprised when I told him that he indeed would—that the paperwork was already going through. It was an offer Paul may never expected to get—and wouldn't have gotten except for the tricky Bob Noyes. Alan's and my paper was eventually accepted, no thanks to Paul.

* * *

The American Academy of Arts and Sciences was a peculiar Boston institution that had been founded by John Adams, John Hancock, and others in the years between the signing of the Declaration of Independence and the writing of the United States Constitution. The American Academy was not as great a success as those other projects. Intended to compete with the great learned academies of Europe (the Royal Society or L'Académie Française), it was eclipsed in the mid-19th century by the *National* Academy of Sciences, whose location in Washington and historic charter signed by Abraham Lincoln—that NAS should be the source of science advice to the federal government—put it on top. From then on, election to the NAS was a big deal, while election to the American Academy was a small honor, a curiosity. By the 1950s, its active membership consisted mainly of Harvard and MIT professors, plus a few others willing to make the trip to Boston for its meetings. It had improved since that time, but it still had many members of no great achievement.

Steve Weinberg was certainly not one of those. He was a member of both academies. He was also a member of that unofficial club of a couple of dozen physicists worldwide who were waiting—some more patiently than others—for Nobel Prizes that might never come. Steve's was on hold, rumor had it, because laboratory experiments at Seattle and Oxford on bismuth atoms showed discrepancies with the Weinberg-Salam theory's predictions. The Swedish Academy dreaded the possibility of awarding a prize for work that later proved to be wrong. This had happened before. The 1926 prize in medicine was awarded to Danish scientist Johannes Fibiger for showing that roundworms caused cancer in rats. It was actually vitamin A deficiency (creating DNA damage). Even Enrico Fermi's 1938 prize was flawed. While Fermi certainly made several Nobel-worthy discoveries, his actual citation was for "demonstrations of the existence of new radioactive elements produced by neutron irradiation." He had even given these elements the names Ausonium and Hesperium. They didn't exist. He had unwittingly discovered nuclear fission. His "new" elements were actually fission fragments, radioactive isotopes of the known elements barium and krypton. (Otto Hahn's 1944 prize in chemistry was for the actual, knowing, discovery of fission.)

Weinberg often came to our theory brown-bag lunches at the observatory, and he was not shy about keeping us apprised about the status of Weinberg-Salam. He was particularly happy when an experiment with polarized electrons at the Stanford Linear Acceleration Center agreed perfectly with his predictions. Shortly after that, the bismuth experiments, which were difficult in the first place, were shown to be wrong. That was enough for the Swedish Academy, and Weinberg and Salam shared the Nobel Prize in 1979.

One month, because Willy Fowler was the guest speaker, Steve and Louise Weinberg invited Margaret and me to one of the American Academy's monthly dinner meetings, which spouses also attended. Louise was a lawyer, an academic expert in admiralty law. That specialty commanded few tenure slots in law schools, and Louise was underemployed, an irritation that may have been responsible for her reputation as a dragon lady. We found her charming. The Academy was at this time housed in Brandegee House, a lavish Belle Époque mansion in Brookline, surrounded by formal gardens and marble statues. Margaret and I were seated at opposite ends of the head table, she with Willy and Steve, I with Louise and Bernie Burke, the MIT radio astronomer. After dinner, we moved for the talk to a gilded salon with white marble columns. Viki Weisskopf, president of the Academy (he still my fan for the Size of Man paper) introduced Willy, who gave his standard talk with its standard jokes: His doctor told him to soak his sprained ankle in warm water. "But doctor, my mother always told me to use cold water." "Your mother was wrong. My mother told me to use warm water."

Jumping out of time-order, in 1982, Louise and Steve Weinberg moved to the University of Texas at Austin. The Harvard Law School had appointed as its first tenured woman professor my Faculty Row neighbor Martha Field. It was the position that Louise had coveted, and she was passed over. Steve was by then a Nobel Laureate, but, oddly, it was Louise who was first offered a chaired professorship in admiralty law at the land-locked Texas law school. Steve's offer followed a long month later. This careful orchestration was the work of a certain still-new Texan—John Wheeler. Nostalgic for the American Academy meetings, Steve and Louise founded their own monthly dinner society, Austin Tuesday Club, to which couples were elected as a unit. Another time jump: When my non-Ph.D. spouse and I moved to Austin in 2007, we were soon elected to the Tuesday Club. We still attend faithfully.

A few weeks after Willy's talk, I got a letter from Viki Weisskopf. The American Academy published its monthly talks as little pamphlets. Willy had submitted a manuscript so slapdash as to be unpublishable. Viki wanted me to quietly rewrite the whole thing. I did so, leaving out the jokes.

36. Academic Politics

In Washington, my father had worked his way into the Carter inner circle of advisors, if not as inner as those from Georgia. He attended Cabinet meetings, sitting with the other senior advisors in chairs against the wall of the Cabinet Room, not at the table with those of Cabinet rank. When a decision was technical, rather than political, Carter would often dismiss the matter peremptorily, saying, "Let Frank handle that. I'll go with what he decides." The politicos of course took note. Frank led a delegation to Moscow that included a private meeting with Kosygin at which he delivered a message from Carter, that the U.S. government would not attempt to moderate the outrage of its scientists at the treatment of Soviet dissident and refusnik colleagues. "And I say this with full knowledge that you consider the dissidents to be criminals who have broken your laws," Frank said. Kosygin listened grimly, but made only a muted response. Although he was still powerful in some areas, including science, Kosygin had by this time been largely eclipsed by Brezhnev as the dominant Politburo member.

The still-new Harvard-Smithsonian Center for Astrophysics (or CfA) lurched from crisis to crisis. George Field was an inexperienced director. I knew little then about the nuances of leadership and management; but it was clear that George didn't know much about them, either. His strength was as a statesman of science, the calm spokesperson for astronomy as a discipline. Plucked from full-time research and teaching at Berkeley, and suddenly in charge of a complex three-hundred-person institute, he was foundering.

George had a knack for making people feel badly about their interactions with him. Brian Flannery was an assistant professor on the Harvard side who had applied for a Smithsonian "permanent" position. He got a form letter, telling him that he was on the "short list"—this, a routine move made by search committees to keep their top candidates from prematurely accepting jobs elsewhere. Brian scheduled an appointment with George to chat, expecting to be courted. "Honestly, Brian, you're not my number one choice, and I can't even tell you that you are my number two," George told him. It was so unnecessary: George could have said, "Brian, I'm blown away by the quality of the top applicants, including you." It wasn't just Brian. Seeing me as their ally, virtually all of the astronomy junior faculty came to me with similar stories.

I complained about George to anyone who would listen. He was inaccessible. It took days to get an appointment to see him about even the smallest issue—the kind of thing that today would be a one-line email—and there was no informal or alternative channel. I needed to ask whether I should explore with Chandrasekhar his spending a three-month sabbatical at CfA. Two days later, George kept me waiting for twenty minutes in his outer office, then, his secretary: "Shall I send Bill Press in now?" "No, I'll come out." The inner-office door opened. "Hello Bill!" I asked my question. "Yes, good idea." I turned to go. "Bill, is that *all* you have to say? I pushed forward my whole afternoon schedule for you!" Another time, I met with George to complain about the shoddy way that incoming theory postdocs were being shuffled around into bad offices. "Bill," he told me, "I really do have more important things to concern myself with." On another occasion: "You are a brash young man, Bill, and I hope that you won't keep causing me difficulties." I probably deserved this, but he might have put it differently.

In the Declaration of Independence, after the famous part that everybody knows ("When in the course of human events…"), there is what I think of as the kvetching part, a bunch of grievances about King George, many of them individually petty, but in toto indicative of—as it turned out—a really big problem for the king. "He has called together legislative bodies at places unusual, uncomfortable, and distant…." "He has erected a multitude of New Offices, and sent hither swarms of Officers to harass our people, and eat out their substance." This was the problem with our King George's administration at CfA: He didn't seem to understand that an accumulation of "little" things set the tone of the place, and could be as important as big things.

Riccardo Giacconi, meanwhile, had his own problems with George's administration. Riccardo's X-ray Astronomy Division was the largest division in the CfA in people and dollars. He had been recruited earlier than, and somewhat independently of, Field. In Riccardo's world view, X-ray astronomy was the center-of-the-Center for Astrophysics, and everything else should revolve around it. This manifested in some peculiar ways. Smithsonian positions had to be advertised publicly before being filled, and there had to be a search committee, even though the final decision would be made by George. Riccardo always had a candidate to push for these positions, generally someone from his own group no longer useful to him whom he wanted to outplace. When there was an open associate director position for ground-based optical astronomy, it turned out that Herb Gursky (a part of Riccardo's team as far back as their 1961 rocket launch) had a previously unknown secret attraction to ground-based astronomy. Riccardo met with George, threatened him with something (his usual mode), and Herb got the job.

One year, I chaired the graduate admissions committee and in due course presented our ordered list of ten recommended admits to the astronomy faculty. "I'm voting against *all* of these," Riccardo announced. He wanted to offer admission to the next ten on the list. "They're not as good, so they will be more grateful to us, and work harder," he said, with a certain logic. What he didn't say was that he was having a hard time attracting the best students to work in his group and felt that he could better find cheap graduate-student labor among a less-exalted cohort.

Riccardo's view of his colleagues, indeed of the whole scientific enterprise, was entirely transactional. His job was to discover X-ray sources in the sky. That was his only job, and he would allow nothing to get in his way. Occasionally his approach happened to align with the goals of others. Affirmative action—specifically the push to hire women—was to Riccardo a distraction that could easily be dealt with. He decreed that one out of five postdoc positions in his group would be filled by a woman. When it turned out that there were no acceptable female applicants, he ordered his people to assemble a list of the best woman graduate students and postdocs in all fields of astronomy, ordered this list by his own estimate of their abilities, and started down the list writing personal letters offering immediate employment with the chance to switch into X-ray astronomy under his supervision. He did not have to go far down the list before getting someone.

Besides the young people's revolt and Riccardo's ambitions, George was also embattled on a third flank. The astronomy department chairmanship had just passed from Alex Dalgarno to Al Cameron. Both had Dean Rosovsky's ear. Both felt that George—wearing his hat as director of the Harvard College Observatory—was trampling on prerogatives of the senior faculty. Our Harvard salaries, and those of our assistant professors, came from the HCO budget. In appointment discussions at senior faculty meetings, George was an oracle. Can we afford to hire X? The oracle would speak: Yes. Can we afford to hire Y? The oracle: No. We never saw an actual budget, nor did we believe that the oracle's answers were budget-based. George had been promised dictatorial powers at a time when the department was (in the Dean's eyes) in receivership. Dalgarno's and Cameron's view was that it was time to end that receivership. Specifically, they wanted transparency and faculty influence on the HCO budget.

More than anything in the world, Al Cameron loved to pontificate. He was a relaxed, but always pompous, pontificator, who could hold forth for hours at a time. Once he took over from Dalgarno as astronomy chair, our department meetings at least doubled in length. It was a combination of Al's self-indulgence and his canny tactics. Our meetings started at four p.m. Items of business on which Cameron had strong views were always last on the agenda. Around six, people would start sneaking out of the meeting. When Al sensed that the remaining few were in agreement with him, matters would come to a vote. When, by chance, the remaining few were opposing him, he would note the lack of a quorum and postpone the matter to a future meeting. I called this form of governance "sedentocracy"—the rule of those who could sit the longest. Unsurprisingly, the impatient George Field was frequently one who left. He had more important things to do.

To his credit, George did try one creative approach. Perhaps if Al could get some pontificating out of his system *before* the faculty meeting, then.... George created a "Committee on Future Automation of Secretarial Services by Computer" that met for two hours just before faculty meetings, with Al as its chair. The rest of the committee were all secretaries, who were in effect paid by the hour to attend. Listening to Al drone on about the future of computers was preferable to doing secretarial work, which in those days was mostly typing. Unfortunately, this tactic only seemed to get Al warmed up for the faculty meeting that followed.

These may sound like three entirely different, and mutually inconsistent, attacks on George's administration of the CfA. By now, however, when I have seen deans and directors and university presidents succeed or fail in many different institutions over many decades, I understand that all the attacks on George had a common root cause. He wasn't *leading*. Leadership doesn't just happen. It takes conscious work. It takes thought. It can be learned. They do a good job of teaching it in the military academies, for example, and in some executive training programs. George didn't communicate to his "troops" any coherent picture of the future of the CfA. Nor, if not so inclined himself, did he put in place mechanisms for surfacing such a vision from below (committees, town-hall meetings, strategic plan, etc.). His fear of losing autocratic control blinded him to the greater power of soft control. In fact, George did command such respect intellectually as an astronomer that he could have steered things his way and built consensus at the same time. Unfortunately, he instead communicated disdain for the people under him; he could simply have hidden this. He didn't have to *like* all people, but he had to *pretend* to like them—and he didn't. There were half a dozen other lessons from Leadership 101 that he just got wrong. Some will come up later in my story, when *I* got them wrong (or sometimes right).

Henry Rosovsky, as Dean, could not avoid the CfA problem forever, though I'm sure he would have liked to. Riccardo, Alex, and Al were each requesting multiple meetings with him; I was writing angry letters. Henry did what deans do: He appointed an outside visiting committee. It was chaired by Dave Heeschen, director of the National Radio Astronomy Observatory and a Harvard Ph.D. from the 1950s. The membership was all Establishment. Maarten Schmidt from Caltech, who discovered quasars, was on it. Individually or (for junior faculty) in groups, we were scheduled to testify to the committee. In advance of that date, I convened an off-the-radar meeting of our junior faculty and younger Smithsonian researchers in my office. From my activities in Washington, I had by now learned that talk—testimony—was lighter than helium. It just floated up and away. Written reports, on the other hand, were denser than osmium (the densest substance on Earth). They could weight down a committee's agenda with issues that had to be addressed. Our cabal divided up the work and wrote a thirty-page white paper, everywhere moderate in tone, mostly just explanatory, but damning in the volume of its evidence. When I talked to the committee, they all had copies in front of them.

A couple of months later, George Field confided to me that Rosovsky had received the review committee report and had asked him to stay on as Director. I wasn't surprised. No one was trying to get George fired, only (by various definitions) tamed. Surprisingly, no one had leaked our white paper to George, although he knew about it. He asked me what we recommended, and I told him. About 75% of our recommendations were accepted by the committee, he thought. The main outcomes of the whole revolt were that George became slightly less imperial; the HCO budget became slightly less opaque (though never actually transparent); a regular outside visiting committee was appointed to come once a year to hear us complain; and Riccardo gained control of the entire CfA Engineering Division. I had not even known that this was one of his demands. It may have just occurred to him on the spot.

Somewhat after that, Henry Rosovsky scheduled George, Riccardo, Al, Alex, and me to meet in his office. I think he just wanted to verify that we were all now playing nicely together. It happened to be the first Tuesday in October, the day that Nobel Prizes in Physics were always announced. Henry had known for some time that he might, one year, have a problem on that day. Steve Weinberg and Shelly Glashow, both Harvard physics professors, were leading contenders for related, but different, work. Rivalry between these two went back to the Bronx High School of Science, of which both were 1950 graduates. All but overtly, they hated each other. (In 1983, two years after Weinberg had moved to The University of Texas at Austin and finally resigned his Harvard position, Glashow argued and voted against a resolution of thanks to Weinberg for his long service to the physics department.) Rosovsky knew that if one of them ever won a Nobel Prize, the other would be furious—and would take it out on him. Deans were inevitably punching bags for wronged prima donna faculty.

"I was shaving this morning when the radio said, 'a Harvard professor has won the Nobel Prize in physics,'" Henry told us. "I had almost decided to slit my throat then and there when it said, 'in fact, *two* Harvard professors,' and I then felt able to live." The prize went to Glashow, Abdus Salam, and Weinberg, the order alphabetical. Steve came as usual to our astronomy theory brown-bag lunch that noon. We had a bottle of champagne, his third or fourth that day. Have you thought about your acceptance speech? we asked. "I've been thinking about it for five years."

He checked to be sure that he knew everyone present before continuing. "I wasn't exactly surprised by the award, you know." He still had a copy of a Stockholm newspaper article five years previous: "Weinberg and Salam win Nobel Prize". It was an error, but the leak must have come somehow from the Nobel Committee. Steve repeated his standard story about Salam, that he (Steve) had quoted from the Old Testament, "For everything there is a season…," only to have Salam say, "Yes. This is also written in the Koran." "But we published first!" Steve snapped back.

Weinberg and Glashow managed to coexist, but their mutual animosity was barely suppressed. Sidney Coleman was the third big wheel at Harvard in theoretical particle physics. He was the smartest—and by far the funniest—of the three, much admired in his field, but never with a big enough discovery to become a public figure. Sidney managed to get along with both Steve and Shelly, but could be caught in the middle, as in the matter of Ed Witten. Witten was a Junior Fellow (like Larry Smarr), Harvard's kind of exalted assistant professor, free of all teaching duties. Ed was modest and soft-spoken, and (everybody knew) a superstar. String theory (so-called) was a new and promising field, and he was its genius. When Ed's fellowship ran out, it took our physics department about five minutes to promote him to associate professor—without any outside letters or a search committee. But associate professors at Harvard, unlike elsewhere, were untenured. Matters came to a head when Princeton offered Ed a tenured professorship. The Harvard department had an available faculty line. It was obvious that Ed should be promoted into it.

Glashow, however, was pushing his own protégé, Howard Georgi, a good physicist but not in Witten's class. That meant that Weinberg would reflexively support Witten, and it put Coleman right in the middle. He flew back from his sabbatical year at Stanford to negotiate back and forth across the second-floor hallway in Lyman Lab. At the crucial department meeting, Weinberg surprised everyone. "I cannot possibly support Witten," he announced. "In fact, I would oppose his appointment even if there were a majority of the department in favor of it." There was shock. Did we hear him correctly? Yes, and Howard Georgi got the professorship. What happened, apparently, was that Sidney, tired of being go-between for two prima donnas, told them that if they didn't agree on a candidate, he was not going to come back from Stanford at all. And, after that, Steve (who did harbor some doubts about Witten—mistaken as it turned out) blinked first.

I had a tiny role to play. All three senior particle theorists told Ed that he would be a shoo-in for the next open line, a year later. He came to me for advice. I think he thought of me as more senior than I actually was because I had shared an office with his father that long-ago summer at Livermore Lab. "Should I wait for a year, or should I go to Princeton?" "Go to Princeton," I told him. "Don't look back." He took my advice. Not too many years later, Sidney Coleman, who had in effect negotiated Witten out of Harvard, was quoted about him in a New York Times article: "He's simply smarter than anyone else. He's brought light where there was darkness. Everything he does is golden. If you go to any theoretical physics department in the world, you can see that people are touched, and touched deeply, by Ed's work."

37. Committees

Murph Goldberger once saw me knock on wood talismanically, and he had taken me to task. I explained that I was not superstitious, but that superstitions worked even if you didn't believe in them. That wasn't his issue. I was doing it wrong, he said. I was knocking the way one would knock on a door, with my mid-finger knuckles (technically, the proximal interphalangeal joints) and with a downward motion. "That way is completely ineffective," he said. The right way to knock was with an upward motion of the knuckles closest to the wrist (the metacarpo-phalangeal joints) against the underside of a table. His mother had taught him this.

I was reminded of this encounter when I started getting appointed to national scientific advisory committees. The Old Boy Network operated effectively whether you believed in it or not. My being invited to JASON was not unrelated to my two years at Princeton. On JASON, I got to know Peter Carruthers, who happened to be chair of the NSF physics advisory committee, and I was soon a member of that, too.

Carruthers was a theoretical physicist high on brilliance and low on achievement. His 1961 Ph.D. was under the great Hans Bethe, who then prevailed on the Cornell physics department to award Carruthers a faculty position. Years later, Carruthers was a full professor, but he was made to feel unwelcome in his department when he took up in quite a public way with the wife of a colleague. Pete left Cornell and became head of theoretical physics at Los Alamos. He was an unabashed romantic, who played violin for two hours every day and chased women. One of Carruthers' outstanding recruitments to Los Alamos was Mitchell Feigenbaum. While at Los Alamos, Feigenbaum discovered the basic principles of chaos theory, including what is now called Feigenbaum's constant. But not long after this discovery, Carruthers took up with Feigenbaum's wife. There was a certain consistency in Pete's life. He was later department chair at the University of Arizona, where, he once told me with much anguish, he had no choice but to terminate the emeritus privileges of a ninety-four-year-old retired professor who, from his wheelchair, had suddenly reached up and grabbed a nearby woman's breast. "I knew exactly what he was thinking," Pete told me. "'*This is my last chance.*'"

The NSF advisory committee met a few times a year with NSF physics director Marcel Bardon. Marcel was a forceful, Gallic character, always on-stage with the exaggerated facial expressions of a mime; witty, manipulative, and an operator par excellence. The NSF physics staff sat around the periphery of the room. At each meeting, we reviewed one subfield of physics. Marcel was on trial. Was he over- or under-funding that subfield relative to its intellectual merit? Were his half-dozen program directors awarding grants to the best investigators? At my first meeting, the subfield was experimental nuclear physics. Coming out of World War II and the atomic bomb project, nuclear physics was a darling field of the new NSF. But that was then. By the 1970s, the best physics departments (Harvard and Princeton, for example) had given up on the field entirely. Caltech maintained an effort, but devoted exclusively to measuring parameters of interest to astrophysics. The Caltech instrument was an old Van de Graaff generator originally built by Margaret's father. But NSF was still supporting more than fifty other nuclear physics centers, the majority at small universities. These had obsolete technologies, attracted few graduate students, and produced little new science commensurate with their large drain on the NSF budget.

The committee's process was that a subset of members—the field's proponents—would write a draft report that the full committee might accept, reject, or rewrite. In this case, the nuclear physicists on the committee submitted an outrageous draft, asking for an increase in funding to *all* of the centers—except Caltech, which was singled out not by name but as an unnamed "applied" facility. Willy Fowler had gotten wind of this, and was attending our open meeting as a member of the public. He was not supposed to speak, but no one dared stop him. In discussion, the full committee (plus Willy) was clearly minded to reject the report, and to recommend the phased closure, over a period of years, of most or all of the existing centers. The two committee members who were the most vocal about this—they happened to be me and Hans Frauenfelder—were assigned to work overnight and bring in a rewritten draft the next morning.

Frauenfelder, who was about my father's age, had genuine credibility. He was Swiss and spoke with the compelling logic of a precise Swiss-German diction. "You know, *life* is a natch-er-al part of ee-vo-lu-tion, but so is *death*, and some-times it is time for a field of sci-ence to *die*." He had been an experimental nuclear physicist for the first half of his career (mostly at the University of Illinois), then decided that nuclear physics had become uninteresting and changed fields, to biophysics.

Hans treated me to dinner at a Washington restaurant far above my price range. At the end of an amazing meal, he asked the waiter to invite the chef to our table, so that he could compliment him personally. "I'm sorry, sir. The chef is very busy, but I will give him your compliments." Hans persisted: "Please tell the chef that I am the great-grand-nephew of Escoffier." He actually was, or was by marriage. The chef came to our table. He and Hans had a long conversation about food and wine. Hans and I worked long into the night eviscerating experimental nuclear physics, and, the next morning, the committee accepted our report.

I came to know Hans well over the next thirty years, and he became one of my favorite people. His wife, Vreneli (familiar for Verena), was the daughter of the wealthy Swiss hotelier, Albert Hassler. The Hotel Hassler, at the foot of the Spanish Steps in Rome, was one of the family properties. Vreneli was a famous beauty in her youth, the 1920s and 30s, and was painted by several notable Swiss and German artists. In 2001, when we (by now, my wife Jeffrey and I) lived in Los Alamos, the whole town lay in the path of an advancing wildfire and was completely evacuated. Hans and Vreneli, by then vigorous septuagenarians, lived in Santa Fe. They welcomed us to their modern, in fact rather grand, house. Because of the apocalyptic circumstances of our stay, they brought out for us the best wines from their cellar. The paintings on their walls—some of the young Vreneli, some from her family—were worth millions of dollars. We asked about the cost of the insurance that allowed the art to be kept out and not in a vault. They had no insurance, they told us. Their fatalistic view was that, if the paintings were stolen, they would take it in stride as just another event in their long lives together.

The subfield taken up by the NSF advisory committee at its next meeting was gravitational physics. Rai Weiss from MIT and I formed the drafting subgroup, with three outside relativists as consultants. We recommended that the NSF for the first time support an experimental effort in gravitational physics, a ten-year program that would lead to the construction of a gravitational wave detector. Our report was accepted by the full committee. This was the start of what became the LIGO project that ultimately won Rai, Barry Barish, and Kip Thorne their shared Nobel Prize.

Another committee that I was on was chartered by NASA head Bob Frosch. Since the end of the Apollo program a decade earlier, NASA had been drifting without any marquee headline projects. Frosch wanted to shake things up. Our "Informal Ad Hoc Advisory Sub-Committee for the Innovation Study" was named by NASA's lawyers so as to sidestep the requirements of public notice and open meetings. The committee was supposed to be populated with people whose creativity outran their practicality, or whose ambitions outran then-current NASA budgets. It was a big group, about twenty people. Five of us were JASON members: Bill Nierenberg, Freeman Dyson, Jonathan Katz, Rich Muller, and me. Other members included aerospace engineer Ivan Bekey, astronauts Judy Resnick and Joe Kerwin, Hewlett-Packard's Barney Oliver, physicist Harris Mayer, Avco Corporation founder Art Kantrowitz, and balloon-launch scientist Ed Ney. George Field and Riccardo Giacconi were also included.

Some of these people had long histories of advocacy. Oliver had long campaigned for a giant radio-telescope search for extraterrestrial intelligence. Bekey championed space tethers, potentially thousands of kilometers in length, that might one day be "space elevators" lifting things into orbit. Harris Mayer and Freeman Dyson had both worked on Project Orion, an interplanetary space ship that (on paper) weighed ten thousand tons and was to be propelled by the sequential detonation of eight hundred small nuclear bombs. Slightly more achievably, Kantrowitz proposed gas-dynamic laser propulsion in space.

Muller, Katz, Resnick, and I were the youngsters. We were supposed to invent craziness in real time. Judy was a Carnegie-Mellon electrical engineer with an entertaining sarcastic streak. She died tragically in the 1986 Space Shuttle Challenger explosion. Later additions to our committee were Dennis Flanagan, the publisher of *Scientific American*; and Jim Michener—the novelist James A. Michener whose weighty potboiler novels were always best-sellers. His 865-page *Chesapeake* had just come out; his new book, to be written over the next four years, was going to be about space. NASA was delighted to bring him into its tent. Jim paid for everyone's drinks when we bar-hopped after meetings. Over drinks, I bet Joe Kerwin ten dollars that the much-delayed Space Shuttle would still not have flown in a year. (I lost the bet.) Frosch recounted that by executive order President Carter was giving to Space Shuttle commanders the police powers of maritime captains—up to and including performing marriages, arrest, trial, and (if required for safety) summary execution. Frosch envisioned a day, centuries from then, when this precedent might actually be needed.

Compared to NSF's austere digs, NASA headquarters featured conference rooms with upholstered chairs, thick carpets, and richly wood-paneled walls. Appropriately, we had richly wooden briefings by senior NASA officials. These were exactly the people Frosch hoped that our committee would inspire—but more likely scare out of their wits. There were also talks by outsiders that were supposed to prime us. A physiologist was studying the mind-body problem in beagles—how were their muscular decisions affected by breaking various neuro-feedback circuits. "You mean you anesthetize a nerve pathway in the dog?" someone asked. "Not exactly. We cut off their heads." When the Washington weather was too hot, our committee met at the National Academy's study center in Woods Hole, Cape Cod. (Frosch had a summer house in the area.) A clambake was provided. In all, we met five times over more than a year. I am not aware of any actual NASA program that came from this endeavor.

Still another committee was the theory panel of the decadal astronomy survey. Brigadoon-like, this major activity came alive once every ten years to make prioritized recommendations to NASA and NSF for those agencies' new-project starts. Astronomers were held in awe by more fractious (and fractionated) disciplines like physics for being able to reach a consensus with winners and losers. Even more remarkable was the fact that the funding agencies actually followed the decadal's recommendations. This decade, George Field headed the effort; a decade later, it would be John Bahcall.

Theory was always the poor stepchild in the decadal surveys. We weren't a big project that needed years of planning, like a satellite or large ground-based telescope, and we were cheap. In the horse-trading, theory tended to end up as a low priority. This decade's theory panel was chaired by Dick McCray, but the driving force was Willy Fowler, who worked himself into a high dudgeon. "Those observer bums are spending millions of taxpayer dollars, looking at any damn thing that comes into their heads, and it's about time to force them to hire some theorists to tell them what it all means—if it means anything at all!" We sent McCray back to the main survey committee with a subcommittee report that made him look like a rabid dog.

Word must have gotten out that I was a reservoir of this rabies. At an MIT colloquium, radio astronomer Bernie Burke took the opportunity to tell me how useless theorists were; how they could, by definition, never discover anything because discovery, by definition, meant observation; how everything that a theorist did was stolen from some observer or another, etc., etc. He then offered to drive me back to Harvard in his new Chevy Citation, which he wanted to show off. It was one of the first U.S. cars with front-wheel drive. In the end, our theory panel was no more successful than previous or subsequent decadal theory panels. Theory remained a low priority.

On these and other committees, I traveled to Washington at least a couple of times a month. I usually took Sara, who was not yet in first grade, taking a taxi from National Airport and dropping her off with my mother for the day or two I was in D.C. Both Sara and I were in Washington, at the National Academy building, for the dedication of the large Einstein statue that subsequently became a popular tourist attraction on the National Mall. My mother was with us. John Wheeler was the dedication speaker. I knew, from him, that there had been an artistic SNAFU. The polished granite circle around the statue, thirty feet in diameter, was studded with three thousand stainless-steel pegs—a precise map of the night sky. But the night sky *when*? Stars are relatively fixed in position, but the planets move. So, from the position of Venus, Mars, Jupiter, and Saturn, one could determine the exact epoch that was represented. It was, by choice of artist Robert Berks, the day of Einstein's birth. Alarm bells went off within the Academy. The National Academy of Sciences would be seen as legitimizing *astrology*! All that could be done was to uproot the planets' pegs and move them to a different date, one chosen by John Wheeler: the date that Einstein's first paper on relativity was published. Billie and Sara edged forward through the crowd so that Sara could be the first child to climb on the statue.

A conference on cosmology ("The Universe at High Redshift") took me to the Niels Bohr Institute in Copenhagen for a week. It was inconveniently during a JASON summer study, but all the bigwigs were going, and I wanted to present a paper on some ideas about creating cosmological perturbations—these would later form galaxies—by a particle-physics phase transition. I had talked to Sidney Coleman about this. He thought it crazy, but not laughable. Less than a year later, Alan Guth would be thinking along the same lines and synthesize his famous "inflationary cosmology" from ideas that I knew about and might have connected, but, sadly, didn't.

Then, I became an immediate evangelist for the Guth model and gave a dozen or more talks about it at conferences, lasting until I was overtaken by the hordes of other physicists who adopted and extended Guth's insight. His was an interesting case in the advance of science: He was at the time a postdoc at Cornell, and considered not very promising. His fusion of ideas already "out there" uniquely advanced the field and made his career. Guth spent the next forty years working on various aspects of inflationary cosmology, was elected to the National Academy of Sciences at the young age of 42, and eventually became the Victor Weisskopf Professor at MIT. It is something that scientists love to debate in private: When the trick is indisputably great, should science give lifetime honors to its one-trick ponies? My answer has always been an enthusiastic yes! How wonderful to be on the carousel of a profession where you only have to grab the gold ring once!

In Copenhagen, I was called out of the conference by Aage Bohr's secretary. Aage was no longer director of the Niels Bohr Institute, and (people said) had become a recluse. I had last seen him at Margaret's father's memorial, in 1973, six years before. "Aage wants to see you, but he does not yet know where or when." Then, later, I was told to go to his house, and given an address. Aage was outside trimming the hedge. "Oh, yes, come inside, we have so much to talk about, so very much to talk about." I could not imagine what he meant. In the kitchen he said, "Yes, we have so much to talk about, but don't start yet, no, no, first we must have some strawberries and cream, would you like that?" I felt as if I was in a Bergman film. Aage had about two liters of strawberries in the refrigerator. In total silence we washed them and took off the stems. On the patio, he ate strawberries and reminisced about visits with Tommy and Margie, and about his wife Marietta, who had died a year before.

The next day, I got another secret message from Aage. We would have dinner together. He would pick me up in his car, but not at the Institute or at my hotel. Was there any place in Copenhagen that I could find my way to? Kongens Nytorv, I offered. I knew it was at the end of the Stroget. The Hotel Angleterre was there, his secretary told me. "He will pick you up in front of the hotel at seven on the dot. He won't be able to wait."

No longer Bergman, I was now in a James Bond movie. Aage drove up. I jumped in. We raced away before…before what exactly? We went to a fish restaurant near the Christenborg, the seat of the Danish parliament. Aage pointed out the nearby building where Niels Bohr was born. The conversation was more professional than the previous day over strawberries. After the main course, Aage asked the waiter, in English, "What do you have for desserts?" The waiter looked surprised. "But Professor Bohr, you are here so often that you know our five desserts better than I do!" "Well, yes," Aage replied reasonably, "but *he* doesn't know them."

I made discreet inquiries at the Institute. It was generally believed that Aage had had a nervous breakdown. When I saw him two years later in Pasadena, he had remarried and was his previous self again. By then, Margaret's mother had also remarried, to Bob Leighton, another Caltech physics professor, recently divorced. Margaret acquired two new step-brothers.

A different conference took me to Cambridge, England, for several weeks. I took Ethan Vishniac, my new graduate student. Ethan's father, Wolf Vishniac, a microbiologist, died in a freak accident in Antarctica, where he was testing an instrument that would fly to Mars to look for life there. A crater on Mars is named for Wolf. Carl Sagan appropriated the tragedy for a segment in his television series, using (Ethan told me) photos from the camera recovered with his father's body. Wolf's father, Ethan's grandfather, was the photographer Roman Vishniac, who became later celebrated for his photographic archive of the vanished Jewish shtetls in Eastern Europe. Roman was also a commercial photographer and photomicroscopist. Ethan had a low opinion of his grandfather. He related that Roman's commercially most famous microscopy, showing Prell shampoo fixing split ends, was in the Prell ad played backwards. In the actual, forward direction, the shampoo *caused* the ends to split.

At lunch in the Cavendish cafeteria with Martin Rees, Ethan complained about the bad English food. I told him that it was a mistake to complain about anything British in the presence of a Brit. Martin agreed. His analysis was that British academics reacted to criticism of Britain the way that American factory workers did to criticism of America, but completely different from American academics. "American academics feel estranged from the American ideal and don't mind having their country criticized. Not so the British academic, who is a patriot." I remembered this conversation many years later when we had lunch with Martin, by then Baron Rees of Ludlow, in the House of Lords dining room.

Despite his faux pas, Ethan later became a professor at Johns Hopkins University's Space Telescope Science Institute. His wife, Ilene Busch-Vishniac was dean of JHU's school of engineering, and was later provost at McMaster, then president of the University of Saskatchewan. Ethan followed her to both places, then both returned to JHU. At time of writing, he is editor in chief of *The Astrophysical Journal*, the position once held by Chandrasekhar.

38. JASON

I can best describe the typical six-week JASON Summer Study as a composite of the first few that I attended. La Jolla, at this time, was not the crowded summer resort destination that it later became. There were single-story houses and bungalows within just a block or two of the La Jolla Shores beach that could be rented for no more than our government per diem. In later years, these would be replaced by multi-story beach apartments, or bulldozed and rebuilt as McMansions. I tried to find a place close to the beach. Most mornings, I ran the length of the beach before work, on the sand and in my swim suit, as I had done on Venice Beach in L.A. It was about a mile from the Beach and Tennis Club at one end to the Scripps Pier at the north; so, a circuit touching both ends was two miles.

I'm sure you know the train-fly math puzzle: Two trains, 20 miles apart, approach each other on the same track, each going 10 mph. A fly alternates back and forth between them at 15 mph, finally getting squashed between them when the trains collide. What distance does the fly travel before its demise? The wrong way to do the problem is to work out the infinite series of individual flight segments and then sum the series. The right way is to observe that the trains will collide in one hour, so the fly will have flown 15 miles. John von Neumann, the Princeton mathematician, famously came up with the answer in only a few seconds—having worked it the *wrong* way and summed the series in his head.

I had my own version of the puzzle. Running was terrifically boring, but I could stay interested by running behind any attractive woman who was also running on the same beach—at a respectful distance, of course. Warming up, I waited for one to pass. That happened fast since this was a Southern California beach. (Not for nothing had I grown up in Southern California). As I ran, a women might pass in the other direction whom I thought more attractive than the one I was running after. When that happened, I reversed direction. And so on, recursively. I had worked out in detail the math that proved that the process was convergent, and that I would be running, on average, only slightly more than a simple round trip.

The JASONs at this time were all male, and most brought their wives and children to La Jolla. Margaret and Sara came for a month, leaving our Cambridge house in the care of student house-sitters. (One of these was Amy Aquino, who later became a film and television star.) There was an active JASON social life, starting the first week with a dinner reception. Rich Muller and Henry Abarbanel, two of the younger JASONs, organized a Wednesday dinner club that visited a different San Diego restaurant every week. Margaret and I were charter members. There was a JASON theater group organized by Rip and Carolyn Perkins that mostly went to musicals in the open-air Starlight Theater in Balboa Park.

In a typical JASON summer, there were generally two or three major studies going. These might be worked on by a dozen people or more. And there were another handful of small studies involving two or three people. Members voted with their feet on what to work on; but it was understood that we should pitch in if a study was understaffed.

A major study several summers running was the proposed M-X missile, and specifically how it should be based. The M-X was to be a MIRV (Multiple Independent Reentry Vehicle) missile carrying ten separate nuclear bombs. The Air Force wanted it for the sheer firepower that it would provide against thousands of Soviet targets. The problem, as articulated publicly by the missile's many critics, including JASON members Sid Drell and Dick Garwin, was that it was strategically destabilizing: Its very concentration made it an inviting surprise first-strike target for the Soviets. That, in turn, made it useful only as a first-strike weapon for the U.S. First strike weapons that give victory to the side that first initiates a nuclear war are, by definition, destabilizing. During the Cold War, these kinds of strategic issues frequently pitted the military and civilian leadership against each other. It often seemed that the issue of strategic stability was one that only civilians cared about.

JASON was asked: Was there a way to base the M-X so that it could reliably ride out a Soviet first strike? That would be stabilizing. The Air Force had a proposal, briefed to us by General John Toomey, at 6' 7" the tallest general in the Air Force. They would build ten thousand underground silos, spread over an area in the west the size of Connecticut. Then, they would shuttle a thousand M-X missiles around, constantly, like a shell game. The Soviets wouldn't know which silos to target. This idea was dead on arrival. If the Soviets ever did ever find a way to figure out the missile locations—even just once—the temptation to strike right then would be too great. And anyway, if you were going to have ten thousand silos, why MIRV at all? Why not have ten thousand much-cheaper single-warhead missiles? It was strange that the JASONs, who as individuals were almost all against the M-X, were trying to find a way to save it. But, to be more precise, we were trying to think of every possible way of saving it, while hoping that they could all be proved unfeasible, thus dooming any large-scale deployment of the M-X. Only about fifty M-X missiles were ever deployed, and these in ordinary "non-survivable" Minuteman missile silos. They remained as monuments to government waste until the 2000s, when they were gradually deactivated.

There is the well-known guillotine joke: An engineer, about to be executed, lies looking up at the blade. They pull the cord, but the blade jams on the way down, sparing him. They try again, and the same thing happens. Then a third time. The engineer says, "Just a minute. I think I see what your problem is." That was the JASON M-X project. We wanted to tell them what their problems were. One JASON idea was to build small diesel submarines, basically World War II models, each capable of carrying a single missile. They had to be diesel: any nuclear-powered submarine would cost vastly more than the missile it carried. We located a retired Navy diesel submarine commander whom Admiral Hyman Rickover had rejected for his nuclear submarine navy and brought him to La Jolla. We finagled a tour of the privately owned *August Picard*, a 90-foot, 100-ton submarine that happened to be transiting San Diego on its way to the Panama Canal. When the Navy showed no interest in the JASON idea, Garwin proposed that the submarines be crewed by the Air Force. That idea didn't fly. I did learn something about submariners: They are all small of stature, affable, and great storytellers. There isn't much else to do on long deployments in tight quarters than be affable.

I tried to contribute. I wrote a piece that attempted to quantify the fragility of deterrence based on deception (like the M-X shell game). It was full of clauses like, "while they may think that we think that they think…." Sid Drell read it and explained in a kindly way that he would not let it go into the report—but that I should not get discouraged. It sometimes took years, he said, for new JASONs to learn the ropes. In the meantime, we were paid the same as others who did contribute. This was an important part of the JASON ethos: Collectively, we were paid by the job (that is, by individual government sponsors for individual studies). But individually, we were paid by the hour or day. You did your best, enjoyed the beach, left editing the final report to the study leader, and didn't let your ego get bruised.

Still, I did contribute something to a smaller study, led by Hal Lewis, a physicist from UC Santa Barbara. Hal was a former chair of JASON, a confirmed skeptic, acerbic to and critical of his colleagues, yet somehow never overtly unpleasant—in the way that Garwin could be, for example.

Hal was a mentor for new JASONs. In a kind of resigned, apologetic whine, he explained why we should not believe what we were told. Lewis got his Ph.D. under Robert Oppenheimer. In the late 1940s, he was considered one of the brightest of the young physicists—on a par with Feynman even. But he accomplished almost nothing of note in physics. Mostly, he served on important and lesser national advisory committees. (A few years later, my father, giving me what he thought of as gentle career advice, said, "Watch out, Bill! You don't want to turn out like Hal Lewis.")

Our study was called AIMVAL/ACEVAL (acronyms for Air Combat Evaluation/Air Intercept Missile Evaluation). The Air Force had done a series of controlled live-flight experiments pitting a "blue" force of highly capable F-14 and F-15 fighters against a somewhat larger "red" force of less-capable, cheaper F-5Es, surrogates for the Soviet air-combat force. The intended purpose of the experiments was to justify the development of a new air-to-air missile for the blue side. To the dismay of the Air Force, the cheap F-5Es usually won the mock battles—with or without assumed new missiles. There were various theories, including that the reds, who were a resident counterforce for training at Nellis Air Force Base, were just better combat pilots ("had superior situational awareness"). Hal told me to go learn about an obscure topic, Lanchester's Law.

Frederick Lanchester (1868-1946) was a British polymath, a pioneer in the development of the internal combustion engine for automobiles, boats, and airplanes. During World War I, he posited a simple differential equation for the attrition with time of blue and red forces that were each able to inflict casualties on the other. Lanchester's Law included parameters for both the quality and quantity on each side. The equations showed that quantity trumped quality. As an example, if blue-force fighters were outnumbered two-to-one, they would need *four* times more firepower per fighter than the red force, just to have an equal chance of winning. I crunched the data and wrote a report showing that, yes, the reds were somewhat better pilots, but that the main advantage was the greater number of F-5Es. At the end of the summer, I briefed my findings to Bill Perry, then the Undersecretary of Defense for Research and Engineering—and thereby JASON's boss. I may have thus contributed in a small way to the Air Force's eventual decision to purchase cheaper F-16 fighters instead of a smaller number of expensive F-15s. Similarly, the Navy bought cheaper F-18s instead of F-14s.

Perry, trained as a mathematician, had a unique career that rotated among Stanford professor, government service, and venture capitalist.

He served as Secretary of Defense 1994-1997 under Bill Clinton. He was briefly a unique member of JASON because, when he was in the private sector, a misguided government bureaucrat tried to expire his security clearances. Making him nominally a JASON member allowed them to be maintained.

The Department of Energy was formed by Jimmy Carter in 1977 by the consolidation of a number of smaller agencies. James Schlesinger, who had already been Secretary of Defense, CIA Director, and Chair of the Atomic Energy Commission, was the first Secretary of Energy. Within JASON, Gordon MacDonald (an older member) and Henry Abarbanel (a younger one) spearheaded a succession of pioneering studies on climate change, work sponsored by a young staffer in the newly formed DOE named Ari Patrinos. Unlike the above two examples, these studies were entirely unclassified. They could have been done by meteorologists and oceanographers in academia. (There were then no "climate scientists" as such.) These were exactly the people who came to brief us—they would say "lecture," not "brief"—and many of them resented that the DOE was listening to us and not just reading their published works. But that was what we did: We digested other people's science and synthesized conclusions and recommendations for our government sponsors. It was something that academia didn't do very well—still doesn't today. My contributions to these climate studies were negligible, but I went to all the briefings and learned a lot of what came to be known as climate science.

Ari Patrinos, our sponsor, was a type of government bureaucrat that I didn't know existed—young and creative. Ari was raised in a Greek family in Alexandria, Egypt, then trained as a mechanical engineer in Athens, Greece, only then emigrating to the United States for his Ph.D. The new DOE was his opportunity, the right place and time. He had a talent for navigating within its new bureaucracy. Invisibly, and almost singlehandedly, he established climate science as a main scientific thrust area. Then, twenty years after the events that I am describing, he pulled off the same kind of coup again: The Department of Energy, improbably, was the first U.S. government agency to embrace and fund the sequencing of the whole human genome, in large part due to the invisible and effective Ari. He and his boss Charles DeLisi sold the project within DOE, and also convinced influential Senator Pete Domenici (whose New Mexico constituency included the Los Alamos and Sandia DOE laboratories) to facilitate matching funds to the National Institutes of Health. NIH was the agency that should have started the project in the first place, but it was hidebound and bureaucratic. Human chromosome number 16 was first sequenced at the DOE's Los Alamos nuclear weapons laboratory.

I accumulated a list of hundreds of acronyms and idioms that littered the briefings to JASON—this in part why it took years to learn to contribute. At one talk, someone said, "We can't fund that. We'll have to do a mipper." I wrote down the phrase, only later decoded to MIPR, a Military Interagency Procurement Request. For lunch, we walked to local restaurants. My favorite was Clay's Texas Barbeque. Mr. Clay (as we addressed the owner, although Clay was probably his first name) was an unlikely transplant to La Jolla from a Black neighborhood in Houston.

JASON members occasionally gave lunchtime talks on work they did outside of JASON. Gordon MacDonald gave a talk on Soviet economy and demographics. It was interesting then, but proved uncannily prescient later on. Gordon's analysis was in response to what he thought were naïve CIA analyses. These simply extrapolated trends with straight lines. He, by contrast, took note of a major demographic trend that would hit the Soviet economy by the mid-1980s, a shortage of workers in the Russian Federation (that is, in the Russian regions of the U.S.S.R.). This was partly the generational echo of the missing millions of Russian men lost in World War II, and partly the result of a sharp decline in the birthrate of ethnic Russians. It was exacerbated by the regime's prohibition of internal immigration into Russia—the Moscow region especially—from the other Soviet republics, particularly majority-Muslim ones. MacDonald's analysis predicted a collapse of the Soviet planned economy by the late 1980s.

Until then-recently, there had been Soviet efforts at partnering with foreign companies to develop oil and gas fields. American oil company CEO Armand Hammer was an enigmatic figure in these efforts. Hammer (1898-1990) had lived in the then-new Soviet Union for most of the 1920s, knew Lenin and Stalin, and maintained his contacts with Soviet leadership throughout his life. In the late 1970s, Occidental Petroleum was active in negotiating deals in the U.S.S.R. Brezhnev provided Hammer with a luxurious Moscow apartment. Hammer's private airplane was specially equipped to do the Houston-Moscow flight nonstop, and was one of very few private Western aircraft allowed into Soviet airspace. MacDonald speculated that such anomalies were evidence that some in the Kremlin recognized the coming crisis. That these efforts had been squelched, indicated, Gordon thought, the aging leadership's intention to just tough it through.

MacDonald's predictions were largely borne out by subsequent history. When Mikhail Gorbachev came to power in 1985 the Soviet economy was in collapse. Gorbachev was at the time the youngest member of the Politburo. The old leadership was in disarray. Gorbachev's *perestroika* was supposed to be about economic restructuring and only incidentally about political liberalization. The latter, however, became unstoppable, resulting ultimately in the complete dissolution of the U.S.S.R. at the end of 1991.

Dick Garwin was JASON chair for two of these summers. The term of the chair was normally three years, and usually renewed for another three, but Dick resigned as chair after two summers. The JASON membership had rejected his attempts to instill discipline, he complained. One attempted Garwin innovation was that, at the beginning of the summer, we were each required to write down (and turn in) what we hoped to accomplish during the six weeks. Most JASONs, the first summer in Dick's term, actually did this. Then, at the end of the summer study, we were supposed to grade ourselves as to whether we had failed, achieved, or exceeded our objectives. Dick proposed to award financial bonuses on the basis of the evaluations. I had no problem with this and turned in a paragraph saying that I had exceeded all my objectives—in fact, substantially exceeded them. Sadly, I was the only JASON member to turn in anything at all, and I didn't get a bonus. This may, however, have been the start of Dick's and my warm, mutual friendship lasting more than forty years.

After he was no longer chair, Dick went back to doing what he did best, which was pointing out every error, however small, in every briefing by every outside speaker. Dick's criticisms were inevitably correct. "I ought to know," he would announce. "It's my field." This, no matter what the field of the talk actually was. Enrico Fermi, Dick's thesis advisor at Chicago just after the War, may have said that Dick was the only true genius he had ever met. Dick was never the one to repeat this quote, but, as I write, it has been repeated for several decades at least. The not-very-good biography of Dick by Joel Shurkin is titled *True Genius*, and the phrase similarly found its way into Dick's citation for the Presidential Medal of Freedom awarded to him by President Obama in 2016.

A full Garwinian outburst was a pleasure to behold. First, he chopped off only a toe of the speaker and gave him a chance to recant. Then a foot, then both feet, then both legs off at the knees. Another chance to recant was offered. Then, Dick worked upward until only the speaker's detached head was left, perched on the viewgraph machine and often still talking, if things had gone this far.

39. Green Apples

I was well out of my Princeton research slump and was publishing, I thought, pretty good papers at a respectable rate. I was interested in two different "big" problems that were of then-current interest, "big" in the sense that each had many researchers pursuing different approaches. The first was the origin of galaxies and clusters of galaxies. The Hot Big Bang was by now the universally accepted cosmological model. Fred Hoyle, the only prominent doubter, had been exiled from academia, retiring to England's Lake District. By itself, the Hot Big Bang did not produce galaxies. Its expanding mixture of primordial hydrogen and helium would simply cool uniformly. By today, the universe would be cold, still expanding, a uniform density of diffuse gas at the temperature of the cosmic microwave radiation (which we actually do see) just three degrees above absolute zero. No galaxies, stars, or planets.

To make those, one needed an additional feature built-in from the start, some kind of matter density perturbations that could later become amplified by their own gravity. What were these? How were they generated? How did they evolve while they needed to be described by general relativity? Or later, when they could be described by conventional astrophysics? (The technical jargon is that perturbations "outside their horizon" require general relativity, while perturbations "that have come into their horizon" can be adequately treated by Newtonian physics.) What was their mass spectrum? Meaning: Were the perturbations bigger on large scales, so that clusters of galaxies formed first and then fragmented into galaxies and stars? Or were the perturbations bigger on small scales, so that small units like globular star clusters formed first and then aggregated "bottom up"? My 1974 paper with Schechter addressed a piece of this, but more was known now. I took up other parts of these puzzles in work with Alan Lightman and with my student Ethan Vishniac. On my own, I worked on exotic ideas like the one I talked about in Copenhagen, for publication in less-rigorously-refereed conference proceedings.

A completely different subject was the "solar neutrino problem". Neutrinos are fundamental particles, produced in nuclear interactions, that interact hardly at all with normal matter. They can easily pass through the Earth or the Sun. Although predicted to exist by Fermi in the 1930s, they were first actually observed only in the 1950s, after nuclear power reactors (strong, close-up sources of neutrinos) existed. Neutrinos are massless consequences of "weak" nuclear forces, similar to the way that photons are massless manifestations of stronger electromagnetic forces. The Sun is a vastly stronger cosmic source of neutrinos than any terrestrial source, but of course much farther away. Starting in the late 1960s, Ray Davis constructed a massive neutrino detector in the deep Homestake Gold Mine in Lead, South Dakota. The idea was to use the Earth itself to shield the experiment from everything except solar neutrinos—from all other forms of cosmic rays, for example. John Bahcall was from the start Ray's theorist partner.

A small number of computational astrophysicists (some at Los Alamos and Livermore) were able to do accurate computational models of the Sun. By force of personality, John organized them to do the required computer runs, reconcile their differences by finding and fixing bugs in their codes, and provide him with the data for a firm prediction of what Davis should see. By the late 1970s, there were solid, but puzzling, experimental results. Homestake (and, later, other experiments) saw only about a third as many neutrinos as the predictions. Either the experiments were faulty, or else conditions in the deep interior of the Sun were not what theorists calculated.

There was also a third possibility, that Fermi might be wrong. If neutrinos were not exactly massless, they could decay in the eight minutes it took them to reach the Earth from the Sun. Or, even more exotically, they might "oscillate" among different types—an idea that came to the fore only later, in the mid-1980s. The nuclear reactions in the Sun produced one type of neutrino, the type that Davis' experiment was designed to detect, but any other types would go undetected. Positing a change in fundamental physics just to solve an astrophysics problem was an unlikely and ugly solution that most physicists rejected. Surely either the experimental result or the theoretical calculations must be wrong.

My niche in this was to explore possibly overlooked hydrodynamic phenomena that might be happening in the solar interior. These might change Bahcall's models. Just a small change in the assumed central temperature of the sun would be enough to reduce the neutrino flux by two-thirds. I had the idea that internal waves, a hydrodynamic phenomenon that I had learned about in JASON, might carry energy away from the solar center. George Rybicki was the more-mathematical partner whom I needed to do detailed calculations, and we wrote several papers together. Rybicki played the role with me that Saul had previously done. George was a genial man and held a permanent Smithsonian position.

He was a decade older than me. His work was in a narrow area of the mathematical theory of radiation transfer in gasses; he had been a protégé of Max Krook. But, unsung, he was a superb applied mathematician generally. As before in general relativity, now in hydrodynamics, I could *consume* sophisticated applied mathematics, but not *produce* it.

For a while, Rybicki and I had difficulty getting our work accepted for publication. Solar physics was a backwater. The few theorists who worked in it were certain that the standard models were correct—so any newcomers with a different view must be wrong. We received scathing referee reports. Oddly, one day, I got a call from Roger Ulrich, a leading solar theorist. Would I be willing to join the solar panel of the decadal astronomy survey already underway? The panel was supposed to chart creative new directions for the field. How did he even know that I was working on the Sun, I asked? "Well…I'm your referee." I told him that he couldn't have it both ways. I couldn't be both an unpublishable hack and also a creative new spirit. We reached an agreement. He would pass my solar physics papers as referee, and I would serve on his (yet another!) Washington committee.

Once, visiting John Bahcall in Princeton, I was introduced to another visitor, whom John described as his wife Neta's cousin from Israel. The three of us got into a discussion of Israeli politics. Neta's cousin prefaced his every remark with, "First, I give you the *facts*. And then, after, I give you the *assessment*." Presently, I said to him, "You must work in Israeli intelligence." He spoke exactly the way CIA analysts did, when they briefed JASON. The visitor and John exchanged knowing looks. I had guessed right. "Is he really Neta's cousin?" I asked, later. John waved a hand dismissively. "Everyone in Israel is Neta's cousin."

I mention this because, regarding my scientific work in these years, first I have given you the facts. And now, after, I give you the analysis.

Science is sometimes the blind search of a green, densely foliaged apple tree for its one hidden apple. You can find the apple only by feeling for it, and you have to climb out onto every branch until you, or another climber exploring a different branch, succeeds. How should you feel when the apple turns out to be on someone else's branch? Disappointed, obviously. But not necessarily too disappointed. You couldn't control where the apple was. You could feel good if you had explored branches likely to bear fruit, unlike, say, your less talented colleagues. The right question was, *if* the apple had been on your branch, *would* you have found it?

There was also the question of beauty. Most physicists believe that, among possible alternatives, Nature chooses a beautiful one. Einstein, Dirac, Witten, and others have written about how the search for beauty led them to professional success. When you don't succeed in grabbing the apple—winning a big prize—your work will be judged by, was it at least *pretty*.

I was pretty good at writing pretty papers that *might* have found hidden apples; but the hard fact is that, in the cases just mentioned, they didn't. My batting average turned out to be zero. (I'll presently make the case that I did just a bit better later in my career.) In the case of the solar neutrinos, the correct answer turned out to be the unlikely option three: Neutrinos were not massless. They oscillated into two-thirds non-detectability on their way through the Sun. Our understanding of the fundamental physics had been wrong. The Homestake experiment was correct. So were those models of the stodgy solar theorists of whom I was so dismissive. Ray Davis, without Bahcall, eventually shared the 2002 Nobel. The other half of the 2002 prize went to Riccardo Giacconi for his work in X-ray astronomy. Evidently the Nobel Committee wanted to knock off two unrelated cosmic particles, neutrinos and X-rays, in a single year. Bahcall was disappointed not to share with Davis. The community belief was that these two were indivisible, as the hands and brain of the Homestake experiment, although I never heard John say this in so many words. Art McDonald, from Canada, and Takaaki Kajita, from Japan, later shared a prize for the definitive experiments observing neutrino oscillations.

For the case of galaxy formation, I was closer, but still missed. What about those necessary cosmological perturbations? Guth's inflationary cosmology, which I had missed only barely, was the key; and my ideas about decaying quantum particles were (as Coleman told me) not completely laughable. Less than a year after his first paper, Guth and others showed that quantum fluctuations in inflationary models generated cosmological perturbations very naturally. That idea, with some later improvements, survives today. And what about how galaxies and clusters of galaxies form from those perturbations? There, I (and everyone else) was missing one astounding and unexpected fact about the universe. To get to it, we need to introduce Vera Rubin.

Rubin was a pioneer woman astronomer. She was of my parent's generation—the generation of my older colleagues at Harvard. I knew her as a Jewish mother type, but more maternal, nicer and less edgy than my own mother. Vera had earned a bachelor's degree in Astronomy at Vassar (a class of one), and was the first woman graduate student in astronomy at Cornell. She dropped out to follow her husband Bob to Washington, finally earning a Ph.D. from Georgetown University in 1954. Now, she held a permanent position at the Carnegie Institution in Washington.

The 1979 annual meeting of the American Astronomical Society was held in Mexico City, and it included a several-day side-trip to the (astronomically interesting) Maya ruins in the Yucatan. Among forty astronomers and spouses on the bus, Margaret and I, and the Smarrs, were "adopted" by Vera and Bob. We ate together and clambered up and down pyramids. I liked Vera a lot. Her life's work, with her longtime collaborator Kent Ford, was the patient accumulation of measurements of the rotation curves of galaxies. Galaxies rotate, but not like rigid objects. Typically, the speed of rotation at first increases as one moves out from the center, then eventually turns over and starts to decrease. From this "curve," one can deduce where the mass in the galaxy resides. It was not in Vera's nature to overtly claim to be a great astronomer, but she was clearly proud of making difficult observations that no one else could do.

The Harvard-Smithsonian Center for Astrophysics in 1980 was home to three outstanding early-career optical observers: Marc Davis (my one-time collaborator with Price and Ruffini), John Huchra, and Margaret Geller. In the astronomy department, we had one open tenured position in that area. It was obvious that we should promote one of these three, but there were positives and negatives about each. A search committee had been formed: Field, Dalgarno, Giacconi, Gursky, Liller—all male (how else?) and all in their fifties. They recommended unanimously that we pass over the three younger candidates and recruit…Vera Rubin. All of them had known Vera for twenty-five years. She was exactly their age. Against all the rules, they had contacted her and confirmed that she would accept.

I was surprised and then angry. This was exactly the appointment that we didn't need. What promising young person would ever again come to Harvard? I made it my personal mission to stop this appointment. And I did. In the end, Vera was defeated by an unholy alliance of me, Al Cameron, and the four deadwood who were misogynists. There was so much bad blood that the tenured position remained unfilled for three more years.

What does this have to do with galaxy formation? The astounding and unexpected feature of the universe that was necessary to explain galaxies and clusters of galaxies was what we now call *dark matter*. Only about fifteen percent of the mass in galaxies and clusters consists of atoms, nuclei, protons, neutrons, or electrons—the stuff of ordinary matter. Eighty-five percent is in some other, completely unknown, form. We know that it is cold, and subject to gravity, but nothing else. Loose talk of dark matter dates back to the 1960s or before. (Zwicky, as always, had some crazy ideas even earlier.) The observations that nailed it, that made dark matter an inescapable conclusion, were the galaxy rotation curves of Rubin and Ford. Vera is now considered one of the great astronomers of the 20th century. She would have been a feather in Harvard's cap. The existence of dark matter was afterwards convincingly confirmed by X-ray observations of clusters of galaxies by Riccardo Giacconi's enterprise.

As a young scientist, I was never sure who was right about science, Wheeler or Chandrasekhar. Did the most important science require theatrical Wheelerian ripostes, or did science advance by deep mathematical insight in the manner of Chandra? These were very different styles, but each had intellectual depth, elegance, and beauty. It never occurred to me that neither might be right, that the best approach might be to throw away these aesthetic touchstones and adopt the no-nonsense, brute force approach of simply grinding out data, or (for theorists) calculating only the data's obvious consequences, the way Lyman Spitzer (whom I disdained) did. Patient accumulation of data was what Rubin and Ford did. And, with a bigger budget and a lot of bluster, it was what Giacconi did. Over the decades, I've come around to thinking that they were right, and that I was wrong. The cosmos is beautiful, but astrophysics does not seem to reward elegance or beauty in theoretical work the way other areas of physics do. Luckily for me in the 1980s, the world had not come to this conclusion.

Besides blocking Vera, I found other issues on which to throw around my newly self-discovered weight. I succeeded only a fraction of the time. I failed to block the reappointment of Herb Gursky as Associate Director for Optical Astronomy. I even tried the John Wheeler method of silently going to the blackboard and writing down my list of desirable characteristics for the job—all of which Herb lacked. I was surprised when my list was unanimously adopted, and then surprised again when a large majority cynically voted for Herb anyway. Still, I was discovering that I could have *heft*. When George Field offered me the position of Associate Director for Theory, I turned him down.

The position had no real power. The other associate directors were (like Gursky) a sorry lot. I suggested that he appoint good-natured George Rybicki. "But I wanted someone more aggressive," Field said plaintively, but he did appoint Rybicki.

Richard Isaacson was my program director at the National Science Foundation. I wrote one proposal annually, and Rich always found a way to fund it, even as my research moved out of his area of gravitational physics and into astrophysics. Now, Godfather-like, he wanted a favor. Up to now, computers used by NSF researchers were, by definition, the mainframes in their universities' computer centers. NSF provided funds in grant budgets to pay for such machines by the hour of use. Now, suddenly, every university researcher wanted to buy his or her own minicomputer, like the Data General Nova 3s that Al Cameron and I had each bought. Rich thought this a bad trend. True, it increased computing "capacity" for researchers; they could use their own little machines twenty-four hours a day year-round. But it decreased computing "capability," because, individually, these small machines were slower and had less memory and storage than the mainframes they replaced.

Isaacson had the idea that NSF should fund a few large, national computer centers that specialized in high capability for university researchers. The word "supercomputer" was not yet in general use, but that was the idea—computing on a scale then done only at Livermore, Los Alamos, and (it was believed) NSA. Rich was a mere program director in one NSF division (physics), so he could only charter a committee of *physicists* to recommend that there be a national *physics* supercomputer center. A skillful bureaucrat, he understood that this would be viewed by upper NSF management as so absurdly parochial that they would have to either forbid it altogether, or else expand it to include all the scientific divisions NSF-wide. He hoped to force them to the latter. He wanted me to chair this physics committee, which he populated with proponents of more-and-better computing from all subfields of physics. Some were heavy hitters, like Ken Wilson, who had won every physics prize except the Nobel—which he won a year later, for his development of renormalization group theory. Larry Smarr, who was by then at Illinois, was a member.

I had never chaired a national committee before. I needed to get tuned up on the leadership thing and separately consulted George Field, who predictably told me to take the high road (whatever that was!), and Al Cameron, who merrily shared his repertory of devious tricks. One was: Go around the table and let each person make an opening statement *of any length*, no matter how long it takes, and no matter how bored everyone else is. This lets people get things off their chests, and it exposes the crazies for everyone to see. When I did this in Washington, it took most of the first day of our three-day meeting. It also showed that, if Isaacson had intended to stack the deck for supercomputer centers, he had failed. The committee was about evenly divided on capacity versus capability. I let people spend the rest of the day arguing, then announced that, since we had been so unproductive, a rump subcommittee would meet after dinner and into the night to draft "the non-controversial parts of the report." Anyone was supposedly welcome to join this group, but, not coincidentally, it turned out to be just me, Larry, and Ken—and we were all supercomputer proponents.

The next morning, I faced a demoralized committee. How would we ever reach any agreement? "Let's get started writing," I said. But what were our recommendations going to be? "They'll emerge," I said. "Everyone will be happy with the consensus." I had no idea if this was actually true. I assigned the best and fastest writers to make the argument for supercomputers, and appointed a slow and methodical "capacity" person to lead that side's drafting effort. We had not been asked to prioritize, so everyone got to have their say. Everyone signed off, but no reader could mistake our report for anything but a strong endorsement of supercomputer centers. Isaacson's plan took longer than he expected, but it succeeded. Five NSF supercomputer centers, at Princeton, San Diego, Illinois, Cornell, and Pittsburgh, were chartered in 1984. Four of the five were founded by members of my committee. Larry Smarr was the first director at Illinois.

Al Cameron was up for reappointment as astronomy department chair at Harvard, so I was not surprised to be summoned to Henry Rosovsky's office. We had all written our mandatory individual letters to him, as Dean, on the subject. Mine skewered every possible candidate, including Al, but then concluded that he should be reappointed. I assumed that Henry was going to chide me for my usual immoderation, then tell me that Al was in for another three years.

"How is everything with you *personally*, Bill?" Then, without waiting for me to answer: "I want you to be the next chairman of astronomy." He went through the logic: Gingerich, Whitney, Layzer, Liller, Lilley were gently eliminated as being "not close enough to the frontiers of astronomy," Krook would be retiring. Field was CfA Director. Dalgarno had been Chairman already. Henry knew that Giacconi was an empire builder who couldn't care less about the department. So that left just Al Cameron and me.

Henry didn't say why he didn't simply reappoint Al, but it may well have been the latter's "Declaration of Independence". At a previous meeting of the department, when George Field happened to be absent, Cameron had announced that he was adjourning the department meeting and reconvening everyone present as the Harvard College Observatory Council—a body that had never before existed. "Our first order of business is to elect a Chair, and I am willing to serve." Bemused, we elected him. We didn't know that he was going to write Henry, demanding that George Field surrender the Observatory's financial records and announcing that he, Al, now served at the elected pleasure of the Astronomy faculty, no longer at the pleasure of the Dean. I had gotten a phone call from Paul Martin, my physics senior colleague in Henry's inner circle. "What the hell is going on there up at the observatory?!" "It's just Al being Al," I had explained.

Now, in person, I convinced Henry that he should reappoint Al, but also announce that I was the heir apparent and would become chair in two years. "Yes," he mused. "I've done that in other departments. It gives you the power now without the responsibility." It would also keep Cameron in check—no further Declarations. I made a beeline back to Al's office. Henry had already called him. Cameron and I shared the (albeit rather limited) powers of a department chair—now indistinguishable from the Harvard College Observatory Council—for the next two years, and even afterwards when I was officially chair and he wasn't.

Not long after my regency was announced, I got a phone call from John Bahcall, who told me, in confidence, that I would be that year's winner of the Warner Prize of the American Astronomical Association, the AAS's top award to an early career theorist. "I suppose that you nominated me," I said. "Not at all," John said. "I called up George Field and dictated *his* nomination letter. My only role was to chair the selection committee." Steve Adler, another professor in the School of Natural Sciences at IAS, called me to say that I wasn't supposed to know that John put him up to calling. Would I consider coming back to Princeton as an IAS professor? I told him (lying) that such a position was my fondest dream, but that (closer to truth) as department chair-elect, I couldn't leave Harvard for several years.

There it was. Without ever being an astronomer—I still considered myself a physicist—I was anointed as a member of the astronomy establishment: Harvard astronomy department chair (to be), American Astronomical Society prize medalist, and job offers seemingly pouring in—well, one, anyway. I did radio interviews, one time telling a BBC interviewer whose questions were dragging: "Just wind me up, and I'll spout cosmology into your tape recorder for twenty minutes. You can edit the tape later and make up the questions."

40. JASON Again

Margaret and I separated in the fall of 1981 and were divorced in 1982. Sara, then age six, made the difficult adjustment of alternating between two households. I have written elsewhere about these events and my adjustment to single life; but this work is supposed to be a professional, not a personal, account. So we must pass on to other things. Strangely, the mood cycle that for many years had given me a couple of wonderful manic days every month or two, followed by the predictable crash to a few days of black depression—my mild bipolar disorder, as we would now call it—completely went away. I missed it, sometimes.

JASON continued as a major part of my professional life. Individual JASON members tended to gravitate toward a single—their favorite—sponsor's projects, but these sub-divisions were not rigid. During my first half-dozen JASON summers, I rotated among several groups. I was pleased that my first JASON work on air combat tactics had gotten assimilated as an anonymous, tiny part of a tectonic shift within the Air Force. Combat pilots from the Vietnam War were just now taking over the senior Air Force ranks and re-examining air combat doctrine, some left over from World War II's massive air raids. Col. John Boyd, a charismatic figure distrusted by the establishment, gave JASON his day-long take on the history of warfare through the ages, emphasizing the importance of fluidity, speed, deception on the part of the winners; and confusion, surprise, panic on the part of the losers. We cheered him on. Many of Boyd's ideas, especially the so-called OODA Loop, eventually took hold across the military. OODA stood for "Observe, Orient, Decide, Act," but the OODA Loop more generally represented the idea that tactical decisions in battle should be made at the lowest possible level and not referred up the chain of command. Twenty years later, the First Gulf War exemplified their application. That short war managed to be a military validation and political disaster simultaneously. In the later Iraq War, Army captains (a low rank) were given vast discretion over large territories and expansive missions—something inconceivable in World War II or the Vietnam War.

Jeremiah Sullivan, an experimental particle physicist from Illinois, usually led studies for the Defense Nuclear Agency, the part of the Pentagon that dealt, not with nuclear weapons per se, but with so-called weapons effects—what would actually happen if the weapons were used. In the early 1980s, a problem of concern was the seemingly basic question, how big a crater did a nuclear bomb make? A lot depended on the answer: Offensively, how many missiles did we need to take out "hardened" Soviet missile silos? Defensively, how vulnerable were our own missiles? Despite forty years of nuclear testing, the data was uncertain by as much as a factor of ten. In theoretical calculations, the size of the crater was enormously sensitive to the height-of-burst, even the difference between a few meters above or (for a penetrating weapon) below ground level. This was "dirty" physics in every sense, with many competing materials and radiative effects. In this way, it was a lot like astrophysics. Jonathan Katz, who joined JASON just before me, became JASON's expert on dirt.

The JASON Navy (as we called them), led by Walter Munk, Ed Frieman, and Roger Dashen, dealt with much cleaner physics. The equations of undersea hydrodynamics and acoustics were well-posed and encouraged beautiful, mathematically sophisticated work. The important problems of the 1980s were almost always in the duality of "SSBN security"—how to keep our missile submarines from being found—and anti-submarine warfare—how to find the Soviets' missile submarines. A particular set of issues was the Soviet Navy's development of long-range active sonar. Sonar, the use of underwater sound to find submarines, can be either passive or active. Passive means just listening for sounds coming from the target—engine noise, reactor pumps, "hotel" noise. Active means sending out a loud *ping* and listening for its reflection back from the target—as so often depicted in World War II submarine movies. The problem with active is that you, the hunter, give yourself away by your loud pings. Active sonar had therefore been largely abandoned by both the U.S. and Soviet navies. But, with much lower sound frequencies and much longer ranges, the Soviets were reviving it.

At JASON meetings, Walter Munk and I sometimes had dinner together in one or another Georgetown restaurant, typically after other JASONs had organized dinner groups and we were the two left out—the implications of this not lost on either of us. Appropriately, we didn't even like each other. Walter considered me young, mercenary, and arrogant; while I found him old, disingenuous, and arrogant. But he was always socially polished and urbane and European. There was some bad blood between Walter and my father going back decades to when Frank was a young scientist—and probably, in Walter's view, young, mercenary, and arrogant. I never got to the bottom of it. In Walter's last years—he died in 2019 at age 101—he often asked me solicitously, "and how *is* your father?" He was gloating. He knew that Frank, a decade younger than himself, had progressive dementia while he, Walter, commanded nearly his full faculties.

Within JASON, apropos of SSBN security, I had the idea of using "active cancellation". Perhaps one could arrange hydrophones and transducers around the hull and use them to make the submarine acoustically invisible. The hydrophones would detect the incoming sonar pings and (after some signal processing) the transducers would emit an exactly cancelling wave, so that there would be no reflection. The general idea of active cancellation was not new (and by now is commonplace in noise-canceling headphones), but no one in JASON had ever heard of it being applied to submarine. There was a lot of mathematical calculation that could be done. I got Freeman Dyson interested in the idea, and then Henry Abarbanel, Sam Treiman, and Doug Eardley (the latter a new JASON member whom I had helped recruit). It looked very promising. We worked on it and kept it very hush-hush.

JASON itself did not have classification authority, the legal ability to declare something an official secret. When we invented things, we guessed how they should be classified, marked and handled our reports accordingly, and sent them to our government sponsors for review. With its potential to counter the whole Soviet sonar program, we thought that our report on active cancellation should be Top Secret. The Navy sent it back to us marked Unclassified. We, the younger JASONs, were astonished. Older JASONs explained: It might mean either of two things. Either the hidebound Navy had no interest in so radical an idea, or else there was actually, somewhere, already a *super*-secret Navy effort in active cancellation—something far above Top Secret. In that case they might declare our report Unclassified to mislead the Soviets into thinking that the U.S. had no interest in the subject. This was typical of U.S. Navy deviousness. I never found out what the truth was.

Dick Garwin sent me to the library to look for any signs of active cancellation in open Soviet scientific journals. Yes! There were at least a dozen papers on the topic, concentrated in a couple of Moscow institutes, with revealing terms like "spheroidal boundary conditions" (to me, the submarine hull) and "infinite half spaces" (to me, the ocean surface). I was reminded of Zel'dovich's textbook on shock-wave physics, so obviously derived from his work on nuclear weapons, yet with the same kind of semantic fig leaves. Soviet scientists wanted their names on open publications, and they played cat-and-mouse games with their own censors to get things approved.

Active cancellation proved to be my entrée into, not Navy circles, but CIA. Don Levine, JASON's year-round Executive Director based in Washington, led me through my first visit at Langley. At the entrance booth off Route 123, the guard looked up our names and gave us the CIA's peculiar parking pass—no words or numbers at all, but just a stylized color map showing unlabeled roads and wooded areas, with an arrow pointing to a particular lot. A mark of status was to get to park in the small VIP lot right at the front entrance, watched over by an elderly man in civilian dress who saluted as you drove in. Once, on a later visit by myself, the same ancient guard not only saluted, but said to me, "Welcome home, sir. It must feel good to be back from *there*." "Yes," I said, wondering who was my doppelgänger and what was he up to somewhere else in the world.

Walking to the CIA front entrance, you passed the statue of Nathan Hale. I sometimes confused Nathan Hale with Benedict Arnold, but my CIA friends told me that it didn't matter: Both were admired as professionals. Then, inside, it was exactly like in the movies—the big eagle seal in the floor and, to the right, the wall of anonymous stars memorializing fallen CIA officers. Those would be Directorate of Operations (DO) people—the Clandestine Service. The DI (intelligence analysis) and DS&T (science and technology) people, when occasionally posted abroad, tried harder not to get memorialized. The next status check was the kind of badge you got from the visitor's office. Best was a *No Escort* badge that gave you the run of the building, although, in practice, its only use was to get you to the cafeteria and the gift shop where you could buy CIA coffee mugs and sweatshirts. It seemed strange that the gift shop was the only place in the building where the name "CIA" was so indiscreetly displayed.

Don and I explained our mission to the two analysts in the submarine warfare office who had been ordered, rather against their will, to meet with us. "We don't try to follow the open literature," they told us. "There is just too much of it." I asked if they could check my list of Russian names against their secret databases. "Well, OK," they said. I expected them to type the names into a computer, like in the movies; but one of them disappeared and came back with a thick book of computer printout, which they flipped through. Don's and my status rose as they found that multiple of my names were known to have KGB connections, and were associated with an institute known to be KGB co-administered. "Like IKI, the Space Research Institute?" I asked. I knew about IKI's KGB masters from my trip to Protvino with our minders 007 and 008, later confirmed by my parents' friend Velodya

Keilis-Borok. "We've never been able to confirm that IKI is KGB. It's an open question." So I knew more than the CIA about IKI. That didn't inspire my confidence in U.S. intelligence.

I learned over time that S&T analysts, as well-meaning and patriotic as they might be, were rarely at the cutting edge of their fields of science. For a time, the exception was the separate office that interpreted the data collected from Soviet missile tests. Those people were at the top of their game, until most of them were fired in a budget cut and reorganization. JASON did its best work for DI when their analysts came to us with well-developed "enigmas". These were Soviet facilities or programs known to exist—as supported by satellite photography and signals or human intelligence—but whose purpose, or state of development, was unclear. Famous within JASON was an enigma from the early 1960s, a manufacturing facility surrounded by triple barbed wire fences of a kind normally used to protect nuclear weapons. The Russian codeword for the place transliterated as "Zeros," and it had something to do with the element selenium. The JASONs sat in silence, stumped, until Dick Garwin announced, "they are making *Xerox* machines." It was obvious. The high security was because copying machines were the source of *samizdat* forbidden books, like *Dr. Zhivago*.

JASON's work for CIA's DS&T fell into two categories. The first was work under the codeword "Byeman". This wasn't for CIA at all, but for the National Reconnaissance Office, NRO, the separate government agency that developed, launched, and operated U.S. spy satellites. In the 1980s, NRO was still an unmentionable "black" agency. Its public cover was the office of an Undersecretary of the Air Force who occupied a whole corridor of the Pentagon, labeled with a single, now-famous, room number 4C1000 (decoded as "Pentagon, 4th floor, C-ring, Room 1000). And there was a back door at 4C1052. But, historically, NRO was an offshoot of CIA's DS&T; and many of its personnel were CIA career officers on assignment. So, JASON's classified NRO work was funneled through DS&T.

The second category of work was for the CIA's own research arm, the Office of Research and Development (ORD) and, separately, for another office, the Office of Technical Services (OTS). ORD managed a portfolio of large institutional projects—those that had not been given over to NRO. OTS housed the people who built gadgets—spy gear—for the DO, the equivalent of James Bond's "Q". For all intents and purposes, OTS was part of DO, except that its staffers, mostly engineers, were not case officers—not formally sworn into the Clandestine Service. This was a bone of contention. The OTS people were full of stories about case officers, liberal arts majors chosen for their people skills, sitting safely in the car outside while OTS engineers did the dangerous work of installing the whatever-it-was inside someone else's building overseas.

Four of us in JASON, all younger members, were drawn to OTS work: Rich Muller, Paul Horowitz, Doug Eardley, and I. Within JASON we became known, sarcastically, as the boy spies. Paul was my colleague in the Harvard physics department, and I had proposed him for JASON membership. He was an electronics genius and, like me, had been appointed a professor on the say-so of Ed Purcell. Beyond building gadgets for the experiments of colleagues, and some work in the exotic field of SETI—the astronomical search for extra-terrestrial intelligence—Paul accomplished only little in academia. In JASON he was a star.

OTS and ORD were often at each other's throats over jurisdictional disputes, and JASON was sometimes called upon to unofficially mediate. Typical was a case where an OTS gadget was supposed to transmit data by radio to a larger ORD system, and they could not agree on the communications protocols. At a meeting in a neutral location in Washington, I suggested a compromise that both sides, after some sniping, agreed to. But I had noticed something. This presumably sensitive data—like, from a spy, maybe?—was being sent *unencrypted.* Shouldn't it be protected? A silence fell over the table until someone suggested that we break for coffee. Pulling me aside, one representative from each camp explained to me that, by Presidential order, only the National Security Agency was allowed to approve any use of encryption. They hated NSA even more than they hated each other. They suggested that JASON design an encryption mechanism for the system. To keep it legal, it would be called a "privacy," not a "security" feature. On JASON's behalf, I declined.

Rich and Paul were the most enthusiastic of the boy spies. Some projects, I wouldn't work on with them. One was to design, using only parts available in a Radio Shack catalog, something with the specifications of a timer for an explosive device. The catalog furnished to us was in Spanish. I was not surprised when it came out that, in 1984, the CIA had mined harbors in Nicaragua. Another project I steered clear of was for the British MI-5, who visited us—with a CIA minder—in the wake of the early '80s IRA bombing attacks. The British team members wore ill-fitting suits and looked and acted like thugs. Three of them gave their names as just "Tim". We thought this might be a joke, modeled on the Monty Python skit where a group of rough, beer-drinking

Australians sitting around a table are revealed to be all named Bruce. But no, these professionals were sticking to their story. They didn't seem like humorous types. They seemed like killers.

41. Chairman

When George Field announced that he would be stepping down as CfA Director a year later, in fall 1982, a search committee was formed. Alar Toomre from MIT was its outside chair. Jim Gunn from Caltech was an outside member; Paul Martin represented Harvard physics. From CfA, Harvard and Smithsonian were equally represented. I was a member. Unsurprisingly, Riccardo Giacconi insisted on having one of his people on the committee: He wanted the director job. It couldn't happen. Sufficient, I knew, was that Henry Rosovsky didn't trust Riccardo. Our committee met many times, took many votes, and eventually recommended John Bahcall, at IAS, as the top candidate and Irwin Shapiro, at MIT, as the second choice. Neither was perfect, and the votes had tilted back and forth.

In John's case, I was never sure, despite knowing him so well, that he actually wanted the job. John was a poker player. Literally so: He had supported himself in college at Louisiana State University playing poker professionally. It could easily be that he wanted the offer as a bargaining chip with Director Harry Woolf at IAS, or for Neta in the Princeton astronomy department. Irwin was an outstanding scientist—intellectually John's better—but he was a fussy, jealous, and emotional individual. He had already turned Harvard down two years before for a professorship and now told us that he had done so because his mother was then dying—at which point in the interview he broke down crying, and we had to take a break. John's interview wasn't that great overall, but at least he didn't cry. Paul Martin drafted our committee's recommendation to Rosovsky (for Harvard) and David Challinor (for Smithsonian), unambiguously favoring John Bahcall.

A formality was the necessity of approving Bahcall as a Harvard professor. An Ad Hoc meeting was as hurriedly scheduled, of time-necessity with only local, Harvard members: Paul Martin, Mike McElroy, Shelly Glashow, Roy Schwitters—not a single astrophysicist. President Bok was out of town, so Rosovsky presided. It was an ambush! Except for Henry, everyone on the Ad Hoc was a friend of Irwin's—Paul, who always had Henry's ear, was Irwin's roommate in graduate school, no less. Several mentioned that Irwin, an experimentalist, would fit well as a joint appointment in the physics department. The purpose of this committee was not to approve John. It was to overturn the search committee recommendation and appoint Irwin as professor and CfA Director.

And it did. A week later, George Field and I were summoned to Rosovsky's office. "I know I should also have invited Al Cameron, as the present chairman," Henry said pseudo-sincerely, "but I somehow don't feel comfortable with him." He had never forgiven Al for the astronomy Declaration of Independence. The purpose of this meeting was to get our assurance, George's as outgoing Director and mine as incoming department chair, that we would work with Irwin for a smooth transition. As I was leaving, Rosovsky called after me, "Remember, Bill, *no surprises.*"

Not long after I actually became department chair, I hosted my first Visiting Committee meeting. Most universities have visiting committees of independent outside experts, convened to examine and report on the health of each department to higher university administrators. Because of its unique governance structure, in this, as so much else, Harvard was different. At Harvard, deans and vice-presidents reported directly to the president—there was at this time no position of provost. The president reported to the Harvard Corporation, a self-perpetuating group of five individuals—at this time all men and mostly Bostonians—who quite literally owned the university. Only in later decades did reforms increase the size of the Corporation and bring to it greater diversity. Quite independent of the Corporation, then and now, there was also a Board of Overseers, about thirty in number, elected by the hundred thousand or so living alumni of the university. The Overseers didn't have much to do—all the real power was with the Corporation—but they did oversee the visiting committee process.

Each visiting committee's membership was a balance among tradition, money, and expertise—the categories not mutually exclusive. For astronomy, the scientists (with expertise) were Herb Friedman (chair), Mal Ruderman, Wal Sargent, Gene Parker, Bernie Burke, and Art Walker, all of whom I knew well. New money was represented by Ken Olsen (the founder of Digital Equipment Corporation), An Wang (the founder of Wang Industries), and James Michener—whose book *Space* has just come out the previous week. Old Harvard, and old money, was personified in A. Lee Loomis, Jr. (money), Charles Francis Adams (money, family, and accomplishment—he was the retired Chairman of Raytheon), William Golden (Jewish money, and he was unofficial science advisor to Harry Truman), and Marie Bullock. Bullock was a Sorbonne-educated socialite and poet who, at age twenty-two, in 1933, had married into the wealthy Bullock family. She was the donor, founder, and life-long president of the American Academy of Poetry, and was on the astronomy committee because she thought the stars poetic. Lee Loomis, by Harvard's way of thinking, somehow embodied his recently deceased father, Alfred Lee Loomis, the investment banker and gentleman scientist whose Tuxedo Park, New York, mansion was famous for its development of radar in the early days of World War II.

The visiting committee banquet was held at the Faculty Club. Ruth Mandalian, the astronomy department's long-serving secretary, had at first ordered the cheapest available wine, an amazing $1 a bottle. This was how it was always done, she told me, anyway since the time of Leo Goldberg. I made her change the order to $5 a bottle (about $15 today). Ruth was a middle-aged woman, a teetotaler, who had never married. The day I formally became department chair, she took me aside. "I've seen chairmen come, and I've seen chairmen go," she said ominously, with no further explanation. She always reminded me of Mrs. Danvers, the housekeeper at Manderley in Alfred Hitchcock's version of *Rebecca*.

Martha Minow, then a young law professor (much later, Dean of Harvard Law) was my date and co-host for the banquet. We had become friends after being introduced by our respective parents. Martha circulated, chameleon-like, among the committee members. To Mrs. Bullock, she was the sweetest little girl; to Lee Loomis, a fellow lawyer; to Bill Golden, the daughter of a prominent Jewish family. At dinner, she sat between Charles Francis Adams and Jim Michener. The latter gave the after-dinner speech on behalf of the committee. He talked like he wrote: stiff, unadorned, humorless prose, but full of facts, facts, facts and a lot of feeling and awe.

Harvard put up its visiting committees at the Charles Hotel, whose ground floor hosted some expensive fashion and jewelry boutiques. Mrs. Loomis, accompanying her husband, regularly charged thousands of dollars of purchases to their room account, which Harvard paid. A couple of weeks after the visiting committee met, Loomis would send a check—a contribution to the astronomy department—for some even thousands, rounded up from his wife's purchases. It was a tax dodge. Harvard was paying for her clothes and he was getting a tax deduction. But it was not my job as chairman to question the arrangement.

Department chairs were required to attend monthly meetings of the Faculty of Arts and Sciences in University Hall, in the Faculty Room, under the oil-portrait gaze of all previous Harvard presidents. In principle, any faculty member could attend these; few scientists ever did. The meetings' only purpose seemed to be to provide a forum for a self-selected group of humanist and social science professors to debate inconsequential issues. The quality of debate was always high, however, and these people were always very well dressed (by scientists' standards). Two hours might be spent discussing a proposal for a technical change in the way that grade averages were rounded in the determination of honors for graduating seniors. (That particular proposal pitted the penunanimous recommendation of the elected Faculty Council—Professor David Layzer dissenting—against the unanimous recommendation of the appointed Committee on Undergraduate Studies—as in that committee proposed by Professor David Layzer.) Memorial minutes (so-called) about deceased faculty members were read. George A. Buttrick, late preacher to the university, was known for his long sermons. A well-wisher had once quipped, "After a sermon like yours, Dr. Buttrick, no man need be afraid to die." (It much later became known that Buttrick, had officiated the marriage ceremony of the parents of President Donald Trump in 1936. If only he had instead counseled them not to marry!)

At another meeting, I happened to be sitting next to William Alfred, whose Beowulf course I had taken as an undergraduate, and heard him whisper to the person on his other side, "What is the best biography of Queen Isabella? Faye is going to play the part and asked me to inquire," the reference to his friend, Faye Dunaway, the actress.

As much as I enjoyed these meetings, they became occasions of uneasy reflection on my professional state. Not yet thirty-five, I was already becoming "mid-career." I was starting to spend more time "advancing the scientific enterprise," as older scientists liked to say, than actually doing science, that is, discovering things. I was much in demand as a colloquium speaker, a dozen or more such trips a year around the country; but my talks were increasingly reviews of work by others. And, unlike many mid-career scientists, I had little appetite for home-institution empire-building. I never took on more than two Ph.D. students at a time, and the postdocs with whom I interacted were generally funded independently by CfA, not hired by me directly.

As one exception, I did use my Livermore/Teller connection to raise money to recruit Julian Krolik, who was a postdoc at MIT, as a CfA research associate—one level up from postdoc. Unlike the rigorous process of writing a grant proposal to, say, the National Science Foundation, to be peer-reviewed and then funded (or not) a year later, my "process" was to visit Livermore Lab with Julian in tow. I was betting that his area of astrophysics, dealing with hot matter and radiation diffusion, was close enough to bomb physics that my friends there would see him as a kindred spirit. Julian and I sat around chatting in John Nuckolls' office with John, George Zimmerman and Tom Weaver. Zimmerman had been a Caltech graduate student in my year who had escaped the draft by going to Livermore—as it turned out, permanently. Now a brilliant computational scientist "on the inside," he was invisible in the academic universe. Weaver, also a Lowell Wood protégé, was moving in the other direction; his "inside" work on supernova explosion modeling was becoming appreciated as far ahead of anything in academia on the "outside".

As our scientific discussion was winding down, I asked Nuckolls for $50,000 to support Krolik at Harvard for a year—salary, overhead, research expenses. "Yeah, he sounds like a good guy," he responded. "Write up a one-page proposal. Don't use the words 'matter' or 'radiation'—those only attract the attention of the security people who want to classify everything. It doesn't really matter what you write. We'll send the money."

These were the Reagan years of expansion of the nuclear weapons program. Money was pouring into the nuclear weapons labs. Fifty thousand dollars was a significant grant for university-based theoretical astrophysics, but at Livermore it wasn't even loose change. Krolik was aghast at this reality—but not about to turn down the support. He and I struggled to write even one page on astrophysics without mentioning matter or radiation. Those, basically, were the entire content of the field! We settled on the words "magnetic fields". In later years, Krolik became a professor of physics and astronomy at Johns Hopkins and at the Space Telescope Science Institute, much respected for his work in magneto-hydrodynamics.

A three-day conference at the University of Illinois at Urbana-Champaign (UIUC) on numerical astrophysics (then a novel term) seemed designed to make me confront my ambivalent past—was I more a product of Caltech, or more one of Livermore?—with my growing mid-career introspections. The conference was organized by Larry Smarr, in honor of Jim Wilson's 60th birthday. It was thirteen years since Wilson and LeBlanc—the Professor and Captain Nemo, as I still thought of them—had initiated me, a new summer student, into numerical modeling as an intellectual discipline and supercomputing as a mode of scientific research. Then, these were black arts practiced with any real scope or sophistication only at the nuclear weapons laboratories, Livermore and Los Alamos. In the intervening years, in no small way due to Wilson, Tom Weaver, and my other Livermore friends, this specialized knowledge had spread out from the labs into academia. It was now taking root.

My computer work was, in a small way, part of that spread. As a matchmaker, I had introduced astrophysicists, including Smarr, to the nuclear weapons crowd. And, the NSF committee that I had chaired with such inexperience two years previously had led to other committees, other reports. It now seemed likely that NSF would soon charter half a dozen national supercomputer centers, located at universities, each with a $10 million supercomputer. Larry's not-so-hidden agenda in organizing this conference was to establish him and his university as an obvious heir, within academia, to Jim Wilson and Livermore.

For this conference, Larry was well-liked enough, not to mention persuasive, to have turned out a roster of speakers and attendees that included the best of the profession: Kip Thorne, Martin Rees, Charlie Misner, Roger Blandford, Saul Teukolsky, Stu Shapiro, Fred Lamb, David Pines, Bernard Jones, Jimmy York, Bryce DeWitt, Dave Arnett; from Livermore, Jim Wilson, Jim LeBlanc, Mike May, Bruce Tarter, Tom Weaver. That would signal Larry's university to take him seriously.

Larry was now his university's rising star, their wonder boy. He was building on this to create an academic empire of students and postdocs, and, if NSF cooperated, bricks, mortar, and supercomputer. At his house, we talked into the night about how a new national computing center might be structured, and how to get UIUC to ante-up the necessary resources before any NSF funding might be available.

I wasn't an empire builder, but I did think that I was ambitious in my own way. We compared our respective outsized ambitions to those of X, an attendee at the meeting. X was the perfect scientist: he had no major achievements in his career, nothing to make him famous or fulfill his promise, but he just kept plugging away on dumb little problems, done crudely, just plugging away through his 20's, his 30's, now his 40's, and we presumed for the next twenty years. He accomplished nothing important, but he never doubted who he was or what his role in the world was. Perhaps his way was best. But neither of us was likely to adopt it.

Jim Wilson was not well suited for the role of computer hero in which the conference cast him. The best-kept secret at the meeting was that, apart from Jim's ability to mind-meld with Cray supercomputers, he was not anyone's favorite collaborator. He was genial, often enthusiastic, but disorganized, inarticulate, dyslexic, and incomprehensibly laconic. His scientific conference talk was terrible. His wandering after-dinner remarks were slightly better. At Los Alamos during the Manhattan

Project, he had been entrusted with the world supply of plutonium (a piece the size of a pea) and asked to determine whether it was malleable. He took it into his office and, when no one was looking, whopped it with a hammer. "It's malleable," he reported.

Jim LeBlanc (to me, Captain Nemo) gave a poorly attended talk. He had been Wilson's assistant, interpreter, Boswell through all the years. It was not difficult to read between the lines of his talk to find an element of suppressed resentment that it was Wilson, the supposed genius, who got all the credit, while he, LeBlanc, who did most of the hard work, got only an hour, just before lunch, and with a lot of conversational hubbub intruding from the lobby outside.

Back at home, I was less worried about empire-building than I had been in Larry's living room. I had two graduate students, Robert Brandenberger and Ron Kahn. Brandenberger, born in Switzerland to an American mother, grew up there, and was an undergraduate at E.T.H. (the Swiss MIT). After his Ph.D., he did postdoctoral work with Stephen Hawking, and became a professor of physics at Brown University and, later, McGill University in Montreal. Ronald Kahn, after getting his Ph.D. in physics, became one of the "rocket scientists" who brought advanced mathematical modeling to Wall Street. He wrote an influential book on portfolio management and became eventually Global Head of Systematic Equity Research at BlackRock.

I enjoyed the minutiae of being department chair. I was trying to work cooperatively with Irwin and to keep my *no surprises* promise to the dean, at least for a little while. I had a clear idea of what I wanted to accomplish as chair—a generational turnover of the senior astronomy faculty. I had started courting Bob Kirshner, whom I knew since undergraduate days in Dick McCray's course. Bob was now himself a boy wonder, having risen at the University of Michigan by age thirty-three to full professor, department chair, and observatory directory. I knew that a Harvard professor offer would have a special allure to him as a Harvard alum—I had experienced the same.

I taught the graduate relativity course in physics, and also a "light" after-dinner course in North House: "Time Travel in Physics and Literature." At Chicago, Bob Geroch, a mathematical relativist who worked with Chandrasekhar, had developed a course for non-scientists and written the cleverly named textbook, *General Relativity from A to B.* My course added H.G. Wells' *The Time Machine*, Twain's *Connecticut Yankee*, and science fiction novels by Heinlein and Poul Anderson. The students and I had a good time.

42. Numerical Recipes

In the fifteen years or so from 1970 to the mid-80s, the constituency for scientific computing broadened from a few high priests in certain subfields to virtually all researchers in large areas of science. Technology moved from punch-card decks and mainframes to time-sharing systems, minicomputers, and—just then nascent—the PC. My career rode on that wave. However, through this period, nearly all scientists enjoyed a blissful, almost touching, ignorance of numerical analysis—the rigorous branch of mathematics that underlay scientific computer calculations.

As graduate students at Caltech, Saul Teukolsky and I were only slightly less ignorant than most. When we ran into difficulty integrating a so-called ordinary differential equation, we consulted Herb Keller, known to be a distinguished professor in numerical analysis. "What method are you using?" Keller asked us. We looked at each other blankly. Did he mean that there was *more than one* numerical method? "Runge-Kutta, of course," we answered. Carl Runge and Willhelm Kutta were German mathematicians who had developed this method in 1901. At Caltech, it had been taught in a first-year graduate physics course since time immemorial. "Runge-Kutta! Runge-Kutta!" Keller shouted, banging his head with his fist. "That's all you physicists know!" Keller told us that we should be using the newer Bulirsch-Stoer method. Roland Bulirsch and Josef Stoer were German mathematicians who were alive, at least.

But this story has a twist: It turned out that Bulirsch-Stoer was even worse than Runge-Kutta for our problem. So, the story illustrates not only our ignorance of numerical analysis, but also that when scientists consulted numerical analysts, either in person or through book or journal articles, they were not infrequently disappointed. Numerical analysis was a field with its own fads and fashions, often poorly matched to practical scientific work.

In the couple of years that Brian Flannery was an assistant professor in astronomy at Harvard—before he left precipitously for a position at Exxon Research Labs at twice the salary—he developed a popular graduate course in computational astrophysics, one that emphasized the practical. When Brian left, I inherited the course and taught it a couple of times from his lecture notes. Independently, Saul was doing a similar course at Cornell. We were by no means unique. Such courses, emphasizing methods that could actually be useful, were springing up in physics and astronomy departments around the country—never in mathematics departments. (Computer science departments mostly didn't yet exist.) Because there were no good textbooks, these courses were typically taught from the instructor's lecture-note handouts. On his way out the door, Brian suggested that he and I turn his notes into a textbook. The need for a practical textbook on scientific computing was very obvious. Conventional wisdom was that there was no real money to be made in writing a textbook—at least not on an arcane subject like ours, and at the graduate level. It would have to be a labor of love. I wasn't interested.

So, the idea lay fallow until I became astronomy department chair and realized that textbook-writing was in many ways the perfect complement to department-chairing. My duties for the department took perhaps an hour a day, on average; but they usually came in the form of a dozen or more five-minute interruptions. These spoiled the concentration needed to think about deep problems—my ability to stay in the trancelike state that I thought of as "mathland". But writing, at least for me, was an activity that could be started and stopped as easily as picking up or putting down a pencil. Writing a thousand words a day of science fiction as a first-year graduate student (albeit unpublishably bad), Kip's firm tutelage in scientific writing, plus my lengthy daily dairy—I thought of myself as a writer, even if no one else knew it. Also, I had decided that the conventional economic wisdom might be wrong in this case: A book on scientific computing that could be both a text and reference book might have a big audience.

Both administration and book-writing were activities looked down on, if not actively scorned, by upcoming scientists. You became a department chair—or wrote a technical book—when you ran out of ideas for your own research. I didn't think that I was out of ideas yet; but, in becoming department chair, I had already crossed this Bridge of Scorn. So why not double down? My intention was to do one three-year term as chair. If, at the end of that, I also had a book to show for it, then all the better.

Brian was contributing his lecture notes, but as now an industrial scientist, he didn't have much time to work on the project. If this was to be a two-year, and not five-year, effort, I needed at least one co-author. Paul Horowitz's textbook on electronics had recently been published and was an immediate best-seller (as these things go) due to its clear, chatty prose style. "But I don't know anything about numerical analysis!" Paul objected, when I visited him in his lab and proposed collaborating.

I didn't see that as an obstacle. I would select algorithms for the book, write out their equations, and explain them to Paul, who was always much smarter than he pretended to be. He would then do the writing. He actually did submit to this for a couple of months, until, seeing clearly that he would end up doing ninety percent of the work, he delicately withdrew from the project. By then I had enlisted Saul Teukolsky in the project, which I named *Numerical Recipes*. Saul's participation shifted the work assignments. As in all our previous collaborations, I would be mostly responsible for the prose, Saul for the equations.

But who would do the detail work that was at least half of what I had intended to unload onto Horowitz? A rule of thumb in writing a book is that when you think you're half done, it's actually a quarter. When you think you're ninety percent done, it's actually only half. You rewrite one section and realize that now you have to rewrite two others. Editing, first for meaning and style and later, copyediting, are Sisyphean tasks. In technical books, equations have to be checked, not just for correctness, but for notational consistency with dozens of other equations throughout the book. References are a fussy nightmare.

Paul Horowitz suggested the name of Bill Vetterling. That was an interesting idea. Vetterling was Paul's junior colleague in experimental physics. Most of his work was with Bob Pound. Pound had helped in Ed Purcell's discovery of nuclear magnetic resonance (known in clinical medicine as MRI), but did not share in the Nobel Prize given in 1954 to Purcell and Felix Bloch. Pound later measured the redshift (change in energy) of photons traveling up and down a shaft drilled through the floors and ceilings of Harvard's Jefferson Laboratory, a gravitational effect predicted by Einstein's General Relativity. Like Paul Horowitz, Bill Vetterling knew little about numerical analysis. He was shy, high-strung, and came across as nervous and fussy in conversation. These were side-effects of his being a consummate detail person. That was just what we needed on the *Numerical Recipes* team. And, significantly, every senior faculty member in physics knew that Bill was not going to be promoted to tenure—but none of them would level with him on the issue. That meant that all the work that Vetterling was putting into building up Pound's lab would end up being ultimately lost to his career.

Bill and his wife MaryAnne were nonresident faculty associates at North House, so I knew them better socially than professionally. Specifically, Bill knew that I didn't have a dog in his tenure fight. That was my opening. I invited him to meet in my office. I told him that he was not going to get tenure and that none of the other senior faculty were brave enough to tell him this. "They think you are too delicate to take the news. You're wasting your time in Bob Pound's lab. You should be out looking for a job." And, in the meantime, *I had a proposition for him.* Vetterling soon signed on as our fourth *Numerical Recipes* author.

Scholarly publishers worked hard to keep authors (and potential authors) in the dark about the economics of their book publishing. Nearly all of a publisher's costs are fixed costs—salaries and office space—that are independent of how many books they sold. The marginal cost of printing, binding, warehousing, and shipping one copy of a book was at the time perhaps a quarter of its cover price. The typical deal for a graduate-level monograph or text was to offer the authors no advance payment, and a royalty of fifteen percent on net receipts or ten percent on cover price—half that for a paperback edition.

Authors always imagined unrealistically large sales. Publishers knew better. The typical graduate-level monograph could sell five hundred or a thousand copies, mostly to libraries, pretty much independent of cover price—the demand was inelastic, in other words. Publishers stayed in business by setting their cover prices so that these inelastic sales, multiplied by the number of such books they published per year, covered their fixed costs. Anything above that was gravy. This was a very conservative business. Acquisition editors—people who came around to faculty offices once a year looking for book manuscripts—were responsible mainly for bringing in books that wouldn't *lose* money.

We trolled *Numerical Recipes* to two for-profit publishers (Addison-Wesley and Benjamin-Cummings) and two not-for-profit university presses (Cambridge and Princeton) and got expressions of interest from all four. There were important distinctions. The for-profits would invest more up front in advertising and marketing a book, but they were ruthless in abandoning books that didn't sell well enough—often "pulping" unsold stock for a short-term tax advantage. The university presses didn't pay taxes, and could keep in print large backlists of books that sold only a few copies per year. As an example, our *Problem Book in Relativity and Gravitation*, published by Princeton, is still in print after almost fifty years. Most of those years it has sold fewer than one hundred copies.

By early 1984, we had a 90% manuscript—meaning that we were really only half done. We let Ed Tenner, the editor at Princeton, send it for review to Forman Acton, whose own book, *Numerical Methods That Work,* had much the spirit we were aiming for; and to Stephen Wolfram. Acton loved us. Wolfram, then at IAS Princeton and whose transformative *Mathematica* software was still five years in the future, said that our manuscript was entertaining, but sterile. I was later an early beta tester of *Mathematica* and, in 1988, gave it a favorable review in *Nature* magazine, under the title, "Twilight of the Pencil".

Regardless, we by now had a clear idea of what were our goals: We wanted to fill an obvious need for a textbook. We wanted to expand the content of then-standard numerical courses to include methods newer than fifty years old (e.g., fast Fourier transforms, Monte Carlo methods). More ambitiously, we wanted to decrease—surely not close—the huge gap between the "best practice" of the rarified mathematical professional software community and the "average practice" of most scientists and graduate students whom we knew. That gap was huge, even at top places. It was not difficult to find people who thought that determinants should be computed by the expansion of minors, an exponentially slow process. Gauss knew better than this at the beginning of the nineteenth century. There was the story of the graduate student who exhausted his professor's entire year's computing budget by repeated runs, over months, of a Monte Carlo program—each run using exactly the same "random" numbers and giving exactly the same results.

Part of our plan from the start was that our book should contain actual, working computer programs. We planned for a language-neutral approach where we might publish versions of the book in more than one computer language, starting with both Fortran and Pascal. In deciding what computer code to include, we often faced an issue akin to "do you teach to the top of the class or to the middle?" We frequently made decisions that traded an advanced method's last factor of two in performance for an older—but not too old—method that was pedagogically clearer or more concise. "Factor of two, but not factor of ten," was our mantra. In later years, we were often criticized for these choices; but we were firmly committed to teaching to the middle of the class.

It wasn't for their backlist or tax status that we eventually signed with Cambridge University Press. But, there was an interesting tax wrinkle: Under the terms of the U.S.-U.K. tax treaty, I owed full U.S. taxes on my royalties, but C.U.P. was not required to withhold taxes, or even inform the I.R.S. of my earnings. That meant that I always declared significantly *more* income than the I.R.S. knew about from direct reporting to them—which would move me to their lowest priority for any tax audit.

It was C.U.P.'s acquisition editor David Tranah who won us over. David was a compact Englishman with dark hair and beady eyes who had negotiated with Paul Horowitz for his electronics book. Tranah *loved* to negotiate. So did I. He and I negotiated for more than a year on everything from type of binding ("sewn hardcover," not "perfect") to initial cover price. On the all-important question of royalties, we arrived at a sliding scale that paid us almost nothing on the first couple of thousand copies—that let David cover his overhead and sell the deal to his bosses—and then rose rapidly to an unheard-of forty percent at the impossible number of fifty-thousand copies sold.

By the time that *Numerical Recipes* was actually published in January, 1986, it had already sold 5000 copies to a subscription science book club, which raised our royalty rate to 20%. C.U.P. first tried to tell us that book club sales didn't count toward our royalty escalation. "Read our contract," I suggested, since I had written it. David Tranah then visited me in person to wring his hands and explain that they couldn't afford our royalty rates—which would obviously escalate further—and would have no choice but to put the book out of print. This was a bluff akin to the child who says, "if you don't give me what I want, I'll put beans up my nose." In fact, our unique contract provided that if the book went out of print, we authors would recover all rights in six months and could take the book to another publisher.

I'll zoom out to the big picture. *Numerical Recipes* had new editions (varying as to what computer language was used, and later adding new material that almost doubled its size) in 1986, 1989, 1991, 1992, 1996, 2002, and 2007, eventually selling more than 500,000 hardcover copies. The book was translated into Japanese, Chinese (separately for Taiwan and mainland China), and Brazilian Portuguese. Two attempts at Russian translation both failed when their respective publishers went bankrupt. When the Internet came to exist, we started a small web business to sell the software separately from the book, and to license it for commercial use. That business eventually produced more income annually than our book royalties—more on this later. *Numerical Recipes* didn't make me rich enough to retire in luxury to my own Caribbean island. It did, over the years, pay for what amounted to two houses—no small thing for a university professor.

Significantly, our project was never looked on favorably by the subfield of professional numerical analysts within computer science. Despite our half-million users, we never won any medals or awards; instead, we attracted quite a few scathing reviews in numerical analysis journals. The numerical analysts' central dogma was roughly, "Good numerical methods are sophisticated, highly tuned, and based on theorems, not tinkering. Users should interact with such algorithms through defined interfaces, and should be prevented from modifying their internals—for the users' own good." This perspective informed the large, opaque scientific libraries that dominated the 1970s and 1980s, the NAG library (developed in the U.K.) and its American cousin, IMSL. That perspective continues today in dominant scientific computing environments like MATLAB, Mathematica, and (to a somewhat lesser extent) Python.

Cleve Moler, who started in academia and invented MATLAB, was an early critic of ours, although he came around a bit after he and I had served on several committees together. We *Numerical Recipes* authors never felt that we had a quarrel with the other viewpoint. We only felt that we had discovered a large, different community of users whose needs were not being met. Our customers were practicing scientists (along with some engineers, economists, and quant financial types) who *did* like to see the internals of methods, who *did* like to tinker, and who might creatively abandon the numerical analyst's rigid and opaque interfaces. We were pleased when a reader pointed out to us that NUMERICAL RECIPES had the anagram UNPRECISE MIRACLE. Some decades later, when I had moved to Los Alamos, Cleve Moler retired to Santa Fe. He gave me a free MATLAB personal license and invited my wife and me to parties at his mansion near the Santa Fe Opera. His guest list was *le tout Santa Fe.*

43. Nukes and Spooks

Back in 1969, after my Green Badge, signifying a "Q" security clearance, had finally come through at Livermore, I made a special trip to the Lab for a private tutorial in nuclear weapons design from John Nuckolls. John was considered as one of the greatest designers in (U.S.) nuclear history, so this was like a beginning student winning a master class in violin with Heifetz or Paganini. I was thereafter familiar with what nuclear weapons looked like—but only in blueprint drawings and photos, not in real life.

The first time I saw—put my hands on—a real nuclear weapon, the genuine article, was in 1977, on a JASON field trip to Los Alamos. Until 1992, when the United States terminated its nuclear testing program in anticipation of a comprehensive nuclear test-ban treaty, the final assembly of weapons—referred to as "devices"—destined for testing in Nevada was done at the lab responsible for the test, Livermore, Los Alamos, or Sandia. Sandia, the third nuclear weapons laboratory, was responsible not for the "nuclear" part of the weapons, but for all the other parts: arming, fusing, security mechanisms, and so on. At Los Alamos, our JASON group first toured the high-explosive (HE) milling facility. We had to empty our pockets of all metal objects that could drop and make a spark. We watched operators machining, from big blocks of HE, the geometrically precise pieces of explosive that would fit together with the "physics package," and, in operation, implode the plutonium "pit" to produce a nuclear explosion—termed its "yield". The physics package was the object that encapsulated a fission weapon's (A-bomb's) plutonium core or "pit"; or, for a fusion weapon (H-bomb) both the "primary" pit and the fusion "secondary" that the primary would implode and ignite. The HE machining stations were spaced far enough apart that, if one exploded, its shock wave would not set off the HE at the others—though it might still kill everyone in the room. Humans were more fragile than blocks of HE.

Next, we toured the plutonium facility, where the pits for the test devices were actually made. The security envelope around plutonium was more rigorous. It took us half an hour to go through a series of document checks and personal searches, and we were assigned "watchers," enough to keep eyes on every JASON at all times. Plutonium was handled (melted, cast, machined, etc.) only in glove boxes, workers inserting their arms into thick rubber gloves sealed to openings in the wall of the boxes, reaching in. The boxes could be quite large—two long armlengths across and, with multiple glove stations, of any length. The quantity of plutonium in each box was strictly limited—and checked and double-checked—to be sure that there was never enough to make an explosive critical mass.

At the end of the manufacturing process, when the plutonium was hermetically encased in inert metal, the finished pit could be taken from the glove box. Our guides led us to a rack of several finished pits, picked one up in both hands, and handed it around to us. It was warm to the touch from the continuing radioactive decay inside. We worried briefly about dropping the pit onto the floor. We knew that without high explosive it couldn't "go off," but could it break open and spread deadly plutonium particles into the room? Our hosts assured us that pits were engineered to survive much worse accidents than this. However, a pit was worth a hundred thousand or a million dollars (depending on how you did the accounting of the whole nuclear complex) and, if dropped, would have to be completely torn down and remanufactured. In later years, common sense prevailed, and visiting VIPs and even Lab senior managers were no longer allowed to handle pits so casually.

Most controlled of all was the isolated facility, a room the size of a high-school gymnasium, where the high explosive parts and the plutonium pits were brought together for final assembly into a testable nuclear device. Although unlikely at any calculable probability, an accidental nuclear explosion could theoretically occur here. With luck, it might kill only a fraction of the population of the small town of Los Alamos. Two nuclear devices, in different stages of assembly, were allowed to be in the room at a time, each on a waist-high steel rack. At the far end of the room was an assembled device, ready for shipping, watched over by two armed guards. The device at the near end, close to us, was not yet fully assembled. It had its nuclear and HE components in place, but not its detonators. That meant that we could approach it at close range—carefully touch it.

I was struck by the moral neutrality of the artifact itself. Nuclear weapons don't seem to be machines in the usual sense, since they have, on any human-sensible timescale, no moving parts. That is an illusion. In the milliseconds that it takes the primary to implode, followed by the even shorter timescale of the secondary, "parts" are moving with an exact, elegant precision. The function of these machines was simple and almost astronomically pure: to transform themselves into very large amounts of instantaneous energy. Their design embodied that purpose and had, innately, no particular knowledge of the self-destructive tendencies of the human species. These were not guns, whose purpose—killing people—was inherent in their design; or chemical-warfare agents, whose chemistry was highly specialized to disrupting human physiology. The abstract technological neutrality of nuclear weapons set them apart. They became evil only by reflecting our own evil back on us. As we JASONs examined the object in front of us, it was easy to disregard the fact that it could kill a hundred thousand souls, give or take.

Each of the three weapons labs had a classified museum with full-scale models of weapons. The models substituted inert materials for the high explosives and nuclear materials, but were otherwise life-like. Some had been built as engineering prototypes; others were built for advertising the lab's capabilities to visiting military and civilian officials. Los Alamos' Blue Room was, in effect, its sales showroom. My first visit to the Blue Room, in 1977, was not long after Jimmy Carter had been elected president on a campaign platform of cutting defense budgets and promoting nuclear nonproliferation. Los Alamos didn't expect to be building new weapons any time soon. The models were arranged as a thirty-year chronology of weapons development, starting with World War II's Fat Man and Little Boy—think of them as Model T and Model A Fords—up through the most recent weapons. These were not very recent. They were designed in the 1960s—think of them as Pontiac GTOs, Dodge Chargers, or similar muscle cars. The history shown was one of incremental improvement.

I next visited the Blue Room six years later, during the Reagan nuclear build-up. History was now relegated to a small corner of the room. Los Alamos and Livermore were in full sales-and-marketing mode, their respective showrooms touting not just stockpile weapons, but new designs that they wanted the military to buy. These were not Pontiacs. They were Ferraris and Lamborghinis. In twenty years since the stockpile was last rebuilt, there was a revolution in weapons design. More kilotons were packed into less space and less weight. When we, JASON academic physicists, looked at these newer designs, it seemed impossible that they could work. Components seemed to be the wrong shapes, or in the wrong places. Things were packed together like the parts of a mechanical Swiss watch. The magic of these designs was that in the first milliseconds of their explosion, they in effect reconfigured themselves to a more physics-friendly design. It reminded me of the Transformers toys that were just then becoming popular.

Livermore and Los Alamos had competing stories as to which lab was more responsible for the amazing advances. A dispassionate historian would find, I think, that Livermore—from Teller on—generated the most important new ideas, but lacked the engineering discipline to, on its own, make them work. Many rounds of competition between physics-first Livermore and engineering-first Los Alamos forced each lab to improve its weaker side. Today, the majority of weapons in the nuclear deterrence stockpile are Los Alamos-designed. But the majority of *ideas* in them arguably come from Livermore.

A different JASON field trip that same winter was to F.E. Warren Air Force Base in Cheyenne, Wyoming, headquarters of the 90th Strategic Missile Wing that commanded two hundred operational Minuteman III missiles. The base was originally a cavalry fort on the Oregon trail. F.E. Warren's daughter married John J. "Black Jack" Pershing. The senior officers now lived in brick mansions dating from the 1880's. The widely separated launch control centers (LCCs) and unmanned missile silos were as much as a hundred miles from the base, scattered into Wyoming, Colorado, and Nebraska. Our escorts, a major and a young lieutenant, drove us in a blue Air Force van for an hour on back roads. It was a sunny day, though with a strong, cold wind. There were few other vehicles on the road. Some that passed us were pickup trucks or vans with two-man combat crews headed back from their 24-hour alert duty. This was big country: brown, flat land spreading as far as the eye could see, intersecting dark blue sky on the horizon in all directions but due west, where one saw what looked like low, white, jagged hills, that were actually the snow-covered front range of the Rockies fifty miles away.

Our lieutenant read his map wrong, so it took us a while to find silo "Quebec Three," seen from outside as a nondescript 11-ton slab of concrete, the blast cover, surrounded by a fence with anti-intrusion alarms and dire warning signs. In a nuclear war, the slab would be explosively blown off. Before the cover landed 200 yards away, the missile would have launched. Each missile carried two 170 kiloton warheads. Each warhead could obliterate a medium-sized city. The 90th Wing targeted four hundred geographical locations in the Soviet Union. Each missile was pre-programmed with its target. It took only a "go" command to send the missile off. That command would come from one of the LCCs, each controlling ten silos. Later, as the Cold War wound down, the targeting information and nuclear unlock codes were, as a safety measure, removed from the missiles. Instead, the missiles would get their targeting coordinates only by Presidential order in a time of crisis.

We drove half an hour to LCC "Quebec". The LCC's "Topside" was a frame building with an attached garage, inside a chain-link enclosure fence. Our lieutenant recited his memorized code-of-the-day into the phone at the gate, speaking directly to the two-man launch crew, officers, who were sealed in their underground capsule. They unlocked the gate for us. The Topside crew—all NCO's, not officers—were the security force. They couldn't lock or unlock the gate. This and other measures were designed to discourage any fraternization between the security force and the missile crews. If an arriving officer stumbled with the code-of-the-day, or if there were any other anomaly, the security force would execute a forceful (supposedly non-lethal) takedown, ending with the officer sprawled on the floor with weapons aimed at his head. To the delight of the security crews, this protocol was randomly exercised: A missile officer at base HQ would be sent to an LCC with intentionally the wrong code. The wing commander later related to us that an Air Force I.G. had once shown up in his office, told him to pick up his phone and issue a series of off-normal commands. He complied. Twenty minutes later, his deputy commander, smiling because he guessed that it was an exercise, showed up, backed by armed security police. "I relieve you of your command, sir." Introductions were made, the exercise was declared over, and all (except the MPs) went off to lunch in the officer's club.

LCC Topside, inside, reminded me (discounting the racks for rifles and machine guns) of an observatory dormitory: a kitchen, a meal schedule around all 24 hours of the day, rooms with bunks, a common room with a pool table. We went through another checkpoint, then down an elevator, then through the outer blast door. The two-man crew had opened their inner blast door and came out to meet us. The commander was a soft-spoken 26-year-old lieutenant. His deputy was a balding captain, some years older, a late transfer into the missile force. Both wore sidearms so that each could shoot the other, if necessary, to prevent a rogue missile launch. A staple of movie scripts, this had never actually happened. The capsule was about 8 x 40 feet in size. Once we were inside, they closed and secured the blast doors.

Where Air Force pilots wore their aviation "wings" badge, Air Force missilemen (later renamed gender-neutral "missileers") wore a badge with a missile on it—inevitably known as the "pocket rocket". The badge had a weighty subtext, however. The U.S.A.F. was (still is) schizophrenic in its view of aerospace missions other than flying manned airplanes. One the one hand, post-World War II, the service was aggressive in claiming such missions—missiles, satellites, space launch, airborne electronic warfare, (later) cyber warfare—as belonging to it. On the other hand, Air Force traditionalists viewed officers in these non-flying career tracks as distinctly second-class citizens, notionally because they didn't put themselves in harm's way as the pilots did, but actually because these mostly technical fields didn't share the Air Force's flyboy culture. As a result, a missileman could be promoted through the ranks up to lieutenant colonel or colonel, but would then hit a glass ceiling. It was indicative of an internecine battle within the Air Force. It was particularly galling to pilots that missilemen, comfortably underground in Wyoming, were credited as doing combat duty. But—the missilemen countered—with their own fingers on the nuclear trigger, how else could it be?

Air Force Systems Command most embodied this battle for the soul of the Air Force. Chartered in the Kennedy era by Robert McNamara's "whiz kids," Systems Command controlled the acquisition of all future Air Force systems—including the airplanes that the pilots would fly. Its culture was high-technology, not aviation derring-do. It had the clout to get rapid promotions for its own best officers, and to get promotions even for officers who, while in the Air Force, had gone off and earned Ph.D.s in physics or engineering. Previously, a Ph.D. would have marked a career as terminal. The cultural conflict came to a head when President Jimmy Carter appointed General Lew Allen as Chief of Staff (that is, head) of the whole Air Force. True, Allen had earned the wings of a bomber pilot just after World War II, but his career was otherwise anathema to the flying, fighting Air Force. He had never served in an overseas or combat assignment. He had instead earned a Ph.D. in nuclear physics from the University of Illinois, did a tour of duty at civilian Los Alamos laboratory, and was promoted within Air Force units that would become part of Systems Command, rising to commander of Systems Command itself. Along the way, he had been director of the National Security Agency, a suspiciously purple command assignment, "purple" referring to a joint command, i.e., one with personnel from Army, Navy, and Air Force combined. The metaphor was that mixing the colors of all three uniforms was supposed to produce the color purple—a doubtful proposition.

Allen served as Chief of Staff for the statutory four years until 1982. By then, Carter's Secretary of Defense (or SECDEF) Harold Brown, a physicist and former president of Caltech, had been replaced by Reagan's Caspar Weinberger, a lawyer-politician who had served as an infantryman in World War II. When Cap Weinberger spoke at a JASON meeting, he told us that the Soviets were evil people who lied and murdered and didn't honor treaty obligations, while we were good people who didn't do those things. Therefore, it was not so important that they should have confidence in our not attacking them—because we knew that we would never do any such thing. On the other hand, we could never be confident that they were not planning to attack us. We listened in stunned silence to this 5th-grade, *Reader's Digest*, view of the world.

Under Reagan and Weinberger, the Air Force found that it didn't need physicists after all. After General Allen, every Air Force Chief of Staff but one has been a fighter pilot with combat experience. Allen retired from the Air Force to become director of Caltech's NASA JPL Laboratory. Air Force Systems Command limped along until 1992, when it was merged with Air Force Materiel Command. After 1982, no full colonel in Systems Command was ever promoted to general officer.

I got to know Lew Allen a bit when he was JPL director; but I later got to know well two of his successors who, years apart and after each retired from active service, became presidents of the Institute for Defense Analyses, of whose board I was a member. General Larry Welch was the first Air Force Chief of Staff to have started as an enlisted airman (that is, not an officer). I'll have more to say about Larry, from whom I learned much about leadership, later on. Normally analytical, thoughtful, unrufflable, the only time I ever saw him become visibly angry was when I asked him about Lew Allen. "Worst mistake a president ever made in appointing a service chief," he said. The other Chief of Staff I got to know well was General Norty Schwartz. He was the one exception to being a fighter pilot. He had made his Air Force career in the special forces, flying black transport planes and helicopters. His view of Allen was similar to that of General Welch.

The battle for the soul of the Air Force continues even today, with the question of whether the pilots of remotely piloted vehicles, "drones," increasingly an important part of the Air Force mission, should be considered as the equals of the pilots of manned aircraft. There is a Remotely-Piloted Aircraft Pilot Badge, but it may be a long time before there is Air Force Chief of Staff wearing those wings.

These matters were playing out far above our heads and our pay grades as we JASONs crowded into the underground missile-launch capsule, which was jammed with racks of electronic equipment and the two-man launch console. Two safes, one for each crew member, contained the two keys necessary for launch. There was a cot where one crew member at a time could nap, and a small TV set for recreational use, then tuned to a daytime quiz show. The launch system was constantly being exercised. In a 24-hour shift, a crew might receive half a dozen Emergency Action Messages (EAMs)—seemingly orders to launch their missiles. One of these came in while we were there, clattering on an ancient teleprinter, and also announced via a voice channel. The message was DELTA BRAVO YANKEE ALPHA TWO THREE North NINE TWO. Sitting at their console, each officer consulted his loose-leaf notebook for the decode. With that day's code, DELTA BRAVO indicated (to our relief) that this was only the lowest level of test. Sometimes, they told us, the test indicator would come only later in the message, after specifying that they should unlock their safes and retrieve their launch keys.

The quiz show was still chattering in the background. What would you do, we asked them, if on an ordinary day like today—no international crisis going on or big news of any kind—you were watching TV, and a valid launch order came in—not a test. Would you launch? They looked at us like we were crazy. "Well, obviously, we'd get on the phone and try to figure out what was going on before we started a nuclear war." We thought this was a good answer, but it left many open questions. Safeguarding against rogue or accidental nuclear war was a topic of perennial interest to JASON, one in which older JASONs had made some important contributions. Touring an Ohio-class Trident missile submarine in Bremerton, Washington, we posed the analogous question to the ship's radio chief, a noncom chief petty officer. All communication into the submarine was hardwired through his duty station. We asked him, what would you do if the captain was preparing to launch missiles—not a drill—and you *knew* that there had been no incoming EAM? "I would go to the ship's master-at-arms, check out a weapon, and we would then shoot the captain," he said. But he said this with a certain smugness, as if it were the answer to a well-known question in a training refresher course, as it might well have been. We weren't left with any assurance that he actually took the hypothetical seriously. The Air Force junior officers who were going to "figure it out" actually earned our greater confidence. I'll have more to say about the Navy later on.

We averaged one or two JASON field trips a year. AFTAC, the obtusely named Air Force Technical Applications Center, was the organization that monitored the globe for surreptitious nuclear weapons tests, using sensors in space, air, land (seismographs), and water (long-range hydrophones). Saving it for the end of our tour, they played for us an audio recording of the 1968 Scorpion submarine tragedy. As the submarine sank, taking forty-five seconds, you could hear the compartments imploding one by one under the pressure. Most compartments had people in them. "We had a two-star admiral in here," they said with Air Force nonchalance, "and he broke down. He just couldn't take it." On the same trip we visited Cape Canaveral and got to kick the tires of the Space Shuttle *Columbia*. Literally—it was the only part of the shuttle that they would let us touch. The Columbia flew 27 successful orbital missions. On January 16, 2003, its 28th mission, it disintegrated on re-entry, killing its seven crew members.

Our CIA friends took us to a nondescript downtown-Washington office building where their document forgers, formally the Graphics and Identity Transformation Group, worked. A rotund, jolly woman who was one of their "artists" was forging signatures on a pile of documents, but took time out to greet us. Rich Muller handed her his driver's license as a challenge. She studied it for half a minute, then took a piece of paper and wrote out half a dozen perfect Richard A. Muller signatures, no two exactly alike, but all clearly a match to the license exemplar. Rich kept the paper as a souvenir. There was nothing classified about it, because no one would ever believe that he had not signed the page himself. Next, they took us to a room of file drawers labeled by countries of the world and invited us to open a few. Inside were blank passports, identity cards, and other artist-ready documents. Some were outright forgeries; others were genuine and had been "acquired".

On a similar trip to the FBI Labs in Quantico, Virginia, we saw a wall of thousands of key blanks—every known type—the raw materials for, among other things, the Bureau's so-called black-bag break-in jobs. A visit to the CIA training camp, "The Farm," whose location was (and is) the worst-kept secret in Washington, gave us the opportunity to see the carpentry shop where they did "concealments" in furniture, and gave each of us the chance to traverse an underground replica escape tunnel. Thirty-six inches wide, eighteen high, and two hundred feet long, dimly lit, you lay on a flat little cart and propelled yourself in the dark on tiny rails by pulling on a rope. Our hosts were impressed by our enthusiasm. It was as much fun as solving partial differential equations.

The fact that JASON worked for CIA was widely known but, for some obscure bureaucratic reason, at the time, classified. At the annual JASON cocktail reception for sponsors, the CIA people wore name badges said "USG," meaning U.S. Government—and they were the only sponsors with that label, so it was not very good cover. On one occasion, Diane Eardley (Doug's wife, an immunologist) said wistfully to a man she had been chatting with, "I've always wanted to meet someone from the CIA, but there aren't any here." He was Julian Nall, the CIA Chief Scientist. He didn't enlighten her.

In fact, I had many pleasant interactions with Julian Nall over the years. He was a Southern gentleman who took it as his mission to teach manners, by example, to rude Yankees. When you visited his large office at CIA, he would take your suit jacket and hang it on a hanger in his closet. He would invite you to sit, ask if you wanted a cold Coca-Cola. Then, there would be a mandatory period of social conversation. He always remembered your wife's name and the ages of your children; I eventually knew that his wife was Christine. When some invisible timer reached five minutes, you were allowed to start talking business. At the end of the meeting, the whole process was reversed, finishing with his helping you into your jacket. When we moved to Austin, Texas, in 2007, we learned that Julian's grandfather had been an important local figure, and that the Nall family was still held in high regard by old-line Austinites.

44. Sea Stories

My initiation into administration as Astronomy Department chair inevitably pulled me away from research. So did the *Numerical Recipes* project, and so did my continuing work for JASON, where I was by now a member of the governing Steering Committee (known within the group as "steerers"). JASON was only the beginning of an explosion of committee-type work that I took on in the later 1980s: National Academy of Sciences studies; board of advisors to the Institute for Theoretical Physics at UC Santa Barbara; Los Alamos theory division advisory committee; scientific policy committee of the Superconducting Super Collider; Herman Feshbach's Annals of Physics editorial board; the U.S. Defense Science Board. Some details about some of these activities are illuminating.

Reports with the imprint of the National Academy of Sciences are conducted by its operational arm, the National Research Council. NRC committees can include people who are not elected members of the National Academy. My first National Academy of Sciences study was on the subject of so-called nuclear winter—the hypothesis that an all-out nuclear war between the U.S. and U.S.S.R. would cause an immediate and catastrophic ice age. It all depended on how extensive were the fires that nuclear explosions would cause, and how much sunlight-absorbing soot these fires would raise into the stratosphere. There were two necessary elements of context: First, we were still in the middle of the Cold War, so that an all-out nuclear war was possible. Second, the dinosaur-extinction hypothesis of Luis Alvarez and his son, geologist Walter Alvarez, had caught the popular imagination, as dinosaurs always do. I knew Luis well from JASON, where he was our only active Nobel Laureate. JASON's other Nobelists remained on the masthead to lend us their prestige but were rarely seen at summer studies: Murray Gell-Mann, Steve Weinberg, Josh Lederberg.

Walter Alvarez was a friend of Rich Muller's, and I knew him somewhat. The Alvarezes' hypothesis was that the Earth had collided with a small asteroid, lofting enough opaque dust into the atmosphere to cause an immediate cooling, global and long-lasting enough to explain the Cretaceous-Tertiary mass extinction event of sixty-some million years ago. Today, this is considered established fact. An asteroid 5-10 miles in diameter impacted at Chicxulub, on the Yucatan Peninsula, 66 million years ago, with a blast equivalent to about a billion nuclear warheads.

Carl Sagan, whose *Cosmos* television series had made him a celebrity of science popularization, was among those who thrust nuclear winter into the public view. As if a nuclear holocaust that might kill a billion people was not bad enough already, nuclear winter was supposed to make such a war even more unthinkable. Nuclear winter was supposed to drive world opinion toward nuclear disarmament—in Carl's public view. The NRC study committee that was chaired by Harvard applied scientist Professor George Carrier didn't include Carl, but did include three of his younger colleagues, Rich Turco, Brian Toon, and John Birks. For balance (rather obviously), it included three younger JASONs, Jonathan Katz, Doug Eardley, and me. Our membership in JASON branded us to the others as nuclear warriors. A half-dozen establishment figures were supposed to be "in the middle": Jack Ruina, Gene Shoemaker, Spurgeon Keeny, and others.

I thought that this whole adversarial framing was bogus. It was as if we younger JASONs were supposed to wear black hats and be *in favor of* nuclear war, while the Sagan henchmen, in white hats, should favor *peace*. That didn't seem like science to me (or Doug or Jonathan). Whether an all-out nuclear exchange between the U.S. and U.S.S.R. could cause the climate to change, and, if so, how much and for how long—those were valid scientific questions. At our first meeting in Washington, my naïve expectation, that a distinguished NRC committee would approach those questions objectively, didn't survive for long. Turco, Toon, and Birks presented their calculations. At every turn, they assumed the most extreme possibilities. Not only would forests and crops burn, but so would the very topsoil. Not only would the structures in cities burn, but the asphalt in streets would ignite. Back-of-the envelope calculations suggested that these were at best implausible guesses. While the assumptions could be replaced by better calculations and straightforward small-scale experiments, the nuclear winter folks had no interest in our committee's recommending such. They believed that the existential threat of nuclear war made it the duty of scientists to endorse the most extreme—not the most likely—scenarios. Via our committee, they wanted the support of the National Academy of Sciences in their mission.

Nuclear winter was my first introduction to a knotty issue that I have never been able to resolve. What *should* scientists do when their research leads to conclusions with urgent or controversial policy implications? Inevitably, those implications will cloud the objectivity of their research. Sagan, Turco, et al., slipped from science into advocacy. Scientists are citizens, and of course they should pursue the political implications of their work. But how? What processes of science and advocacy can arrive most rapidly both at accurate facts and at wise policy decisions? There is no simple answer.

Our committee met several times and wrote a draft report that was not quite as extreme as the Saganites, but it wasn't good science. NRC reports are supposed to be, if possible, unanimous; Jonathan, Doug, and I signed the draft under pressure and without enthusiasm. We were lonely voices who wanted a more evenhanded report—even if that meant admitting that the key question—was nuclear winter possible?—simply could not be answered. NRC reports are sent out for independent review to as many as a dozen reviewers. In our committee's case, the result was unusual: A preponderance of reviewers rejected the draft. Its assumptions were unjustified, the reviewers said; its advocacy didn't follow from the science. Within the committee, we three right-wingers turned out to have been in the center after all. The report as finally issued was a mish-mash of on-the-one-hand, on-the-other-hand that left everyone with a bad feeling.

A few years later, I had a call from the MacArthur Foundation, asking if I would support Turco's being awarded a MacArthur "genius" fellowship. I said yes. He deserved the award for advocacy in the public sphere—just not, please, for his science. He did get the award.

I am dwelling on this one report because it presaged the evolution of my views on the more important—and more real—subject of human-caused climate change. JASON dipped a toe into climate studies before the field of climate science existed in any organized sense. As that field developed, I was often left unimpressed by the quality of its work. This was brand-new science. It had people entering from many different fields. Sometimes the first to enter a field are not the deepest or best. The variability was great. Echoing my experience with nuclear winter, many climate scientists had very strong and visible policy agendas.

In later years, JASON became notorious as a hotbed of climate-change skeptics—deniers, even—including Bill Nierenberg (who, while director of Scripps Institution of Oceanography, was also JASON chair), Freeman Dyson (who wrote widely and influentially on the subject), Will Happer (who served in high government positions), and Rich Muller (who later underwent a self-publicized conversion from doubter to believer). In the 1980s, I may have been somewhat aligned with them. But by the 1990s, I felt that the situation had changed. It was no longer fruitful to debate the quality of the earliest climate science. The field had come into its own as a self-correcting branch of science: Dubious publications evoked rapid countering responses and healthy debate. The vast preponderance of new evidence supported both the hypothesis of global warming—as virtually established fact—and the hypothesis that climate change was largely human-caused—this as something between 95% and 99% probable. The resulting policy implications were well-founded and necessary, both for mitigation (that is, actions to slow the change) and adaptation (that is, surviving it).

Climate science had thus become, for me, a poster-child success. In the end, its science and its policy recommendations were concordant. Any lack of objectivity from its early days had been overtaken, as it should be, by a preponderance of solid scientific evidence. Dyson and Happer never modified their skeptical views, and they came to be seen as pariahs. A charitable interpretation is that they expected climate science to meet the standards of the best experimental physics, where conclusions could be validated with 99.999% certainty. That was never going to happen. My personal standard was that climate science should be "good enough," that is, convincingly self-correcting and with a growing corpus of consistent data.

Not yet tainted by his climate change views, Bill Nierenberg's term as JASON chair was nevertheless fractious. In his first couple of years, he replaced virtually all older JASON steerers with newer JASONs, including me. Steering Committee meetings became monologs of Bill talking to us, his supposed acolytes, at a speed of several hundred words per minute. Man, was he a talker! It was a total immersion, a drowning baptism in verbiage. I was one of few—possibly the only one—who enjoyed these meetings. Bill's themes and topics ranged wildly and unpredictably, but the bit-rate (to borrow computer terminology) was always high. I always learned a lot, though only rarely anything relevant to the task at hand.

Up to this time, no original JASON member had ever been kicked out of the group. Some had drifted away for their own reasons, but, for those remaining, JASON membership was viewed as a lifetime perk. Nierenberg determined to change that. It was hard to tell whether his motives were pure (i.e., generational transition and getting rid of deadwood) or were purely personal—that he was finally able to act on long-held grudges. The process was simple enough: Bill would propose an action to his now-captive steerers and then talk until we concurred. Rich Muller's analysis was that JASON now consisted of old farts and young pricks; and that we young pricks needed to decide whether it was in our interest to let the old farts kill each other off. You could argue it either way.

Joe Chamberlain, an atmospheric scientist from Rice University, was an easy case because his contributions to JASON in recent years were zero, and because he had an evident daytime alcoholism problem. Next was Hal Lewis, one of my original mentors within JASON. Nierenberg's overt indictment was that Hal had repeatedly bad-mouthed JASON while serving on the prestigious Defense Science Board, causing JASON to lose business; but it was really that Hal's summer-study office at the opposite end of the hallway from Bill's had become a kind of JASON government-in-exile where old farts and young pricks gathered to conspire against Nierenberg's management. I warned Hal that Bill was gunning for him, and that he should lower his profile. He responded by demanding an appearance before the Steering Committee, without Bill, for the purpose of clearing his good name. I was out of town that day, luckily. After listening to him, the steerers immediately voted him out of JASON. It was the classic case of someone who almost shot himself in the foot, but aimed so badly that he shot himself in the head instead.

Gordon MacDonald's was an unhappy case, because Gordon's career (inside and outside of JASON) combined periods of brilliance—he was elected to the National Academy of Sciences at age 32—with tendentiously argued contrarian views. As a young scientist, he had made a name by disputing the overwhelming evidence for plate tectonics, then called continental drift; as an old one, he insisted that climate change, while possibly real, had astronomical, not human, explanations. That Gordon had become a close scientific advisor to Bill Casey, the notoriously tricky CIA Director, endeared him to no one. Within the CIA bureaucracy, the long knives found their target when a personal indiscretion of Gordon's came out in a security re-investigation. We in JASON never learned the details. CIA suspended MacDonald's clearances, Nierenberg took the opportunity to suspend him from JASON, and steerers concurred. I was a direct beneficiary: I became JASON's designated representative to CIA for what amounted to (but was never called) sales and marketing, a role that lasted for a decade. Unlike Gordon, I made friends rather than enemies of the career spooks, to my and JASON's benefit. It helped that I already had some connections, one, for example, with a former graduate student of Kip's, my contemporary, who had become a distinguished CIA analyst. Over the years, he and I shared hobbyist computer hacks.

A year or so after Hal Lewis was kicked out of JASON, I was visiting Santa Barbara as chair of the visiting committee of the Institute for Theoretical Physics. ITP was at the time an unusual experiment, a place funded by the NSF specifically for hosting semester-long, topical programs, attended by ten or twenty sabbatical visitors from other institutions. One semester the program might be a topic in condensed-matter physics; the next, one in astrophysics, and so on. Over time, ITP's was recognized as a successful model and is now replicated at half a dozen centers. I became chair of its visiting committee unexpectedly. The position was scheduled to rotate to an astrophysicist on the committee. There were two of us. Stirling Colgate, as the more senior, was the presumed choice. He was the after-dinner speaker at the meeting banquet. His speech was so offensive that I was elected instead of him.

Stirling was a remarkable individual, an heir to the Colgate-Palmolive toothpaste fortune. He had graduated early from the Los Alamos Ranch School in 1942 when it was seized for use by the new Manhattan Project. After the war, and with a Ph.D. in physics from Cornell, he was an early recruit to Edward Teller's new Livermore Laboratory. His career there, and later at Los Alamos, was a mixture of bomb work and astrophysics. Colgate was famously the bad boy of astrophysics, both intellectually—where his work ranged unpredictably from insightful to crazy—and in his personal behavior. He was president of the New Mexico Institute of Mining and Technology (New Mexico Tech) from 1965 to 1974, resigning under the cloud of accusations of sexual and other misconduct.

Hal Lewis discovered that I was in Santa Barbara and insisted that I come flying with him the afternoon after our ITP meeting. This was his first contact with me since being kicked out of JASON. He was a longtime private pilot, but his invitation seemed odd, peculiarly insistent. His plane was a Piper Comanche, a fancy single engine low-wing with retractable gear and a variable pitch propeller. His pre-flight inspection of the plane seemed to me cursory.

It crossed my mind that JASON had been a big part of Hal's life—perhaps now a life not worth living. He could kill himself and in the process take one of those JASON steerer bastards (me) with him. It was only an idle thought until, halfway down the runway on our noisy takeoff roll, he took his hands off the wheel and shouted, "You take it, Bill!" (He did know that I had flown with Margaret.) I managed to take off, and I was able to get to stable, level flight with everything trimmed. The plane's feel was not too different from Margaret's plane, in which I had had as much as ten or twenty minutes of hands-on piloting experience. Hal was silent. I flew over UCSB, then over the 3000-foot mountains behind Goleta (near Ronald Reagan's ranch), then up the valley and back through San Marcos Pass. Then out to sea, aiming for the oil platforms, and maybe the Channel Islands visible in the distant haze. Hal still said nothing. Some minutes later he spoke: "Bill, are you *sure* that you want to fly out over twenty-five miles of open ocean at an altitude of only 4000 feet?" I knew then that he wanted to live!

Hal likely had a role in my being invited to join the Defense Science Board. Perhaps my courageous piloting helped. The DSB had a history going back to 1956 and was by charter "the senior independent advisory body" to the Department of Defense. On paper, we reported to the Secretary of Defense, at the time Reagan's appointee Caspar Weinberger. "Cap" Weinberger was a right-wing Republican politician with little interest in science, a far cry from his predecessor under Jimmy Carter, Harold Brown. That didn't matter to us, for two reasons. First, we only met with him ceremonially, about once a year, never for more than fifteen minutes. "He spits a little holy water on us, and we kiss a little ass," someone said privately. Second, the DSB didn't in any real sense report to him, nor did it have much to do with science, at least not in the way that JASON did. DSB was pretty much its own freestanding institution within DoD. Members like me, with four-year terms, came and went, thirty of us at any given time.

The invisible governance of the DSB was a clique of former members, labeled "senior consultants," career defense individuals who included Gene Fubini, Johnny Foster, Al Flax, Sol Buchsbaum, and John Deutch, supported by a DoD civilian executive officer and three mid-career military assistants, Army or Air Force lieutenant colonels, or Navy commanders. Fubini was a former Assistant Secretary of Defense. Foster, a protégé of Edward Teller and an early director of Livermore Lab; Flax, a director of the National Reconnaissance Office (while it was still an unacknowledged "black" agency) and later president of the Institute for Defense Analyses. Sol Buchsbaum was a senior executive at Bell Laboratories and chaired the White House Science Council under President Reagan. John Deutch was Undersecretary of Energy in the Carter administration (and a buddy of my father's), later CIA Director and Deputy Secretary of Defense.

The DSB's overt product was its reports, four or five a year, mostly forgettable but occasionally influential, written by "task forces" of board members. There were five such task forces when I joined. I was assigned to "Technology," which was a ghetto reserved for the half-dozen actual scientists on the Board, at the time Lewis, Deutch, Fred Brooks, Ed Frieman, Norm Hackerman, Anita Jones, Josh Lederberg, and me. Notably, four of us were JASONs. The other task forces, mostly comprising defense business executives and retired senior military and civilian officials, the DSB's large majority, were, I soon learned, the more important ones: Management, Strategic, Tactical, and C3I (which meant Command, Control, Communications and Intelligence). In addition to monthly meetings of each task force or its report-producing study group, there was a two-week summer study at which all the task forces met to write or finalize reports. But not exactly that. Except for our technology group, the people who actually wrote the reports were not Board members, but their staffers—lower-level executives and engineers from the members' home corporations. In no sense independent advisors, these workers always kept the interests of their companies foremost. DSB was my first experience with the phenomenon, common in government, of figurehead committees or commissions, whose distinguished membership delegated everything to staff; and also my first experience with the pervasive, institutionalized conflicts-of-interest that characterized the DoD and its contractors.

DSB membership had some lesser perks. Our lunches in the Pentagon seniors' "Blue Room" were served by enlisted Filipino stewards in dress uniform, on china with the seal of the Secretary of Defense. (Oddly, a year later, all the Filipino waiters were suddenly replaced by enlisted African-American waiters—in dress uniform.) We were issued permanent passes that let us roam the Pentagon and all Washington area military installations. We had the protocol rank of one-star generals, could eat in Officers Clubs or stay overnight in Bachelor Officers' Quarters (known as B.O.Q.s.) on any military base. There were sometimes interesting after-dinner speakers invited to the dinners. Paul Nitze described a British ambassador's view of negotiating with the Soviets: "It's a bit like dealing with a defective vending machine. You put your coin in, pull the lever, and nothing comes out. If you take hold of the machine and give it a big shake, then that can sometimes do some good. But for heaven's sake don't bother talking to it!"

When I was in Washington on JASON or other business with a couple of free hours, I often used my Pentagon pass to access the bookstore in the Pentagon's main corridor. This stocked a vast collection of books on history (especially military), biography, intelligence (including spy novels), computers, technology, diplomacy, and anything even faintly related. Before the web and Amazon.com existed, this store was a completely unique resource—more concentrated and browsable than any library that I knew of.

If reports were the Board's overt product, its less-visible function was to provide a venue for schmoozing. Under cover of giving independent scientific advice to DoD, senior executives from defense aerospace companies could have conversations that their company lawyers would never have allowed. I almost fell off my chair when, at lunch early in my term, two CEOs were chatting openly about the circumstances under which their respective companies would bid on a large defense contract. I was no lawyer, but I was pretty sure that this was an illegal, anti-competitive conversation. When Will Taft IV, who was deputy SecDef, gave an after-dinner talk to DSB on his office's efforts to ferret out wasteful and corrupt contractor practices, he was met by a stony silence.

It wasn't just my imagination that the DSB industrialists were, overall, a rum lot. At a DSB banquet, my date and I were seated at a table of them. She was socially very adept and charmed them all, but, afterwards, she was livid. How could such stupid, self-centered, arrogant male jerks be the captains of industry who were running the world? I wondered myself. At lunch one day, one of the industrialists picked up on the fact that I was an astronomer and asked me to explain something: He had bought his teenage son a little telescope, and they looked at this very bright star and found that it was crescent-shaped, "like the moon is sometimes." Could I explain that? "That's Venus," I said, mentioning that Galileo in 1609 had beaten him to the discovery. I demonstrated, with salt shakers and ashtrays representing the planets and Sun, how the phases work. "That's just fascinating," he said, not quite getting it.

OK, to be fair, not everything about the DSB was rotten. I came to admire a few of my new colleagues. Norm Augustine, at the time CEO of Martin-Marietta Aerospace, radiated the charisma of leadership that I had seen only a few times in my life. When I mailed him a copy of *Numerical Recipes*, he responded with a handwritten note of thanks and, at the next DSB meeting, made a point of commenting on several sections in the book—that he had actually read. After his retirement in 1997 (as by then CEO of Lockheed Corporation, which had acquired Martin-Marietta), Augustine devoted himself to public service. I again came into contact with him on National Academy committees and when I was a member of President Obama's Council of Advisors on Science and Technology. Personally lobbying Congress, Augustine played a big role in the passage of the America COMPETES Acts of 2007 and 2010, which supported science and science education in multiple federal agencies.

George Heilmeier of Texas Instruments also impressed me as a person of integrity. Anita Jones and I were almost always on the same side of issues. She was a pioneering computer scientist in compilers. We formed a lifetime bond as easy co-conspirators, including while she was Director of Defense Research and Engineering in the 1990s. It was on DSB that I first got to know (and appreciate differences among) retired senior military—generals and admirals. Sober leaders included General Russ Dougherty, who had been commander-in-chief of the nuclear-armed Strategic Air Command, and General Leonard Chapman, former Commandant of the Marine Corps.

Admiral Bobby Inman was the military equivalent of a badly behaved child. He had risen nevertheless to director of the National Security Agency and deputy director of CIA, and was later nominated by President Clinton as Secretary of Defense. His nomination was withdrawn after he gave a press conference that was a virtual temper tantrum, with rants against Senators Bob Dole and Trent Lott, and New York Times columnist William Safire.

Much later, when I arrived in Austin in 2007, Inman, by then in his 70s, was an adjunct professor in UT's LBJ School of Government. With computer science department chair J. Moore, Inman and I conceived a plan to appoint Fred Chang, a former NSA Research Director, to a well-deserved professorship. The effort failed after Inman insisted on addressing our faculty and then, figuratively wrapping himself in the American flag, lectured us on the importance of U.S. intelligence gathering against foreign threats. It was not a pitch that our department, about half born in those same foreign countries, took to.

At DSB meetings, the retired military tended to sit together on one side of the horseshoe table. As a small show of independence, I often picked a seat in their midst. "Are you a briefer?" Admiral Ike Kidd, a former commander-in-chief of the Atlantic Fleet, asked me the first time I did this. "No, I'm one of *us*."

Kidd himself defied characterization. His father (of the same name) was the Navy admiral killed on the bridge of the battleship *Arizona* in the Japanese attack on Pearl Harbor. Our Admiral Kidd knew nothing about science, but had made his way onto the DSB by having been a science groupie on his way up through the ranks. The older JASONs remembered him as—yes—Captain Kidd. Ike always greeted me with "a hearty hello to my favorite astronomer." He was full of naval aphorisms: "Gentlemen, it's time to take the storm warnings out of the flag bags," and "Liberty, but no boats," meaning that the captain of the ship had granted the crew shore leave, but there was no way for them to get to shore. At one meeting, a gung-ho Army major briefed us about an automatic gun system that identified targets, tracked them, and prioritized them for "servicing" in an optimal order. "Young man," Ike interrupted, "I have a question about your use of the English language. When you say 'service', you really mean 'destroy', is that right?" "Yes sir, Admiral." "Young man," Ike intoned, deadpan, "don't *ever* try your hand at animal husbandry."

There were wheels within wheels within the DSB. "Stealth" had the specialized meaning of design measures that reduce the detectability of aircraft by radar, typically by a combination of shaping, materials, and tactics. The much-guarded secret (and big surprise) of the 1970s was that many orders of magnitude reduction in radar cross section could be achieved. When I was assigned to a study on U.S. air defense, I did my own back-of-the-envelope calculations of how future foreign stealth technology might in the future change things. The existence of a U.S. stealth program, but no details about it, had been revealed by SECDEF Harold Brown on his way out the door at the end of the Carter Administration in 1980; and JASON's work on active cancellation—the work that the Navy had refused to classify as secret—also gave me insight. It seemed to me that our air defense committee was woefully uninformed. Not quite: There turned out to be a separate, undisclosed, DSB panel on low observables (i.e., stealth) that I was quietly invited to join.

That group was chaired by Bill Perry and his longtime aide, Col. Paul Kaminsky (ret.), these two the godfathers of stealth in the Carter years. Everyone on the panel except me was a former high government official. I had the impression that the committee's sole raison d'etre was to provide an official way for Perry and Kaminsky to keep their hands in the stealth program. I was there not for my expertise, or as window dressing, but basically to shut me up everywhere else. When I walked around the table to introduce myself, Gene Fubini was amused. "Bill, you've reached a level where you don't have to say who you are. Just sit down and say, 'General, I am glad that you can be here with us today.'" The Air Force had a labyrinth of security compartments with layers of fictitious cover stories. The main evidence that our panel was not just in support of some cover story was that I couldn't think of any *higher* former officials who could be window dressing on the *real* panel. But I was never 100% sure. There were stealth jokes that I was never sure were jokes: That they checked the pilots to be sure that they had no metal fillings in their teeth, but that the pilots' lucky felt dice in the windshield were deemed OK.

With the sole exception of stealth, the JASONs had, in my experience, access to much higher levels of secrets than the DSB. We in the DSB did once get a briefing on the SIOP, the Single Integrated Operational Plan for waging nuclear war, however. It was chilling. You've got your DGZs (designated ground zeros), the briefer explained, and your laydowns, and your land-based versus SLBM waves of attack. The SIOP was a fossilized relic of the 1950s Cold War, still worshipped as if it were the relic of some saint from the Crusades, a shinbone, maybe.

My disdain for the DSB must have been evident to its leadership. After four years, unusually, I was not renewed for a second term.

The Superconducting Super Collider project, directed by my former Harvard physics colleague Roy Schwitters, embodied U.S. competition with Europe's monumental Large Hadron Collider at CERN in Geneva. The SSC was billed as able to achieve energies and temperatures "last seen in the Big Bang," so its scientific policy committee rated a token astrophysicist—me. The committee's roster was Roy's attempt to balance the field's opinionated prima donnas (Sam Ting, Steve Weinberg) with sensible people (Leon Lederman, Jerry Friedman). Before the project was canceled in 1993, fifteen miles of tunnel out of a planned ninety-mile circumference were bored near Waxahachie, south of Dallas, in the Eagle Ford shale and Austin chalk geological strata. We walked a mile of the completed twelve-foot diameter tunnel, 150 feet underground. For me, a highlight of our meetings was always the progress report on the design and fabrication of the ten thousand required superconducting magnets, each about the length of an ICBM missile and involving nearly impossible stretch technology. Appropriately, the magnet project manager was hired out of defense industry, where he had been project manager for Trident submarine missile manufacturing. He frequently got the words "magnet" and "missile" confused in his briefings.

The SSC project was plagued by huge budget overruns. Roy became the scapegoat, but the real offender was Admiral James Watkins, then the Secretary of Energy. Schwitters submitted accurate budget estimates to Watkins at DOE who dishonestly trimmed them for submission to Congress, betting that overruns were more likely to be funded year-by-year than the truthful numbers all at once. Watkins lost the bet. Schwitters lost his job, ending up a professor at the University of Texas at Austin—and a pariah in the physics community. I recruited him into JASON, where he flourished.

Herman Feshbach's *Annals of Physics* editorial board, which met monthly at MIT, was service "lite". *Annals* was a second-tier journal founded by Phil Morse, Feshbach's mentor. The names Morse and Feshbach were linked for at least two generations of physics students by their two-volume standard textbook on mathematical physics, the eponymous "Morse and Feshbach". My copy was the one that Frank had used in the 1950s. *Annals of Physics* once had pretensions of becoming important, but Feshbach was too genial, too nice, one might say too lightweight, to make it happen. At our meetings we quickly triaged the small number of received manuscripts into fair, poor, and abominable. The first group, we accepted immediately; the second, we sent out for further reviewing. After this sorting, we sat around and traded Harvard-MIT gossip.

Once a year, we had a visit from the New York based publisher. Then, instead of our usual desultory mode of operation, we loudly "consulted" each other. We "raised important matters for discussion." We "made difficult and measured judgments," "debated" and "reached a consensus." It was like a Monty Python skit. In return, the publisher took us to dinner at a fancy Boston restaurant. A different highlight was when Eve Sullivan, the journal's editorial assistant, brought sheet music from her church choir, and, instead of triaging manuscripts, we all sang four-part Thanksgiving hymns. Another time, Eve had cold-called Tom Lehrer and invited him to lunch, then described it all to us. Lehrer, famous for satirical songs that he mostly wrote as a Harvard graduate student in the early 1950s, was listed in the Cambridge phone book. He happily accepted Eve's invitation after establishing that she would pick up the check.

I served on the *Annals* board for seven years and was paid $2,000 a year, even then not a lot.

45. The Grove

Within JASON, Rich Muller and I were (what was later called) frenemies. We constantly worked together on spooky intelligence problems and constantly needled each other. We collaborated on an elaborate musical skit, *JASON's Golden Fleece*, performed by younger JASONs lampooning our elders at the JASON 25th Anniversary banquet. I was jealous of Rich's being awarded a MacArthur "genius" Fellowship. I ascribed his selection entirely to the influence of his mentor-hero Luis Alvarez. Rich, on the other hand, was jealous of my being a Harvard professor and department chair. His primary academic appointment was not in the prestigious Berkeley physics department, but up on the hill at Lawrence Berkeley Laboratory.

On a line through Alvarez, who had been Ernest Lawrence's protégé, Rich considered himself heir to the famous Lawrence legacy in experimental physics. As far as I could tell, no one else, except possibly Luis, thought this. Yet no one could doubt that Rich was very smart, very creative, and an obnoxiously ambitious scientific and social climber. "I don't want to rush into print with things that would have been discovered three months later by some perfectly respectable plodder," I said to him. "I can't imagine anything more rewarding than a career of being three months ahead of the competition," he countered.

In explaining mass extinctions, dinosaurs the prime example, Muller went beyond Luis and Walter Alvarez's theory of asteroid collisions. He embraced the weak evidence that extinction events were *periodic*, occurring every 65 million years. He proposed that the periodicity was due to an unknown distant object orbiting the Solar System. He named it Nemesis. Unlike the Alvarezes' theory, which was backed by Walter's discovery of an iridium-rich thin geological stratum, Rich's theory was purely speculative. His Nemesis paper in *Nature* attracted only brief attention. I suggested, somewhat sarcastically, that he write a popular book and take his case directly to the people. I offered to help find him an agent, for which he agreed to pay me one percent of any advance obtained. John Brockman, whom I introduced him to, got him an advance of $262,500, about $750,000 in 2020 dollars. I was again jealous—and Rich owed me $2,625. To this day, the issue of periodicity is not settled. The periodicity, if present, is now known not to be exact;

but neither do the mass extinction events seem to be spaced entirely randomly.

Rather than paying me in cash, Rich suggested that I be his guest—a comparable cost, he assured me—at the Bohemian Grove summer encampment. This was pure frenemies: San Francisco's Bohemian Club was the snobbiest and most exclusive Old California private men's club, and Rich (thanks to Luis) was one of its youngest members. Attending with him would raise my jealousy to new heights, he knew. He noted that, by club rules, no individual guest could be brought to the club—by anyone—more often than once every two years.

The Bohemian Club was founded in 1872 by San Francisco journalists as a counter to the stuffier Establishment men's clubs of the time. A century later it had become the pre-eminent social gathering of the California business elite, who defined themselves as rich, WASP, Republican, and in charge of the country. These were the club's "full members." They included Ronald Reagan, Gerald Ford, George Schultz, Caspar Weinberger, Thomas J. Watson, Edgar F. Kaiser, Edmund G. Brown, Robert Chandler, and on and on. The waiting list was about 15 years long, and it was one of those WASPy things that you never, ever, actually got in by waiting out your time in line. Instead, you had to get quietly promoted to the top of the list, say by becoming CEO of a Fortune 500 company, or being elected Governor or Senator, or being appointed an ambassador or Cabinet Secretary.

Wisely, the full members had long ago realized that a club with only themselves in it would be boring. There were therefore additional special categories of membership for famous musicians (who would perform, gratis), artists, licensed pyro-technicians (gratis fireworks), playwrights and theater professionals (to assist, gratis, with the club's amateur theatrics), and even a category for university professors. Even by 2020, the club has no women members and few female employees. In the 1970s, it successfully defended a court case brought by a woman who applied to be a summer waiter at the Grove. The club lawyers presented evidence that, by long tradition, members habitually urinated on the Grove's giant redwood trees; that, because their age, members needed to pee frequently; and that, because they were gentlemen, they could not do this in front of women. Ernest Lawrence was a Bohemian member, followed generationally by Alvarez, and then Muller. Alvarez was himself from a patrician California family. With his being a Nobelist, he might have qualified as a full member. But the annual fees for special membership categories were much lower, and Luis was always very conscious about money. Lawrence, before him, fit right in with the

Bohemian elite. He cultivated contacts with industrialists who thought on the same grand scale that he brought to experimental physics. As Lawrence's guests, the S-1 Committee, which initiated the Manhattan Project, met at the Grove in September, 1942; and the new Atomic Energy Commission held key private meetings there in summer, 1947.

The Grove comprises a thousand acres of redwood forest bordering the Russian River, two hours' drive north of San Francisco near the town of Monte Rio. The valley floor of the Grove is an old, cut-off bend of the river. The redwoods are 300 feet tall and 10 to 20 feet across. Sunlight and rain filter down to the valley floor in the most hushed and muted way. The rustic cabins are insignificant features under the giant trees. The forest up the valley walls is still young—it had recently been logged when the land was acquired by the Bohemian Club in the 1870s. There were no protesters at the Grove front entrance when Rich and I arrived in his car, but there were police barricade sawhorses stacked at the side of the road in anticipation of the usual demonstrations by liberals, women, and other malcontents. The entrance arch bore the slogan, "Weaving Spiders Come Not Here," meaning that business dealing was forbidden inside. Many people interpreted the club motto also to mean "no Jews". Except for a very small number of high-profile Jewish individuals, this was the case.

On peak weekends, the "midsummer encampment" attracted 1,800 Bohemians to the Grove, lodging in their respective "camps," about fifty in number. Rich was a Pelican. Pelicans took pride in the fact that they didn't take their breakfasts and lunches in the raucous outside dining area. Instead, they had their own little glassed-in dining room and their own longtime Filipino chef. Most camps were surrounded by rough board fences with decorative gates. Near the "civic center," where Pelican was located, the camps were pretty much crowded up against each other, with dirt trails between. Within the Pelican perimeter, perhaps a hundred feet on a side, there were rustic, permanent buildings for kitchen and bathroom facilities, the dining room, and platforms for semi-permanent sleeping tents. The tents were equipped with electricity, fresh linens daily, and electric blankets. This was not roughing it. The center of the Pelican camp was an open lounge area under a stand of hundred-foot trees. There was an elaborate open bar, a large stone fireplace, and wickerwork rocking chairs and tables.

Now I have to start shamelessly name-dropping. As a quintessential part of the Grove experience, it simply can't be avoided.

At breakfast, I met camp members Harry Grey (board chair of United Technologies), Chuck Percy (recent former U.S. Senator), Aethelstan Spilhaus (science popularizer), and Fred Crawford. Fred was ninety-five and couldn't hear very well, but he had co-founded TRW with Si Ramo and Dean Woolridge when they came to him "with some good ideas." Finding that I was a Harvard professor and, even better, a Harvard alumnus, Fred asked what was my class. "Sixty-nine," I said. It took a couple of seconds for him to figure out that I meant 1969, not 1869. He had known people from the earlier class. His was class of '13.

"I saw Halley's comet from the Harvard Yard in 1910, when I was a sophomore," he said. "I lived on the family farm in Watertown. I was the tenth generation born on that farm. It was right across from the cemetery. What's there now?" I told him that it was the parking lot of the Star Market where I shopped. He talked about growing up in rural Watertown, before the telephone came in, before electricity, before running water, before even the horse-drawn streetcar went that far out. The ice man carried news from house to house.

"Where would a fellow go," he asked, "if money were no object, and he wanted to see Halley's comet twice in his lifetime?" I said the Southern Hemisphere in 1986, and that I would find out more and get the information back to him, which I later did.

"I'll tell you another Sandra story," John O'Connor said to the group. His wife was Supreme Court Justice Sandra Day O'Connor. "You see, the Court is out at a restaurant and the waiter takes Sandra's order first. 'I'll have the luncheon filet with the baked potato.' 'And what about the vegetables?' the waiter enquires. 'Oh, they'll have the same thing.'" Dirty jokes were a big part of Grove life. Walter Cronkite related a shaggy dog story about Olympic competition in wrestling, where the United States beats the Soviet Union despite an unexpected new Soviet move, the Pretzel. The punch line, delivered in the famously resonant Cronkite voice, was, "Coach, you have no idea how much strength a man gets when he bites his own balls." Cronkite asked me what I thought of the Reagan Strategic Defense Initiative ("Star Wars"), and listened thoughtfully to my answer, with occasionally the famous avuncular chuckle.

That day's lunch was a special occasion in honor of camp member General Jimmy Doolittle, who at age 88 had been awarded an honorary fourth star by President Reagan. For the event, guests were invited from other camps. Treasury Secretary and former White House Chief of Staff Jim Baker read a letter from fellow Bohemian Reagan. Baker related that

President Jimmy Carter, sleepless during the 1980 hostage crisis, had asked the Presidential portraits in the East Room for advice. Teddy Roosevelt said, "Go to war and lick the bandits." Woodrow Wilson said, "Be a man of peace and convince them of the right." President Lincoln said, "Go to the theater." Television personality Art Linkletter made a little speech about what the children of America thought of General Jimmy Doolittle. (Linkletter's television show featuring interviews with children had last aired twenty-five years earlier.) Admiral Jim Stockdale, who was awarded a Congressional Medal of Honor for his heroism in North Vietnamese prison camps, was a speaker. Later, in 1992, Stockdale was the Vice-Presidential candidate on Ross Perot's third-party ticket. Garnering 19% of the popular vote, mostly drawing votes from Republicans, the Perot-Stockdale run was arguably responsible for the win by Democrat Bill Clinton. Fred Crawford responded to Stockdale with a story about botching his own chance to be a hero—finding Earl Warren asleep on the swimming raft out in the Russian River not half a mile from here and failing to push him off the raft, "and hold him under until the bubbles stopped coming out." This was not a liberal group. After lunch, Rich and I walked down to the river, to see the site of Fred's failed heroism. The river was slow and green and muddy, only a hundred feet or so across.

Robert Millikan, Caltech's first president, got along well with the California establishment, so Caltech presidents were subsequently invited to join the Bohemians. All accepted. Weaving spiders notwithstanding, the philanthropic potential of the Bohemians was irresistible. When new Caltech president Murph Goldberger visited the Pelicans—as a prospective camp member—he got into an argument with Jimmy Doolittle over arms control. "That Goldberger fellow didn't treat Jimmy right. He wouldn't fit in here," was the consensus. Charlie Townes, a member, was able to get Murph into another camp, Isle of Aves.

Pelicans was not the highest status camp. That one, everyone knew, was Mandalay, the only camp that barred entry to other Bohemians. There was usually a Secret Service agent stationed at the entrance. Gerald Ford was in Mandalay, as was Henry Kissinger (a rare Jew) and Cap Weinberger (not Jewish, despite his name). The annual Cremation of Care ceremony was held at the lake that evening, an elaborate but sophomoric dramatic event climaxed by a huge fireworks show. After that, the various camps, especially the ones that specialized in performing arts and music, hosted informal cabarets and jam sessions. The music was generally jazz, some very good. Back at Pelicans, the cronies told dirty jokes: the one about the deaf-mute man buying a condom (you can Google for it); and the one about the three priests who agree to the fourth's plan that they should confess secretly among themselves their worldly transgressions. (After lurid details from the first three, the fourth confesses his only sin, that he is an uncontrollable gossip.)

A guest at Sunday's Pelican breakfast was William Ruckelshaus, famous as the first head of the Environmental Protection Agency under Nixon. Ruckelshaus was later FBI director; and then, as deputy Attorney General, famous for resigning, along with Attorney General Eliot Richardson, in Nixon's Watergate "Saturday Night Massacre". At the Grove, Ruckelshaus sat between me and John O'Connor, with Rich and Walter Alvarez across the table. Rich took a small bottle out of his pocket and, without asking, poured what looked like dirt onto Ruckelshaus' bread plate. "This is a sample of the worst environmental pollution ever known on this planet," he said dramatically. Ruckelshaus recoiled. It was a sample (via Walter Alvarez) of the iridium-enriched clay layer—harmless unless it was blocking sunlight in the stratosphere and you were a dinosaur.

At lunch, I sat between O'Connor and Neil Morgan, the editor of the San Diego Tribune. Morgan was the guest of Pelican Bob Chandler, publisher of the *L.A. Times*. You could tell when Chandler was in residence by the thousand copies of the *Times* trucked in overnight for general distribution. Each guest at lunch was called on for a five-minute talk. I gave a condensed version of one of my Astronomy 1 lectures, "How is the ordered state of this redwood tree able to come out of cosmic chaos?" I verbally traced the origin of its free energy (the technical term for the opposite of disordered entropy) backwards to the Big Bang. "You did fine," Rich told me, smugly.

When, a week later, I flew back for a second, final weekend, it was with a sense of purpose. I knew that I was there to collect names for future name-dropping, and jokes for future after-dinner talks to stag audiences—although I doubted that I would have many such opportunities.

Charles Black, husband of Shirley Temple Black. Herman Wouk. General Joe Foss, war hero and former South Dakota governor. Senator Alan Simpson of Wyoming. Tom Clausen, former head of Bank of America, then-head of the World Bank. "Moose" Taylor, head of some big oil company and member of the Smithsonian Board. Arjay Miller, ex-Ford, ex-dean of Stanford Business School. Some ex-president of Citicorp—there were just too many of these to remember one individual's name. Andy Devine's son. George Murphy. Dan Rowan.

After the evening's amateur theatrics, Rich, Walter and I did the parties in the camps along the river road. We were looking for Luis and finally found him, very drunk, sitting on a log with Art Linkletter—quite the odd couple.

Several Camp members went out of their way to give me warm goodbyes, a few making clear that they considered me definitely membership material. The Bohemians were not as bad as I had feared. Rich told me that the first time he went to the Grove (as Luis' guest), he was most impressed by how friendly everyone was to him, how they treated him, from a lower-middle class New York Catholic family, as if he really belonged—made him in fact belong. I liked their jokes and even their mateyness. What I didn't like was the pervasive assumption that the jokes (and mateyness) bonded them as a power elite. I certainly didn't like that women were excluded from this elite. I was also a bit more cynical than Rich about WASP culture. The WASPs treated you like you belonged until they decided that you didn't—as in the case of Murph Goldberger—and thereafter you wouldn't get the time of day from them. The WASPS were cool customers.

In JASON, I was always in awe of Luis Alvarez, but always had vaguely mixed feelings about him as a person. His Nobel Prize was for the invention of the hydrogen bubble chamber, a device that dramatically opened up the field of experimental particle physics. That invention was surely important, but I was more in awe of things that Luis had discovered about the physics universe—facts so basic that they seemed to come directly from God. Tritium, an isotope of hydrogen, consists of one proton and two neutrons. Helium-3, an isotope of helium, consists of two protons and one neutron. Tritium is unstable and radioactively decays into Helium-3, which is stable. Or is it the other way around? "We knew that one of them was probably unstable and decayed into the other," Luis remarked one day at lunch, "but we didn't know which until I measured it in 1939." It blew my mind. It was like meeting the first discoverer that the Earth was round.

But there was also a selfish, grasping aspect to Luis' character. In the course of a JASON study on detecting explosives (e.g., at airports), Luis invented a novel and promising new technique. We described it in our report to the sponsor, the FAA. JASON's work product was, by contract, "work for hire" and was owned by the federal government. So, we were surprised when, a year later, someone discovered that Luis had quietly filed a patent application on the idea in his own name. I was surprised at the unanimity with which the JASON steering committee felt that he must not be allowed to get away with this. My own view was that Luis would be Luis and we must just take the bad with the good. But no, they were out for blood. I did manage to have myself appointed as the JASON ambassador to Luis for purposes of negotiation.

It was an uncomfortable meeting. We wanted Luis to assign his patent rights to the federal government. That would end the matter. This was the one thing that he would not do. Far from truculent, his demeanor was alternately contrite, whiney, and feigning innocence. He would sign any letter of apology that I dictated, he said. He would prostrate himself before the steering committee and ask them for mercy. But he would not assign the patent rights. The most revealing part of our "discussion" was his retelling a story I had heard before: During World War II, he had invented a system for combining radar data with human radio commands, termed Ground Control Approach (GCA), that allowed aircraft to be "talked in" to a safe landing even in zero visibility weather. He had filed a patent and assigned the rights to the government—he thought only for the duration of the war. But, after the war, the government kept everything. "It should have made me rich," he said. In Luis' mind, no matter what our JASON contracts said, this was just a repeat of GCA. He was *not* going to assign any rights.

I drafted Luis' sniveling letter of apology and squirmed at his fawningly hypocritical appearance before the steering committee. He knew that we would not, in the end, kick out of JASON one of our Nobel Laureates—and he was right. His patent eventually issued as U.S. Patent #4,756,866, "Nitrogen Detection," sole inventor Luis W. Alvarez. Luis remained a member of JASON until his death in 1988. His autobiography, unabashedly titled *Alvarez*, had come out just a year before. Alan Lightman reviewed it for the Washington Post submitting a piece that began, "One senses Luis Alvarez' deep regret that he can be praised only by the greatest *living* physicists."

46. Boxes

While I was astronomy chairman, and afterwards through the 1980s, I fought a long series of losing academic skirmishes with Center for Astrophysics director Irwin Shapiro. My diary records page after page of my (I thought) subtle political maneuvering, alliances gained and lost, department meetings turned to shouting matches, appointments with the Dean (by this time economist Michael Spence), conspiratorial tête-à-têtes with colleagues in astronomy and physics. Personalities aside, there were legitimate underlying issues: Irwin's vision of CfA was strictly hierarchical, with no room for faculty autonomy. His top-down priorities aligned exactly with his previous career as a radio astronomer, a meticulous builder of instruments, a reporter of precise data with a minimum of speculative interpretation.

In Irwin's worldview, "theory" ought to mean exact physics, in the mode of Einstein's Theory of General Relativity—the field in which Irwin had made his reputation by high-precision radar measurements of the Solar System. To the kind of looser theoretical papers that I wrote—or for that matter George Field, or Al Cameron, or (at Princeton) John Bahcall or Jerry Ostriker—Irwin gave lip service and no more. "You're being paranoid, Bill," George Field counseled me early in Irwin's tenure. "He doesn't *hate* theorists. We're just not at the top of his priorities."

With today's distance on the events, Shapiro can be credited with bringing to CfA world-class observational facilities at multiple wavelengths: X-ray (Riccardo Giacconi's group), optical (the Multi-Mirror Telescope, MMT, in partnership with the University of Arizona), and—Irwin's own baby—the Submillimeter Array (SMA) on Mauna Kea in Hawaii. What this distant view misses are the nuances that Riccardo was in constant conflict with Irwin; and that Irwin's support for the MMT was grudging and under-resourced, its main impetus coming from the younger optical observers John Huchra and Bob Kirshner, and from the Arizona side of the partnership. Within CfA, the SMA—and the hiring of millimeter-wave radio astronomers—was a giant sucking vacuum cleaner of resources.

By contrast, compared to observation, theory was cheap, and new sources of outside funding were becoming available. The monumental, priority-setting decadal study of astronomy and astrophysics recently chaired by George Field for the National Research Council had made a strong case for theory's robust return-on-investment. NASA was establishing a new, well-funded theory program, with a call for proposals for new centers. We had hints that a theory proposal from CfA, led or co-led by George Field, would be a shoo-in—NASA was notorious for buying prestige and paying off people who had done service for the space program. A group of us, including then-Smithsonian researchers Don Lamb and Alan Lightman, worked for several weeks on a preliminary proposal—that Irwin then didn't allow us to submit. Under Smithsonian rules, he could do this as a matter of his authority; under Harvard rules, he could do it de facto, by refusing to allocate any office space.

As this kind of evidence accumulated, even George—the most genial and forgiving of men—came around to my view about Irwin. "He really does hate us," he eventually admitted. "Us" now included himself, Cameron, Dalgarno, and me, the tenured theory professors. George added, "He dislikes *you* in particular, Bill. I don't understand it." I kept from George my father's opinion: Frank had been Irwin's department chair at MIT, and the two had not gotten along. It was consistent with Irwin's character as Frank understood it that Irwin might now delight in punishing me for the sins of the father. It certainly felt that way.

On the Harvard side, I might not have lost on *every* issue if Mike Spence, as the new Dean, had taken more of an interest in the science departments. Instead, he delegated all matters relating to the physical sciences to my senior physics colleague, Paul Martin—who had engineered Irwin's appointment as CfA Director in the first place and had been Irwin's roommate in graduate school. Neutralizing me with the Dean as a favor to his buddy Irwin was as easy as swatting a pesky fly. Oddly, Paul and I had a perfectly friendly relationship. We found each other amusing and often interacted at weekly physics department lunch meetings. In this case, it was George Field who had the darker view: "Paul is jealous of you, Bill. You've already been on more national committees than he'll ever be on."

Paul overstepped his bounds with the dean only a couple of times. He never scheduled with Mike any discussion of whether I should be reappointed to a second term as department chair, or who my replacement should be. The day my term expired, Ruth Mandalian, our battle-ax department administrator, called the dean's office to find out whom she now worked for. They referred her to Paul Martin's office. Paul told her that, for the foreseeable future, astronomy's "acting department chairman" would be—guess who!—Irwin Shapiro. Knowing that a firestorm would ensue, Ruth gleefully informed the senior faculty. The dean hurriedly appointed Josh Grindlay as my successor.

A couple of years later, Irwin and Paul—without authority and without informing the astronomy faculty—offered an astronomy professorship to Columbia's Patrick Thaddeus, a distinguished millimeter-wave radio astronomer who would advance Irwin's SMA project. That led to an unprecedented excursion by the dean up Garden Street, to actually meet with the astronomy faculty. Looking Irwin in the eye, Spence stated that the astronomy faculty, not the director, was the voice of astronomy at Harvard; and that the director would not take the lead in initiating future appointments.

"We've got him on the run, Bill!" George Field gushed after the meeting. But it was all for show, because Spence did nothing to revoke Paul Martin's day-to-day authority. Furthermore, the Thaddeus appointment, on its merits, was too good for Spence and President Bok to turn down; so Pat—who was indeed a first-rate scientist—joined our faculty. A delightful New York Irishman, Pat sometimes referred to himself as "an old Bolshie"; his grandparents had lived in a utopian Communist community.

Irwin outlasted me. I left Harvard in 1998. He remained CfA director for more than twenty years, until 2002. During his term the only tenured faculty appointments in theoretical astrophysics were one-for-one replacements for the eventual retirements of Field and Cameron, and then for my departure. There was no expanded theory postdoc program, no new theory center or institute—all of which soon happened when Charles Alcock assumed the directorship, and all of which proved very productive.

Thus, during the 1980s and early 1990s, Cameron, Dalgarno, Field, and I just hunkered down. We could each support our own small groups from grants already in place. Irwin didn't interfere with our renewal applications. He did take away our previous internal CfA funding; and he did from time to time reduce our allocations of office space. The office that I was left with for a postdoc had formerly been a storage room behind the women's restroom.

My research group in the late 1980s was me, my student David Spergel, and my postdoc Katie Freese. There were occasional theorists appointed as CfA postdocs, and we interacted with them, Bob Scherrer and Barbara Ryden especially. Spergel later became a professor at Princeton, succeeding Jerry Ostriker as department chair. Spergel is widely considered the top theoretical astrophysicist of his generation. He in 2021 replaced founder James Simons as president and CEO of the multi-billion-dollar Simons Foundation. Freese later became a professor at the University of Michigan, and later the University of Texas at Austin.

In various combinations, David, Katie, Bob, Barbara, and I worked on speculative ways in which exotic particle physics phenomena—themselves conjectural—might manifest themselves in astronomical observations. David and I explored the implications of what we termed "cosmions," hypothetical massive particles that could be the (by that time accepted) dark matter in galaxies. In this we lost the battle of names. These and similar hypothetical particles instead became known as WIMPs—weakly interacting massive particles. Thirty years of large-scale physics experiments have not yet found them. But the missing mass is still there, and so may be WIMPs.

We hoped, mistakenly as it turned out, that cosmions/WIMPs might also explain the solar neutrino deficit. With Ryden and Scherrer, the three of us investigated the effects of so-called topological singularities—cosmic "strings" or "domain walls" that were predicted by some exotic physics theories. Every decade or so, some unexpected and statistically marginal observation hints that cosmological topological singularities may actually exist; but, thus far, none have panned out. I still harbor the distant hope that our forgotten work may someday prove to be a "pre-discovery," like the Press-Gunn paper on gravitational lenses.

On the practical side, I developed a fast method for calculating the gravitational interactions of many particles—a so-called N-body code—by aggregating the positions of the particles into a "tree" of nested boxes. Willy Benz and Al Cameron used my code in several of their papers. A related method later became known as the Barnes-Hut method. I didn't mind Josh Barnes and Piet Hut (both at IAS) getting the credit because they needed it more than I did. These ideas were "in the air," and battling over precedence was a waste of time.

When Supernova 1987A exploded, the closest supernova in 400 years, I was part of John Bahcall's IAS group in analyzing the data, visiting Princeton frequently. Because of the presence of Ed Witten there, cosmic strings were taken more seriously there than at Harvard, a bonus for me. Another bonus was sitting next to George Kennan, who was then in his mid-80s, at the weekly IAS professors' dinner. He talked about his meetings with Stalin. There was now good scholarly evidence, he said, that Stalin was a Czarist informant before the revolution. Kennan had come across the incriminating material some years previous, he said, but decided that, as a former ambassador to the U.S.S.R., he couldn't publish it: It would look like an intelligence operation, a back-channel U.S. attack on the legitimacy of the Communist Party.

I always prevailed on John to put me up at Marquand House at 150 Stockton Street, the official IAS guest house. This was a brick mansion rebuilt in the 1930s with modern conveniences (a separate eight-car garage, for example) that still had the flavor of those times. The books in the paneled library were vintage 1930s and 40s. Princeton's Marquand Park, an adjoining seventeen acres, had once been the mansion's grounds. I usually arrived after the housekeeper Mrs. Moriarty had gone home in the evening, so I was often alone in the house. Sleeping in the master bedroom—which connected by a secret passage to the (empty, alas) mistress' bedroom—it felt like living the first chapter of an Agatha Christie novel, waiting for the scream. "Mrs. Marquand was a Hochschild," Mrs. Moriarty told me when I asked about the house's history, as if this would tell me everything.

Doug Eardley, Paul Horowitz, and I made an effort at getting the supersecret NRO to point a certain large orbiting telescope *up* instead of *down* so as to observe Supernova 1987A. There was no astronomical satellite capable of making the observation. Launch of NASA's space telescope (ST, later the Hubble Space Telescope) was still three years away. Through JASON channels, we managed to get an appointment with General Tom Moorman, who was acting NRO director, but, arriving at the Pentagon, Moorman had been called away and we were stuck with his deputy, a dour civilian. The man listened to our briefing and told us that it was impossible—too much technical risk to a billion-dollar system, too much interruption of its real-time mission. We were about to leave when Moorman breezed in. "Don't repeat the whole briefing," he said. "Just give me the bottom line." Doug spoke up: "Sir, for the first supernova in four hundred years, we want a series of occasional observations from now until NASA's ST is launched." "I hadn't realized they were that rare," Moorman said, and to his deputy, "That sounds like a good idea. Arrange it."

A small number of observations were subsequently made, but they produced nothing useful. The satellite's instruments were too finely optimized for its intended surveillance mission, not for astronomy. After General Moorman moved on to his next assignment (gaining an additional star), the remaining powers-that-be classified the whole effort in a "compartment" so secret that we lost access to it. There was no valid security purpose in this; they were mainly afraid of some higher-up finding out about the "misuse" of national assets and saying, "*you did what???*"

These were also years in which I spent a lot of time continuing *Numerical Recipes*. Saul and I took turns writing a monthly column in the journal *Computers in Physics*, each month's piece destined to be a new section in the next edition of our book. Continuing for four years, the exercise resulted in a 1992 Second Edition of the book.

At Harvard I was hunkered down, with limited resources for my research. In astrophysics generally, I had a niche, but I was no longer the boy wonder that I had once been. But there was still that other world, the one of national security, science policy, and consulting to the government where I was still—relatively—a young fresh face. There increasingly in demand, how could I not be more and more attracted to it?

In the late 1980s, Soviet officials who happened also to be high-level spies for the CIA, numbering on the order of a dozen, began to disappear. Some were publicly arrested, tried, and executed. Others simply evaporated. It couldn't be chance. Clearly the Russians had discovered some new way to identify U.S. agents or their activities. Sensitive U.S. intelligence hinted that new technical means, or completely new technologies, were involved. Both directly as the result of these events, and also indirectly as we developed new expertise, JASON got a lot of new business from the intelligence agencies.

This was fascinating stuff, much closer to real intelligence operations than we previously been allowed to get. We weren't just interpreting overhead satellite photos any more. I was the JASON point-of-contact for all the new work. Rich Muller, Paul Horowitz, Dick Garwin, and occasionally Luis Alvarez—collectively the "boy spies" (title awarded honorarily to the latter, older two)—were always involved. Summer study reports were chartered by CIA, NSA, and FBI. These agencies detested each other. We learned never to mention to NSA that we worked for CIA, never to mention to CIA that we worked for FBI, and never to mention to FBI that we worked for NSA. Curiously, that particular three-circle covered most of the institutional sensitivities.

Our studies looked in turn at several different technical possibilities to explain the loss of our agents. CIA was enamored of "spy dust," a hypothetical substance that, clandestinely spread around the U.S. embassy by the Soviets, would cling invisibly to CIA case officers working under embassy cover and allow them to be tracked—even long after the fact—anywhere in Moscow. For several consecutive summers we invented, and then ourselves debunked, innumerable chemical, biological, and nuclear possibilities. Suddenly, around 1989, spy dust was out. We were directed not to think about it anymore.

It was five years before we learned why. CIA had finally concluded that they were being betrayed by a mole inside their Langley headquarters. The previous intelligence hints about new Soviet technologies were now believed to be disinformation planted by the KGB—layers of cover stories designed to mislead, rather like the U.S. Air Force's layers of cover stories regarding stealth, I thought. That CIA had been taken in by spy dust for so long was a bureaucratic embarrassment. Well-known today as sorry episodes in counter-intelligence history, it took several bungled investigations before the mole, Aldrich Ames, was finally arrested and charged in 1994.

Back at the earlier time, the other technical possibilities mostly involved electromagnetic waves—radar. Perhaps there was some new way of intercepting communications, or bugging shielded rooms in the embassy, or tracking people. The Soviets had been irradiating the Moscow embassy with unexplained radar signals for decades. The famous "Great Seal Bug," a commemorative plaque mounted on the wall behind the U.S. ambassador's desk, had been found to be a bug in 1951. The Soviets tapped it remotely by radar. This famous device was designed for the KGB by Leon Theremin, inventor of the eponymous electronic musical instrument. Our CIA friends located and let us page through the official U.S. Government contemporary file on the incident—a surprisingly thin folder that included letters from all the principals involved. Technical understanding of the device was quite rudimentary, supporting British counterintelligence officer Peter Wright's published claim that it was the Brits who explained the thing to the Americans after the fact.

CIA now did its sensitive work in shielded rooms that electromagnetic waves couldn't penetrate. Or could they—somehow? We JASON boy spies were "read into" the GUNMAN top secret compartment that protected a 1984 discovery that IBM Selectric typewriters in the U.S. embassy contained very clever Soviet radio bugs. We visited NSA, met the NSA engineer whose persistence—but mostly animosity toward CIA—led to the discovery, and examined the actual finds. We felt very special for a few months until Dan Rather described everything on the CBS Evening News. But we *were* a little bit special, actually. The full story of GUNMAN and Charles Gandy was first made public only decades later, in Eric Haseltine's 2019 *The Spy in Moscow Station*. Haseltine was a former director of research at NSA.

By the time that (if unknown to us) the mole theory had replaced other possibilities, the boy spies had, in effect, developed a new line of business for JASON. Collectively, we now knew a *lot* about electronic surveillance, past, present, and—with respect to what the laws of physics might allow—future. Conveniently for us, the Moscow agent catastrophe—never actually a technical issue—was around this time replaced by another issue of national importance, this one undeniably technical. Previously, in 1977, a fire had gutted most of the eighth floor of the U.S. embassy in Moscow, including spaces occupied by CIA. A lot of classified materials were afterwards found to be missing—neither in place nor burned up. Few doubted that the fire had been set by the Soviets. Construction of a New Office Building or "NOB" began in 1979. Work proceeded at a snail's pace, partly because the Russians were in no hurry for the United States to occupy new quarters, and partly because the all-Russian construction force were supposed to be monitored continuously by Americans—to prevent, supposedly, the installation of new Soviet bugs.

In 1985, the State Department "pouched in" to Moscow several tons of apparatus. While diplomatic pouches were once just that, the United States, in the 1980s, unilaterally took the position that large items—sealed pallets or whole shipping containers—were protected from inspection under international law. The impetus for this was in fact GUNMAN's need to ship several tons of typewriters and other office equipment back to the United States for dismantlement and close inspection. In response, some forty countries enacted limits, not always enforced, on the size of diplomatic "pouches," but not the U.S.S.R.—which reciprocally began to ship large quantities of materiel to the United States under diplomatic seal.

Now in 1985, a gamma-ray computerized tomography (CT) machine developed by a Department of Energy laboratory was assembled on the NOB construction site. It was said to be capable of producing detailed three-dimensional images of the *interiors* of concrete beams, pillars, and walls—even the inside of the steel reinforcing rods. Medical CT imaging devices are routine today, but, at the time, this was a novel technology—especially with its ability to penetrate concrete—and, supposedly, it caught the Soviets unaware. The machine soon produced "smoking gun" images. The building's concrete pillars were riddled with hundreds of connected implants of a technology never previously seen. Clearly this was a coherent "system" of some kind, but its nature, mode of operation, and even what were the functions of its individual components, were entirely opaque.

JASON came into this picture serendipitously. Johnny Foster, who knew me from DSB and previously as a distantly orbiting Tellerite, was a member of PFIAB (pronounced "PIFF-ee-ab"), the President's Foreign Intelligence Advisory Board. PFIAB was a vestigial White House office that did little. Membership was a minor political honor given to loyal Reagan supporters, the board met rarely, and, despite its name, it had little traction within the intelligence agencies, who kept it at arm's length. PFIAB members included Anne Armstrong (former ambassador to the U.K. and Republican Committeewoman), former Sen. John Tower, Professor James Q. Wilson (a rare Harvard right-winger), and Leo Chern (a Republican banker and fundraiser). PFIAB had been briefed on the classified details of the Moscow discoveries. Foster, seeing an opportunity for the board to actually do something, convinced his peers to open a formal investigation. He would chair its technical side—and JASON would be brought in to do the actual work.

It was a kick, and I was in charge. The JASON boy spies met a number of times in the ornate PFIAB conference room in the Old Executive Office Building next to the White House. We had White House visitor passes and used them to eat in the OEOB cafeteria. When Paul Horowitz loudly repeated a Fawn and Ollie joke, referencing the Marine Corps colonel at the White House who had been responsible for the Iran-Contra affair, the people at the next table scowled at us. Likely they knew Oliver North and Fawn, his assistant, personally.

PFIAB's executive director was a Navy captain in intelligence, Fred Demech, a short, solidly-built guy who had spent much of his career at NSA and was nearing military retirement at age fifty. PFIAB was a final, honorific, posting for Demech. He took an unexpected liking to us and—by a combination of invoking the White House name and himself knowing where thirty years' worth of intelligence community bodies were buried (figuratively, I hoped) was able to get us into many highly classified "compartments" and summon for us important briefers.

Fred told us that before being offered his job, he was interviewed by a Reagan White House functionary. For the occasion he was wearing his dress-blue uniform with campaign ribbons and full Navy captain's insignia. "Do you support the President politically?" the man asked him, startling him. "I am a serving Navy officer, and he is my Commander-In-Chief," Fred carefully replied. "Yes, but do you support him *politically*?" "Sir, have you ever heard of the Hatch Act?" Fred asked, now curious. "No, what's that?" Now, he felt he could answer the original question: "Yes, I support President Reagan completely."

To us he explained: "There was no way I was going to get through to this guy. And anyway, I didn't want to tell him that I was a registered Democrat."

Charlie Gandy, the original force behind the GUNMAN discovery and now a hero of the intelligence world—except for CIA whose complacency he had exposed—turned out to be a mixture of solid engineer and semi-paranoid nut job. He told us that the Russians had developed a radar that could *read your thoughts*, by picking up subliminal impulses sent to your vocal cords. We did our own back-of-the-envelope physics calculations to assure ourselves that this was nonsense.

Another briefing was by a trio of FBI agents, the most senior of whom was introduced to us only as "Mike, the Silver Fox". He was the Bureau's best break-in artist, with midnight skills honed over twenty-five years inside, perhaps, foreign consulates and embassies the length of Massachusetts Avenue. We learned a lot about black operations, Soviet construction practices, and some crude jokes.

We also learned a lot, not always relevant to our task, from Dave M., a senior FBI guy in counter-intelligence. Dave thought of himself as a kind of movie director, spinning epic fictions for audiences of one. The Soviets didn't understand that ordinary Americans would readily assist FBI counter-intelligence operations. Here changing some specifics, when a foreign embassy employee out alone (in violation of his embassy's rules) sat next to a pretty American woman in a bar and chatted with her, the Bureau followed up by recruiting her as a volunteer. Weeks later, the man drove across town to have dinner alone in a new restaurant. He made the reservation by phone just before setting off in his car.

Caught in an unexpected traffic jam (a road closure), he arrived twenty minutes late. And, surprisingly, the same woman was there, eating alone. In their previous meeting, *he* had approached *her*, not vice versa, and he had done so at random. This, now, could only be coincidence. The two had dinner. It was the beginning of a beautiful relationship—one with FBI counter-intelligence.

"There are no coincidences," Dave made sure to explain. "It's people like me who make things *seem* like coincidences." The sudden road closure was not chance. It was orchestrated to allow the FBI to get the woman to the restaurant before their target. Low-level embassy people didn't think of themselves as important. They never imagined the scale of the resources that would be brought against them if they seemed like good targets. Code clerks were best of all.

NSA, apart from Gandy, was the agency most difficult for JASON to crack open. This, apparently, came from the very top, from DIRNSA (Director, National Security Agency) General Bill Odom. He knew exactly who the JASONs were, and he ordered his people not to talk to us. There was some specific reason, and we never found out what it was. However, Fred Demech's old-boy network had seen DIRNSAs come and DIRNSAs go, and, down in the working depths, they were happy to ignore the order from above. Sometimes these people would tell us anything except their own names. By contrast, our friends at CIA, up to and including Deputy Director for S&T Jim Hirsch, loved us—and we knew their names.

We also had our own channels into NSA. Dick Garwin was a longtime member of—and I had just joined as a new member—the visiting committee to an outlying division of the Institute for Defense Analyses (IDA), its so-called Communications Research Division (CRD), located in Princeton, New Jersey. CRD was, for all practical purposes, a fully owned subsidiary of NSA. In an intentionally academic setting, it coddled the country's best crypto-mathematicians, the makers and breakers of the most sophisticated codes. This was an incestuous world: CRD, and its parent IDA, shared many of the same post-Sputnik forefathers as JASON: John Wheeler, Jack Ruina, Eugene Wigner, and others. General Odom blacklisted JASON, but his staff never pulled the thread on who were the individual JASONs and what might their other connections be.

There may have been something deeper and more specific. Julian Nall, chief scientist at CIA, tried to patch up relations between JASON and NSA by inviting his opposite number K. Speierman, NSA's chief scientist, to meet with me in Nall's office at Langley. The meeting went well, I thought; but David Lieberman, CRD's director, later told me that K. (as he was called) related to him that Bill Press was an arrogant, self-inflated dilletante, "no different than the other JASONs."

By this time I had been promoted to be a member of IDA's corporate board, and I sat next to General Odom at a private dinner for IDA sponsors. "I know exactly who you are," he told me, "You wrote the JASONs' Moscow report. I'd like to get you more involved with several matters at the Fort." We were in a classified setting and he mentioned a few. (Insiders always referred to NSA as "Fort Meade" or "the Fort".) I said that several other JASONs could also be useful to him. "I have no use for the JASONs," he interrupted, adding, "Only people like you and Dick Garwin are useful to me." I never resolved these contradictions. JASON's FBI contact thought that there was a specific JASON security issue, although he never could come up with any details. Dick and I may have been OK because we had taken and passed NSA's polygraph test.

The committee at CRD had been around for a generation and was called FOCUS—no one remembered why. On FOCUS, I got to know well some of the cryptology old hands, especially Lou Tordella, the legendary, then-retired, deputy director at NSA who ran the agency for many years; and John Tukey, the hugely inventive Princeton statistician. Among many more substantial contributions, Tukey first coined the word "bit" for "binary digit". I already knew mathematician Andy Gleason from Harvard faculty events, but only on FOCUS learned about his secret past. Another new committee member was Bob Mercer, then a researcher in speech recognition at IBM Watson Labs. Mercer would later become rich as co-CEO of billionaire James Simons' hedge fund empire, and notorious as one of candidate Donald Trump's largest backers in the 2016 election.

Tordella had an interesting take on Peter Wright's book *Spycatcher*, which had been recently published. In the book, Wright, a retired British counter-intelligence official, revealed without permission many U.K. and U.S. secrets, notably the VENONA deciphering of Soviet communications. According to Tordella, Wright felt that he had been cheated out of his British pension on a technicality. He recorded his recollections in a notebook manuscript whose threatened publication was supposed to be leverage in the pension dispute. Tordella, who knew Wright well, thoroughly disapproved, but believed that the threat was only a bluff. Unplanned for was Wright's progressive dementia. His lawyers gave the notebook to a ghostwriter—and the book was published. In the U.K., its publication was long held up in legal proceedings.

We had another unlikely back-channel into the surveillance business, a Cambridge professional acquaintance (call him Joe) who owned a small electronics design firm that specialized in audio circuits. Audio didn't mean hi-fi. It meant bugs for spying. Oddly and bureaucratically, the physical devices that his company made were born into a kind of classification limbo—unclassified until physically delivered to his government customers. That meant that he could show and explain them to us without, technically, violating any rules. Joe was our Deep Throat and helped us JASONs become experts in the trade.

PFIAB chair Mrs. Armstrong decided that Johnny Foster should brief President Reagan on their (i.e., our) findings. Foster called me. "The President is a show-and-tell kind of guy. I need a visual aid to get his attention." I suggested that he take an ordinary pen out of his pocket and tell Reagan that it could contain a bug that could record and later exfiltrate many hours of speech, and that such pens could be unwittingly carried into the White House by anyone. Johnny told me later that Reagan had been interested, examined the pen, and said, "*This* pen?" "Yes," Foster lied, so as not to embarrass the president. At that time, the claimed technology would have been a stretch. Today it could be realized by a high-school student with ten dollars of parts from digikey.com.

Over two years, we came to a good understanding of how the KGB's NOB "system" was supposed to function. We had high confidence that it could be disabled without tearing down the building, leaving its disconnected, non-functioning components in place to rust away. After PFIAB, we wrote additional reports with detailed recommendations on how to accomplish the disabling. In the end, it was a matter of politics. Given the magnitude of the provocation, the United States couldn't *not* tear down and rebuild the NOB—or at least the top half, where classified work was to be done. It cost half a billion dollars. The building was finally occupied in the year 2000.

47. Tryouts

In my twenties and thirties, my role on national committees, at invitational meetings, and the like, was often to be the token young person. I imagined the backroom deliberations: "We need someone *young*...what about Bill Press?" The Defense Science Board, where my term was winding down, was a prime example; I don't think I was ever taken seriously there as a player. My friends on DSB (Norm Augustine, Johnny Foster, Bill Perry come to mind) likely thought it a good experience for me to have while young, their investment in my later career. The rest of DSB, especially the industrialists, must have wondered why the DSB leadership hadn't found a more *respectful* young person.

Seemingly not very different was the invitation, in 1988, to join the board of the Institute for Defense Analyses as by far its youngest member, this coming after I had been on their so-called FOCUS visiting committee for a couple of years. IDA's CEO at the time, W.Y. (Bill) Smith, a retired Air Force general, attended FOCUS meetings ex-officio, although he must have understood next to nothing of the very mathematical presentations. An Arkansan, Smith had risen in rank through a succession of deputy and chief-of-staff positions rather than direct commands, and had done a stint on the National Security Council staff in the White House. He was a loquacious fast talker who didn't do much editing between thinking and speaking—in this, quite unlike the thoughtful retired four-stars whom I had previously met. That may have been why, at FOCUS, he took a liking to me: I didn't do much editing either.

There wasn't much else we had in common, but Smith must have had a plan to refresh the board—and there I was. I was one of five new board members that year, an unusually large turnover. We "new boys" were general Bernie Rogers (recently Supreme Allied Commander in Europe), admiral Harry Train (recently Commander of the Atlantic Fleet), Lawrence Eagleburger (a Kissinger protege and former ambassador to Yugoslavia), John Palms (then president of Emory University), and me (then supervisor of exactly one postdoc, Barbara Ryden, who, coincidentally, was the niece of business executive Bob Frohlke, another IDA board member).

I liked Harry Train's stories. When he retired from the Navy he "decompressed" by walking the whole Appalachian Trail in one go, completely alone, 1,500 miles from Georgia to Maine, March through August. His wife sent food packages to post offices along the route, but he needed to hitchhike into towns to pick them up. "At first, when I stuck out my thumb, all I could think about was that six months before, I had a *personal* staff of thirty-five—stewards, orderlies, drivers and so on—and a professional staff of a thousand officers. I had a four-engine plane, two helos, and my own personal heavy cruiser. But I found that I got really good at hitchhiking. I would wait for middle-aged women to come along. I got to know the type who would not only wait for me to get my stuff at the post office, she would also drive me to her house and wash my clothes for me, and then drive me back to the trail." He had a lot of charm.

Bill Smith also had interesting stories. He had recently been in Moscow for a conference that brought together the surviving principals of the Cuban Missile Crisis: American, Soviet, and Cuban. In 1962, Smith had been military assistant to General Maxwell Taylor, then Chairman of the Joint Chiefs. In the social functions at the conference, Smith said, the Cubans and Americans got along much better than either did with the Russians. He had always wondered why, in the wake of the Missile Crisis, the Soviets had not taken any decisive action to regain face. He was right to wonder, he learned. There had been a Soviet plan to invade northern Norway, seize and hold a fjord, and then negotiate to trade it for the surrender of West Berlin. Apart from whether it would even have worked, the plan was abandoned when NATO caught wind of it and visibly ramped up logistic preparations for the rapid deployment of anti-aircraft missiles across the North Sea from the U.K.

IDA might have been another DSB for me—I could have been seen as having insufficient gravitas and eased out after a three-year term—but it didn't turn out so. IDA's business as a federally sponsored, not-for-profit "think tank," was to give technical advice to the Pentagon. Its niche was not as technical as JASON's: IDA's divisions did studies on the requirements for large-scale acquisition programs (planes, ships, etc.), on cost-accounting procedures, and on organizational issues within the military. More than half the staff were Ph.D.s. These made presentations on their work to the board. It was considered a plum to be so chosen. The board's composition was, by IDA's bylaws, one-third business executives (Barbara Ryden's uncle was one), one-third retired senior military, and one-third scientists; but the latter group were mostly people whose active science careers were long behind them.

IDA had been originally a consortium of universities, its so-called "members," each one of which appointed individuals to the board of trustees. In the Vietnam years when universities were getting out of the defense business, IDA's universities adopted the legal device of electing their board members as individuals rather than representatives. Thereafter, the board would become self-perpetuating. The bylaws specifying the board composition were adopted at that time.

I was often the only board member to ask a technical question after a staff presentation. But here, nodding as if they actually understood the talk (which I doubted), my elders approved of this behavior. It became my role for, completely unexpectedly, more than thirty years. When I ultimately reached the mandatory board retirement age in 2021, I had aged from youngest board member to oldest, asking technical questions and lacking gravitas all the way.

Not all the IDA board members were complete strangers when I came on. I knew Herb York from JASON, Russ Dougherty (a retired Air Force general) from DSB, statistician John Tukey from FOCUS. New as acquaintances were Stanley Resor, Secretary of the Army under Johnson; and my fellow Harvard professor Sam Huntington, who was influential among neo-conservatives, served in the Carter White House, and later became notorious for his 1996 *The Clash of Civilizations*, which predicted a future apocalyptic conflict with Islamic extremism. Retired army general Andy Goodpaster was the class act on the board. He had been Eisenhower's White House military assistant and later himself served as NATO Supreme Allied Commander in Europe. He was recalled from retirement to rescue West Point from a cheating scandal, voluntarily relinquishing his fourth star to accept the assignment as Superintendent. (When he retired a second time, President Reagan restored him to four-star rank.)

Goodpaster was fond of quoting Eisenhower on all matters. A famous one was, "Plans are worthless, but planning is everything." A less well-known Eisenhower quote was, "When you enter the men's room, get in line behind the youngest man." But what I found interesting was not any particular thing Andy said, but rather the way that he—without saying anything—radiated integrity. This was a new phenomenon to me. It got me interested, intellectually at least, in what one saw very rarely in academia: the intangible aura of *leadership* that made people want to follow. Even more surprising to me was that this quality could, to a considerable extent, be *taught*—that it was a recognized part of young officer training in the military.

Before long, I got to see this leadership-radiation phenomenon at an even greater magnitude. When General Smith announced his plans to retire, I was appointed to the board's search committee to find his successor. Our committee sifted names and dossiers. Historically, the CEO position had alternated between scientists such as Jack Ruina and Al Flax, and senior retired military like Goodpaster and Smith. I pushed ineffectively for a return to that tradition. Still, I was not the only one to be annoyed when Bill Smith burst into our meeting to announce (in his chattering way) that we could simply stop searching. General Larry Welch, recently retired as Air Force Chief of Staff, was interested in the job. But…really? Welch was known as a hardball defender of Air Force interests, a true blue-suiter, dismissive of the other military services even as a member of the Joint Chiefs. IDA worked primarily for the Office of Secretary of Defense, where "jointness" was the religion. Welch seemed like a very poor match for IDA.

We couldn't refuse to interview this famous four-star who had risen from airman (enlisting at age 18) to head the U.S. Air Force. At the appointed time, Larry came into the boardroom and, before anyone could say anything, said, "I know what you are all thinking. You think that I am 100% Air Force, and that I couldn't do the IDA job." He was right in that of course. He continued, "So in the last couple of weeks, I've met with each of the following people…" His list included present and former army Chiefs of Staff, Chiefs of Naval Operations, directors of the National Security Agency, and former senior White House officials. "You can talk to any of them about me. When I was Air Force, I did the Air Force job. These people will convince you that at IDA, I'll do the IDA job." We checked and were bowled over. That a Navy CNO might so highly recommend a rival Air Force Chief of Staff was mindboggling.

The other issue on our minds about Welch was whether he could be respected by IDA's Ph.D. scientific staff. Outside of the Air Force, he had only a high school education. His bachelor's and master's degrees (in political science) were from the service staff and war colleges. "I don't think I'll have a problem with that," he said slowly, after a pause.

Whether inborn or taught, Larry Welch was a preternaturally capable leader. I observed him at close range for what came to be many years. He was a laconic Oklahoman, talked slowly when he talked at all. He was tall, but not particularly handsome, and seemed sometimes physically awkward. It was hard to imagine him fitting into the cockpit of the F-4C Phantom that he flew in Vietnam. He never seemed like the smartest person in the room. Yet, he was someone that you just wanted to follow.

It was a very non-obvious kind of charisma. When an individual at IDA needed to be fired, the person entered Larry's office not knowing what to expect, and he came out smiling—and without a job. "How did you manage that?" someone asked Larry. He answered in his slow Oklahoma drawl, "I led him to see that his future was very, very bright—*just not here.*" It was magic. As he predicted, there was never a problem of respect with the Ph.D.s who worked for him.

I may have been observing him, but he, at board meetings, was also observing me. He took me aside once, as if to impart some great confidence that the other board members shouldn't hear. "Bill, I've noticed that you normally like to sit *there*," he pointed, "but that, for some talks right after lunch, you move to the front of the room, *there*. Why is that?" I explained with some embarrassment that, when I needed an afternoon nap, I positioned myself so that the only person who could see that I was asleep was the *speaker*—because he or she was precisely the least likely person to out me. (This was a trick I had learned from Dave Schramm, who often slept at the front of seminars but, unfortunately, also snored.) "Ahhh!" Larry said, in a way that made me think that his regard for me had increased.

IDA activities increased the frequency of my travel between Boston and Washington. On that route, flying Eastern Air Lines was something of an adventure in the last couple of years before its final bankruptcy. They seemed to be doing no maintenance, and flights were often delayed or cancelled for "mechanicals". On one flight, we passengers worried when the co-pilot came out of the flight compartment carrying a flashlight, ripped up the carpet, opened a hatch in the aisle floor, and peered down. After that, we circled for an hour to burn off fuel in advance of what the pilot announced would be a "land ditching" (that is, a landing without landing gear). I was in first class on a frequent-flier upgrade. The flight attendant explained to us that, as first-class passengers, we were required to perform certain duties. After the landing, we should form a "human barrier" to stop coach passengers from mobbing the forward exits. Then, in the event that the forward flight attendant was incapacitated, we should open the doors ourselves, engage the escape slides and—this was the best part!—assist all the coach passengers off the plane before we ourselves exited last. The man sitting next to me and I looked at each other: We knew that we were going to be *first* off that plane. In the event, it was a normal landing.

When the pilots' and flight attendants' unions went on strike, Eastern kept flying with minimal statutory crews of management and non-union personnel. There were often only ten or twenty passengers, and there was usually only a single flight attendant. Eastern's corporate headquarters kept coming up with bizarre incentives to keep us as customers. On one flight, the management flight attendant came down the aisle and gave each of us a new twenty-dollar bill. On another flight, from Boston, they gave each of us a boxed live lobster from Legal Sea Foods. Rather than keep it in my hotel room as a pet, I found an acquaintance in Washington to give it to. Eastern finally stopped flying in January, 1991, with no advance notice—they were still selling tickets for nonexistent future flights.

Around age forty, I noticed a change in the kinds of committees that I was invited to be on. I was becoming too old to be the token young person on old-people's committees. Instead, I had arrived at a rising middle level of the Establishment pyramid. Never their stated purpose, some activities seemed structured so that we "young comers"—the overripe term was actually used—could meet and network; and clearly also so that soon-to-be-past leaders could form judgments on whose rise within the power structure to facilitate. Sometimes these events seemed like beauty contests where we were supposed to display, if we were able, that intangible *leadership* aura that I had now discovered existed.

An extreme example was a one-day conference sponsored by the Carnegie Commission on Science, Technology, and Government. The Commission itself comprised luminaries a generation older than me: Norm Augustine, Josh Lederberg, Sid Drell, Bill Perry, Andy Goodpaster, Guy Stever, and so on. I knew most of them. I never did figure out exactly what the Commission's business was. It had something to do with science policy, and scientific advice to the government, and science education, and scientific illiteracy, and, well...whatever. The Commission sponsored but did not attend our event. The format, which was not announced before the meeting, was that we went around the table for each of us to give a fifteen-minute off-the-cuff discourse on what we thought about, well...*anything*. To emphasize the seriousness of the event, our remarks were transcribed by a stenographer with a stenotype machine. They would be "made available" to the full commission, whatever that meant. It was total bullshit, but somewhat engaging at the time.

For this particular gathering, the young comers, besides apparently me, were: Barry Horowitz, new president-elect of the large defense think-tank, MITRE Corporation; Danny Boggs, appointed by President Reagan at age 42 to the Sixth Circuit Court of Appeals in Cincinnati; Tom Moss, then at Case-Western, recently a high Congressional staffer; Stewart Personick, an AT&T vice president of engineering; Ellen Futter, then president of Barnard; Craig Fields, newly appointed as director of the Defense Advanced Research Projects Agency (DARPA); Bill Drayton, a Third World advocate and president of the Ashoka Foundation; Granger Morgan, a science-policy type at Carnegie-Mellon; Ash Carter, an academic arms-control pundit at Harvard; and Peggy Hamburg, an assistant director of NIH. We were all in our thirties and forties. I didn't perceive much leadership aura in this group—including me. Several of the group were obnoxiously assertive—again including me—but I had already come to understand that this was not the same thing as leadership.

As I write now thirty years later, these people are near the end of their careers. How did they turn out? Googling, I can't find any trace of Tom Moss. Perhaps I recorded his name wrong. Horowitz and Fields were each fired, or resigned under pressure, from their respective big jobs after brief, stormy tenures. Arrogance, vindictiveness, and the inability to lead were mentioned about both; they could have been twins and, afterwards, both pretty much disappeared. Boggs remained on the Sixth Circuit for his whole career. He was said to be considered by George W. Bush for the Supreme Court, but John Roberts was the actual nominee. Personick and Morgan each became known and respected within their engineering specialties, though not much more broadly; each was in due course elected to the National Academy of Engineering. Drayton remained CEO-for-life of Ashoka and coined the term "social entrepreneur," but was not much heard from after the 1990s.

That leaves three out of ten who could be credited objectively with influential (as distinct from merely solid) careers: Ellen Futter was president of Barnard for thirteen years, later board chair of the Federal Reserve Bank of New York, president of the American Museum of Natural History, and a director of several large corporations. Peggy Hamburg became New York City's Commissioner of Public Health and later headed the U.S. Food and Drug Administration. Ash Carter seemed at the time the least promising in the group. He was a protege of the Harvard-MIT club of ex-physicist arms controllers (Jack Ruina, for example), but his reputation fell off rapidly with distance from Boston. He didn't seem like the smartest person in the room. But, under President Clinton, Ash became Assistant Secretary of Defense for Global Strategic Affairs. Then, returning to government in the Obama Administration, he became in quick succession Under Secretary of Defense for Acquisition, Technology and Logistics; Deputy Secretary of Defense; and then Secretary of Defense from 2015 to 2017. Either there was talent there invisible to me, or else Ash just figured out how to play the game better than the rest of us. Maybe both. He was the most successful of us. He died unexpectedly at age 68 in 2022.

More predictable than the young comers were the career arcs of young-to-mid career academic research scientists. A "First Annual Frontiers of Science Meeting"—"frontiers" was the obscure euphemism for "young"—brought about forty of us across all fields to the new National Academy of Sciences conference center in Irvine, California. By design, none of the invited were Academy members. Thirty years later, virtually all are. Future Nobel Prize winners included Tom Cech, Ahmed Zewail, and Frank Wilczek. The token young economist was Larry Summers, later Secretary of the Treasury and president of Harvard. Judging by the number of participants whom (i) I had never met before and whom (ii) I later worked with at a national level—judging by networking, in other words—the meeting was a success. Here I met Eric Lander, Francis Collins, Greg Petsko, Sylvia Ceyer, and Mike Witherell for the first time, while Larry Smarr, Steve Koonin, Roy Schwitters, Bob Kirshner, Margaret Geller, Rocky Kolb, Rich Muller, Frank Wilczek, were participants I knew already.

A long-lasting activity across fields of science was my involvement with the David and Lucile Packard Foundation. David Packard, co-founder of Hewlett-Packard Corporation (known now as the original Silicon Valley company) was, in his mid-70s devoting himself largely to philanthropy. I joined the selection committee of his newly established David and Lucile Packard Fellowships in Science and Engineering. Eight of us, chaired by Yale applied physicist Allan Bromley, were supposed to cover all fields of science and engineering. The initial roster was: Tom Cech, Millie Dresselhaus, David Goodstein, Phil Griffiths, Allan Spradling, Bill Schowalter, and me. After a few years, the committee was enlarged with John Hopcroft, Stuart Rice, and Skip Scriven. David Packard was an ex-officio member of our committee, as was Franklin (Lynn) Orr, professor of petroleum engineering at Stanford who happened to be David's son-in-law. The fellowships were half-million-dollar research awards to the best new assistant professors in fifty top research universities, among whom twenty were awarded annually.

Somewhat like the NAS Frontiers meetings, the creation of social and professional networks of scientists in unrelated fields was a stated goal of the Packard Fellowships. Until their cumulative number became too great, present and past fellows were encouraged to attend an annual three-day meeting and give short talks. Bromley always gave a pep talk with phrases like "future leaders". I don't think he ever stooped to "young comers," but he might just as well have done. His valid point was that, at the national level, decisions had to be made for the allocation of resources among different fields of science. It was understood that narrow specialists would advocate for their own fields—those voices were discounted. It would thus fall to researchers with broad networks and a broad appreciation of fields other than their own to ultimately influence these key decisions. Spouses and children were encouraged to attend the Packard meetings, which were held in the then-brand-new Monterey Bay Aquarium and included childcare, and also social events in the aquarium after its public closing time.

David Packard was fond of relating the backstory of the fellows program, and also of the aquarium and its affiliated research institute. Sometime around 1985, he, his wife Lucile, and their four adult children were sitting at the family breakfast table in Los Altos when David proffered that the time was right for the family to take on new philanthropic projects. He invited each of the other five to come up with a suggestion for new funding by the existing family foundation. It sounded in tone to us like an ordinary family debating what movie to go to, except that, in the Packard family, the scale was hundreds of millions of dollars. Lucile's project became the Lucile Packard Children's Hospital in Palo Alto. David, Jr., founded the Packard Humanities Institute, devoted to classical studies.

Julie Packard, joined by Nancy, developed the idea of an aquarium and marine research institute. Julie's plan was originally that the aquarium be free to the public, like the Smithsonian museums in Washington. Her father convinced her that people value more what they must pay for. A consulting firm recommended quite a hefty admission price—but with discounts up to 100% for school and other nonprofit groups. The aquarium has flourished with this business model for several decades now and has become an influential institution both among museums and for its environmental work.

Susan Packard Orr, after discussion with her husband Lynn, proposed the fellowship program in science and engineering. I found the Packard family to be the least pretentious billionaires I ever met—or maybe tied for that honor with Cecil and Ida Green, Texas Instruments billionaires who were listed in the phone book, answered their own phone, and used to take young Sara and me to lunch in La Jolla. David and Lucile raised their children with a solid work ethic and a sense that the purpose of the family fortune was the public good, no more and no less. At the annual fellows meetings we got to know Susan, Julie, and Nancy to some degree, and also the next generation, David and Lucile's grandchildren, who were brought up with much the same sense of purpose. A seat on the Packard Foundation board was reserved for grandchildren who had turned eighteen. They rotated through it a year at a time—a kind of try-out for whether they deserved to have a continuing influence on family philanthropy. While David Packard was alive, we received every year in December a package of several pounds of dried apricots from his orchard, along with a nice note.

Giving away fruit is apparently something that rich people do: Decades later, my wife Jeffrey worked with Luci Baines Johnson in connection with the Lady Bird Johnson Wildflower Center in Austin. Every year in December, we got from Luci a jar of orange marmalade, the fruit from the Johnson Ranch.

I was on the Packard committee for twenty-five years! I was not even ten years older than the early classes of fellows; many became friends as well as colleagues, especially Josh Plotkin, Jackie Hewitt, Jennifer Doudna, Dan Jaffe, Shri Kulkarni, Jon Kleinberg, Joe DeRisi, Rob Schoelkopf, Noam Elkies, and Chris Stubbs. Jeffrey and I introduced Frances Arnold to her future husband and, over years, watched her career soar to eventually win a Nobel Prize. Doudna, whom we liked enormously, was awarded a Nobel two years later.

To summarize, by the beginning of the 1990s, my professional life had shattered into multiple part-time careers, each rewarding to me, but each with an entirely different and non-overlapping outside constituency. Packard, Carnegie, and Frontiers were examples of one shard, my continuing education in working across—or at least understanding something about—a range of different scientific disciplines. My investment in these kinds of activities paid dividends only gradually over the following two decades. In this vein was also the NAS Film and Television Committee, which provided scientific advice, sometimes scathing, for a series of PBS documentaries produced by WQED Pittsburgh. Between the scientists and the filmmakers there was a love-hate relationship. Jim Ebert, then president of the Carnegie Institution of Washington, chaired our committee. An even-tempered embryologist, he exploded when the film people opined that we had to be careful "about how we treat evolution as a theory".

Leonard Nimoy, who played Mr. Spock in the Star Trek movies, was traveling around the country picking the brains of astrophysicists for his next movie. Paul Horowitz and I had lunch with him in the Harvard Faculty Club. I thought this a discreet location, but no—we were constantly interrupted by star-struck Harvard faculty seeking autographs.

After George Lucas did *Star Wars*, Paramount woke up to the fact that they owned the *Star Trek* rights and started making the franchise movies. Nimoy was by now in his fifties. He agreed to reprise the Mr. Spock role on condition that they pay him a lot, and also, if the franchise was successful, let him direct several of the films.

Nimoy had decided that the plot of his next film should involve time travel back to the 20th century—but for what purpose? For *lost arts*, I suggested that he pick an occupation that we take for granted now but that becomes extinct in the future. As an example, I let loose with the improv of a cursing, tobacco-chewing, wildcat oil roustabout who knew everything about oil-well drilling. When, a year later, Sara and I went to see the movie, *Star Trek IV*, my oil roustabout had become a beautiful female marine biologist. "You don't have any whale biologists in the 23rd century. You need me," she says in the film, but she doesn't spit.

JASON and IDA were the centers of my continuing involvement in defense science and related policy. In a generational transition, Bill Nierenberg had been eased out as JASON chair by a palace revolution in which I played a role. Feuding with several of the older JASONs, Bill had packed the JASON Steering Committee with younger members whom, he thought, would be his respectful supporters. But when we noticed that we constituted a majority of the committee, we voted him out of office. He went almost gracefully. Elsewhere in the national security world, I was a consultant to the theory division at Los Alamos, and I did the rounds of two- or three-year terms on the scientific advisory boards of several three-letter intelligence agencies.

A particularly interesting one was a committee that for a number of years supervised the development and construction of a highly classified, multi-billion-dollar intelligence project. This was my first exposure to formal project management as a serious discipline, i.e., undergirded by statistical tools and methods: PERTs, Gantts, critical paths, earned-value reporting, and knowing the essential difference between "reserve" and "contingency" funds. While formal project management did not guarantee that a project would succeed in "scope, schedule, and budget" (as the mantra went), its absence on big projects guaranteed almost certain failure.

Watching this secret project develop, I learned that effective project managers were usually hated. The requirements of project management are different from those of leadership. Years later, at Los Alamos, I learned the formal expression of this idea: Every project manager should report directly to a "senior executive" whose job is to provide top cover for the project. The senior executive is supposed to have the leadership skills, ideally, to take the flak and not be hated by everyone. Years later, at Los Alamos, I was the senior executive for several large projects, as will be told.

My professional life at Harvard had no relation to any of the pieces just described. My research was on cosmic strings, domain walls, and similar topics at the interface between cosmology and particle physics. This kind of interdisciplinary work was given public lip service in both the physics and astronomy communities—it was sexy—but, privately, neither group thought highly of it. And I spent about equal time and attention working on *Numerical Recipes*. Saul and I turned out monthly technical columns for *Computers in Physics* magazine that would eventually become sections in the next edition of our book. On their face, the columns were pedagogical efforts, not original research; but we knew that our published computer routines hid many original tweaks that were not known in the literature. It was our private convention not to point these out—except to each other. We wanted to sell books and make converts to our style of scientific programming, not puff ourselves up to no obvious benefit: Clever computer code was simply not something rewarded in academia.

Our programming style was "close to the machine" and "highly integrated". This was right, I still think, for a period that lasted about twenty years; but it was eventually superseded by the development of more expressive programming languages—not at all close to the machine—and by the necessity of well-controlled abstract interfaces in large computer codes—making them not at all highly integrated in our sense. The last, third, edition of *Numerical Recipes* was published in 2007. It still sells a couple of thousand copies a year, but nothing like its former glory, which gradually faded.

Rarely did wires get crossed among my fragmented activities, but one time they did. A proposed column of ours explained how the public Data Encryption Standard could be used to generate random numbers for scientific purposes. Since the word "encryption" occurred, I dutifully sent the draft to my IDA friends in Princeton; they were supposed to obtain clearance for publication from their NSA sponsors, a routine formality. Instead, I received an unannounced personal visit, in Cambridge, from Marlin Wagner, who was the head of all research at NSA. "There is nothing classified here," I protested. "I'm not saying that there is," he said carefully. "I've done nothing wrong," I asserted. "I'm not saying that you have," he said noncommittally. What it came down to was that he just *wished* that we wouldn't publish the column. He was literally wringing his hands. Saul and I were many months ahead in writing columns, so I offered a compromise: What if we didn't publish the piece in the magazine, but still did include it in our new book edition, two years hence? Marlin's whole demeanor brightened. "That would be *wonderful*! We would have no objection to that!" I never did find out why the fuss. There must have been some time-limited sensitive operation whose success our innocent column would somehow have endangered.

I knew that my chances of ever being elected to the National Academy of Sciences were by now nil. The Academy was organized into disciplinary sections, each responsible for the election of its own new members. My separate constituencies in astronomy and physics were each small. My work was at a diffuse interface, not "at the top" of either field. My national security engagement and cross-disciplinary interests were invisible. *Numerical Recipes* was very visible, but not helpful. No one was ever elected to NAS for writing a textbook, especially a very popular one. I could imagine the conversation behind closed doors: "Oh, you know Bill! He's written a very successful textbook. He tells people that he's made enough money to pay off the mortgage on his house. That's not the kind of person we elect to the National Academy of Sciences—a person with no home mortgage!"

I wasn't supposed to know, but John Bacall and Dave Schramm did keep trying to get me elected. Academy elections were deeply secret, but John and Dave were people who just couldn't resist keeping me informed. For several years, around the same time of year, one or the other would call me with the assignment of writing a few paragraphs describing how wonderful I was. These words were provided by them to the people who were supposed to write letters of recommendation for me. It was against all the rules. John always took great delight in political maneuvering. When I visited Princeton, he would take me into his office and close the door. "It's looking good, Bill. It's looking good. We've made it to the next level. It's looking good." And then, maybe a month before the election results were due to be announced, John would go radio silent. He never said, "By the way, Bill, your election failed again this year." But a year would go by, and it would again be, "Can you update your paragraphs?" and then "Looking good, Bill. Looking good." And then radio silence again.

I never felt too badly about this, although I knew that it was a disappointment to my father, who was then in his second, final, term as NAS president. But I had made some career choices very consciously, even as other choices were made for me. Perhaps, if John Bahcall had become director of the Center for Astrophysics instead of Irwin Shapiro, there might have been a flourishing of theoretical astrophysics there instead of the suppression of it. Then, I might have done more (or better) in my disciplinary specialty, and been electable in the NAS astronomy section. But even in such an alternative universe, the siren call of other fascinating activities would probably have instead prevailed.

48. Evangelist

In the 1980s and later, the Alfred P. Sloan Foundation supported a series of commissioned scientific autobiographies. Luis Alvarez wrote vaingloriously about himself. Freeman Dyson's beautiful *Disturbing the Universe* was the beginning of his new, and soon rightly celebrated, career as a writer for popular audiences. In 1991, *Nature* magazine asked me to review Viki Weisskopf's memoir in the series. I had become known as something of a bad boy reviewer, so the *Nature* editors, who loved to cause a stir, often thought of me. "Victor Weisskopf is a fortunate and happy man," my review began, "and he does not mind telling us so."

What struck me about Viki, whom I had encountered on many occasions since our first friendly meeting at Caltech, was that he had become a Great Man in physics without ever having actually discovered anything important. He had been in all the right places: in Göttingen with Max Born, Leipzig with Werner Heisenberg, Berlin with Erwin Schrödinger, Copenhagen with Niels Bohr. At Los Alamos during the war, he was Hans Bethe's deputy in the Theoretical Division. In the 1960s, he was director-general of CERN, the pan-European physics laboratory. He had won many major prizes and medals—but he had no major scientific discoveries to his name. Weisskopf was not without self-awareness on this point. "It is regrettable," he wrote, "that among scientists the presentation of ideas is not as highly valued as the creation of ideas. This is in stark contrast to music, where the performer is a partner equal to the composer." I loved the concept: *Einstein's General Theory of Relativity performed by William H. Press*, I thought; or, *William H. Press conducts a new interpretation of Hubble's Expansion of the Universe.*

In one of his book's rare sardonic passages, Viki wrote about the successive periods in the life of a theoretical physicist. "The first, when one is young, is the time of hard work on new ideas. In the second period, one has become reasonably well known and is invited to give talks at various universities and at conferences. One still contributes valuable original work, but more often with the help of younger colleagues or by suggesting a line of research to them. In the third period one receives invitations to give general talks surveying the whole field through the light of experience."

I saw myself as being in the second period that Viki described, but teetering on the brink of the third—which I was not looking forward to. It was a fate that I knew that I must try to escape, although I didn't yet know how. I was not unlike Viki in often being a communicator of other people's discoveries, especially ones that I thought should be better appreciated. I thought of the role not as performing, but as *evangelizing*. As an evangelist for Alan Guth's inflationary cosmology in the early 1980s, I gave many talks at conferences, perhaps helping to advance his fame. Now, in the 1990s, I was giving talks on wavelets, or, as Viki might have it, *Ingrid Daubechies' Wavelets in an artistic interpretation by William H. Press*.

Some backstory is useful: The vague concept of wavelets, meaning undulatory (that is, wavelike) functions that were localized in space and time—that idea had been around for many decades. In the 19th century, so-called Fourier analysis showed how any signal could be decomposed into sine and cosine waves, but those were not localized. Twentieth century quantum mechanics introduced the idea of "wave packets" that had the desired localization in both position and energy, the latter implying frequency. Eugene Wigner formalized that notion, but, like quantum mechanics generally, it required the introduction of complex numbers. I knew about Wigner's wave packets from a JASON project: Radar engineers had adapted the same idea to something that they called "range-Doppler ambiguity functions". In JASON, I had written a computer program to calculate such functions efficiently. So I was well primed when a visiting DARPA program director gushed about a new fad in French applied mathematics, a new kind of wavelet that required only real numbers and were *perfectly* localized in both position and frequency, a Holy Grail that the mathematicians termed "compact support."

There were only a handful of published papers on the subject. I found them very obscure—and not just because they were in French. One 1988 paper by Ingrid Daubechies, in English, was more comprehensible. Ingrid, who was Belgian, had worked in France with Yves Meyer and Stephane Mallat, then brought their ideas, with many of her own, across the Atlantic—first to the Courant Institute in New York and then to Bell Labs in New Jersey. I knew her from one of the NAS Frontiers meetings of young scientists. Studying her paper, I came to understand that her particular mathematical construction of wavelets, far from being impossibly abstract and weirdly French, was a practical piece of engineering that begged for application in all kinds of areas.

I was an immediate evangelist. I gave seminars and conference talks. I wrote a "Primer on Wavelet Transforms" that was incorporated into an unclassified JASON report and later became a section in the second edition of *Numerical Recipes*. I wrote and freely distributed computer code that implemented a sequence of Daubechies wavelets, which I designated Daub4, Daub6, Daub8, and so on. When I taught wavelets in my courses, I pointed to the function "Daub4" and exclaimed in mock awe, "Can you imagine having something as basic as 'cosine' *named* after you?" To my delight, my "Daub" names soon became universally adopted. In 1994, Ingrid moved to a professorship at Princeton and soon became the first female full professor of mathematics there. Success has many parents, and I like to think that I am one of those for Daubechies wavelets, which are now the standard in many audio and video compression algorithms like JPEG and MPEG.

My only original research on wavelets was a forgettable paper that I wrote with Mike Freedman, a pure mathematician who was briefly a member of JASON. In truth, this was not a paper worth publishing; but I was drawn in by the idea of publishing with a Fields Medal winner and someone whose Erdős number was *two*—so that mine would become a respectable *three*, about the same as most professors of mathematics. The famously collaborative mathematician Paul Erdős had, by definition, an Erdős number of zero. His coauthors are assigned Erdős number one. Their coauthors (on any other paper) have Erdős number two (about 10,000 people at present), and so on. *The role of mathematician will be performed tonight by understudy William H. Press.*

Ten years had elapsed since the National Academy survey committee chaired by George Field had established national priorities for astronomy and astrophysics in the 1980s. I had then been a member of Dick McCray's theory subpanel. Now, a new committee was being formed to recommend priorities for the 1990s. Its chair was John Bahcall. Knowing that he could not completely control the choice of the chairs for the various disciplinary sub-panels—those had to be each field's own big guns—John invented a structure where the subpanel vice-chairs, not chairs, would comprise the main committee. A generous interpretation was that the vice-chairs were chosen as individuals of broader perspective; a more accurate one was that we were mostly friends of John's. I was vice-chair of the subpanel on computing and data processing, chaired by Larry Smarr, and I was thus on the main committee.

Almost by definition, the chair of the decadal Astronomy and Astrophysics Survey Committee becomes the unofficial dean of astronomy in the United States. This was a big deal for John. The implementation of his priorities for a decade would be a career-capper to be surpassed only when he won a Nobel Prize for the solar neutrino experiment—his not-so-secret hope. John was a poker player and viewed the AASC as a kind of poker game. Figuratively, the subfields represented by subpanels would bet on the hands they were holding—their priorities for large optical, infrared, or radio telescopes, satellites, and so forth. Then, somehow dealing himself the best cards, John would *call* and dictate the final results. This was a worldview sure to get him into trouble, and it did, both with his own committee and with the government agencies that it was supposed to advise.

An early battle was with Erich Bloch. Close to the end of his six-year term as NSF director, Bloch was universally hated by the scientific community. He was a Reagan appointee, an engineer, and the first (and to the present, only) NSF director without a Ph.D. Bloch's avowed mission was to direct the NSF toward industrially relevant research. Astronomy was understandably at the bottom of his priority list. He had appointed a loyalist, Pat Bautz, as astronomy division director, her mission to gut the previously planned astronomy program. He was angered when the astronomy division's outside advisory committee petitioned the National Science Board to have Bautz removed for cause. It was that bad.

My father, as president of the NAS and someone who knew both Bahcall and Bloch, was alarmed when Erich announced that, uninvited, he would swoop in to address Bahcall's committee. Frank delegated Phil Smith, his number two guy, to give us a private pep talk just before that meeting. Erich was unpredictable and vengeful, Phil explained with a terrifying serenity. Erich liked nothing better than to goad people and make them say things that they would regret. Our best course of action was to listen to him politely and make absolutely no response, no matter how much he tried to provoke us.

After Phil left by a side door, John kept Erich waiting in the hallway for ten minutes, just for spite. The NSF director then stormed in with Bautz following. The first thing that he said was, "I want to take this opportunity to commend Pat for the fine job that she is doing. You people ought to award her a medal." He then looked around the room, expecting someone to contradict him. Nobody said anything. We were as if drugged, sedated. This was theater of a high order. Eric went on to say more and more outrageous things. No response from anyone. He couldn't believe it. He had heard about John's reputation as a street fighter. He wanted to take on John in single combat, but he instead found himself in a room full of zombies.

Bloch then played his nuclear option. Our committee was irrelevant, he declared. He wondered why he was here at all, given that he and Pat had *already decided* what would be NSF's astronomy priorities in the 1990s—two fairly modest telescopes, one in the Northern, the other in the Southern Hemisphere. Any different recommendation from us would be dead on arrival, he told us. We all glanced sideways at John. He looked ready to explode. "This meeting is over," Erich said. That did it. John did explode. He and Erich yelled at each other for the next half hour while the rest of us watched in awe. But we knew that time was on our side: Bloch's term as NSF director would end in August, 1990, a year before our report was issued.

A different crisis for our committee was secondary to a larger one in space science. The Hubble Space Telescope had been launched in April, 1990. At the end of June, 1990, just before one of our AASC meetings, NASA belatedly revealed that HST's large main mirror had been fabricated incorrectly. HST's images were all blurry. The AASC "policy" subpanel, chaired by Dick McCray, had finished its report before the announcement but, presciently, had taken NASA management to task for its poor oversight of HST's construction. It was proof that the HST disaster could have been—and was—foreseen.

Our main committee had evolved the process of inviting each panel chair to present their report, followed by a debate as to which findings and recommendations should be taken over into the main report and endorsed by the full AASC. In this one case, after McCray's presentation, instead of calling for discussion, Bahcall simply handed around his own draft chapter. Contradicting McCray's, John's chapter read like something written by NASA's public relations department. It was a full-throated endorsement of NASA's management of HST.

We read it and were stunned. I was the one who dared to speak up. "John, I think that there will have to be some compromise made on the issue of criticizing NASA." "*That's the last straw!*" John shouted. "*I don't have to take any more of this crap! I resign! I won't be at the meeting tomorrow! You can just get yourself another chairman!*" With that he walked out. We sat around waiting for him to return, but he didn't, and we eventually all went to dinner. The next day John arrived an hour late, apologized for his late arrival—but for nothing else—and continued as if nothing had happened.

What was really going on? I never found out for sure. John had spent more than a decade lobbying for HST; he was often called its godfather. My best guess is that, before our meeting, NASA associate administrator Len Fisk had gotten to John and convinced him that any criticism of NASA management would politically endanger a now-necessary Hubble repair mission. Fisk probably wrote the laudatory prose himself. John must have known that such a draft wouldn't fly, but he needed NASA's spies to report back that he had gone to the mat supporting it. His next-morning tardiness was probably when he was on the phone with Len with his carefully orchestrated bad news.

The final AASC report terms the HST failure the greatest scientific disappointment of the decade and offers muted criticism of NASA management, exactly the kind of compromise that I anticipated from the start—not coincidentally because I ultimately wrote those paragraphs. The Hubble rescue mission was launched three years later, in December, 1993. Astronauts Kathryn Thornton and Thomas Akers installed a corrective optics package. As I write, HST still holds the record, after 30 years and several additional Space Shuttle upgrade missions, as the U.S.'s most productive space astronomical facility. Fisk was elected to the National Academy of Sciences in 2003. Bahcall wrote his nomination.

Sometimes even I, who knew John very well, couldn't decode his motives. By 1991, LIGO, the Laser Interferometer Gravitational Observatory, was starting to become real. Since my time as his graduate student twenty years earlier, Kip Thorne had dreamed of actually detecting the gravitational waves that were predicted by Einstein's general relativity. Kip was no experimentalist, and he convinced Caltech to hire away from Glasgow Ron Drever—a portly, and very Scottish, physicist in the tradition of Glasgow's famous Lord Kelvin; and Kip negotiated a partnership between Caltech and MIT for a joint proposal to NSF by himself, Drever, and MIT's Rai Weiss. By 1987, a 40-meter-long prototype interferometer had been demonstrated. Caltech appointed Rochus ("Robbie") Vogt to be the founding LIGO director who would shepherd the project into its full construction phase, two geographically separated five-kilometer interferometers that would cost hundreds of millions of dollars. Vogt, a physicist, had successfully managed the construction of Caltech's Owens Valley Radio Telescope and played a leading role in building the Keck Observatory Telescope on Mauna Kea.

The money had to come from NSF, and this would be the largest construction project that that agency had ever undertaken. Big science, particle accelerators and the like, were normally the province of the

Department of Energy. But DOE didn't support astronomy—or general relativity. And, propitiously, an activist National Science Board over NSF was itching to do big science with the big boys. In 1990, the Board approved LIGO construction.

LIGO was an *astronomy* project only in Kip's mind—this going all the way back to the provocative title of his and my 1972 review article, "Gravitational-Wave Astronomy." Everyone else in the world thought LIGO a physics experiment based on an intriguing technology, the utilization of laser quantum measurement principles to detect a tiny, classical Einsteinian effect. Still, Kip had managed to get his entire LIGO trio, himself, Ron, and Rai, onto one of the lesser of the Bahcall committee's astronomy subpanels. These three were canny enough not to push for LIGO be funded with astronomy money. But they wanted the astronomy committee's endorsement of their project.

The astronomers on our main committee would have nothing of it. That didn't surprise me. What did, was John's eagerness to do battle on the issue—and on the wrong side. He could have just let the matter drop. LIGO would have ended up with a subpanel endorsement, and no mention in the main committee's report. No one would have noticed. Instead, John decreed that, since Kip et al. had boldly asserted that LIGO was astronomy, our astronomy committee should rank its priority among all other astronomy proposals for the coming decade. Of course, he knew that we (except for me, Chris McKee, and at most one or two others) would rank it dead last. Such a public repudiation would kill the project, even if it had no astronomy funding.

John's motives were, I think, a mixture of personal and scientific. He resented the way that the Caltech and MIT administrations had used their formidable political influence to support the project—his own small IAS had no such power. He was close enough to the astrophysics to understand that Kip's so-called theoretical estimates of what LIGO might observe were vastly overstated—never quite fraudulent, but close. Painfully, I agreed with John on this. More fundamentally and unemotionally, he argued that the observation of gravitational waves—even if it could happen—would not be an interesting scientific discovery.

That point, which turned out to be completely wrong, deserves some unpacking. By 1990, radio observations of the Taylor-Hulse binary pulsar—two neutron stars orbiting very closely—had indirectly confirmed the existence of gravitational radiation. The two neutron stars were losing orbital energy and spiraling closer and closer together. This was exactly the process whose final, cataclysmic burst of radiation LIGO was designed to someday observe. The pulsar observations were not only qualitative: Energy was being lost at exactly (to within one percent!) the rate predicted by general relativity. By contrast, LIGO's observations, if possible at all, would test relativity with an accuracy of only ten or twenty percent. In John's view, the LIGO project would spend hundreds of millions of dollars for a one-shot, essentially phony, newspaper headline: "*Einstein Proved Right.*"

I thought—still think—that this narrow argument missed two key points. First, physics is fundamentally the most experimental of sciences. As such, it seeks direct, not merely inferential, confirmations of physical theories. That nobody doubted the existence of Einsteinian gravitational radiation was not the point. It had not been directly observed. In mediaeval times, no one doubted the existence of angels. They could be inferred (according to the standards of the time) from the evidence of their works. Something as fundamental as gravitational radiation could not be left in such a state.

Second, completely independent of general relativity, I saw LIGO as a fantastically interesting experiment in fundamental measurement. If successful, it would detect the displacement of a macroscopic object by a distance one ten thousandth the size of a proton—the fractional equivalent of measuring the distance to the star Alpha Centauri with an accuracy the width of a human hair. I believed—still believe—that the most important driver of all scientific discovery was the development of new measurement capabilities. LIGO would cross the threshold of quantum-dominated measurement by many orders of magnitude in one jump. Who could say what might serendipitously be discovered, or what spinoffs of laser quantum technology might arise? This was important in itself.

I also did have a dog or two in this fight. John was a close mentor, but I had been Kip's student. LIGO could be traced, in one interpretation, to Kip's and my original review article. A few years after that article, Rai and I had been the authors the Physics Advisory Committee report that established the NSF's experimental gravitation program, where LIGO was now situated. Rich Isaacson, its program director, also ran the theoretical physics program that still supported my research at Harvard. Rich now weighed in privately to make clear to me his hope that I could get John off his case.

All parties on both sides were soon secretly communicating with my father as president of the NAS. Frank quietly sought my advice on how to get John under control. Eventually, he called John and told him that

LIGO was not within the scope of the AASC committee, an unusually strong position for Frank, who generally liked to be diplomatically above the fray. John then backed down.

That the final report of the AASC made no mention of LIGO could not pass unnoticed, however. That inconvenient fact surfaced in the year's Congressional appropriation hearings—and LIGO fared badly as a consequence. By the following year, however, NSF had completed its objective site selection process for the two major LIGO facilities construction projects, choosing Hanford, Washington, in the district of speaker of the House Tom Foley; and Livingston, Louisiana, the state of Senator Bennett Johnston, who sat on the Senate appropriations subcommittee for NSF. LIGO was back on track.

I'll skip forward in years to finish my version of the LIGO story. In its new wish to manage big science, NSF was seriously out of its depth—Caltech even more so. Robbie Vogt proved a poor manager, bullying and prone to temper tantrums; but Caltech refused to replace him. I blame Kip for this. He was naïve about the management and leadership skills needed for big project management, and he mistook Robbie's bullying for necessary firmness. In 1994, under Congressional pressure, NSF froze LIGO's funding until Caltech came up with an acceptable management plan—a euphemism for firing Vogt. Robbie was replaced by Barry Barish, an even-tempered experimental particle physicist with more experience in the construction and operation of large scientific facilities.

Interestingly, I knew about LIGO's dirty laundry early on, this from newspaper reporter Bob Cooke, who had begun his career at the *Pasadena Star-News,* where he had several times interviewed Saul and me about black holes. Bob was by the time a science reporter at *Newsday*, and we kept in touch. He was tenacious and loved a good story—and LIGO was one. Another example, some years earlier, was when the *New York Times* science department had mounted an elaborate and expensive expedition to once-and-for-all find the Loch Ness monster. (It failed of course.) Cooke got to Scotland a week before the *Times* advance party, hung out in the pubs, and signed up half the town as his paid confidential informants. Anything that happened in the *Times* camp got reported in *Newsday*, under his by-line, days before it ran in the *Times.*

Much later, when I was at Los Alamos, Robbie Vogt was chair of University of California's science oversight panel for Los Alamos, and I was his host and dinner partner. Undiplomatically, I made an oblique reference to the failure in managing LIGO of his authoritarian management style. "What exactly do you mean by that?" he said in his native German accent, suddenly very coldly. I should have shut up, but we had both been drinking too much, and we had just spent the day working together harmoniously on Lab issues. "Well, some people call you a Nazi," I blurted out. He stiffened. "What do you *think* you know about my background, Bill?"

He then told me that, in Nazi Germany, his father had been a newspaper editor who was critical of the Hitler regime. One day when Robbie was twelve, his parents were taken away. He never saw them again. He was sent to a residential Hitler Youth unit, a kind of punitive orphanage for the children of enemies of the regime. The children were given military training and drilled in Nazi propaganda. Still, he said, they knew that the Nazis were their enemies. He was just barely young enough not to be sent to the front as cannon fodder before the end of the war.

After Barish was in charge, it took twenty years more, and several generations of technology improvements, before LIGO first detected gravitational waves. My tiny contributions during that time were as one of the invisible cadre of outside reviewers who recommended to the NSF the continuation of the long project at each stage. I was part of one site-inspection visit at the Hanford facility, and I chaired an outside committee that evaluated the project's plans for data processing. In 2016, after a major upgrade in its sensitivity, LIGO first detected gravitational waves. The event had the clear signature of two inspiralling black holes each of mass thirty times that of the sun, at a distance of four hundred megaparsecs. The detection of other similar events followed. This was a class of events that even Kip's optimism had not anticipated. Even now, there is no accepted explanation of how such massive black-hole systems can be formed. Rai, Kip, and Barry shared the 2017 Nobel Prize in Physics, the prize awarded with unusual speed. Ron Drever had died earlier that year. LIGO is a jewel in the NSF's crown, now often cited as the exemplar of basic science's need for long-term, patient support.

Returning to the early 1990s, my most important evangelism may have been for something more abstract than inflationary cosmology, wavelets, or LIGO. Tom Loredo, a younger colleague of Saul's at Cornell, had made it his mission in life to convert all astrophysicists to the religion of Bayesian statistics.

Why was this a "religion" and not merely a class of statistical techniques? That long story can only be sketched here: Bayesian methods go back to the eponymous Rev. Thomas Bayes (1701-1761) and were in common use until the beginning of the 20th century. At that time, Fisher, Neyman, and Pearson, so-called "frequentists," crusaded for what they saw as greater rigor in statistics. Bayesian methods were accused of having "subjective" elements that needed to be rooted out. This was the beginning of a fifty-year war between Bayesians and frequentists. By the 1980s, professional statisticians had reached a reluctant truce, agreeing that, for different kinds of problems, each methodology had value. Physics graduate education lagged by twenty years, however, so, until the 2000s, most physicists still thought Bayesian methods taboo. I was an easy convert for a couple of different reasons. The physicists' bible for the Bayesian religion was a manuscript textbook by Ed Jaynes that circulated in third- and fourth-generation Xerox copies. Jaynes was a charismatic figure, revered by some, never quite crackpot, but far from mainstream. Harry Sahlin, another not-quite-crackpot, had introduced me to Jaynes' book my first summer at Livermore, twenty years earlier, and I loved it. (The book was finally published in the normal way some years after Jaynes' death.)

What cemented my Bayes conversion, however, was hearing about work at IDA's classified cryptology center in Princeton. The researchers there were Bayesians to the core. They celebrated, sometimes with bottles of champagne, amazing successes that the outside world would never know anything about. There were methods like Markov Chain Monte Carlo that married Bayes with supercomputers. MCMC had been developed at Los Alamos in the 1950s and, although it was not classified, was barely known outside of just two communities, nuclear weapons and cryptology. One of IDA Princeton's resident geniuses, Nick Patterson, educated me on MCMC and other large-scale Bayesian approaches like Gibbs sampling—also unclassified and underappreciated.

David Spergel was by now an associate professor in Princeton astrophysics, and I introduced him to Nick at IDA. David, too, caught the Bayes bug. In 1995, he spent a sabbatical year at NASA's Goddard center in Maryland and soon after became leader of the science analysis team for the Wilkinson Microwave Anisotropy Probe satellite. The scale and sophistication of that team's fully Bayesian approach was a huge leap in the analysis of astrophysical data. Exaggerating only a little, this was like the Emperor Constantine's conversion to Christianity. After WMAP, all astronomers were Bayesians—or else were suitable only for feeding to lions.

Jeffrey, who was soon to become my new wife, was the Harvard assistant dean for finance whose institutional approval was required for my NSF proposal renewals, along with all research proposals from Harvard's Faculty of Arts and Sciences. When we married in 1991, Sara, then fifteen, acquired a new, much younger brother, James. Jeffrey and James were with me for the first two weeks of that summer's JASON study. They then flew back to Boston, leaving me in La Jolla for a month. For some reason, I fell behind on my daily diary, keeping only very abbreviated notes. In August, when I was back home, I expanded my notes into diary entries and got back into synch. On August 17, caught up, I typed the incomplete entry: "I meet with Henry Rosovsky, who is again the acting dean. I requested this meeting, because"

I no longer recall my reason for meeting with Rosovsky. What I do clearly recall is a sudden dazzling awareness: The disembodied reader to whom I had been writing my diary for two decades was now an actual person. She was Jeffrey! She was in the next room. Instead of typing, I could just tell her about my day. I never wrote another entry. I had kept a diary for twenty-one years, seven months, and seventeen days.

49. Global Grid

Until the 1990s, I had little to do with people who called themselves computer scientists. Physicists, even computational physicists like me, looked down on them. A stale physicist joke was that any science that had to call itself one, *wasn't*. Social science, political science, and computer science were then inevitably cited as the examples. Slightly more elevated was a debate over whether a field that proudly termed itself "the study of the artificial" (a phrase coined by Herbert Simon) could be a science. Wasn't science by definition the study of the natural, not the artificial? Mathematics might also be termed artificial, but the mathematics community didn't welcome computer science as a branch of their field. Oddly the issue was not that computer science lacked rigor. It was the opposite: The computer scientists' proofs seemed to mathematicians pedantically rigid, mechanical—soulless. Mathematics and computer science were separated by incompatible aesthetic principles.

Harvard had no computer science department per se. Les Valiant was first-rate, an introverted Brit who had singularly come to rest in Harvard's applied sciences division and worked on abstract complexity theory. Harvard thought so little of him that his office was in the basement of the old computer center. (Valiant was in 2010 awarded the Turing Award, considered to be the Nobel Prize of computer science.) Harry Lewis was my former neighbor on Faculty Row. His strengths were in undergraduate teaching and (later) university administration. When these two plus Paul Martin, then dean of applied sciences, attempted to recruit H.T. Kung away from Carnegie-Mellon, I was put on Kung's schedule of interviews—because there weren't enough actual computer scientists to fill his day. Kung's work touched on both theory and practical computer systems. He was a star. Carnegie-Mellon was a computer science department second only to MIT. What were these three even thinking?

I was Kung's last meeting of the day. I explained that I wasn't actually a computer scientist, and that I wasn't even in the division of applied sciences. I just wanted to meet him, I lied. With a sigh, he told me that he was sorry to be wasting my time—not just mine, but everybody's. Harvard had so far to go in computer science, he said, that for him to come here and improve the place, he would need resources so great—faculty lines, research support, a building—that it would sound ridiculous and grandiose to the people he had been meeting with. I told him that—if we were both going to be frank—I actually didn't much care whether Harvard bettered itself in computer science or not. But I added, "Instead of wasting our—and your—time, why don't you just tell the dean and the president what it would actually take to get you here." He left my office and did exactly that. Jeremy Knowles, the dean, and Neil Rudenstein, the new president, gave him everything he asked for, and he came to Harvard, where he has been ever since. Harry Lewis' connection to Bill Gates, who had been his undergraduate student, delivered on the promise of a new building.

It was also around this time that I was invited by Joe Traub to join the National Research Council's Computer Science and Telecommunications Board. Traub—who, coincidentally, had done famous work with Kung—was the founding chair of Columbia's computer science department. Within the NRC, CSTB was an effort to merge and revivify two boards that had fallen on hard times, a computer technology board whose imagination never advanced much past 1960s mainframes, and a telecommunications board of telephone industry stalwarts. Traub, as the new chair, rotated out the old membership and recruited modern computer scientists (Al Aho, Ruzen Bajcsy, John Hennessy, Raj Reddy, Ivan Sutherland, Bill Wulf, among others) plus a few networking and telecommunications people who were already envisioning a coming revolution. I was the token computational, as opposed to computer, scientist. The distinction is between one who uses computers to advance a field of science (computational) and one who advances the theory or practice of computing or computers themselves (computer). These are very different specialties. The computer scientists' formulation of this was, "Press is the applications guy," as if all applications of computers to all sciences merited only a single seat at the table.

Today's established subfields of computer science (e.g., theory, systems, networking, languages, graphics, security, machine learning, artificial intelligence, human-computer interaction) barely formally existed at the time, except insofar as the work of individual researchers lay in those areas. Traub appointed to CSTB people with a broad range of interests. The computer scientists were, to me, a new and different tribe—or, more accurately, set of warring tribes. Highest in status were the theorists, especially the complexity theorists. Lowest were the people in industry who actually designed and built computers. Academic computer science departments often comprised only theorists. The people who thought about actual computers were in separate engineering departments. Even the high-status theorists, however, recognized that they were looked down on by all the sciences that didn't have "science" in their name. The notes I took at CSTB meetings have entries like, "Computer science: we ain't got no respect—fifteen-minute discussion."

In the Byzantine organization of the NRC, standing boards like CSTB met three or four times a year and generated no direct product. What boards did was to charter ad-hoc committees, usually chaired by a board member and populated by experts in a particular subject. Committees wrote reports with specific recommendations to one or more federal agencies, delivered their report, and then went out of existence. The influence of such reports varied widely. A board with a string of successes—meaning that the recommendations of its committees actually became law or policy, created new national initiatives, influenced funding decisions, or were adopted by academia—those boards became known in Washington as powerful. Such a reputation in turn gave a board's subsequent committees additional heft. Under Traub and his successor Bill Wulf, CSTB became arguably the National Research Council's most successful board.

Soon after I joined, CSTB published the report of a committee chaired by Cornell computer scientist Juris Hartmanis on the future of computer science as an academic field. Given the composition of our board and the committee, it was not surprising that the report recommended that academic departments be more outward-looking. "A broader agenda would legitimize closer couplings to science, engineering, commerce, and industry, encourage greater interaction between research (especially theoretical research) and computing practice." What was surprising was the community's reaction. Stanford pioneer computer scientist John McCarthy led a petition drive demanding that the Academy withdraw the report. McCarthy, a co-founder of the field of artificial intelligence, could not be ignored. Eight hundred self-identified computer scientists—who knew that so many even existed?—signed.

Traub and Wulf kicked the matter upstairs at NAS, so it fell to my father to tell McCarthy that the report would not be withdrawn and, moreover, that it was being warmly received on Capitol Hill. McCarthy's record of past crusades didn't help his case. At Stanford, he had petitioned the administration to cut funding to the Stanford libraries because, in the future, everything worth knowing would be distributed by email long before it could be printed—a position that, ironically, seems much less extreme now than it did then.

Not just Congress, but many university administrations took the CSTB report as a roadmap for the future. Today, all good computer science departments have the kind of connections that were envisioned, although computer science's links to "commerce and industry" have come to vastly dominate those to "science and engineering" in the sense intended by the report.

IDA had become another connection for me to computer science in general, and supercomputing in particular. At NSA's instigation, IDA created a new Supercomputer Research Center in Maryland, not far from Fort Meade. SRC was intellectually and intentionally quite distinct from the Princeton crypto-mathematicians.

These first years of the 1990s were a breaking third wave in the information revolution. The first wave, a century earlier, had been the telephone, which changed social life and the way business was conducted; the second, starting at the midcentury, was the computer, cresting with IBM and Apple personal computers in the 1980s. This third wave was about global digital connectivity. It was *happening*, but barely appreciated outside of a few island communities of academics. Long distance optical fibers, which could carry a thousand times more data than copper, became suddenly possible with the invention of in-line laser optical amplifiers. TAT-8, the first trans-Atlantic optical fiber cable, was laid in 1988. Satellite up- and down-links were moving to more capacious, higher frequencies. The first Block II (non-developmental) GPS satellites were just being launched. The specifics of strong encryption algorithms, previously hidden as classified or regulated trade secret, were coming into public view.

Before 1995, when the Internet was made fully commercial, email was mostly limited to universities—available to the public only via expensive dial-up subscription services. When science journalist Gary Taubes (he happened to be the brother of my transitory graduate student Cliff Taubes) asked me for advice on getting connected in New York City, I recommended the service providers Maestro, MindVox, and Panix, now long forgotten. "For services," I wrote to him, "you definitely want ftp, gopher, and wais, along with telnet and (if you can get it) www (world-wide web) and/or mosaic." Mosaic was the earliest point-and-click web browser, developed by an undergraduate student named Marc Andreessen in Larry Smarr's computer center in Urbana-Champaign. I was an early beta tester. I exchanged emails with Marc on technical points, and with Larry to reassure him that it was a worthy project for him to support. After graduating, Andreessen co-founded the company Netscape and became a billionaire venture capitalist.

It was obvious that there would be big winners and big losers. The landline (formerly Bell) telephone companies were slowly rolling out an anemic and expensive International Standard Data Network to homes and offices. ISDN was designed around a fixed rate of 64 thousand bits per second. The companies never imagined that anyone would want a faster speed. They were wrong by a factor of (at least) a thousand. This failure of imagination provided the opening for one-way cable TV to morph into two-way broadband.

From my several vantage points—Harvard, JASON, IDA, CSTB—it was clear to me that policymakers in the U.S. defense establishment didn't understand these and the many related trends. Outside of DARPA and a few other hot spots, the government was a telephone and paper operation—barely out of the first information wave, in other words. Terminals on the desks of CIA analysts were used for word-processing. NSA had the biggest and best computers in the world, and good technical chops, but its leadership saw this new glass as half empty: More intercepted communications were being encrypted—requiring either huge classified computing resources to decrypt, or impossible to read at any cost. Communications channels long intercepted on terrestrial microwave links were lost as they moved to fiber. Packet-switched networks like the Internet were nearly impossible to read, because the individual packets moved asynchronously along variable paths.

In spring, 1992, I pitched to JASON's CIA sponsors the idea that we should do a broad study of the "Global Grid"—the underlying technologies, the likely societal implications, and what the U.S. government, especially the intelligence community, should be doing in preparation and response. "We are in the middle of a revolutionary change," the pitch went, "as regards the sources, quantities, uses, timelines, and technologies of the U.S. government's information collection in the broadest sense." JASON's CIA point-of-contact was Linda Zall, a robust, vibrant, red-haired woman who seemed to be able to convince her superiors of anything. Zall was later the driving force behind MEDEA, a CIA task force of outside academics (self-consciously modeled on JASON) that promoted the declassification of CIA satellite imagery for environmental studies. Linda added to our pitch the selling point that, aided by JASON, CIA management would be seen as getting out in front of the issue—and the other three-letter agencies.

In JASON, co-authorship is viewed more as collective defense than as recognition. The Global Grid report that we delivered at the end of the summer had fifteen co-authors. Actually, there was no report—just a stack of 140 then-newfangled PowerPoint slides. These were projected from acetate transparencies, because computer projectors or large displays didn't then exist. My Global Grid briefing was two and a half hours long. Intentionally, its tone was the opposite of NSA' s pessimism. The world was going to become completely connected. Information would, or could, flow everywhere. U.S. intelligence collection would be able to touch anything, anywhere. They just needed to learn how—and we had a whole list of specific technology suggestions. HUMINT, a CIA franchise, would be more relevant than ever, we pointed out, with facilities to penetrate, communications technicians to suborn, encryption keys to steal. (Intelligence professionals now sometimes refer to this kind of information as coming from "the bit fairy".) NSA, in turn, would move from passive collection to active intrusion—later called "tailored access operations"—collecting bits at their source, "at rest". This should be viewed as a time of opportunity, we said.

The study hit a nerve in government. There was a presidential election campaign underway. It seemed likely that, with Ross Perot taking votes from incumbent George H.W. Bush, Bill Clinton and Al Gore would win. Career civil servants and non-political technocrats were keen to embrace ideas that might become new initiatives in the next administration. Vic Reis, who had become DARPA director, arranged for me to brief Bush's science advisor Allan Bromley, whom I knew well from the Packard committee, and, a week later, Al Gore's intelligence advisor Leon Fuerth. Fuerth wanted me to brief Clinton, but our CIA sponsors scotched that as bureaucratically too risky, because Clinton wasn't yet elected. Instead, I gave the briefing to the mostly Democratic staffers of the House Permanent Committee on Intelligence. Linda Zall quietly briefed Al Gore, with whom she had already, privately, connected on issues of climate policy.

After the new administration was in place, I repeated the talk to presidential science advisor Jack Gibbons and national security council senior director George Tenet. Before I was shown into Gibbons' ornate briefing room in the Old Executive Office Building, a staffer took me aside. Gibbons suffered from narcolepsy, he said, and would likely fall asleep during the briefing. I should pretend not to notice and continue talking. This indeed happened. I did manage to keep Tenet awake, in part by passing around my own personal handheld GPS device—one of the first available to civilians. I had paid $1,000 for it, largely for bragging rights. The people in the room had never seen one before. Tenet later became CIA director.

Around this time, John Deutch had been nominated, but not yet confirmed, as Under Secretary of Defense for Acquisition. A few days after his nomination was announced publicly, he called me at home late in the evening. "I want you as my DARPA director, Bill," he told me. "Don't tell anyone now, because I'm not confirmed yet. But you can count on it." Jeffrey and I couldn't get to sleep that night. We spent the whole night talking about the offer—which I knew I had to accept. There was a lot to be done, and I could help do it. But, where in D.C. would we live? Could cousins Fred and Sudie help us get James into Sidwell Friends School? Would we rent out both our houses? And so on.

I never heard back from John, or anyone else, about the job. After this experience, we were both more skeptical of idle job "offers" that came my way.

In retrospect, JASON's roadmap for the U.S. intelligence community was mostly the one followed over the next twenty years. While this could be coincidence, or inevitability—success has many parents—JASON did educate a lot of policy people and helped create a conventional wisdom that soon became self-sustaining. What we did was package a social and technological phenomenon, turn it into a consistent story, and provide a vision of the future that was optimistic and actionable. In favoring the enlargement of CIA responsibilities in some areas, our report may have helped goose NSA out of complacency. For the next couple of years, I was invited by CIA to attend some mid-level inter-agency meetings; it was obvious that, although there was no change in NSA's customary arrogance, new winds were blowing there, too.

Linda Zall, as energetic as ever, nominated me for a medal, the CIA's Agency Seal Medallion, given approximately annually to outsiders who made "significant contributions" to the agency's mission. The recipient the previous year was Jeannie Rousseau, a French woman who had spied on the Nazi's rocket base at Peenemünde and been captured and tortured by the Gestapo. I felt I had gotten my CIA medal with a lot less personal risk—a bargain.

For the medal presentation ceremony, Jeffrey, Sara (by now college-age), and I were met at the CIA front gate by a limo and driver—the CIA director's. On the two-minute drive to the main headquarters building we mainly stared at the car's secure scrambler phone and its button for direct connection to the White House. The car pulled into the underground garage, from where we took the elevator that went directly up to the director's suite. In CIA lore, this was the elevator used by spies so secret that no one should catch a glimpse of them. The presentation was made by CIA deputy director Admiral Bill Studeman, whom I had met before when he was the director of NSA. Studeman read a little speech that Linda must have written. A photographer took pictures while they presented me with a substantial gold-plated medallion. A couple of years later, when Jeffrey and I were invited to an unrelated event fancy enough that the invitation called for black tie with decorations, I phoned the CIA protocol office to ask if there was a wearable rosette associated with my medal. There was a long pause. "Mostly, our recipients don't like to advertise the fact," the woman said diplomatically. So the answer was no.

JASON's Global Grid study gained me a lot of political capital. The next summer, when the new administration had been in office six months, I convinced the JASON Steering Committee that we should spend some of it by weighing in on the Clipper Chip controversy. The background: Clinton's presidential campaign had been dogged by his history of avoiding—opponents said dodging—the draft; by his admission of using marijuana—his famous "I never inhaled"; and by the Democratic Party's general vulnerability to the accusation of being soft on crime. Once in office, the new White House overcompensated with an uncritical receptiveness to tough initiatives in law enforcement and national security. One such initiative culminated in the 1994 Crime Bill, later seen as validating mass incarceration of minor drug offenders, especially African-Americans. The Clipper Chip, more arcane and technical, grew from the same soil.

The idea that cryptography should be regulated by government was longstanding. Cipher devices like the famous World War II German "Enigma" machines were classified by the United States as munitions and banned from unlicensed export. With computers, the ban was extended to software—even when the cryptographic algorithm was the Data Encryption Standard promoted for general commercial use by the U.S. government itself. Except for the most visible products—Microsoft Windows, for example—this ban was practically unenforceable; and, anyway, products with strong encryption could be freely *imported*. NSA's position on these issues was both secret and generally muddled. The agency had conflicting responsibilities for both "signals intelligence," which was hindered by encryption—one of the half-empty glasses addressed by our Global Grid study—and "communication security" which encryption aided. Historically, the SIGINT people generally trounced the COMSEC side in what were called "equities" debates at NSA, but there were occasional exceptions, such as the release of the Data Encryption Standard in the 1970s.

NSA saw key escrow, so-called, as a way of satisfying both missions. The goal was that encryption products would ultimately be banned *unless* the encryption was done by a government-supplied hardware chip, the Clipper Chip, whose encryption would have a built-in "back door" allowing the government—but only with a valid warrant, it was claimed—to read the messages. NSA enlisted an enthusiastic FBI to sell the idea to the new administration, and the plan was announced in April, 1993, in a breathtakingly disingenuous White House release, most likely originating the office of NSA's general counsel Stewart Baker—someone whom I came to know. Encryption would never be banned, the release said. The use of the chip would be completely voluntary—this despite bills that were already being introduced in Congress that were quite the opposite. Because the back-door key could be released only under warrant, the chip would *protect* the American people from unauthorized wiretapping—but everyone knew that NSA's SIGINT collection was exempt from any requirement for warrants.

Reaction against the Clipper Chip by the academic computer science community was swift and fierce. The Electronic Frontier Foundation, Mitch Kapor's previously obscure civil liberties organization, spearheaded the fight. NSA's long checkered history of straying from its mission and spying on Americans made it an easy target. FBI was not helped by its clumsy attempt to prosecute Phil Zimmerman for "exporting" (by putting on the Internet) his popular Pretty Good Privacy encryption program. Zimmerman had responded by publishing the full PGP source code in a hardcover book. Published materials that are "open and available to the public" in "libraries or other public collections" are not subject to export restrictions. In the original statute, this provision was known informally as the NSF exclusion, because that agency insisted on its inclusion, out of fear that the open publication of a large body of scientific work could be criminalized. (Ruth Greenstein, whom I knew as IDA's long-serving general counsel, wrote the text of the provision as a young attorney at NSF.)

Still, once committed, the White House had to defend the Clipper Chip. This hot potato was delegated to the Office of Science and Technology Policy, and there to Mike Nelson, who had been a junior staffer of Al Gore's. We in JASON had interacted with Mike the previous summer in the Global Grid effort.

Although JASON members individually opposed key escrow on civil libertarian grounds, our report took the simpler tack that, independent of any politics, the proposal was simply unworkable. This was deliciously easy, since the Clipper proposal was half-baked from the start. Each chip was going to cost a prohibitive twenty dollars, and the proposal specifically excluded the possibility of manufacturers incorporating Clipper functionality on their own chips—the underlying algorithm being in fact classified. Computer scientists soon showed that the law enforcement access mechanism could be negated by simple software patches that criminals would no doubt use. There were many other things wrong with the proposal.

That fall, I briefed our JASON "Encryption and Privacy" study around town. George Tenet at NSC in the White House listened noncommittally—his hands were tied. John Deutch, by now Under Secretary of Defense for Acquisition and Technology, loved the report and arranged for me to brief Bill Perry, who was by now Deputy Secretary of Defense, and whom I knew from my DSB days. "Government stifling private sector innovation" was a headline that resonated well with our friends in DoD, which didn't otherwise have a dog in NSA's or FBI's fight. Gary Denman, the new DARPA director, had me brief his whole group of Office directors. At CIA, Admiral Studeman was surprisingly enthusiastic—for a former director of NSA—and wanted me to brief "the board of directors at the Fort," i.e., NSA senior management. He assigned someone to make this happen, and, in October, I briefed NSA director Admiral Mike McConnell.

McConnell was polite, but not very interested: By the end of 1993, my sense was that NSA "seniors" were no longer betting on this lame horse. Stewart Baker left government to return to private practice in mid-1994. FBI was, as usual, left dangling in the wind. FBI assistant director James Kallstrom, whom we knew as one level up from JASON's point of contact at FBI made increasingly desperate statements. In February, 1994, the White House issued a face-saving "approval" of the use of the Clipper Chip by U.S. government agencies—none of whom actually ever did—along with streamlined regulations to "help American companies sell their products overseas," notably products with non-Clipper encryption. In 1996, the Clinton administration transferred control of encryption export restrictions to the Department of Commerce, signaling that encryption was a trade commodity, not (so much) a national security concern.

This was all fun, but, this time, I don't think that JASON in fact made a difference. CSTB did its own anti-Clipper report, as did the Association for Computing Machinery and industry groups. The Clipper Chip proposal would have collapsed without us. However, the crypto wars (as they are known) have continued to the present. As I write, Apple's decision *not* to know or keep a record of the user encryption keys of its iPhone users—intentionally making Apple unable to assist law enforcement—is being litigated in court and in the court of public opinion. FBI says that law enforcement is "going dark," exactly the same language as Kallstrom's thirty years ago.

Kallstrom later became a right-wing commentator on Fox News, but I came away from Clipper with increased respect for Stew Baker and Mike Nelson, with each of whom I interacted on other issues in later years. Each was only doing his best to argue a bad brief. I had less respect for Kallstrom's take-no-prisoners and often deceptive approach, which I would later see as typical FBI tactics.

50. Far Out

By far my greatest scientific achievement was the discovery by Paul Horowitz, Jacqueline Hewett, and me of radio signals from intelligent extraterrestrial beings—an event said by many to be the greatest scientific discovery since Galileo. What, you never heard about it? Let's see, there must be a reason for that.

Throughout the 1980s, Paul, my colleague in physics at Harvard and also in JASON, had been doing SETI (the Search for Extraterrestrial Intelligence) by means of radio observations. Harvard owned an obsolete, but functional, 84-foot radio dish at what was called Agassiz Station, a hill in the small town of Harvard, Massachusetts. Anyone choosing to work on SETI did a utility calculation: There was approximately zero chance of success multiplied by approximately infinity payoff; so you could get any utility answer you wanted. Paul loved SETI because it gave him the excuse to build exotic state-of-the-art electronics, and because the subject was so *cool* that it brought him into contact with other cool people. His effort at Agassiz was privately funded by Carl Sagan and Steven Spielberg, the latter of whose interest stemmed from the success of his 1982 movie, *E.T. the Extra-Terrestrial.* The search concept was to search for radio signals that were highly coherent, like a perfect bell or tuning fork but orders of magnitude better. No natural process, but only an engineered radio transmitter, it was thought, could generate such signals. To filter out human-produced, terrestrial sources, a putative signal had to have exactly the tiny time-varying Doppler shifts that the Earth's rotation would produce for a celestial, but not terrestrial, source.

In the 1980s, my contribution to Paul's effort was to suggest that the message signaling an actual detection, automatically displayed, be changed from "one channel greater than 20 sigma," to "POSSIBLE SIGNAL OF EXTRATERRESTRIAL ORIGIN! NOTIFY OPERATOR IMMEDIATELY!" This was enough to get me a favored position at the 1985 opening ceremony, smiling into the cameras from just behind Spielberg's left shoulder. Then, in 1987, my all-caps message was actually triggered by a series of detections. They all occurred when the telescope was at its extreme pointing limit, close to the zenith, and were probably some kind of glint off the metal support structure. The other sure sign of their terrestrial origin was that their frequency (removing the Doppler) was 1420.00000 megahertz. All those zeros meant that the extraterrestrials would have to be measuring time with exactly our definition of the second, which didn't seem likely. Paul and I discussed driving around Boston trying to find the actual source of the signal, until we realized that the only radio receiver on Earth sensitive enough to see it at all was his 84-foot telescope at Agassiz.

One night in October, 1991, the all-caps message again displayed, this time with several sequential detections at precisely the same location in the sky. There was no coincidence about the length of the second this time. These were exactly the kind of signals that Paul's experiment was designed to detect. On subsequent nights he pointed at the same location—and he never again saw anything. The mantra, "Extraordinary claims require extraordinary evidence," had been popularized by Carl Sagan in his *Cosmos* television series. Paul told me about the signals, but only in strict confidence, because he didn't regard a single, unrepeatable night as the required extraordinary evidence. Ironically, he especially didn't want Carl to find out about the signals, for fear that Carl would abandon his own television mantra and immediately call in Spielberg and the *New York Times*.

I did, however, get Paul's permission to tell Jackie Hewitt. Jackie was a young MIT radio astronomer and Packard Fellow, with whom I had analyzed radio data to extract the time delay between the two images of gravitational lens 0957+561. That work had regrettably turned out to be, pun intended, star-crossed. Jackie had beautiful data, and I contributed sophisticated statistical methods. We got a clear answer, 540±12 days with a 95% confidence level. We had many consistency checks on our correctness. We published two papers that were, at the time, admired. But, over the next several years, it became clear that our result was just wrong. The actual value obtained by others with new data was close to 420 days. Going back, we never found an error in our analysis. I've published many papers that turned out to be irrelevant, or misguided, or have small, correctable errors. These were the only two (as far as I know) that were just incomprehensibly *wrong*.

In these pre-web days, Jackie knew how to really search the literature. In a 1986 catalog of hundreds of radio sources in the Galactic plane by Canadian radio astronomer Phil Gregory, whom she knew to be a careful observer, she found a radio source at close to the position of Paul's signal. Gregory had found only a few sources that changed with time, and this was one of them. He saw it on July 15, 1978, but, despite thirty-six later attempts, never again. The one time he saw it, it was one of the strongest sources in his whole catalog and it "did not exhibit the characteristics of interference." There was no known bright star or galaxy at its location in the sky.

Interestingly, Gregory's receiver was not sensitive to the kind of coherent signal that Horowitz detected. Gregory could see only broad-band sources, those with natural causes—or, one might imagine, those carrying the E.T.'s broadband communications, which might be interspersed with coherent, monochromatic finding beacons for Paul to find. Jackie and I wanted to publish a short note on all this with Paul. We wouldn't claim to have found extraterrestrials, of course, just put out the facts and seeming coincidences so that others could make further observations.

But the data belonged to Paul, and he vetoed the plan. Over the next three years, Jackie found ways to bootleg telescope time at Haystack, MIT's 120-foot radio telescope, to look for the Gregory source. Once, she thought she saw it; but she couldn't publish bootleg data, or even formally apply for telescope time, without revealing the connection to Paul's SETI observations; and Paul still adamantly refused. His own paper, with Carl Sagan as its co-author—because Carl always insisted on authorship whether he did anything or not—appeared in late 1993. Their paper took the tack that, since none of the largest signals in five years had survived *re*-observation on a different day, upper limits could be set "on the prevalence of supercivilizations." Jackie and I never got permission to tell our version of the story. E.T. may still be out there, but by now his (her, their?) research grant may have run out, so their experimental transmissions in our direction may have ceased.

While this was all happening—or not happening—I was more prosaically working on an invited review article on the so-called cosmological constant for *Annual Reviews of Astronomy and Astrophysics*. Such invitations were considered plums, and I had last been invited to do one, on black holes, in the 1970s. I was comfortable with the general relativity part of this new assignment, but I needed help on other angles, so I invited as co-authors Ed Turner at Princeton for his observational expertise, and Sean Carroll, a graduate student of George Field who could write about the exotic connections to particle physics.

The cosmological constant was a hypothetical term that could be added to the Einstein equations and (depending on its sign) enhance or inhibit the predicted expansion of the universe. Einstein needed the cosmological constant for consistency with the static universe that was at the time believed. After Edwin Hubble's discovery that the universe was not static, but expanding, Einstein had famously called the cosmological constant "my greatest blunder."

Or did Einstein actually say that? While there is little doubt that Einstein came to see the cosmological constant as unnecessary in his theory, the "my greatest blunder" quote is known to us only as reported by George Gamow, who, as a popularizer, was not above making things up. Later, John Wheeler claimed that he was also a witness, that he had been walking past Einstein's office door at IAS *at the exact moment* that Einstein made the remark to Gamow. Such a coincidence is hard to fathom, especially by those of us who knew well Wheeler's delight in pulling people's legs. My guess is that Wheeler thought that Gamow made up the quote.

In any case, by the 1990s, the cosmological constant was an additional parameter in the wide-open, contentious debate about the mass density of the universe, and in what form that matter was: atoms, photons, Vera Rubin's dark matter, or something else. Particle physicists viewed the cosmological constant as a kind of matter, "the energy density of vacuum." They couldn't compute its value to within the laughable factor ten to the power one hundred; but they saw no reason that it should be zero. Most astronomers, agreeing with Einstein's later view, thought it ugly and unnecessary, and therefore probably zero.

I don't think that the editors of *Annual Reviews* had any particular dog in this fight; in the debate on cosmological mass density, they wanted reviews covering all the bases. Carroll, Turner, and I wrote: "An epochal astronomical discovery would be to establish by convincing observation that the cosmological constant is non-zero. An important physics discovery, on the other hand, would be to adduce a convincing theoretical model that requires it to be exactly zero." Allan Sandage, a referee, marked up our manuscript with many angry comments that we pretty much ignored, although I sent him a fawning reply. Sandage was the dean of his generation of cosmologists, the intellectual heir of Edwin Hubble. Dennis Overbye's then-recent book, *Lonely Hearts of the Cosmos*, cast Sandage as its anti-hero, lonely, pompous and out of touch. Overbye's heroes in the book were my contemporaries, Dave Schramm and his Fermilab crowd.

In part because of our review, in part because I had written a couple of papers that touched on the value of the Hubble constant, which measured the rate of expansion of the universe, I found myself invited to talk at conferences on cosmological observations. At the weeklong Aspen Winter Physics Conference in January, 1993, I was a part of the opening panel discussion on Monday morning, "Importance of the

Distance Scale," and I gave a talk Friday morning on the 0957+561 gravitational lens time delay, which could give information about the Hubble constant. On the days between, I listened to sessions on other topics. Thursday's session was about Type Ia supernovas, and the speakers included Saul Perlmutter, Craig Wheeler, and Bruno Leibundgut.

Cosmologists loved "standard candles," defined as classes of stars, or events like supernovas, that all shone with the same luminosity, no matter where in the universe they happened to be. If two objects were standard candles, then the ratio of their apparent brightnesses was the inverse ratio of the squares of their distances—the famous inverse square law. Put differently, if by other measurements you knew the distances to some close standard candles, you could *infer* the distances of more distant ones—even ones at cosmological distances far across the universe. That gave a distance scale on which the Hubble expansion could be measured. More subtly, the effects of relativity implied small deviations from the Newtonian inverse square law. Measure those, and you could infer a value (or possibly zero) for the cosmological constant. At the time of the Aspen Conference this was thought to be an impossibly futuristic goal.

Supernovas, the violent explosions of some stars at the end of their lives, could be classified by type on the basis of their optical spectra. They were bright enough to observe at cosmological distances. There had been high hopes that the Type Ia supernovas would prove to be standard candles. However, several of the speakers dispiritedly showed data indicating that they were not standard at all, but varied in brightness—and even in the shape of their light curves—their brightness—over time. I raised my hand and said, "Well then, you should use the observed shape of the light curve to *calibrate* the brightness, and that way get a *standardized* candle." It was pretty obvious from the data they showed that the light curve shapes could be described with one or two parameters. In fact, jumping ahead, this turned out to be almost exactly right.

My recollection is that Saul Perlmutter later asserted that his group independently had the idea of calibrating the light curves, but, if so, he certainly didn't mention it at this meeting. He likely first heard it from me. Bob Kirshner was at the time more interested in a completely different approach, the so-called expanding photosphere method for Type II supernovas. But he must have filed away my remark in his head, because, back home, he assigned graduate student Adam Riess to work with me on developing the idea.

Adam was not the most fun graduate student I ever worked with, but he was tough. Once Adam got his teeth into a problem, he would shake it and not let go until he broke its neck. Kirshner and I became Adam's co-supervisors. Adam and I developed a unified statistical procedure for best estimating the calibrated "templates" for light curves that standardized them. At the same time, with Bob, Adam was part of the loosely organized group of observers at multiple places who were discovering new supernovas. In the period 1993-1994, most of the available data came from a group in Chile at Cerro Tololo observatory. The Chileans (as we called them—they were actually a multi-national group) gave us access to their data in raw form, before they published it, a big favor. By the end of 1993, Adam and I had refined our standardization method. We could apply it to the Chileans' data and get beautiful results. By early 1994, we had written the draft of a paper on the Hubble constant. Soon after that, we had an additional application: The more distant supernovas established a "rest frame" with respect to which we could measure the velocity of the Earth, with a result that agreed with measurements of the cosmic microwave background—the striking confirmation of an expected, but never-before demonstrated, result.

The problem was that we couldn't submit these papers for publication without permission from the Chileans. The essential data belonged to them. They first told us to wait until April. Then it was wait until July, when they would submit their own paper (and data) to *The Astronomical Journal*. We offered to make them co-authors of our papers, but they declined. After their paper was submitted, they changed their minds again and told us to wait until it was actually accepted by the journal. This finally happened in September, and we submitted our papers to *The Astrophysical Journal* the next day. The Chileans still would not be co-authors. They didn't like the style of Adam's and my newfangled, mathematical approach to the data. They were young in age, but they were very traditional astronomers. In what came to be a neck-and-neck competition between two groups, Kirshner's and Perlmutter's, the group in Chile could never have been competitive on their own.

During 1994 and 1995 it became clear that the precision of our method was enough to enable a measurement of the cosmological constant—if only enough data from high-redshift (that is, very distant) supernovas could be had. Adam developed a multicolor generalization of our analysis method specifically for that purpose. I gave a lot of advice, but Adam did all the work—definitely an example of Weisskopf's "second period." Saul Perlmutter's group at Lawrence Berkeley Lab had a head start on discovering high-redshift supernovas. Their well-organized project utilized a custom-designed camera (and matched software) that had been built by him and others in Rich Muller's eclectic research group. Saul was Rich's graduate student and then postdoc, and I had met him many times. From the beginning, Perlmutter's intention was to be first to measure the value of the cosmological constant. He, like most astronomers and physicists, expected it to be zero.

Adam wrote to Saul proposing that we all collaborate, but the vague response could only be interpreted as negative. A couple of months later, Kirshner heard Saul speak at a conference and reported back that, although the Berkeley team's instrument was indeed good at discovering supernovas, their photometric measurements and data analysis seemed to him shaky. These events are significant because they gave Bob conviction that there was time for him to catch up. His "High-Z Team" came together as a group of observers at Harvard (him, Adam, and Brian Schmidt), the European Southern Observatory, Hawaii, and, yes, Cerro Tololo. Their effort depended more on manpower than on custom hardware and software, but it had access to major telescopes and, in due course, many orbits of Hubble Space Telescope time.

I knew that it would take Bob and his collaborators at least a couple of years to amass the necessary data for Adam's and my kind of precision analysis. Adam and I wrote two further papers together, with Bob as last author. Bob liked that order of our names because our last-name initials, R, P, and K, were his own, "Robert P. Kirshner." These papers became part of Adam's Ph.D. thesis. Adam took his final Ph.D. exam in May, 1996, and soon went off to a postdoc position in Berkeley—but in the astronomy department with Alex Filipenko, not in the Perlmutter group.

From this point forward, my involvement in the "epochal Astronomical discovery" of the cosmological constant (our 1992 review's foreshadowing phrase)—today referred to as "the accelerating universe"—was slight. But, having started the story, I'll finish it, jumping forward in time as necessary.

Adam and I kept in touch by email, and I of course saw Bob frequently in our department at Harvard. I was not an official part of the High-Z Team, but I expected that I would be brought back into the fold when there was a sufficient accumulation of data. In September, 1997, I got an email from Adam with a question about how to further generalize our standardization method. It was clear that he was hard at work on a large corpus of data. This seemed like the right time for me to rejoin the team, and I asked Bob about the mechanics of doing so. He responded with a long, lawyerly email on the team's governance and, in particular, all the objections that would be raised by the other team members if he proposed to them that I join. It came down to: Adam was enough of a theorist for them. In other words, I was *out*.

At this time, Adam had not yet made what one might consider the actual discovery. Both competing teams were capable of discovering a nonzero value of the cosmological constant, but it was clear from their informal conversations and conference talks that both expected an answer of zero—that the race was over who would set the smallest upper limit. In such a situation, where a null result was expected, being *first* was less important than being *best*. The field would progress by a succession of better and better upper limits, closer and closer to a value of zero. Already in 1996, the Perlmutter group had touted their effort, and signaled their expectation, by publishing a very preliminary measurement centered approximately on zero—but, as they correctly noted, with still a broad range of possible values.

If the cosmological constant was *not* zero, however, discovery credit would go the other way: The *first* claim capable of passing rigorous peer-review would be the one to go into the history books. In early November, 1997, Adam recorded in his lab notebook an analysis of the then-available data showing that, if one assumed zero cosmological constant, then the implied density of ordinary matter was negative—an impossibility. A few days later, he realized that things must be the other way around: If he assumed a positive mass density, then, with high statistical significance, the cosmological constant *could not* be zero. However, he did not have sufficient confidence in his own analysis even to tell his collaborators about this, until January 9 or 10, 1998, when he was about to go away on his honeymoon. That date became significant in later disputes over priority, because on January 8 the Perlmutter team had orally presented a poster at the annual meeting of the American Astronomical Society in Washington, D.C, that, as reported the next day in the *San Francisco Chronicle*, showed evidence of a nonzero value. There is controversy over who on the Kirshner team knew about this before Adam felt secure in communicating his own results. There were no better angels here: Subsequently, each team darkly hinted that the other's confidence in their results was solely due to purloined knowledge of their competitor's work.

I never took this time-priority dispute seriously. The teams had independent data that, when analyzed correctly, would inevitably demonstrate the astonishing fact of a nonzero cosmological constant, implying that the expansion of the universe was accelerating. At the core of both analyses was the standardizing of supernova light curves. My 1993 remarks in Aspen were likely the original injection of this essential notion into both the Kirshner and the Perlmutter camps; but there were other people in the world with similar ideas. Mark Phillips and the Chileans had a version, for example. In the end, Perlmutter's group used a very simple method. They simply stretched each light curve to match a template, and standardized by the amount of stretch. However—here my modest priority claim—they came to this method only after Adam and I had published our more general analysis showing that the data *allowed* a one-parameter fit. Had nature not been so kind, the Perlmutter method would have failed, while ours, which was markedly more statistically sophisticated than its competitors, would still have succeeded. It is now generally accepted that the single important parameter corresponds to the mass of the exploding star. However, at the time, we thought that other physical parameters might be important: the star's rate of rotation, or the amount of interstellar dust nearby. While these effects are real, they proved to be small.

Kirshner's team submitted their definitive paper to peer review in March, 1998. Adam was the lead author. Perlmutter's group lagged with their formal submission by six months, but both groups were cited as *Science* magazine's "breakthrough of the year" for 1998. Still, each group kept trying to find flaws in the other's analysis, so as to knock them out and claim sole credit. Kirshner was quoted in *The New York Times*: "Hey, what's the strongest force in the universe? It's not gravity, it's jealousy." Several conferences organized debates or panels at which the two sides confronted each other. I was the moderator for more than one of these sessions, because I may have been the only person in the world still on friendly terms with both Kirshner and Perlmutter. Despite the obvious hostility between the teams, I thought that these soberly structured face-offs made for good science, since there were still issues that needed to be exposed and settled. The two teams never seemed to understand that they needed each other. No less than for SETI, extraordinary claims required extraordinary evidence. That two fierce competitors with different data and different analyses had gotten the same answer contributed to the necessary *extraordinary*.

It remained possible that both groups were getting the same *wrong* answer because of a common systematic error. We all knew that data from the future Microwave Anisotropy Probe and Plank satellites could expose such an error. That data was still five years away. In advance of one of their debates that I was moderating, I wrote to Adam and Saul that it was important to achieve well-controlled, believable results so that MAP and Planck would not "blow the whole supernova method into the dustbin of astro-history." In the event, the supernovae and satellite data proved to be not only consistent, but highly complementary. Each depended on the unknown cosmological parameters in a different way, allowing for the resolution of fitting degeneracies in both. David Spergel, my former student, led the MAP data analysis (the satellite by then renamed WMAP after the late Dave Wilkinson). For this work, David in due course shared a Shaw Prize and a Breakthrough Prize (each more than a million dollars); but it was universally recognized that credit for *first* discovery of cosmic acceleration belonged to the supernova groups.

Everyone knew that this discovery of an accelerating universe, if correct, was worthy of a Nobel Prize, but a prize could be divided at most among three individuals. Kirshner, Riess, and Schmidt were clearly the leaders of the High-Z Team. Perlmutter, with his colleague Carl Pennypacker, led the other team. Counting on the fingers of one hand, it was obvious that there could be no prize unless one or the other teams' claim was found faulty or dishonest. This fact prolonged the unseemly animosity for years.

A decade later, in 2007, Kirshner posted on his web site an essay with his version of events. His main point was that scientific priority had for a century been assigned by date of acceptance in a peer-reviewed journal, and that, by that measure, his High-Z Team had been first by many months. This plausible, or at least arguable, point was, however, accompanied by darkly insinuated ad-hominem attacks on Perlmutter and his team. Already, a year before Kirshner's posting, Perlmutter, Riess, and Schmidt had divided a Shaw Prize, which also could be awarded to no more than three. The consensus view at the time was that the Shaw selection committee—who were individually known to be supporters of younger scientists—had punted on the priority issue by honoring the less senior leaders on both teams. Bob's web post called the Shaw award, "a great thing," adding vaguely, "I am very glad to see that cosmic acceleration is being recognized."

I laughed when I read Bob's screed. He was clearly oblivious to the fact that it could be interpreted in two opposite ways. His priority argument might be trying to knock Perlmutter's team out of the running, leaving the Nobel to himself, Riess, and Schmidt. Or, by his endorsement of the Shaw award, he might nobly be stepping back from the discovery—giving the credit to the more junior Adam and Brian, so that a Nobel to those two plus Saul could follow. I asked several colleagues what they thought. Bob was well-liked—in person, he was the funniest astronomer in the world. Because of this, perhaps, everyone I talked to interpreted Bob's intent as unselfish—he was signaling that his students should have priority for the prize. The 2011 Nobel Prize in Physics was indeed awarded one-half to Saul Perlmutter, and one-quarter each to Brian Schmidt and Adam Riess.

A couple of weeks after Bob got back from the award ceremony in Stockholm—he was there as Adam's and Brian's guest—he and his wife Jayne visited Jeffrey and me in Austin where, by then, we lived. Our dinner foursome in a fancy restaurant was a kind of somber celebration. Bob and I had both *not* won a Nobel Prize for the most important discovery in cosmology since Edwin Hubble. Admittedly, he had come hugely closer than I had; but he nevertheless kindly presented me with a gold foil-wrapped chocolate Nobel medal. He had brought back from Sweden a supply of these.

Thanking him for the chocolate medal, and mentioning his web post, I consoled, "It was a generous thing you did, stepping back." By the nature of the occasion, we were all four practically in tears. "What are you *talking* about?" Bob responded. "I never intended *that*."

Pursuing, for a moment, a similar narrative thread, ten years later, in a completely different field, I was positioned to watch at close range another historic example of Nobel-seeking miscalculation. CRISPR, an enormously powerful technology for editing and changing genomes, was invented in 2012 by Jennifer Doudna at UC Berkeley (a friend since her time as a Packard Fellow) and her French collaborator Emmanuelle Charpentier. Soon after, Feng Zhang, at the Harvard/MIT Broad Institute, demonstrated CRISPR's feasibility in human cells. While Zhang's claim to a Nobel was the weakest of the three, it was widely expected that the three would eventually share a Nobel prize. In 2016, Eric Lander (whom I was then close to on PCAST) published an article, "The Heroes of CRISPR," in the prestigious journal *Cell*, his own history of the discovery of CRISPR. Eric was director of the Broad, so it was not too surprising that Zhang emerged as the main hero.

What *was* surprising was that Doudna and Charpentier had been almost completely written out of Lander's history. Well, not entirely surprising: Broad and UC were fighting over CRISPR patent rights. It was a grand miscalculation. In 2020, Doudna and Charpentier alone shared the Nobel Prize in Chemistry. Zhang's omission was widely believed to have been furthered by Lander's shabby article. At time of writing, the patent dispute is ongoing; but, in the meantime, the independent development of many CRISPR variants, not covered by the early filings, has diminished the value of the original patent claims.

Backtracking now to the 1990s, and to my own story: In April, 1994, against all odds, I was elected to the National Academy of Sciences. John Bahcall's and Dave Schramm's politicking had finally succeeded. I was elected in the astronomy and physics sections jointly. Kip surely also helped. Indirectly, I learned that Vera Rubin had argued forcefully for me. I had pleasant memories of climbing pyramids in Mexico with Vera in 1979, but she surely also knew that I had helped block her appointment as a Harvard professor in 1980. She was just a good person.

It would make narrative sense to report that it was my work with Adam on calibrated light curves that pushed me over the threshold for election to NAS. But actually, in 1994, that work was still completely bottled up, waiting for permission from the Chileans for us to publish or even talk about it. No Academy members knew of it. Another theory might be that my election in 1994 was influenced by my father's retirement as NAS president in 1993. His successor, Bruce Alberts, was a controversial choice—in some ways representing an anti-establishment backlash against Frank. Perhaps my election was helped by some number of votes of nostalgia for Frank. Or perhaps members' worries about nepotism had held up my election *until* Frank retired. These are possible theories, but I think it more likely that I was just lucky. My interests were drifting away from astronomy and thus NAS electability. The work with Adam was the exception. Bahcall and Schramm had persisted on my behalf and succeeded in the nick of time. Many years later, at a time that I could not have been elected, this bit of luck would much influence my professional life.

A year later, in 1995, I became the JASON Chair, succeeding Curt Callan. Jeffrey and I joked to each other that, between JASON and NAS, I had achieved my last two professional ambitions. What was unsettling was that this wasn't completely a joke. I was forty-seven.

51. Computers

Our second editions of *Numerical Recipes* in the computer languages Fortran and C, thirty percent larger than their respective first-edition counterparts, were published in 1992. Sensing success, Cambridge University Press did a lot of direct-mail and print advertising. The books sold 40,000 copies in the first year and 20,000 more in each of the next three years. We sent gift checks of $500 each to half a dozen C.U.P. staffers whom we had found supportive—after verifying that the Press allowed them to accept gratuities. I had already paid off the mortgage on my house from *NR*'s first-edition royalties and now paid off the mortgage on Jeffrey's house in Weston. Apart from the book, my main computational interest was still big data (although that specific term was not yet in use). George Rybicki and I analyzed the publicly available optical spectra of high-redshift quasars with new "fast" algorithms. The cleverness was mostly his; the analysis mostly mine. We showed that with more data we might be able to make interesting statements about galaxy formation. But that data didn't yet exist.

I didn't yet know that I would not have access to the High-Z Team's supernova data, but I was also unable to get access to another new big data set that cried out for statistical analysis, the Center for Astrophysics Redshift Survey of 18,000 galaxies being done by my colleagues Margaret Geller and John Huchra, whose offices were steps away from mine. With an earlier survey of a thousand galaxies, Geller and Huchra had made the major discovery of large-scale structure in the positions of galaxies—that the galaxies lay on "sheets" separated by "voids". Margaret Geller's and my relationship, personal and professional, was always fraught; but Huchra was a close friend. Jeffrey and I had introduced him to his wife, MIT economist Rebecca Henderson, and we often socialized with them. John told me that Margaret would oppose my seeing their data, but that I might have a slim chance if I could proffer some new, useful computer algorithm to help their analysis.

Over several months, using published spectra, I developed an automated method for extracting the redshift (that is, the distance) of a galaxy from its noisy spectrum, a task that, by hand, took five or ten minutes per galaxy. My computer program could do 20,000 galaxies overnight, and with greater accuracy than the graduate students who were usually assigned this boring work and could do only a few dozen per day. It was not enough to sway Geller. Her students continued to work by hand, and I never got access to the data. I gave seminars on my method at various places, never published it, but gave my code away freely. Variants of my method found their way into the Caltech Deep Redshift Survey, the Two Degree Field survey (250,000 galaxies), and the Sloan Digital Sky Survey (4 million galaxies). Only the Sloan team ever made an honest man of me: Jill Knapp and Robert Lupton at Princeton sponsored me to become a full member of the SDSS consortium, *with* data rights. Ironically, I never used my SDSS data rights because, by the time the data were available, I had already begun a new career at Los Alamos.

Scientific supercomputing—what my CSTB computer-science colleagues spoke of so disparagingly as "applications"—had a simple storyline in the 1980s: This was the decade of dominance by Cray computers. After he left Control Data Corporation to start his own company, the Cray-1 was Seymour Cray's first design. Expected to sell only a dozen or so, these machines were priced accordingly, in the many millions of dollars. It was no surprise that the first two were bought by NSA and Los Alamos. Then, to everyone's surprise, almost a hundred Cray-1s were sold over the next few years, to other government agencies, NSF supercomputing centers, and industry. The Cray-1 was followed by the Cray-2, then the X-MP and Y-MP designed by Steve Chen. "MP" meant "multi-processor" but only in the sense that they could cram up to four logically separate machines into one circular arrangement of cabinets—about ten feet diameter—plus the required entire room of cooling equipment.

Cray's talent was for designing and integrating very fast circuit board-level subsystems that were made of discrete components—not the integrated-circuit processors that were starting to make personal computers possible. Those PC chips had more transistors on them, but were at the time slow. Cray computers achieved their power both by being fast, and by being heavily vectorized. The latter meant that they could do an operation—a multiplication, say—not just on two numbers, but on two *lists* of numbers, multiplying all the corresponding items on each list simultaneously. This speedup was termed "data parallelism" or, more technically, SIMD—single instruction, multiple data, pronounced "simdee". It was up to the programmer to figure out how to make use of this feature by structuring the desired calculation to use these "vector" lists. That was the art of supercomputer programming at the time.

As single-chip integration got faster and cheaper in the early 1990s, it became possible to build data-parallel SIMD computers less expensively and on larger scales. Cray Research was hidebound in its designs, however, and began to lose business to startups like Thinking Machines Corporation, an MIT spin-off founded by Danny Hillis, whom I knew from CSTB. Jill Mesirov was a mathematician, an early TMC hire whose job was to evangelize for scientific applications. She was recruited from IDA's Princeton center, and she was married to Harvard mathematician Dick Gross, so I knew her well and visited her at TMC.

TMC's "Connection Machine 5" could broadcast an operation to 1024 separate processors, each of which performed it on one item in a vector. The vector was spread out over all the processors; but the programmer could still think of the vector as a unified logical entity. Actually, each of the processors in the CM-5 was capable of doing its own thing, not in lockstep with all the others, something termed MIMD—multiple instruction, multiple data ("mim-dee"). The CM-5's processor chips were in fact each capable of powering an entire Sun Workstation (a high-end personal computer in those days). That notwithstanding, the CM-5 by design *simulated* a SIMD machine: Vectorization was a familiar concept to scientific programmers. Telling them, as with MIMD, "here are 1024 processors running amok—you figure out how to corral them" was considered a commercial non-starter.

When I lectured in the 1993 Trieste summer school, another lecturer was Mike Metcalf, a Brit who had worked for many years at CERN in Geneva. Mike was a world authority on SIMD programming. He and a few others had convinced the international Fortran standards committee to undertake a major overhaul of that language—basically, to turn it into an efficient vehicle for data parallel programming. The result would be called Fortran-90. I was not surprised to learn that my friends at TMC also figured prominently in the design of the new language, since it was perfect for their machines. Sitting in on Mike's lectures, I came to appreciate that SIMD wasn't just a matter of grubby details of computer architectures. It was a whole different approach to algorithms—thinking not just in time (program step following program step), but in "space" as one laid out the algorithm across vectors. "Parallel thinking" was less like writing the lines of a poem and more like designing a garden landscape.

Gyan Bhanot, a postdoc of Steve Adler's at IAS in Princeton, contacted me soon after I got back from Trieste. Gyan knew Mike Metcalf, he knew Jill Mesirov, he had worked at TMC for several years, and—this was the interesting part—he had translated about half of *Numerical Recipes* into a prototype version of the still-unreleased Fortran-90. We *NR* co-authors liked to brag that *Numerical Recipes* had "narrowed the gap between average practice and best practice" for our then two-hundred thousand scientist-readers. It now seemed written in the stars that we should introduce our readers to parallel thinking. We already had an actual (part-time) employee, Seth Finkelstein, an MIT ex-undergraduate who functioned as our master coding and quality-assurance guru. Now, we also hired Gyan as a part-time consultant. Mike Metcalf accepted a nominal honorarium to become our senior advisor.

Over more than a year, we studied Gyan's code line by line, argued with him, wrote our own code, argued with each other, and corresponded with Mike. Fully implemented Fortran-90 compilers didn't yet exist. We connected with compiler development teams at Microsoft, DEC, NAG, and Intel. With nondisclosure agreements to each, we got their unreleased builds, ran them on several different kinds of computers, and sent back many bug reports. They all loved us for this. Without violating our secrecy agreements, we were, in effect, bringing them all up to the same level—to Mike Metcalf's level, in fact. Our typical email to Mike would list a single line of code, followed by a laconic, "DEC allows, NAG rejects, who is right?" Sometimes these were ambiguities in the standard itself, which Mike brought back to the standards committee. We relied on Seth for a lot of detail work. Any task I sent to him before I went to bed usually got done by the time I got up the next morning. I learned more about computing that year (parallel thinking, but also compilers and more) than in any year since my own undergraduate time as an all-night hacker.

Around this time, I happened to make two trips to Australia. I had never been to the Southern Hemisphere before. For a July, 1995, symposium of the International Astronomical Union, I was invited to give the conference summary talk, "Prognosticating the Future of Gravitational Lenses." I liked seeing for the first time the Southern Cross and Alpha Centauri (the nearest star), and Phar Lap, the stuffed racehorse in the Melbourne museum. But I was painfully aware of entering Weisskopf's third stage as a scientist with "invitations to give general talks surveying the whole field through the light of experience."

Then, in January, 1996, I was again in Australia for a series of lectures to a graduate summer school at the Australian National University in Canberra, on parallel computing and Rybicki's and my fast algorithmic methods. Mount Stromlo observatory was only twenty minutes away, and I gave talks there on quasar spectra and automatic redshift determination. Although I avoided Weisskopf's stage three on this second trip down under, my distress this time was that I was so clearly "all dressed up and nowhere to go," that is, all algorithms and no access to important data. Between them, these two trips to Oz underscored my professional frustration.

Numerical Recipes in Fortran-90 was published with much fanfare in 1996. And it flopped. Over five years, it sold only ten thousand copies, a success by academic standards, but far from our dream of bringing parallel thinking to the masses. With a nod to Phar Lap, we had bet on the wrong horse. Danny Hillis's certainty that programmers could not master the complexity of MIMD—lots of processors doing independent unsynchronized operations—was not quite right. MIMD was beyond *most* programmers, but the few who could master it could and did write libraries of reusable code that hid most of the details. These Message Passing Interface libraries were ugly. You didn't really have to "think parallel" in any elegant way. You just had to chop up your problem into a bunch of little "threads" and send them to the MPI.

Data parallelism never completely died. Many aspects live on inside today's GPUs—graphics processing units. SIMD might even someday return in processor chips with very long register lengths. But MPI became the dominant paradigm for the next twenty years. Thinking Machines Corporation went bankrupt. Sometimes the prettiest horse doesn't win the race. *Numerical Recipes in Fortran-90* has some of the best code I ever wrote. C.U.P. keeps it in print, but barely: In 2020, it sold exactly twenty copies. Jill Mesirov moved to MIT's Whitehead Institute, later the Broad Institute. She was later my colleague on the IDA board.

Later in the 1990s, Saul and I started learning C++, which, according to our publishers, was going to be the next big thing. Our continuing connections with compiler vendors got us test copies of their latest, or yet unreleased, products. They were all full of bugs. None implemented the full language standard, which had been around since 1990. The C++ project got put aside, to be restarted only in the mid-2000s.

Our *Numerical Recipes* business did have its less intellectual side. In early 1994, our post office box received an anonymous "whistleblower" letter from an individual who said he was an unemployed former MIT student. Desperate for money, he had signed up as a contract programmer with not one, but three, well-known, large software companies, promising each its own proprietary library of scientific programming functions. The smallest of the three (the only one I am allowed to name even today) was Lotus Development Corporation, the makers of then-famous "Lotus 1, 2, 3". Call the other two companies X and Y. What our whistleblower actually gave to his three unwitting employers was the copyrighted computer code from our book. Although his letter to us was couched as a confession, you could read between the lines to see that what he really wanted was for us to pay him to identify himself and provide the documentary evidence that would allow us to sue the three companies that he had already cheated. A quirk of copyright law is that an infringer is liable for damages, even if the infringement is unintentional or unwitting. In principle, the infringing companies could then in turn sue our whistleblower, but there was no blood to be squeezed from that stone.

I bought a copy of Lotus 1, 2, 3, and played with its scientific functions. For such functions, any particular computed result is not exact. The true mathematical answer might be an infinite decimal, but the computer rounds all of its calculations to, say, fifteen decimal places. Because of roundoff in the intermediate steps of the calculation, the very last of those fifteen digits of the final answer might be wrong in either direction by one or two units. That is referred to as roundoff error, an accepted fact of life. What I found was that, trying by an automated computer program thousands of different examples, *Numerical Recipes* and Lotus 1, 2, 3 always made *exactly* the same roundoff errors. It was better than a fingerprint. It was a smoking pistol.

Lotus was headquartered right in Cambridge. It was a surprise that the Lotus general counsel, to whom I had to explain roundoff errors, didn't just blow me off. It was obvious that we poor academics could never afford the cost of an actual lawsuit. Perhaps Lotus was actually an ethical company! Or perhaps they thought that, without a lawyer of our own, we would settle cheaply by their standards. They paid us $100,000, plus they agreed to provide the identity of the whistleblower—they owed him no loyalty, after all. He was surprised, a few weeks later, when I phoned him and invited him to lunch at the Legal Sea Foods restaurant near MIT. Over a plate of fried scrod, I told him that we had no intention of suing him, which would be pointless, but that it was still an open question whether we would decide to pursue *criminal* charges against him. We hoped that he would cooperate with us as we approached companies X and Y. He readily agreed.

Company X, located on the West Coast, was famous for their scorched-earth approach to business. "Take no prisoners," was practically their motto. I reasoned that if we could get a settlement from them, the remaining Company Y would easily fall in line. But I knew that Company X's legal department would never give me the time of day. I needed fancy, expensive lawyers whom I could never afford. Somehow, my network of friends led me to Greg Moore, a very junior associate at

Ropes & Gray, the fanciest of white-shoe Boston law firms. Before going to law school, Greg had been all-but-dissertation in graduate school in astrophysics. In my free, pre-engagement, half-hour consultation with him, I laid out the evidence—roundoff errors (which he completely understood), the name and confession of the whistleblower (and his address, and his *parents'* address), and the Lotus settlement letter.

Greg set me straight on the hierarchy of law firms. Ropes & Gray never did contingent fee arrangements—that was for ambulance chasers. I would need at least $100,000 up front. It was like a TV game show: Should we take our $100,000 from Lotus and bet it all on suing Company X? I consulted my co-authors. We decided no. Greg was disappointed. Our case was *so good*, and Company X was widely considered *so evil*. Ropes & Gray had a specialty in plaintiff's intellectual property cases, and winning publicly against X would be terrific publicity for them. Greg took our case to the firm's senior partners. Perhaps for the first time since the firm's founding by two Harvard Law graduates in 1865, a contingent fee arrangement was agreed to.

We did still get to make a heart-stopping game-show decision: In infringement cases, a plaintiff can file for a preliminary injunction against the defendant. The standard of proof is fairly low—our settlement letter from Lotus would likely be enough. An injunction would immediately stop all sales of the infringing product. For Company X's flagship product, that amounted to millions of dollars *per day*. The catch was that, if we eventually lost at trial, we would be liable for the full amount of their lost sales. It could easily be hundreds of millions of dollars—maybe even billions if a trial dragged out! What would really happen was that all four of us authors would have to declare bankruptcy and lose our houses and life savings. We could keep our wives and children, dressed in rags, if they would still have us. I polled my co-authors on whether they wanted to go this route. They all said yes. We were all-in.

When Ropes & Gray informed Company X that they were preparing to file for a preliminary injunction, Company X flew their General Counsel and several lesser lawyers by overnight company jet to meet with Greg Moore and me in a Ropes & Gray conference room. The General Counsel, who was a famous figure in high-tech, gave us his take-no-prisoner talk. Could we appreciate that Company X dealt with more than eight hundred frivolous infringement claims every year? "We *never* settle," he said. "That would just open the floodgates to more claims. We allocate the necessary resources to win our cases in court, no matter what the cost." Because they didn't know about our unlikely contingent fee arrangement, X assumed that they could easily outspend us. But we New England puritans were made of sterner stuff than these West Coast sybarites. Harvard, for example, had been one of Ropes & Gray's main clients for more than a century. So, in answer to this tirade, we just sat, saying nothing.

"And when we *do* settle," X's General Counsel said presently, ending the long silence, "we *never* pay more than a million dollars." That was the beautiful music that we wanted to hear. The terms of our settlement with Company X, if any, remain under a secrecy agreement, as also for our settlement with Company Y subsequently. The whistleblower got off scot-free.

Around the same time that we were bottom-feeding off of these reputable software companies—no, I mean, defending our valid intellectual property rights—we first registered our dot-com domain, "NR.COM". This was 1993, and it cost $35. While the Internet dated back to the mid-80s, it had taken a decade for it to be opened up to any and all comers—not just large companies that did business with the government. We were part of a rush of smaller tech companies to register—new entrants into the Global Grid. At the time, there were about 25,000 domains. The supply of two-letter domain names like "NR" was obviously limited, but they were not particularly in demand. Companies tended to register their full names. The earliest dot-com domains (years before we small-fry were allowed in) were for the likes of "APPLE.COM," "IBM.COM," "XEROX.COM," and so forth.

Jump forward now twenty years to 2015, when there were more than 200 million registered domain names. ICANN, the Internet Corporation for Assigned Names and Numbers, by now had deprecated two-letter domain names. Any whose registration lapsed would not be reissued. But, in deference to a thriving world market in trading registered domain names, ICANN still allowed operating two-letter domains like ours to be bought and sold. I regularly got emails offering to buy NR.COM for thousands or tens of thousands of dollars. Most of these emails were ungrammatical. I mentally classed them as a variation of Nigerian email scams.

An email in good English from one Andy Booth that offered $200,000 did get my attention. Booth invited me to check out his bona fides on the web: He was a famous English footballer, playing for Huddersfield and England International, who, after retiring in 2009, had become a high-value Internet domain investor. Who knew that such people existed? The second time that Booth wrote to me, he raised his offer to $450,000. I was motivated to do some research and easily discovered a whole subculture devoted to domain-name trading: investors, speculators, brokers, dealers, escrow companies, journalistic websites, sober or flaming bloggers, and so forth. The name Andrew Rosener was often mentioned as a particularly successful broker. I contacted him. He immediately offered $500,000 and said that he was located in Panama and that, for safety, we should communicate only by phone, not email. Just hours later, I got an email from footballer Booth saying that he heard that I was talking to Rosener, and this meant that he couldn't do business with me. As a parting piece of advice, he wrote, "ask for a lot more." Clearly these were deep waters!

On the phone, Rosener told me that he had relocated from the United States to Panama for business reasons. He had been a seafood broker, buying and selling the cargos on large fish-processing vessels while they were still at sea, and sometimes redirecting them to different ports. When he was stuck once too often with the cargo of a ship whose refrigeration had broken down, he switched from fish to Internet domains. I told him that we wanted more than $500,000. "You're in luck," he said with the total sincerity of fish brokers world-wide. "I just got off the phone with an investor who will offer $675,000. He wants a clean, private transaction, with no publicity."

When you sell a high-value domain name, what you actually transfer is intangible—the password to your account with a domain registrar. The buyer uses it to log in, transfers the domain to his own registrar, and, finally, points the acquired domain to his own web and email servers instead of yours. Every buyer fears transferring money only to receive a bogus password; and every seller fears giving up the password only to receive bogus funds—a forged check or its electronically stolen equivalent. These are transactions, often international, with little documentation, and with no practical recourse to any country's legal system. Caveat emptor and caveat venditor, both.

For these reasons, transactions are done with intermediaries. As broker, Rosener was supposed to reassure both parties that the transaction was legitimate, an assertion backed only by his reputation in the blogosphere. Money and password would actually be sent to an online escrow company, also trusted on the basis of its online reputation. The escrow company would validate the funds and the password before releasing either.

We were supposed to close in forty-eight hours, but the funds didn't arrive. Rosener was vague; the deal was starting to smell, well, fishy. By now I had figured out that the buyer was in China. As in our legal dispute with Company X, I decided that I now needed an expensive lawyer and (after some online research) contacted Wiley Rein LLP, a D.C. law firm with a specialized practice in Internet domain transactions and disputes. Their lawyer laughed. He had done deals with Rosener and vouched for him—up to a point. As for taking our case: "We assist in these transactions for big corporate clients. We'd have to charge you $10,000 for almost no work. You don't need us." I engaged him anyway. Because of the repeat business angle, Rosener was less likely to cheat Wiley Rein LLP than to cheat me. Also, I wanted the lawyers' due diligence to make sure that the transaction was in all ways legal and that tax forms were properly filed. All the money to us four co-authors should flow through the law firm.

A week later, the funds arrived at the escrow company, but not the authorization to release them. We then got a communication from an email address and name that we had never seen before. The sender assured us that he was our same buyer and asked us to instruct the escrow company to release the password to *him*. Our lawyers thought that the buyer was probably just trying to hide the transaction and his identity from Chinese authorities—not against U.S. law—but they advised us to say no anyway, which we did. We thought that the buyer might back out at this point and were surprised when, after another week, the funds were released to us from escrow.

The buyer soon indeed changed the password. We no longer controlled the NR.COM domain. But, strangely, when you typed "NR.COM" into a browser, it continued to point to our website. And so it remained for four years. We moved to a new domain, "numerical.recipes," cutely among mostly cookbook sites. As far as I can determine, the buyer has never made any use of his $675,000 purchase.

What was going on? Best guess is that a rich Chinese family wanted to transfer wealth abroad without leaving any fingerprints. Maybe they were planning to move to Canada or Australia. Once there, and carrying nothing but a memorized password, they could sell the domain for cash—maybe again through Andrew Rosener in Panama. In fact, in 2020, NR.COM was transferred to a new registrar. Perhaps it was quietly sold. It doesn't lead to our website any longer; it now points *nowhere*, an intangible liquid asset that exists only in cyberspace and can cross national boundaries without taxation at will. I read somewhere that Rosener, still in Panama, now also deals cryptocurrency, and invests in legal cannabis startups.

52. Sea Duty

About the same time that I became JASON Chair, I was invited to join the Chief of Naval Operations Executive Panel, the senior outside civilian advisory group to the Navy. Somewhat like the Defense Science Board, the group met several times a year and divided into subgroups to write reports. Where DSB was always dominated by industry executives, CEP (as it was known) took pride in its roster of strategic thinkers. The Navy considered itself the most intellectual of the services. A CNO—the Navy's highest-ranking admiral—might not be an intellectual himself, but he was expected to be able to hold his own with them, much as, I am sure, he was expected to be good at ballroom dancing. CEP members were drawn from both inside and outside government. Andy Marshall, a member, was the archetype. Then in his seventies, Andy was famously director of the Pentagon's Office of Net Assessment, the obscure name of a small, prestigious outpost of strategic thinkers within the DoD.

When I joined CEP, its fifteen or so members included Josh Lederberg, Steve Trachtenberg, Richard Solomon, Charlie Herzfeld, Robert Murray, Walt Morrow, Paul Bracken, Alf Andreassen, and Frank Fernandez. I knew most of its members from JASON interactions, but only a couple of them well. Lederberg was a member of JASON and Nobelist. Trachtenberg was president of George Washington University. Solomon was a former Assistant Secretary of State and president of the United States Institute of Peace.

Charlie Herzfeld was a former DARPA director and Director of Defense Research and Engineering. We had previously bonded over the fact that his uncle Karl Herzfeld was John Wheeler's thesis advisor. Murray was a former Under Secretary of the Navy. Morrow (1928-2017) was director of MIT's Lincoln Labs. Paul Bracken was a professor of political science and business at Yale, and a disciple of Herman Kahn. Andreasen and Fernandez were key figures in some highly compartmented Navy programs that JASON had worked on. Fernandez was later DARPA director.

Within CEP, I was assigned to a task force labeled "Hedging," whose charge was to recommend strategic directions that might be robust in the face of future uncertainties; and later to "Littoral Surveillance". This was indeed a time of uncertainty for the Navy in general and thus the CNO in particular. After the fall of the Berlin Wall and the end of the Cold War, defense budgets plummeted. The Navy fleet shrunk from nearly 600 ships to less than 350. Difficult times called for new thinking, and, in 1994, President Bill Clinton appointed Admiral Jeremy Boorda as the twenty-fifth Chief of Naval Operations. Although Boorda had checked the required boxes at sea for promotion (CO of a destroyer in the 1970s, for example), he had risen in rank mostly in the land-locked Bureau of Naval Personnel. Boorda was the first CNO who was not a graduate of the United States Naval Academy—he was a high-school dropout and the product of an enlisted-to-officer program in the 1960s—and the first who was Jewish.

Admiral Boorda championed unconventional ideas about the future of the Navy. He thought, for example, that towed floating platforms (like oil platforms) might replace aircraft carriers; that ships could be made largely autonomous, with much smaller crews; that the future of the Navy lay in littoral, not blue-water capability. Such thinking was anathema to the Navy's seagoing, fighting admirals, especially those who had been passed over for the CNO position by the draft-dodging President Clinton. The Navy establishment hated Boorda with a barely concealed rage. More than previous CNOs, Boorda turned to our civilian CEP for strategic and technical advice. He spent time with us at most of our meetings.

To civilians, we were known as the CEP, but within the Navy our staffing unit was designated N00K. The N signified Navy. The double zero—zeros, not letter O's—signaled to all that we reported directly to the CNO. Like every other Navy organization that I ever encountered, N00K was lavishly overstaffed. The office was run by full Navy captain (rank O-6) Robby Harris. There were always several commanders (O-5s), typically new in rank and marking time until the availability of their first shipboard command. Ditto for the several lieutenant commanders (O-4s) awaiting assignments as XO on a small ship or department head on a larger one.

Seeing me as inexperienced, and always trying to further their own careers by ingratiating themselves, the junior officers took an interest in explaining to me how the Navy really worked. Importantly, they said, I needed to understand that, although I was a civilian, I had the protocol rank of "commodore." That rank had not officially been used since 1899, but was the equivalent of a one-star admiral. Genuine one-star admirals were titled "rear Admiral, lower half" and automatically promoted to two-star rear admirals after six months. We civilian commodores could never get promoted. Still, I must conduct myself according to my ersatz rank, they told me. After that lesson, I never had to take a taxi from Naval Intelligence headquarters in Maryland (where we sometimes met) back to my hotel in Washington. "You wouldn't happen to be driving downtown after the meeting, would you?" I would ask the most junior officer during a coffee break. "Sir! It's right on my way home from work! May I give you a lift?" The other junior officers smirked.

More than just commandeering rides, my rank empowered me to request "familiarization tours" of Navy facilities and commands. The N00K staff loved arranging these, because it gave them the opportunity to flaunt their organization's "double zero" status. The more junior officers loved *going* on these tours, because they provided opportunities for glad-handing and self-promotion—essential, I learned, to a Naval career. When any CEP member requested a tour anywhere, we were all invited to participate; but there were usually only one or two of us on each. After I did land-based tours of Norfolk Navy Base and several of the commands at Hampton Roads, I raised my sights and requested an actual "drive" on a nuclear attack submarine.

In JASON, we had toured various submarines at the dock and once, in 1992 in San Diego, had a day cruise on the ballistic missile nuclear submarine USS *Alaska* (SSBN 722). Its Ohio Class were the largest submarines ever built in the U.S. Navy. While underway, we toured its Trident missile tube spaces, reactor control, officer cabins (tiny and shared, except for the skipper), crew berthing, and wardroom. We spent the most time in the command center and adjacent navigation center. There, the junior officers enjoyed training us in protocols like: "Officer of the Deck! Sir! Civilian William Press requests permission to take the rudder controls *under supervision*, Sir!" "Permission granted." Smiles all around. Steering a huge underwater submarine with twenty-four intercontinental missiles, each carrying fourteen operational nuclear warheads, was a thrill. The *Alaska* conducted some simulated missile launching drills, very professional but nonetheless sobering in their implications. Surfacing just outside San Diego harbor took more than ten minutes, because it was not career enhancing for the CO to surface right under, or in the path of, a sailboat—these generally invisible to sonar and barely visible by periscope. On the surface, going in, I was given the honor of riding solo at the top of the conning tower outside, waving to the pleasure craft and pretending to be driving the ship.

This time around, with my N00K lieutenant commander in tow, I met up with the Los Angeles Class USS *Alexandria* (SSN 757) at the dock in New London, Connecticut. From the outside it seemed tiny, and inside it was tinier still. Submerged, the *Alexandria* displaced 7,000 tons; the *Alaska* displaced 19,000. The mission of SSBNs like the *Alaska* was to be slow and quiet and invisible for months at a time, ready to launch missiles in the unlikely event of a nuclear war. Attack submarines, by contrast, played cat and mouse games with (formerly) Soviet subs, and they sneaked into unlikely places to perform intelligence missions. The character of the crews was also entirely different. SSBN officers were sober, seasoned professionals, who could deliver earnest lectures on the theory of strategic deterrence. SSBN COs were full Navy captains. The CO of the *Alexandria* was Mike Klein, a new O-5 who had maybe just turned thirty. His XO was an O-4, a couple of years younger. The other dozen officers were lieutenants and lieutenants junior grade, not long out of the Naval Academy and nuclear submarine school. The crew comprised about twenty petty officers and a hundred or so eighteen- and nineteen-year-olds.

We transited the Thames River on the surface, submerged as soon as it was deep enough, and charged at high speed for an hour out into the Atlantic. I was taken around the ship and shown every nook (N00K?) and cranny of the boat, these excursions separated by coffee breaks in the wardroom. The wardroom was perhaps eight by ten feet in size. A full complement of officers could barely fit in around its table. But an interesting effect was that every time we returned there, it seemed larger. In fact, over the course of the day, the whole submarine got larger. The perception of space adjusts—the submariners told me that it was a familiar effect.

These were wild and crazy guys. "What would you like to try?" CO Mike asked me. Diving drills? Find a ship and trail it for a while? Then simulate shooting torpedoes at it? High speed turns—no, we weren't in deep enough water for that. This nuclear submarine was basically their toy. Training drills were an excuse for—no doubt within limits—showing off and having a good time. We got back to the dock late in the afternoon. Protocol required that I buy a souvenir crew hat and coffee mug for $11, the proceeds going to the crew entertainment fund.

It wasn't necessary for me to exercise my synthetic rank to get a familiarization ride on an aircraft carrier. World events brought me the invitation. On November, 21, 1995, the Dayton Peace Accords were agreed to, ending the Bosnian War. In advance of the final signing ceremony in Paris on December 14, the U.S. Navy conducted a major exercise to signal to allies and adversaries alike what U.S. military reengagement might look like if the peace plan failed. Specifically, the exercise simulated a full-scale carrier land attack on a supposed transgressor. A hundred VIPs—journalists, politicians, and various other thought leaders, including the whole CEP—were invited in rotation to observe the exercise firsthand from the aircraft carrier USS *George Washington* (CVN-73), two hundred miles at sea. The GW's carrier battle group consisted of the carrier, a cruiser, several destroyers, a couple of invisible attack submarines—and a full carrier air wing. The exercise's twenty-four hour flight operations simulated continuous attack on a small country—that role played by Fort Bragg in North Carolina.

At Norfolk Naval Air Station a dozen of us boarded the twin-propeller C-2 Greyhound for the forty-five minute flight out to the carrier. During the flight, they briefed us on the tailhook landing. If we didn't catch the hook, the pilot would accelerate at full throttle and attempt to take off before we fell into the ocean. It was exciting.

The GW's CO, Captain Malcolm Branch, met us and escorted us to the large officers' wardroom. This was where we would get briefings and take our meals. Crew on board numbered about 6,000 and were divided into departments (engineering, communications, supply, operations, deck, training, etc.). Nominally, the CO's job was to have the ship in the right place at the right time so as to service the air wing's operational mission of bombing targets. In practice the work of driving the ship was done by junior officers on watch. The CO's substantive job was more like being the mayor of a small city. The ship had the equivalent of (their pamphlet told us) five large restaurants, two department stores, a bank with ATMs, a post office, laundry and dry cleaners, a police department, two barber shops, eight gyms, TV and radio stations, medical and dental clinics, a jail, and on and on.

That first afternoon, we flew from the GW by helicopter to the USS *Mount Whitney* (LCC-20), the command ship for the Atlantic Fleet, one of two such active fleet command ships in the U.S. Navy. In principle, the fleet commander (a four-star) would direct the war from here. After touring the *Mount Whitney*, we flew to the USS *Guam* (LPH 9), an amphibious assault ship that was carrying 1,200 marines—who, in this exercise, were not actually going anywhere. We stayed on the *Guam* overnight.

In the morning, back to the GW, where our "VIP quarters" were undecorated, steel-walled cabins, each with four bunk beds. Dick Solomon was my roommate. Our cabin was located directly under the flight deck and separated from it only by some layers of steel plate. Every ninety seconds or so there was a huge THUMP-BANG as an aircraft landed and caught the arresting cable; or else an equally loud

BANG-WHOOSH as a plane took off on the steam catapult. It was obviously impossible to sleep here, and we had the unsupervised run of the whole ship, excluding the nuclear propulsion spaces, so I resolved to make the most of 24-hour days.

Wandering around this floating city, I got good at grabbing the handle above bulkhead doors and swinging myself through them, Tarzan style; and at scuttling down ladders. I encountered the CO several times in different parts of the ship. "Management by walking around" was his style, and I never saw him on the bridge. The bridge was, however, an interesting place to hang out and watch flight operations, especially in the middle of the night. Anyone could get food, "mid rats," throughout the night in any of the messes.

The Air Wing Commander (CAG, for Carrier Air Group), Navy Captain Ronald McElraft, was co-equal to the CO and did not report to him. Both reported to the commander of the battle group, the two-star admiral aboard. Admiral's spaces in the *GW* (easily identifiable by their wood paneled walls—the rest of the ship was bare painted steel) included the Combat Information Center (CIC), a large operations room from which the admiral (in this case Rear Admiral Henry C. Giffin, III) and his staff actually directed the war. We VIPs could come and go freely.

The big situational awareness display showed our battle group operating in a fictitious enclosed sea, surrounded by equally fictitious land, both superposed on what was actually the open ocean of the Atlantic. The sea opened toward the north, with the enemy (red) country on the west and a friendly (blue) country on the east. It was hard to miss the point that this was exactly the Adriatic Sea between Bosnia/Croatia and Italy, but turned upside down. The locations of our battle group ships, plus other friend, foe, and neutral ships (some real, some notional) were shown. From satellites, radars, sonar, and so forth, our side knew exactly where everybody was. From the bridge all you could see was empty horizon in all directions. Carrier group tactics were to spread out with no more than one ship in a single horizon patch, so as to mask the battle group's presence and operational intent. We, the blue team, knew the big picture but, supposedly, the red-team enemy didn't.

The admiral's staff, ship's officers, and crew all tried hard to make us feel welcome. After all, our presence was, at least indirectly, the whole point of the exercise. Not so the air wing officers. After a single punctilious briefing, we had only frosty interactions. "Pilots are like that," the more friendly ship's officers told us. The air wing lived and ate separately from the rest of the crew on board. On land, by analogy, the carrier would be a base, with a base commander. The air wing would be a tenant command. Typically an air wing didn't join a battle group ("embark") until the latter's naval vessels were on station. The air wing would then fly in, a logistically complex operation that might require mid-air refueling.

I learned by observation that, except for maybe the casual "hi, there" in passing, officers never spoke to enlisteds directly. Rather, officers talked to chiefs (Chief Petty Officers) and chiefs talked to enlisteds. This would play out even when all were standing within a few feet of each other. During our visit, the XO had the notion that, since we were civilians, we should be exempt from this protocol. It was decided that our CEP group would eat a lunch in an enlisted mess. They briefed us very carefully. We should casually walk in, go through the cafeteria line, and each sit at a table with an unsuspecting group of sailors. I did this. "Hi, fellas," I said. "Sir, what are you doing here in the enlisted mess, sir?" "Oh, you don't have to call me 'sir'. I'm a civilian. Just visiting. I'm interested in getting your take on what's good and bad aboard the GW." Boy, were they not going to buy this! I had never seen kids gulp down their food so fast, excuse themselves, and be out of there, leaving me alone at the table!

We witnessed one deviation from the exercise plan. Alarms went off and an announcement was made throughout the ship: The exercise was temporarily suspended. I made my way to the CIC and learned that an incoming plane had been waved off because its landing gear seemed to be not properly extended. It was instead going to fly back to Norfolk to land there, hopefully safely, but with emergency crews at the ready. This necessitated an aerial refueling and the repositioning of other aircraft for possible search and rescue in case the plane ran out of fuel. Everything else stopped for more than an hour until the plane landed safely.

Takeoff in the C-2 for the trip home was as exciting as landing, but differently. The steam catapult doesn't quite shoot you into the sky. It only gets you going fast enough that, when you go off the front of the flight deck, the falling toward the ocean is only temporary, until you gain additional speed. Back in Norfolk, I was awarded my official "One Trap One Cat" plaque (meaning that I had experienced one arrester gear landing and one catapulted takeoff), engraved with my name. Mounted on the plaque are the two halves of a broken catapult tension (or "hold back") bar, whose function was to prevent a plane from moving until the catapult had built up enough steam pressure to strain the bar to its calibrated breaking point.

In JASON, I had from time to time been "read into" a few Navy compartmented "special programs"—those much more highly restricted than Top Secret. In my experience, these programs were amazing without exception—most amazingly good, but some amazingly bad. The bad ones were often programs championed by senior Navy officers who may have lacked technical training but made up for this with commanding demeanor and wishful thinking. I suggested to CEP executive director Captain Harris that a few of us should personally do a quick technical scrub of *all* of the Navy's black programs. My written proposal made clear that we would make no recommendations other than suggesting that, just perhaps, some programs might deserve a further technical evaluation. This was an unusual request to say the least. In part, I actually wanted to be useful; in part, I was testing the power of "zero-zero". Harris was skeptical, but forwarded my request to Admiral Boorda's staff. We were surprised—no, flabbergasted—when it was approved.

Navy special programs were overseen by a two-star admiral, at the time, Richard Riddell. This was a rotating job for newly minted admirals, who rarely much engaged in the technical details. As a captain, Admiral Riddell had been the last CO of the USS *Nautilus*, the world's first nuclear submarine, on its way to being decommissioned and scrapped.

I now know much more about Dick Riddell's naval career than most people. A couple of years ago while looking for something else on bookfinder.com, I happened on his memoir *Through My Periscope*—self-published and apparently intended only for family and friends. Riddell's book is fascinating and artless at the same time. He is the unabashed hero of every happening in his own interesting life, and he can't resist including every joke he had ever told. "Just like me in my memoir!" I thought. I sent a letter of appreciation to his last-known physical address, but I never heard anything back. He would have been eighty.

But in my CEP days, Riddell commanded a suite of offices on the sixth floor of the Pentagon, 5D660. Appropriate to the Navy's cult of secrecy, few even knew that the Pentagon *had* a sixth floor—these were converted attic spaces that could be reached only by one staircase. The real authority in the special programs command was in the hands of a Navy captain, Earl DeWispelaere, who was said to be the longest serving captain never promoted to admiral in the U.S. Navy. Long past mandatory retirement as an O-6, his waiver for continued service was endorsed by each rotating two-star in turn, because he was thought to be the only person who knew about all the secret programs. Captain DeWispelaere also knew how to delay. Despite a letter signed by the

CNO, or at least officially rubber stamped by his staff, my permission to visit his attic office kept being delayed indefinitely.

In May, 1996, Admiral Boorda shot and killed himself on the lawn of the CNO's residence at Washington Navy Yard. He had come under attack for wearing small brass "V" (for valor) devices on the battle ribbons of two of the decorations that he had won during the Vietnam war. The V's were customarily awarded for service in combat. While his citations mentioned combat missions, they did not specifically authorize the V's. His suicide note, addressed to "the sailors of the Navy" called it his honest mistake. After the suicide, retired admiral Bud Zumwalt, who had been CNO at the time of Boorda's citations, said in an interview that, in his judgment, Boorda was indeed entitled to wear the V's. Zumwalt may have been sympathetic to Boorda for several reasons. It was Zumwalt who, as CNO, had eliminated the requirement of a major sea command as prerequisite to the rank of admiral. Also, although Zumwalt was raised as Christian, his mother was Jewish. Zumwalt, like others, may have perceived an anti-Semitic element in the Navy establishment's hatred of Boorda. The long knives in the Navy were out to get Boorda, and Boorda knew it. If it wasn't the V's, it would be something else. The tragedy was that, unknown to anyone, his mental state was so precarious that he could neither fight back nor just stick it out, either of which would have been possible.

Art Cebrowski, then a two-star admiral, was in his way even more radical than Boorda, and had been adopted by him as a protégé. Cebrowski's first talk to the CEP made the hairs stand up on the back of my neck. He was a zealot on a subject I knew well and thought of as *my* Global Grid—even if that wasn't the term he used. It was all there: satellites, reconnaissance, GPS, fiber, bandwidth, wavelength-division multiplex, undersea cables, computing power, the Internet—all now wrapped up into a military-resonant term, "information supremacy". Cebrowski wasn't influenced by our JASON report—he had never heard of JASON. This was convergent evolution. He and I had arrived by very different routes at the same place. I was the technologist. Cebrowski was the naval aviator who had commanded a carrier air wing, an aircraft carrier (the next-to-last CO of the USS *Midway,* decommissioned in 1992 to become a museum in San Diego), and a carrier battle group. In matters of the Global Grid, he was entirely self-educated.

Perhaps because Cebrowski had come up through the officer ranks as a blue-water warrior, his ideas, which came to be known as "network-centric warfare," survived Boorda's unhappy end. Over time, the main idea—that information superiority, not sheer military mass, most determined military success—was embraced first, surprisingly, by the Navy and later the other services. In 2001, SECDEF Donald Rumsfeld appointed Cebrowski (by then a retired three-star) to head a new Office of Force Transformation, reporting directly to SECDEF. Cebrowski died in 2005 at age 63. The Cebrowski Institute for Military Innovation at the Naval Postgraduate School is named in his honor. As much as feasible, the name Boorda has been erased from U.S. Navy history.

Boorda was succeeded as CNO by his Vice CNO, Admiral Jay Johnson. Despite a series of shore-duty assignments, Johnson had risen to command a carrier battle group, so he didn't share Boorda's taint. Admiral Johnson held a formal reception for the CEP at his residence—now moved without explanation from the Navy Yard to an historic house on the hill overlooking the State Department at C Street. Spouses were invited—an innovation—and Jeffrey flew down for the novelty of it. Admiral Johnson was straightforward and soft-spoken, and he appeared to have little interest in the CEP beyond hosting receptions.

Nevertheless, once in motion, the wheels of the Navy grind exceeding fine. Many months later, I was notified of a scheduled meeting with Admiral Riddell himself in his Pentagon attic office, and, after that, with the notorious Captain DeWispelaere. No other CEP members were approved. I made several trips there, up the secret staircase, for briefings by him and his staff. All those secret codewords notwithstanding, the information I got was far from satisfactory. I couldn't figure out whether the staff were hiding the technical details from me, or whether, at this command level, they truly didn't understand their own projects—that they were all drinking the Kool-Aid offered by the corporate contractors doing the actual work. In that case, who *did* actually understand what was going on, enough to be critically in charge? Perhaps only DeWispelaere, or, more likely, no one. It was likewise clear that Admiral Johnson, as new CNO, had no interest in any report from me. After typhoon Boorda, the Navy's priority was to sail in calm seas. Nor could I report to my CEP colleagues, because they weren't cleared for the compartments. My task was quietly dropped from the CEP's list.

Similar to the practice at the DSB, some CEP members stayed in its orbit for decades, while others were asked to serve only for a single term. That decision process was entirely opaque. I didn't think myself any less intellectual or strategic than my CEP colleagues, but it is probably true that I never learned to present myself as the Navy's kind of strategic intellectual. In 1999, by which time I had left Harvard for Los Alamos, I was not renewed. The N00K staff never had the opportunity to evaluate my ballroom dancing, so it couldn't have been that. But I never smoked a pipe, nor did I pause in midsentence to wag my finger at some abstract intellectual concept. So, maybe it was that gravitas thing again.

53. Arrived

With an acronym soup of activities in Washington and elsewhere—JASON, IDA, N00K, NAS, CIA's STAP, NRO's TAG, even the legitimately astronomical SDSS in Princeton—I was flying somewhere for a couple of days almost every week, generally in blue blazer and khaki slacks, then practically the uniform of academic defense consultants. In contrast, my Harvard-based career in astronomy was as if stuck in amber—visible, shiny, decorative, but immobile. I was an entertaining speaker and got invitations to talk at conferences and department colloquia. I gave talks on Bayesian statistics, supernovas and the Hubble constant, cosmic strings, parallel programming, fast algorithms, high-redshift Lyman-alpha spectra. I sometimes took my old talk on wavelets out of mothballs. The reality hidden behind the curtain was that none of these topics were going anywhere as my further research.

Still, this visibility—the feeling of having *arrived*—was addictive. I made it onto the list of people (as a guess, some hundreds, or even thousands) invited annually to nominate for the Nobel Prize in Physics. For any candidate to advance, there had to be a consensus of multiple nominations. So, an effective nomination was one that guessed how others would nominate in that and future years. I nominated Joe Taylor a year before he and Russell Hulse won for the binary pulsar; and I nominated Smoot, Mather, and Wilkinson for COBE, the first cosmic microwave satellite. There, I was both too early and too late: Wilkinson died before the prize was awarded to Smoot and Mather in 2006.

My trip to Sweden had nothing to do with the Nobel. I was appointed a member of the U.S. delegation to a meeting of the International Union of Pure and Applied Physics. This was my first experience with the governance of international bodies. There was very little content to the meetings, but many bureaucratic things that needed to be debated and voted on. It always went like this: On every motion there was a perfectly reasonable U.S. position, which would be introduced into the debate by our delegation head, Hans Frauenfelder (in his thick Swiss accent). Then, delegates from almost every other country would take the floor to denounce the U.S. position. I initially found this alarming and asked Dave Schramm what was going on. "Just wait," he said. Eventually someone called for a vote, and someone else called for a secret ballot. When the votes were counted, the U.S. position always prevailed.

Associated with the meeting was a tour of the Swedish nuclear power plant at Forsmark. In 1986, the West's first indication of the Chernobyl disaster had been when the plant's radiation alarms went off—and only after an operator opened an *outside* door to go to the restroom. I was only barely interested in the plant. I had already toured many nuclear plants, starting with a high-school trip to the San Onofre construction site in 1964. Much more interesting was a tour of the adjacent Swedish national radioactive waste disposal site, comprising a series of parallel tunnels in saturated sediment sixty meters below the bottom of the Baltic.

The gimmick was that, although the Baltic was salty, the saturating water below was fresh—paleo-water from glacial melt at the end of the last ice age. Carbon-14 dating showed that this water had no molecular exchange with the modern, salty Baltic over many millennia. So, the theory went, even if the storage tunnels eventually flooded, their radioactive contents wouldn't migrate anywhere. From land, we walked down a sloping tunnel, soon below sea level, to the site's operational center a kilometer offshore. The tunnel walls dripped water. Our tour guide collected some in a cup. "Who would like a taste?" People recoiled: What? Drink runoff from a radioactive waste disposal site? I was the only volunteer. I had confidence in the Swedes as good engineers. As advertised, the water tasted fresh. They told us that it was twenty thousand years old.

Another lesson in national character was provided by a Dahlem Conference in Berlin. This was a European conference, with only three of us invited from the United States. For meals and social activities, participants tended to group by country. Faced with the prospect of spending time with two boring Americans, I asked Friedel Thielemann, whom I knew well from when he was at CfA, whether I might be adopted by the half-dozen Swiss. They welcomed me, even to the point of speaking English or standard German—not Swiss German—when I was with them. My standard German was good enough to understand mealtime chit-chat, and, if I looked lost, they switched to English. On the last day, the Swiss organized to share rides to the nearby airport. "How long before your flight time would you like to leave the hotel?" they asked me deferentially. I said three hours. It was the right answer—the *Swiss* answer. They had discussed among themselves whether I might be one of those Americans who liked to catch planes at the last minute. That would have been a problem for them.

In December, 1997, I was in Pune, India, for GR15, the 15th International Conference on General Relativity and Gravitation. I had never travelled to India. That was one reason for going. The other was simply nostalgia for the international general relativity community that had once been my tribe. Cosmology was within GR15's scope, so I still had a thin excuse to go. Jeffrey came too.

Pune was a university town of "only" a million people. The Pride Hotel was a modern high-rise near the confluence of the Mula and Mutha rivers. We watched women doing laundry in the muddy river, beating clothes with a stick. We only realized that they were doing hotel guests' laundry after a batch of my white shirts came back gray and gritty—although beautifully ironed. We traveled within the city by gasoline-powered, three-wheeled autorickshaws driven by (it seemed) suicidal daredevils. Other than traffic accidents, however, it was said to be safe to go anywhere, even looking like an American and carrying in cash more than people earned in a year. Conference sessions were at the university, within the pleasant, walled compound of the Inter-University Centre for Astronomy and Astrophysics. IUCAA's director was Jayant Narlikar, who had been Fred Hoyle's longtime collaborator at IOTA in Cambridge.

Associated with GR15, Narlikar had arranged for Roger Penrose to give an evening public lecture in the Balgandharva Municipal Auditorium, which could seat a couple of thousand people. Penrose's popular-science book *The Emperor's New Mind* had just been published in India. We conference participants were taken by bus from the conference banquet to the auditorium. The crowd that showed up for the lecture numbered more than five thousand. They surrounded the auditorium and shouted to be allowed in—physically impossible. The situation was near riot. Police were called. Our bus couldn't get through the crowd. We were led through the chanting crowd as a tight group, to a side door. "Kindly make way for our distinguished international guests," our organizers shouted. The angry crowd parted for us. We had reserved seats in the front three rows. The auditorium was packed, including standing room and all the aisles, probably to twice its safe capacity. Once we were seated, all fire exits were chained shut, to stop people inside from opening doors to the outside crowd. We imagined a headline like, "Distinguished International Guests Die in Fire," with perhaps a subheadline, "Two Thousand Local Residents Perish."

Roger's book (and talk) put forward the absurd thesis that human consciousness was fundamentally quantum mechanical in origin, and that it could not be explained by the equations of chemistry and physics, but required a special intervention of quantum in the collapse of the probabilistic wave function—something (in Roger's telling) both enabling of and uniquely enabled by the conscious mind alone. Other physicists (including, unfortunately, John Wheeler) had entertained similar notions, but Roger took things to new speculative heights—and sold a lot of books. To be fair, he was likely sincere in his belief. More mathematician than physicist, Penrose could harbor such quasi-religious thoughts without cognitive dissonance. Unexpectedly, Penrose's thesis touched a nerve of Indian spirituality. Unwittingly, he had become a celebrated physicist guru mystic. That he spoke in the accent, and with the diffidence, of a British colonial administrator, might even have helped his fame. The vast crowd for his lecture came by train from all over India and were mollified only when the police announced that Penrose would repeat his lecture the next day in a larger venue.

I was appalled by Penrose, profoundly depressed at the hopelessness and misery of India, and, by conference's end, saddened that theoretical general relativity (as distinct from its offshoot, relativistic astrophysics) had progressed so little since the phenomenal 1960s and 1970s. Email into India was sporadic. On the last day of the conference, as Jeffrey and I were packing for thirty-six hours of incommunicado travel home, word came that Dave Schramm had been killed. He was piloting his own plane, en route from Chicago to his vacation house in Aspen, where Judy and the family awaited him for Christmas. There was bad weather. He should have waited a day. The plane went into a stall and crashed. He was 52. That was all we knew, and all we could think about, on our trip home. At JFK in New York when we arrived, there was a different blizzard. All flights to Boston were cancelled. We rented a car, the last one on the lot, swearing falsely that we would return it to the same location the next day, and, conscious of the irony, drove six hours home unsafely through a blinding storm.

This should have had nothing to do with my deciding to leave Harvard. A travel adventure to India had turned into a disappointment for a spoiled, first-world brat—me—coincidentally the same week as a friend died in an accident. But, now looking back, it did have an effect. Mortality, misery, stalled areas of physics, all seemed blended into a single gestalt. Schramm was one of my most valuable mentors, he more than the others egging me on to be ambitious—like him, in other words. Now, he didn't have to grow old; but I still did.

If I left Harvard, the obvious question was whether I should stay in astrophysics, but refreshing my career in a more fertile setting. Princeton was a place that I still visited for several weeks a year. A decade earlier I had deflected the offer of an IAS professorship, but Phil Griffiths was director now. He and I were friends on the Packard Fellowships committee. My mentor John Bahcall and friend Freeman Dyson could come up with an offer there, I was sure. But I was also sure that I didn't want to go back to Princeton. I would never get over how unhappy I had been there the first time around.

What about government service? In 1996, when he was appointed DARPA director, Larry Lynn had asked if I was interested in becoming his deputy. That wasn't attractive enough and I said no—of course politely. A few months later, after Bill Clinton's re-election, Josh Lederberg wrote to me asking for suggestions to replace Jack Gibbons as science advisor—not excluding myself. It was unclear whether Josh was actually a part of the transition team, or was just freelancing. I wrote back that I didn't think that I could compete successfully for that job, but that I did want to do a stint in government someday, for example as DARPA director, or Deputy Director for Science and Technology at CIA. By this time, I had known a succession of incumbents in both positions, and I certainly didn't feel outclassed by them. "Someday" was not exactly throwing my hat into the ring.

But, by the beginning of 1998, I was looking for a new job more actively. Sid Drell forwarded to me the job posting for deputy to Los Alamos National Lab director John Browne, who had just succeeded long-serving director Sig Hecker under unusual circumstances. In a year that the Lab's budget was cut, Sig had ordered a RIF, "reduction in force," of two hundred people (out of about 12,000). A process was put in place to figure out who should be fired, based on people's documented performance. But that process had gone off the rails in neglecting an important boundary condition. Almost all the people selected were Hispanic. Almost none were Anglo (as the term was used in New Mexico). This didn't go over well with New Mexico's political and community leaders. Sig was soon on his way out. John Browne, an experimental physicist, was a man known for his personal warmth. As new director, his job was to regain the confidence of the workforce, and to repair the damage to community relations.

I checked with people I knew, then wrote back to Drell that, yes, I thought that I could be a terrific LANL deputy director, but that the acting deputy director, Pete Miller, was an old buddy of Browne's and (my sources said) sure to be selected for the permanent job. I didn't want to be the outside candidate who very publicly didn't get the offer. That would really make me look like deadwood at Harvard. Still, I was now ready to be more aggressive in searching. I heard that Ruth David might be stepping down as DDS&T at CIA, and now asked John Deutch to nominate me for the position. This was itself iffy, because John had resigned as CIA director just a year earlier, under the cloud of an investigation into whether he had ignored the rules for handling classified documents—and we who knew him had little doubt that he had. But he was my best connection.

Deutch recommended me to CIA director George Tenet and deputy director John Gordon, but soon wrote back that, according to his source, Ruth was not leaving her job. Deutch added that he was also nominating me as Assistant Secretary of the Navy for Research, Development, and Acquisition. In due course, I traveled to Washington for interviews with Under Secretary of the Navy Jerry Hultin (whom I already knew from the CNO Executive Panel), and then with Secretary of the Navy John Dalton. Thirty years earlier, while I had been avoiding the Vietnam draft, Dalton and Hultin had both been serving naval officers. I didn't get the job.

Stepping down as DARPA director, Larry Lynn recommended me as a possible successor, but Frank Fernandez quickly got that job. We in JASON knew and liked Frank. But, after failing to get these job offers, I was starting to reconsider Harvard in a new light. Some are born deadwood, some work to achieve deadwood status, but I might have to have it thrust upon me.

In that vein, Jeffrey and I applied to become Co-Masters of North House, which had recently been renamed as Pforzheimer House. In our favor was strong support from outgoing masters Woody and Hannah Hastings, they who had first welcomed me to Faculty Row twenty years earlier and were only now stepping down. Harvard house master was a job that one could do for a long time. This fit my deadwood plan perfectly. Woody and Hannah's recommendation got us invited to interview with the advisory committee of undergraduates. But against us was the fact that the final choice would be made by undergraduate dean Harry Lewis. I didn't know for a fact that Harry disliked me; but it seemed unlikely that, in twenty years, my low opinion of him hadn't somehow gotten back to him.

And then, our interview with the undergraduate committee did not go well. Only four of twelve students bothered to show up—an indication that we were already dead meat—and they found us, as we later learned from Jeffrey's inside sources in the dean's office, cold and unsympathetic. Harry sent a sincere handwritten note thanking us for our interest and telling us that we didn't get the job.

54. Uprooting

The Los Alamos National Laboratory, LANL, had at this time about 12,000 employees in divisions of average size 500. Divisions comprised between five and ten groups, with each group leader responsible for 50-100 employees, a scale familiar to me as that of an academic department—I had been chair of one of these. The whole of LANL's management was something like 10 senior executives, 25 division directors, and 200 group leaders. The Lab's budget was about $2.2 billion per year.

In January, 1998, in my response to Sid Drell, I had rejected the idea of becoming deputy to the new director, John Browne. As it turned out, that future was not so easily dismissed. An April email from Hans Frauenfelder told me that the position was still open and that he, Hans, was on the search committee. A couple of weeks later, at the NAS annual meeting in Washington, Hans told me that Browne was inclined to appoint an outsider—this quite the opposite of what my sources had told me—and he was likely to choose from the search committee's top couple of candidates. Back home, I sent him my CV. A week or so later, Browne called to assure me that Pete Miller didn't want the job at all—he was serving as acting only as a favor to John. The position was wide open.

Later, after we did move to Los Alamos, I learned the more complete story: Browne and Miller were longtime collaborators and buddies. Browne was an experimentalist, an accelerator physicist who had been at Livermore for a decade before moving to Los Alamos, where he rose through the management ranks. Under Sig Hecker, John held a series of associate director posts as a kind of management utility player. In the latter years of his reign, Sig had infamously "flattened" LANL's organizational structure, so that he sometimes had as many as twenty associate directors and division directors reporting to him, changing frequently in number, person, and title. Pete Miller was a West Point graduate, one of the tiny number of such African-Americans before the Vietnam War. He had been a Ranger in Southeast Asia, left the army, earned a Ph.D. in nuclear engineering, and came to Los Alamos. Like John, Pete also rotated in and out of transient associate director positions in Sig's "flatland" (it was derogatorily called). Part time, Pete was also a professor of nuclear engineering at Berkeley. So, compared to the many inbred Los Alamos lifers, John and Pete were, in different ways, both relatively cosmopolitan.

A year before, when Sig had resigned in the wake of the RIF, Browne and Miller had for fun compiled a private list—a kind of betting line—of possible candidates for the director job. There were some obvious external candidates. Roy Schwitters and Steve Koonin, both JASONs, were given favorable odds. Roy visibly wanted the Los Alamos job. He made several trips to Los Alamos, interviewed favorably, and seemed to be the top candidate. For a while, Steve also wanted the directorship. He had been a postdoc at Los Alamos. Now provost at Caltech, he had been appointed by president Tom Everhart; but Everhart was succeeded by David Baltimore, who was difficult to work for—and Steve now wanted to be top dog somewhere. His candidacy ended, when Laurie, his wife, mentioned that if he moved back to Los Alamos, it would be without her.

There were also internal candidates for Lab Director, but all were flawed—longshots at best. Pete and John agreed that the Lab could use some new blood, but—here showing that they did have provincial sides—they thought that bringing in an outsider as lab director would be going too far, akin (as Pete later told me) to the Vatican announcing that the next Pope would be a Muslim. Over beers, John tried to convince Pete to become a candidate. He was respected, capable, a Berkeley full professor—and his being African-American would send the right signals nationally. Pete adamantly refused. What came out of that drinking session was that it was John who became a candidate—and he got the job.

Unlike Sig's flatland, John's org chart had three deputy directors: a principal deputy (that became me) and two others for operations and business administration, respectively. Three associate directors would be line managers respectively for nuclear weapons, threat reduction (meaning other classified work), and supporting research (meaning unclassified science). To demonstrate his commitment to winds of change, Browne determined to hire his principal deputy and two of the associate directors from outside.

Traveling to Los Alamos, I met with the search committee in late May. The badge office gave me a green (i.e., cleared) badge: I was somehow in their system as having a green badge at Livermore—not for me to tell them that it had expired five years earlier. As a matter of process urged by the committee's HR minder, the committee posed an identical list of questions to every interviewee. All the fixed questions had "right answers," and none really probed the candidate: How would you characterize your leadership style? The right answer was: Listen well, communicate often, delegate much. What would you like to accomplish in your first year? The right answer was the same: Listen well, communicate often, delegate much.

In what little time was allotted to freeform discussion, I came up with some other right answers by accident. I didn't then appreciate the extent to which Los Alamos scientists hated all management above their own group leaders (the first level of managers). Even group leaders came under suspicion if they evidenced any management ability or ambition. I impressed the search committee as knowing nothing about management, and this was exactly what they wanted to hear. At heart, I would be one of them, a scientist—they thought.

More of an inquisition was a separate meeting with Steve Younger, whom Browne had appointed as the associate director for nuclear weapons. Nuclear weapons were the Lab's principal mission and accounted for about half of its funding. Younger came from X-Division (nuclear weapons design). Browne came from P-Division (physics) and was not at heart a weapons guy. So, while Browne's org chart slotted Younger as "only" an associate director, Steve would in practice have autonomy over fully half of the Lab's budget. Meeting with me, he needed to be sure that I would not be an impediment to the real business of the Lab—his business—nuclear weapons.

Younger was a thin, intellectual, nuclear warrior, slightly paranoid and always deadly serious. At our meeting was also Jas Mercer-Smith, his deputy and a rising star in nuclear weapons design. I had known Jas some years before as Al Cameron's postdoc in nuclear astrophysics at CfA. Jas spanned the two cultures, academia and weapons, so he could function as a translator. My nuclear weapons apprenticeship with Edward Teller, Lowell Wood, and now-former Livermore director John Nuckolls added to my credibility. When Steve started telling me about his Jesuitical education, and how his weapons work was influenced by his close reading of St. Augustine's *Confessions*, I guessed that I had passed his test.

When I visited the Lab a second time in mid-July, John Browne invited Jeffrey to come. She went on a tour of the town while I had meetings with Lab senior leadership (then still mostly acting, while Browne filled positions), prominent Lab scientists, and a representative group of division directors—LANL's middle management. Everyone seemed to think that I would be offered the job.

The most interesting meeting was with Pete Miller, whose job I would be filling. "Do you have any idea what you are getting into?" he asked. I shouldn't expect any privacy or anonymity. When I went to the supermarket, I should expect to be buttonholed by Lab scientists who wanted to give me a piece of their mind. Everyone would recognize my car; if I was ever speeding in town, they would call the police—just for the fun of cutting the deputy director down to size. He asked me what Jeffrey thought about moving here. So-so, I said. "Someday," he said, "you'll be ready to leave this place, and you'll find that she has put down a tap root twelve-hundred feet into the mesa and won't be willing to leave." I smiled politely. (But, as we will see, ten years later, his prediction turned out exactly true.) John and Marti Browne took us to dinner at Gabriel's, a touristy restaurant on the highway to Santa Fe. John didn't quite offer me the job, but made it clear that he was going to. There were a few loose ends he needed to tie up.

The Lab had arranged for us to spend the next morning with a local realtor, Ardie Hafer, to get an idea of the housing situation. We saw several unsatisfactory houses. There were beautiful houses in Santa Fe, forty-five minutes away, but we wanted to live "on the Hill," close to the Lab. When Ardie showed us a newly listed, attractive house on a unique lot, a pine-forested, three-acre property, we decided on the spot to buy it. How much would we lose if we put down a deposit and later had to back out of the deal—for example if I didn't get the job offer, we asked. Ardie told us that she was a past president of the New Mexico Realtors Association. It would be unethical of her to let us make *any* offer without waiting a day.

But hey, she said, you might be the new deputy lab director. The normal rules don't apply. She told us to write a check for $1,500. In Massachusetts, the deposit for a binding offer would have been in the tens of thousands of dollars. Are you sure of that amount? we asked. "Oh, if you don't move here, I can probably get you your $1,500 back," she said. It was a small town and everybody owed her. We handed her the check.

A day later, I got a phone call from a very annoyed John Browne. "I guess you don't understand small towns, Bill," he said. He was right of course. Within hours of our making the offer on the house, the whole town knew that I was coming to LANL as deputy director. It was a job offer that John intended to make, but it was embarrassing to him because there were other candidates, including internal candidates, that he had been intending to talk to before any announcement was made. My explanation, that we had put down the offer on spec, and that we were at risk at most $1,500, and possibly nothing, barely mollified him. There was a great amount that we would have to learn about being highly visible in a small town. But we had a house.

Back home in Cambridge, we mapped out whom we must tell and in what order. I needed to do a delicate dance with my dean, Jeremy Knowles. Harvard long had a policy of granting two-year leaves (without pay) for government service. In case Los Alamos turned out to be a mistake, I needed this two-year safety net. I would come back with my tail between my legs and revive my plan of actively becoming deadwood. But Los Alamos was technically not government service: LANL was managed for the government by the University of California. How would the dean feel about a nuclear weapons laboratory? What if he said no? But this thinking was all unnecessary; my leave request was granted without a murmur. Knowles was probably happy to get rid of me.

When Los Alamos put out a press release about my appointment, and word spread among my colleagues, I got many emails with pro-forma congratulations. Most expressed surprise, but there were a couple of exceptions: John Bahcall was surprised only that I had kept him in the dark successfully. Martin Rees wrote that he was not surprised that I had decided on a move, a judgment about my future in astrophysics, perhaps; but also perceptive of how blocked I was at Harvard. As soon as Irwin Shapiro heard about my move—not waiting for me to actually leave—he cut off salary support for my half of George Field's and my secretary, putting her part-time hours below the threshold for her to earn benefits. The message couldn't have been clearer.

During the fall, I made half a dozen trips to Los Alamos, a few days each. Pete Miller stayed as acting deputy until January 1, but John Browne was building out his so-named Senior Executive Team (soon known as SET), and there were meetings that I needed to be at. In Cambridge, we were busy with the tasks of uprooting ourselves after twenty years—no easier for Jeffrey, who had lived in Cambridge long before I knew her.

The moving van picked up our Cambridge goods on December 9, the same day that my Mazda car ad ran in the *Boston Globe* classifieds. At this time, before national companies like CarMax or AutoNation existed, local dealers offered only about half as much as you could get in a private sale. I got my first and only call on the second day of my ad. The would-be buyer sounded fishy. He didn't want to see the car. Instead, he wanted me to meet him, with the car and its title papers, at an address in a run-down part of Boston where, he said, he would pay cash. I refused.

As an alternative, I agreed to meet him at his local bank branch. He withdrew the cash, handed it to me, and watched me give it back to the bank in exchange for a cashier's check. Now less afraid of being mugged, I let him drive me to a nearby subway station. He seemed like a perfectly ordinary guy, or maybe a drug dealer.

I flew to Albuquerque on a one-way ticket two days afterward. Jeffrey and James joined me a week later. Harvard was now in my past.

55. Harmony

Tourism, federal spending, and oil and gas were New Mexico's three major industries. Including the military facilities and the two national labs (LANL in Los Alamos and Sandia in Albuquerque), an eighth of the state's population was on the federal payroll. The carefully staged "Land of Enchantment" seen by tourists portrays New Mexico as a happy place where Hispanics (about half the population), Anglos (about 40%) and Native Americans (about 10%) live in harmony.

My education to the contrary took place in stages. The Lab's scientists and engineers ("technical staff members" or TSMs) were virtually all Anglo and lived either "on the Hill" or in Santa Fe, a forty-five-minute commute. The Lab's support staff (technicians, crafts, administrative) were predominantly Hispanic and lived in Española or one of the other nearby Hispanic towns in the Rio Grande valley, typically a half-hour commute.

An early lesson in ethnic politics was hiring my senior "admin". The Lab had a rigid ladder for support staff. John Browne's admin, Cecelia ("Cec") was the only rank OS-8. Admins of deputy directors were supposed to be OS-7; assistant directors, OS-6; division directors, OS-5, and so forth. Theoretically, managers at these levels hired their own admins, advertising positions and interviewing in the normal way. In practice, there was an admin deep state, controlled at the top by Cec. It was not coincidence that all the OS sevens and sixes, and almost all the fives, were Hispanic. While I was commuting part-time that fall, Cec had assigned someone to my work, her protégée, a sweet young OS-6. Cec's plan was for me to hire this woman, automatically getting her promoted to OS-7.

John Brown had designated an experienced Lab physicist, George Kwei, to help in my transition to Los Alamos. George and his wife were both Californians with unclouded views of New Mexico ethnic politics. Cec's protégée was too gentle for the job, Kwei insisted. I needed someone more assertive—and also loyal to me, not to Cec. Kwei insisted that the OS-5 admin in Physics, Trish, be added to the interview list, which consisted of Cec's protégée and two non-starters also chosen by Cec. Trish, an Anglo, was a hard scrabbler. She (like the other candidates) had only a high-school education. But she was smart and an autodidact. She knew how to butcher large animals—literally: She was on the county's list of volunteers to call when deer or elk became roadkill. After Kwei and I selected her, it took a month of fighting to get her promoted to OS-7 over Cec's objections. Trish worked for me for five years, during which she was shunned by all the other admins. They ate lunch in a group; she, alone. They spoke to her in cold, formal English; to each other, warmly, in Spanish. (In Northern New Mexico, the Hispanic population is bilingual from birth. English is the language of education and business, Spanish the language of family and social life.) "Does it bother you?" I asked her. "I ignore it. I'm better than that," she told me. Welcome to New Mexico, the Land of Enchantment.

Not just in the Lab were there hierarchies that I had to learn, and pitfalls to be avoided. We had a hard time finding someone to clean our house. Local cleaners would work only for cash, with no paperwork or tax filings. Because of my visibility, I couldn't do this. We were lucky to find Helen and Phil Archuleta, a couple in their sixties, recently retired after twenty years in support positions at LANL. Their housecleaning business—including Phil's by-the-book tax accounting—was as much social as economic. Phil cleaned the house while Helen chewed the fat with Jeffrey. It was Helen who set us straight: "We are the descendants of the *conquistadores*," she said. She didn't mean all New Mexico Hispanics, but only the northern New Mexico families—Baca, Chávez, Sánchez, Maestas, Martinez, Villarreal, Archuleta, Lucero, Salazar—whose ancestors settled the valley in the 1600s. The *patrones* who comprised the power structure—local politicians, land and business owners—these were all descendants.

Representing the Lab, Jeffrey and I one time attended a banquet in honor of the visiting Spanish Crown Prince (later King) Felipe. Guests filled the large Albuquerque hotel ballroom, and there were endless speeches—all in English, curiously—by *patrones* of all stripes. Felipe's speech, also in English, was even more curious: "Your mother country has not forgotten you," he declared. "Someday Spain and New Mexico will be reunited as one." Even allowing for some hyperbole, this irredentist royal declaration could not possibly have been approved by the Spanish Ministry of Foreign Affairs! I think that Felipe just got carried away after meeting so many self-professed descendants of *conquistadores*.

Farther down the social ladder were Hispanics whose families had arrived in the recent two centuries. These people would not be invited by the archbishop of Santa Fe to participate in his annual guided tour of Spain, as Phil and Helen had been (all profits going to the Church). The archbishop knew them by name, they told us. It didn't seem to matter that the archbishop, Michael J. Sheehan, was actually Irish-American. Far, far down the social ladder—not on the ladder at all—were recent Mexican immigrants, legal and otherwise. These were tolerated in northern New Mexico only as day laborers working for local Hispanic employers. In later years, we read about incidents where the rented houses of legal Mexican immigrants attempting to settle in northern New Mexico were burned to the ground—the arsonist's message likely directed more at their landlords than at the tenants.

Openings for scientists and engineers at the Lab were advertised nationally and hired from the national candidate pool. Given our location and local workforce, a special effort was made to recruit native Spanish-speakers. These usually ended up being from Puerto Rico; or individuals from Mexico or Chile who had gotten Ph.D.s in the United States. When I asked Helen where these people fit into the Hispanic hierarchy, she only shrugged. They weren't even on the radar. Nor was there a single unity of Anglos. At the state level, the Anglo power structure was dominated by the farming and ranching families who had settled in the West before New Mexico's 1912 statehood; royalty were the Udalls and Kings. In the decades after World War II, the Anglo population quadrupled with new arrivals from "back East". We were the parvenus. Los Alamos County, the smallest (though not least populous) in New Mexico, carved out especially for the Manhattan Project, was all parvenus by definition.

Soon after I arrived, at the instigation of the Lab's government relations office (which reported to me), our state representative Jeanette Wallace introduced in the legislature a resolution welcoming me to New Mexico, as new LANL deputy director. Such resolutions were introduced all the time and passed by unanimous consent. On the happy day, our government relations people drove me to the Roundhouse, the state capitol. There happened to be another honorific resolution on the docket, one congratulating the archbishop—the man who greeted Phil and Helen by name—for some recent success. Jeanette (a tiny woman) escorted me to the platform; another representative escorted the archbishop. Our respective resolutions were adopted simultaneously to scattered applause. I turned to the prelate and extended my arm to shake hands. He stared at me, frowned, turned, and walked off the platform. It was another New Mexico lesson. He and I had non-overlapping constituencies. There was no profit for him in even the smallest friendly gesture.

But I was not quite as foreign to New Mexico, it turned out, as the archbishop thought. Several times in my first few months, Trish came in to tell me that groups of administrators from elsewhere in the Lab wanted to meet me—and they didn't want to give a reason. We soon learned the pattern. A delegation of (typically) young women would file into my office. We exchanged hellos, and they told me where in the Lab they worked. They stared at me. Then, the bravest would say something like, "Are you really Jewish?" I wasn't by any means the only Jew at LANL. The town had a modern Jewish Center with an ordained part-time rabbi, Jack Schlachter, who was also a scientist in Physics Division. But, to the Hispanic workforce, I was the *visible* Jew.

My visitors were always quick to explain that they were not anti-Semitic. Far from it. Among the *conquistadores*, there were *conversos*, Jews who were forcibly converted to Catholicism during the inquisitions of the 1500s, but were expelled from Spain nevertheless; and *marranos* (literally, pigs), Jews who had pretended to convert for just long enough to seek safety in the New World, where they secretly practiced the old religion. This 16th century history was remembered in the families of 20th century Española, New Mexico, USA. In every group there was at least one woman who told me seriously that her grandmother had always lit candles on Friday night and would never explain why.

I nodded avuncularly. I never expressed skepticism. Sometime later, I mentioned these *converso* visits to an astrophysics postdoc from Spain in T-Division (theory). With him, I must have let my skepticism show, because he responded by telling me with great certainty that my name, Press, probably came from Pérez, a common converso name. Spanish Jews whose Hebrew names were Peretz and Perutz often became Pérez. So, I too was a *converso*. I didn't tell Helen; and Jeffrey didn't start lighting candles on Friday nights. I did start giving my name as Pérez at Mexican restaurants outside of New Mexico, when the line was long.

I found out much later that the New Mexico converso (also known as crypto-Jew) stories had been long debunked. Many were traced to New Mexico state historian Stanley Hordes' popular accounts—and, in some cases, seemingly, fabrications. Hordes lectured widely in New Mexico and firmly established the converso story as part of Santa Fe's tourist-serving narrative. It seems likely that the sincere visitors to my office had been exposed to the convincing Hordes tales. New Mexico is a small place. We knew Hordes socially as a perfectly nice man.

Elmer Torres, in the community relations office, was in charge of the Lab's tribal relations. Northern New Mexico had its Eight Northern Pueblos, southern New Mexico eleven more pueblos plus three Apache tribes. Far to the west was the Navajo Nation. "Are you Native

American?" I naively asked Elmer, who to me looked Hispanic. "You're from back East," he said. "We don't say 'Native American' here in New Mexico. We're *Indians*." His pueblo was San Ildefonso, just down the hill from the Lab. Much of Los Alamos County was land appropriated from the San Ildefonsos for the Manhattan Project. In fact, this history was reflected in the deed and title insurance on our house, which based our title not on documented historical transfers, but on specific federal legislation when the land was privatized in the 1960s. The San Ildefonso pueblo had a fifty-year record of failed lawsuits trying to get their land back. In the unlikely event that they ever succeeded, we might lose our house, but we would supposedly be compensated from the U.S. Treasury.

After I had been at LANL for a time, Elmer arranged for Jeffrey and me to tour the eight northern pueblos and meet their governors and lieutenant governors—all in a single long day. Each pueblo had a population of one or two thousand, with a small fraction of residents living in the traditional adobe structures and most in small cinder-block homes or manufactured houses and trailers. Economic poverty was everywhere evident. The titles of governor and lieutenant governor were not traditional, "from time immemorial." The tribes had been forced to select governors by the conquering Spanish, who needed someone in each pueblo to hold accountable. Governors served one-year terms. The Spanish gave each tribe a ceremonial cane as the governor's symbol of authority. During the Civil War, President Lincoln replaced the Spanish (and later, Mexican) canes with silver-headed canes symbolizing (depending on whom you believe) either the pueblos' independent sovereignty, or else the Union's authority over them. "Before the Spanish, what was your traditional governance?" I asked at one pueblo.

That was my first experience with what I came to think of as "the mask". There were questions—many—that tribal Indians in New Mexico simply wouldn't answer. They didn't temporize, or explain, or change the subject. Instead, a figurative mask would descend, and I would find myself looking at a blank, silent face—staring into the face of time immemorial, as I thought of it. The silence might continue indefinitely, until I changed the subject. It was hard to know in advance when one would encounter the mask—but sometimes obvious in hindsight. "What is your pueblo's language?" was OK. (Taos and Picuris spoke Tiwa, the others Tewa, but I gathered that they were mutually comprehensible.) But, "What is the name of your pueblo in Tewa?" was definitely *not* OK—and I remembered from my undergraduate Harvard folklore course that if you knew the true name of a people, you could cast spells on them. Assigning saint names to the pueblos was purely a Spanish intervention. In 1599, the conquistador Juan de Oñate put down what became known as the Pueblo Revolt and, at San Juan Pueblo, decreed that all Indian males over age 25 should have one foot cut off. In 1999, on the four hundredth anniversary, by dead of night, a foot of the triumphal bronze statue of Oñate in Española was sawed off, presumably by Indian activists. That about summed up 400 years of Indian-Hispanic relations.

The reason that governance questions about the pueblos were off limits—anything beyond the governor, lieutenant, and the small pueblo offices that dealt with practical matters—was that the invisible, actual government was a religious theocracy—and *everything* about traditional religion was off limits. We got a hint of this on our tour, at lunch in Taos Pueblo's coffee shop/diner with the pueblo's governor. Taos, the most northern of the eight pueblos, was somewhat different from the others, having assimilated cultural elements of the plains Indians. They kept a sacred herd of buffalo, for example. In our Formica-tabled booth, we were chatting with the governor when he glanced at the restaurant entrance and visibly stiffened. An older man, dressed like everyone else in jeans and a long-sleeved cowboy shirt, had just come in. The governor leaned toward us. "That is the *War Chief*," he said softly. He paused for a second. "I'll go ask him if he is willing to meet you," he said. Unlike the archbishop of Santa Fe, the Taos War Chief proved willing to shake hands with me—but that was about it. The War Chief was responsible for the sacred buffalo herd; that was all that we were allowed to know.

Anthropologists in the 1920s and 1930s had done better at documenting New Mexico pueblo culture; and, in our limited observation, little had changed. Each pueblo comprised two clans or moieties, their names varying—summer vs. winter, north vs. south, turquoise vs. squash. Supreme authority vested in a kiva leader or *cacique*, whose title also varied (in Taos, the War Chief). Tribal decisions—including the selection of a new governor every New Year's Day—were made by the male elders, who gathered in the kiva, smoked peyote, and then listened to the guidance of the spirits. While the governor was the public-facing chief executive, behind the scenes he was entirely subordinate to the religious leader, who had lifetime tenure.

A little-remembered piece of New Deal legislation, the Indian Reorganization Act, became law in 1934. At that time, each pueblo was given the option of adopting "new-style" governance, with an elected tribal council and universal suffrage (including women). Most pueblos retained old-style kiva governance. Santa Clara Pueblo was one adopting the new system; when we visited, this was still being talked about six decades later, as a matter of pride. But when Jeffrey, as an avid gardener, asked our friendly guide at Santa Clara what she grew in her garden, the mask dropped: her plants were used in sacred ceremonies (Elmer later told us), and the subject was taboo.

At LANL, Earl Salazar was a senior manager in business, the Lab comptroller roughly, whom I interacted with and enjoyed talking to. Earl lived in San Juan Pueblo, had a college degree and an MBA. His father, Nick, was a prominent state politician who, born in San Juan Pueblo, had defied convention by marrying a non-Indian and managing to build a political constituency crossing Indian-Hispanic lines.

San Juan Pueblo, like several of the pueblos, had a casino. Theirs, apparently, was not doing well. Earl had not been especially active in tribal affairs, but, one New Year's Day, he was informed by a knock on his door that, in recognition of his acumen in business affairs, the spirits in the kiva had chosen him as the new governor. Specifically, the spirits wanted him to restore the casino to profitability. Earl was apologetic in telling his Lab colleagues that he would have to leave for a year—he had no choice. LANL had a special program, approved by the Department of Energy, that granted one-year paid leave to employees who were elected to tribal office.

Why the federal government would approve spending Congressionally appropriated taxpayer money in this way was also a part of my new education—a one-hour lesson in Indian law from the LANL general counsel's office. The U.S. Constitution's "commerce clause" gives Congress power to regulate foreign and interstate trade, and also—mentioned on an equal footing—trade with Indian tribes. These few words came to be interpreted in law as implying, first, that the tribes were sovereign nations; second, that they could be regulated *only* by Congress, and not (for example) by the State of New Mexico or Santa Fe County. And third—most obscure to me—that the federal government had a "duty to protect" the tribes, this implying a whole slew of legislative and executive authorities that have generally harmed Native Americans far more than protected them.

So, if each pueblo was a sovereign nation, then who in the U.S. government was responsible for formal diplomatic relations with them? The obscure answer was that, depending on circumstance and location, each tribe was assigned to a federal agency. As a legacy of the Manhattan Project, the pueblos in northern New Mexico were assigned to the Atomic Energy Commission, which became the Department of Energy, which in turn delegated responsibility to the Director of the Los Alamos National Laboratory, who in turn often (when he had other things to do) delegated it to me. On my probably fictive Hispanic side I might be just another *converso*, but to the Indian tribes, I was—occasionally, when they wanted to make a point of it—a plenipotentiary ambassador. A few times a year, the Director's Office would get word that the pueblos wanted a formal sovereign meeting. Then, John Browne and I, or just I, would show up at one of the pueblos at the designated time, usually 10:00 a.m. "Bring something to read," John advised me. We needed to be respectfully prompt, but the pueblo leaders were sovereign; it might be noon before anything actually got started.

These diplomatic meetings always began with prayers in Tewa, or Tiwa, or both. A prayer would go for fifteen or twenty minutes. No translation was provided, but the tone was sometimes quite vehement and accusatory. Sometimes, there was then actual business to be discussed (in English), such as the distribution of LANL funds for scholarships and community services. Most of the time, however, the purpose of these meetings was for the Indian leaders to lecture us on their grievances great and small—many real, but few remediable: Number one was always that we had stolen their land. After that, our employees were speeding on their roads. We were poisoning their sacred herbs with radioactivity. (Not true as far as we knew, but they would not tell us which herbs or where they were gathered.) We were giving jobs to Navajos or, worse, letting Pojoaqueans self-identify as Native American for affirmative-action. My guess was that the Tewa prayers probably had covered the same material—prayers for the return of Indian land, for relief from speeders, and so forth—a kind of agenda-setting.

Over time, I came to understand that many Indians did expect that their prayers would be answered. They had been on the land from time immemorial in the past, and would be there for time immemorial to the future. Someday the white man would be gone and it would again be as it was. This was a spiritual, not necessarily an intellectual, belief. I knew Catholics who truly believed in the transubstantiation of the Eucharist, and Jews who truly believed that the messiah would come. It was like that.

Fred Begaye, a Ph.D. physicist at LANL, was emissary between the Lab and the Navajos. Dwarfing the pueblos in size, the Navajo Nation had a population of more than a hundred thousand, its own police, courts, social services, and representative government. Two hundred miles west of Los Alamos, the Navajo reservation covered 27,000 square miles. Its sovereign relations were not assigned to DOE or LANL, so

Fred's diplomacy was only unofficial, but he was well wired into the Navajo power structure—the Begayes were Navajo royalty. One day he brought me a white-paper proposal. The United States was spending billions on the development of an underground nuclear waste disposal facility at Yucca Mountain, Nevada. The alternative to underground waste storage was above-ground storage that was permanently monitored and guarded. Above-ground storage was less expensive and could allow for the development of future, better storage technologies, or even for new uses for nuclear waste.

But it was dropped from consideration when Congress ridiculously legislated that any form of storage must be engineered to last for 50,000 years. The Navajos' proposal was straightforward: By treaty between the sovereign Navajo Nation and the United States, and in consideration of a large lump-sum payment (to be invested in a permanent fund), the Navajos would store and monitor all nuclear waste shipped to their reservation for the next fifty millennia. Fred, an educated man and a scientist, had no doubts that the Navajos could keep their side of such a bargain. The white man would be long gone, but the Navajos would still be there to guard the white man's radioactive legacy. I forwarded Fred's white paper to Ernie Moniz and a few others at DOE in Washington, where it disappeared.

Pojoaque (pronounced po-WA-kee) particularly stuck in the pueblo leaders' craw. Officially, it was a federally recognized tribe. In practice, it was a small, prosperous Hispanic town, well-located on Highway 84/285, with two gas stations, an RV park, a casino, a liquor store, and a hotel. The Indian leaders conceded that there had once been a Po-eh tribe. In their telling, however, the Po-ehs had become extinct in the late nineteenth century, long before two Hispanic families, Gutierrez and Viarrial, showed up in 1934, claiming to be Po-eh descendants. Under the Indian Reorganization Act, the two families were granted land and federal recognition as a tribe. In the 1990s, a profit-making cultural center and museum opened with—depending on whom you believed—lovingly recovered, or cynically invented, historical exhibits. In later years (here jumping ahead) the Pojoaqueans continued to thrive. Their Buffalo Thunder Resort, run by Hilton as the largest golf resort in New Mexico, opened in 2008. Its greens are watered from wells that draw on the Rio Grande aquifer in violation of New Mexico water rights law; but Pojoaque's sovereign Indian status prevents any state enforcement. Like many other New Mexico water rights cases in federal court, the dispute will likely continue for decades.

If New Mexico Indian politics had its ridiculous moments, one laughed only not to cry. The pueblo residents were trapped in poverty, holding onto shreds of cultural pride as the excuse for undemocratic, inefficient, and often inept governance within the tribes and—generally even worse—by the federal government's Bureau of Indian Affairs. That BIA's bureaucracy was largely populated with enrolled tribal members only made things sadder.

One time, I met with the governor of San Ildefonso Pueblo to explain how the Lab might pay for scholarships for pueblo kids at any University of California campus. This was already somewhat fanciful, because it would be the rare pueblo kid who could gain admission to UC, given the poor local educational system. "Are you making the same scholarship offer to Santa Clara and San Juan?" he asked. These were the pueblos just upriver on the Rio Grande, the San Ildefonsos' traditional enemies. "Of course," I said. "Then we don't want it," he said.

Besides the Indians, LANL had something akin to diplomatic relations with the British and the French—or at least with their respective nuclear weapons programs. Soon after I arrived at the Lab, I was invited to visit the U.K. Atomic Weapons Establishment at Aldermaston. After World War II—and after the discovery that Klaus Fuchs, a British national, was a Soviet spy at Los Alamos—nuclear weapons cooperation with the British ceased until 1958. Then, a new, Cold-War US-UK Mutual Defense Agreement authorized the transfer between the countries of "nuclear materials and other technology for nuclear weapons." But *not*—the Lab lawyers who briefed me before my trip emphasized—actual weapons blueprints or designs. I could talk to the Brits, and they might talk to me, about general design principles—but not about actual weapons.

When I arrived at AWE, my friendly hosts took me to a classified conference room for a familiarization briefing. The first slide was the detailed cross-sectional design of their submarine warhead. And, to my only partly expert eyes, it sure looked a whole lot like the U.S. W-76, a Los Alamos product. It was obvious that our actual cooperation was not what the lawyers imagined. The British had tested their nuclear weapons at our Nevada Test Site. In October, 1992, they had a weapon "down-hole" and ready to fire when George H.W. Bush peremptorily halted testing. They were given no advance notice and lost the test opportunity, a matter that still rankled—they reminded me of it several times during my visit. They gave me a silk necktie with geometrical figures and equations, which I still sometimes wear.

Our special relationship with the British became even more special shortly after my visit, when the British industrial firm Hunting-BRAE lost the management contract for running AWE to a consortium led by the U.S. firm Lockheed-Martin. John Hopson, a senior Los Alamos weapons designer (and, with his wife Emmy, good friends to Jeffrey and me from the time we arrived) took a leave from LANL, joined Lockheed-Martin, and moved to the U.K. as a senior AWE officer. John was entirely American, yet there he was, practically running the U.K.'s nuclear weapons program. He told me that, as a kind of fig leaf, there was one small safe that only his assistant, a Brit, had the combination to.

The case of the French was rather different. Nuclear weapons cooperation between the United States and France began secretly in the 1970s under Nixon. It was still so sensitive that I wasn't cleared to read the governing agreement. People in the know alluded meaningfully to the period of 1992 to 1996 when U.S. nuclear testing had ceased, but French testing (in the South Pacific) continued. All documents relating to U.S.-French cooperation were isolated in a separate vault room, which I wasn't cleared to enter; and I didn't get a trip to France. I did once get to entertain a visiting French delegation for a social hour. Alcohol was not allowed on-site at the Lab, so Gallic sociability was limited. In my pre-brief for this event, I was told that the French would have green (cleared) badges. We should not be fooled. We were to let them think that they were cleared for everything, but actually tell them nothing.

56. Senior Executive

John Browne's organizational structure looked good on paper, but its limitations soon became evident. Every CEO needs a COO, a Chief Operating Officer who actually runs the place—bashing heads as necessary. John didn't have a COO. None of us three deputy directors—me supposedly the "principal"—had the necessary authority: structurally because the three associate directors reported to John, not to any of us; and de facto because none of us were temperamentally suited for the task. I didn't know anything about running an organization of 12,000 people with 2,000 buildings and structures spread over 43 square miles. Tom Garcia, the deputy for business administration, seemed to see his job as mainly "greening the valley," that is, providing jobs and contracts for New Mexico Hispanics and styling himself a *patrón*. Too zealous at this for the government auditors, he was soon out and replaced by (once again in an acting capacity) Pete Miller.

Dick Burick had the title deputy for operations, so, arguably, he should have been COO. Dick was an engineer who had worked with John over the years. He was an outdoorsman, whose living room walls displayed the heads of deer, elk, and—his proudest—four species of bighorn sheep. Elk chili was invariably served at his parties. Jeffrey and I always left hungry, because Chronic Wasting Disease was already being reported in North American cervids, and we didn't want to take any chances. CWD, to digress just briefly, was a prion disease of the brain, related to mad cow disease and, in humans, Creutzfeldt-Jakob disease. Prion diseases propagate by the ingestion of misfolded proteins that aren't neutralized by heat (e.g., cooking). It became apparent only in the 1990s that animal prion diseases (in Britain, mad cow disease) could infect humans through food.

Strangely, and tragically, a LANL division director was diagnosed with Creutzfeldt-Jakob disease in 2001. His increasing problems with balance went undiagnosed until Browne sent him at Lab expense to the Mayo Clinic in Phoenix. There, they did a workup and told him the good and bad news. The good was that they had arrived at a definitive diagnosis. The bad was that he would be dead in six months. He was. There were 200 or so cases annually in the United States. The unfortunate man was not a hunter and didn't consume deer or elk, but he had lived in the U.K. for ten years during the mad cow outbreak. It was a case of exceptionally bad luck.

Returning to Dick Burick, here was a person who truly believed that there was a rational engineering solution to any problem. Dick failed as a COO amid the multitude of problems that boiled down simply to human conflicts without engineering solutions. Nor was John good at being his own COO. His response to conflicts among members of his Senior Executive Team—most such entirely professional and on issues that simply needed a decision—was to convene the whole SET for meetings over which he would good-naturedly preside. We were expected to reach a consensus. But effective COOs don't govern by consensus. They listen and then decide things, and that wasn't John's style.

In practice, my job, DLDSTP (deputy laboratory director for science, technology, and programs) was a dog's breakfast of responsibilities. I had to learn a lot just to understand what they were—much less do them. LANL had over years evolved into what was termed a matrix management system. A first lesson was to understand the difference between *line* and *program*. Everyone at the Lab had one (and only one) line manager, defining a hierarchy from lab director through associate directors, division leaders, and group leaders (who were the lowest or "first-line" managers) to technical staff members and support staff. Line managers hired new employees, set salaries, wrote annual performance appraisals, and were responsible for safety and regulatory compliance. Crucially, they were also responsible for "making payroll," that is, finding work for their people. And they got no budget for that, zero. That is where *program* came in. Every LANL activity, large or small, had, by definition, a government sponsor who was paying. The human interfaces to sponsors were program managers. These were LANL's rainmakers, in effect its sales force and marketing teams.

Programs were supposed to be executed *across* line organizations, not *down* them, hence the idea of a matrix. A defense program in "weapons aging," for example, might involve materials scientists in the physics division, chemists in the chemistry division, engineers in weapons engineering, and so forth. Program managers disbursed their programs' funds to line managers (i.e., making payroll) and were responsible for the successful execution of programs. A Lab scientist might work, and divide his or her timecard, among several programs. Adding only a small amount of confusion was the fact that program managers, like everyone else, were embedded in line organizations and had line managers. A program manager might report to a group leader or (for a big program) a division leader or associate laboratory director—but that line manager, unless also given program responsibility, wasn't supposed to interfere in the allocation of funds. So, cutting to the chase, power at LANL flowed on two axes: who wrote your performance appraisals (line), and who was paying you (program). Matrix management has its strengths and weaknesses; that is too extended a subject for here.

I had a mixture of line and program responsibilities, along with other responsibilities that didn't fit the matrix at all, probably ending up in my office for just that reason. As a direct line manager, my span of control had organizations totaling 200 people—a drop in the bucket. But some of those organizations existed solely to house programs totaling several hundreds of millions of dollars, and for these I also had programmatic responsibility. My most significant program was Laboratory Directed Research and Development, residing in my Science and Technology Base programs office. All sponsored programs at LANL were taxed five percent to pay for LDRD, which supported Lab scientists in basic research in areas that underlay all of LANL's capabilities. That amounted to about a hundred million dollars a year, and it supported, part-time, more than a thousand Lab researchers. In Sig Hecker's flatland, LDRD had drifted into being a kind of entitlement program for the Lab's most senior scientists. I reinvigorated the peer-review process for scientific merit, mainly by appointing younger scientists to the review committees, especially ones with first-time LDRD funding. Trish, controlling my calendar, metered the flow of meetings requested by irate senior scientists who lost their flexible funds.

Also within my purview were government and community relations and offices handling University of California coordination, educational programs, strategic planning, quality assurance, and the library. Most of these functions reported to me indirectly, through lower-level managers. Besides Trish and my special assistant (a two-year rotating position for a Lab scientist), I had about half a dozen direct reports. I learned that I needed to meet at least weekly with each of them, seem always interested, urge them to think strategically, and—my real job most of the time—run interference for them in their relations to other parts of the Laboratory.

Two of my direct reports were *project* managers. A project was not a program. Programs were level-of-effort activities that continued as long as the sponsor sent money. A project was an activity with a defined beginning and end, a scope, a schedule, and a budget. A project manager was someone personally responsible for delivering successfully on-time and on-budget—ideally, anyway. Paul Lisowski initially reported to me as project manager for the Spallation Neutron Source front-end linac (linear accelerator). SNS was a multi-lab national project under construction at Oak Ridge, Tennessee. LANL's piece was the design and construction of the linac, a $200 million item. Paul was direct with me: "I'm the project manager. You're the Senior Executive. You don't interfere with my management of the project, and, in return, I keep you completely informed about how it is going—the good and the bad—with no surprises. Your job is to keep senior management off my back, or to get me fired if I completely screw up."

Paul was happy that I already knew a little bit about project management. I knew what a *work breakdown structure* was. The WBS was a hierarchical set of boxes that broke down the whole project into a set of tasks, subtasks, sub-subtasks, and so forth. All had to be accomplished for the project to be completed. I understood the idea of the *critical path*, the sequence of events whose dependence on each other defined the minimum time to complete the project. Things *on* the critical path were the highest priority to keep on track, while things *off* the critical path had to be kept off. I had some understanding of the concept of *earned value reporting*, an accounting methodology for verifying that a project stayed on budget.

I did need a lesson in *reserve* and *contingency*. Both described funds that were not yet allocated to boxes in the WBS. Reserve was money that shouldn't ever be needed at all. In fact, DOE projects had zero reserves. If they ever needed them, DOE would have to go back to Congress, ask for it, and take a beating. (This did regularly occur, but was above my, and John Browne's, pay grade.) Contingency was completely different. Contingency was money that *would* be spent—you just didn't know exactly which WBS box to assign it to at the start. The recorded experience of many varied projects produced a pretty good estimate of how much contingency a project would require. For high-tech development projects like SNS, a number like 20% was typical. Key then was who "held" the contingency. The project manager—Paul Lisowski soon moved to a different job and was replaced by Don Rej—held most of the contingency and was empowered to juggle it among the WBS boxes—the project manager giveth and taketh away. I held some of it, a way of guaranteeing that when Rej started running out of contingency, he would have to involve me in decisions. And, above me, the SNS uber-manager in Oak Ridge held a fraction, which he could allocate to (or from) the national labs participating in the SNS project. Don Rej and I traveled to Oak Ridge twice a year for DOE's overall project review, where my job was to take the lashes on my bare back for things, luckily few, where Don had screwed up. In the end, LANL's piece of SNS came in on-time and below budget, and we had contingency to give back to the project.

But if SNS was my education on what good project management looked like, my other big project was an education from the other side. D-Site Prime, also known as Technical Area 21, was a nearly forgotten piece of Lab property at the far end of the eponymous DP Road. On paper, the land was supposed to be given over to the county for business development. But first, there was the not-so-small matter of greenfielding the Old Tritium Building. Tritium is the radioactive isotope of hydrogen and a key ingredient in nuclear weapons. Paradoxically, it is both hellishly radioactive and not very dangerous, the former because it decays with a half-life of only 12.3 years (to be compared, e.g., to plutonium's 24,000 years), the latter because its decays emit only a wimpy beta particle that doesn't penetrate even the dead outer layer of human skin. Still, it would be dangerous for someone to live or work in an environment with a lot of tritium. Its atoms exchange with the hydrogen in water and organic materials, so a person could ingest a significant dose by that route. Thus, we couldn't just bulldoze the old building and not tell the county.

Built in the 1950s, the building had been pretty much abandoned in the 1980s. The decommissioning project that I inherited had been going for ten years. The wall-paint and the cement floors were all contaminated. There was no budget to tear down the building and bury it somewhere as low-level radioactive waste—and anyway, where? No existing waste site would take it. The basic strategy was to keep the interior humidified so that the tritium would exchange with water in the air—and then vent it up the building's stack. This was environmentally permissible. Dispersed as water vapor and diluted in the dry New Mexico atmosphere, the radioactivity at the closest site boundary was well below allowable limits. And the project would eventually succeed, if only because half the tritium decayed away every 12.3 years. But, as project management, it was ridiculous. Every year we missed our milestones, and every year the DOE sent a little more money. I periodically reported the project's failures to John Browne and the SET, as I was required to do. They shrugged. No one, at the Lab or in government, actually cared. It was a phony project.

Anyway, there was another reason that the land wouldn't be given to the county any time soon. Manhattan Project alumni—those still alive by this time were in their eighties—remembered that during World War II plutonium waste had been disposed of at Technical Area 21 by drilling a shaft a few hundred feet into the mesa (intentionally still far above the water table), dumping the waste down the hole, and then filling it in. No one knew where these holes were. While there was likely no actual environmental danger, the public-relations danger was acute. The slow progress on tritium happily kicked the whole TA-21 can down the road.

Safety, in the sense of "safety rules," was not something that I had ever paid much attention to before coming to Los Alamos. But, after a month that saw several small accidents at the Lab, reportable though not life-threatening, safety came very much onto the SET agenda. An expert safety consultant was brought in, and we were told to block a whole day for training. It was more interesting than I expected. The consultant was formerly from DuPont. That company was the poster-child of good safety practice, and many ex-DuPonters did safety consulting gigs—a kind of post-retirement perk.

Reportable accidents, I learned, were those requiring medical care or resulting in lost work time. If you could be treated with a band-aid or aspirin from one of the ubiquitous first aid kits and go back to work, then it wasn't reportable. Fatalities counted against companies only if they were "preventable injury-related deaths." Heart attacks by sedentary office workers didn't count against us—a good thing, because obesity and diabetes were high in the New Mexico workforce. In dangerous industries like mining and agriculture, fatal accident rates were on the order of ten, or even more, per 100,000 work years. We had nothing like that many at LANL, despite our performing quite a few "high consequence" operations—often things involving explosives or fissionable materials. In national statistics, "safe" office jobs had a fatality rate of about one per 100,000 work years—things like fires or falling down stairs. For our size, that rate would imply, on average, about one preventable death every eight or ten years.

"Do you want to set that as your goal?" Mr. ex-DuPont pressed us. It was a trick question. The goal had to be zero. Data showed that organizations that weren't *constantly* working to improve safety were *inevitably* getting less safe. It was the Red Queen in Alice in Wonderland: "It takes all the running you can do to keep in the same place." Also, since the Occupational Health and Safety Act of 1970 that created OSHA, there was officially no acceptably low fatality rate. That sounds like a pious sentiment, but it was in fact at the time controversial. Big industrial projects used to be "costed" in both dollars and expected lives lost. Twenty workers died building the Brooklyn Bridge; five, the Empire

State Building. Post-OSHA, big projects were supposed to be designed with a strong expectation of no deaths at all.

But if the Lab had a fatal accident only every some years, how could we, the SET, ever know whether we were getting better or worse? A single fatal accident might just be bad luck. DuPont had an answer to this. Extensive data existed not just on deaths, but on the whole pyramid of lesser events: hospitalizations; outpatient care; non-reportable accidents. And, importantly, data on near misses, where nothing actually happened—but might have. The rates of lesser events were dramatically higher than for deaths, obviously. For every fatality, there were something like a thousand accidents requiring medical treatment, many more thousands of non-reportable accidents, and at least another order of magnitude more near misses. What the data showed was that the rates of these more frequent events were *predictive* of how often more serious events occurred. There might not be enough cases to prove this for any single organization, but nationwide the pattern was clear. So, we could measure our risk of serious events by less-extreme ones that actually happened.

That brought us to the basics of safety culture. An organization either had it, or didn't, and there were a bunch of indicators. In a safety culture, workers about to begin a task stopped—literally stood still—and mentally identified the hazards and how they were mitigated. If the hazards weren't adequately mitigated, then *work stopped.* A manager who retaliated against a worker who called a safety-related "stop-work" had to be fired. On the contrary, managers were expected to at least thank, and if possible reward in some way, employees who pointed out safety concerns, or who reported near-misses. This sounds fanciful, but it was *measurable*: If we, senior management, didn't see data that documented many more near-misses than non-reportable accidents, and many more of those than actual injuries, then our people weren't reporting—and we didn't yet have a safety culture.

And that brought us to *behavior modeling.* Employees wouldn't accept a safety culture unless they saw management at all levels visibly modeling it, on and off the job. No manager could let pass an unsafe situation, in their own organizations, elsewhere in the Lab, or out in the world. If you encountered a spill in the supermarket aisle, you couldn't just steer around it. You positioned your cart to shield others from the hazard, reported the spill, and *didn't leave* until you saw it being taken care of. In other words, you became a sanctimonious prig on all safety matters large and small. "Get used to it," our trainer told us. "It is a condition of your employment as a manager." John Browne nodded enthusiastically. During our day of training, when people got up for an individual break, they often left their chairs pulled out from the conference table. Without interrupting the flow of his lecture, our safety man would walk around and push in the chair. He never explained or mentioned it. We were supposed to catch on. He was modeling.

Once thus inducted into safety culture, I came to enjoy it with the smugness of sanctimonious prigs everywhere. I already always fastened my seat belt driving, so that didn't count. But I did start checking my tire pressure more often and enjoyed speeding less. I took the aluminum ladder in our garage to the town dump and replaced it with a fiberglass ladder—heavier and more expensive, but not an electrocution risk. As for electrocution, when I replaced light bulbs, I now turned the switch *off*, despite *on* being more convenient for seeing right away that the bulb worked. And I now always pointed out to strangers if their shoelaces were untied and, if feasible, I waited while they tied them. You meet some interesting people doing that.

I did enjoy the time I substituted as acting Lab Director for an out-of-town John Browne in presenting certificates of appreciation (minor, but frameable) to two Lab scientists. The two had been at a scientific conference in Chicago, in a big auditorium, when the next plenary speaker was introduced. On the man's way to the podium, he collapsed, clutching his chest. There was a collective intake of breath by the audience, *but nobody did anything.* Our Lab pair jumped over people to get to the aisle, then ran to the stage. One of them began administering CPR while the other pointed to someone in the front row and instructed him to call 911—now!—and instructed the rest of the audience to stay in their seats and keep the aisles clear. The ambulance arrived quickly, and one of the Lab people rode with the victim to the hospital. "It was our Lab safety training," they said. "We knew that we were *required* to act, so we did."

Safety wasn't the only training that John insisted on for the SET. An outfit called Benscheidt came in to give us a day of media training. Carl Benscheidt was the former West Coast director for the CBS Evening News with Dan Rather. With his wife Lisa, this was a family business. The training was all about having well thought-out talking points, four at most, and sticking to them no matter what. You had to control the interview in every way possible. Beforehand, you could question the setting and background: "I'm just not comfortable doing it here." On camera, you always smiled and projected a positive image. That was what people remembered, much more than what you actually said. Except for the rare live broadcast, you didn't actually have to say anything at all: On questions you didn't like, you just kept smiling—later, an editor wouldn't waste airtime on your silent, smiling face. Or, you could "bridge" with a minimal answer followed by, "but what we *need* to talk about is…" or "but as a matter of fact…" Then, you could pretend that the question was exactly one of your talking points. "I'm glad you asked that," was always good. Control the pace: Slower was better. Don't repeat the negative that you are denying. Counter factual errors immediately: "That's just not true." Avoid ambushes: "How nice to see you! Let's arrange a time to talk." Don't look at an offered document or photo. For print media interviews, take out your Perlcorder (today, cell phone): "You don't mind if I record this also, do you?" These were all good lessons that stayed with me.

I got a lot of good mentoring from John. "Los Alamos isn't Livermore," was a common theme. At Livermore, the accepted way to start a scientific conversation was to go into someone's office and start criticizing them and their work. This would lead to a lot of shouting at each other and, eventually, real progress on the issue. It was a style of science that I also knew from Zel'dovich's group in Moscow—and Zel'dovich had gotten it from the Soviet master himself, Landau. Kip's group was not too different. But the approach wouldn't work at Los Alamos, John said, adding that he was already getting complaints about me. Los Alamos was small-town. The person you insulted might not know about Landau, but they might coach your kid's soccer team. For sure, they were going to tell the whole town about your bad behavior.

There were pitfalls that I could not have anticipated. Jeffrey volunteered with the local search-and-rescue club; to support her, I needed to pass the written multiple-choice test for an amateur radio "ham" license. The test was administered monthly by the local Los Alamos chapter of the American Radio Relay League. I was a physicist and already knew a lot about radio technology: The day before the test, I digested the pamphlet that listed every possible question and its answer. On the day, I showed up, took the test, and passed with a grade of 100%.

Two days later, John Browne called me into his office. "I've received a formal complaint about you, Bill." In Los Alamos as elsewhere, ham radio was a largely blue-collar hobby. In showing up cold to take the test—and especially in scoring 100%—I had insulted the club's membership, most of whom worked at the Lab. Everyone else taking the test had attended six weeks of night courses in preparation. When I showed up, they all knew who I was, but I didn't even have the courtesy to introduce myself or (for example) compliment them on their club's work. It was an outrage. *Who did I think I was?*

On this and other occasions, I had to assure John that I was not blind to the responsibilities of leadership, but only inept. I had observed at close range a natural leader in Larry Welch. I knew that leadership wasn't about being the smartest person in the room, but rather the person in the room that others wanted to follow. I knew that leadership in some form could be taught to those less gifted than Gen. Welch—they gave courses in it at the military academies. "I'm *trying*," I whined privately to John, "but it's not my strength." "Why don't you do a one-week intensive course somewhere," he suggested. He had done one once and had found it helpful.

From various alternatives, I picked a course called "Leadership at the Peak" run by the nonprofit Center for Creative Leadership and, my week, held at the Broadmoor Resort in Colorado Springs. Among its competitors CCL was known for being more data-driven, less kumbaya, and they had been around since the 1970s. My week, the participants were mostly from industry, typically senior vice presidents in their forties being groomed for the CEO position of their medium-sized corporation. One or two were sons of founders of big family-owned businesses being groomed to take over from Daddy. There were a dozen of us.

CCL didn't do things like have us fall out of trees and be caught by our teammates. Instead, three hours a day, we took every known industrial psychological test: tests of abilities, tests of aptitudes, tests of desires and ambitions. Then, afternoons, we met individually with our assigned coach, who would go over the previous day's test results. We also did war games and tabletop exercises, where we were observed through one-way glass. The next day, your coach would critique you on your performance. Generally, I did very well on the objective tests. Jeffrey and I had always joked about our schoolteacher mothers, who were disappointed if we ever brought home from school a score on an objective test of less than 99th percentile. We had both learned to be really good at *test-taking*.

One of the tests was supposed to be a creativity test, whatever that meant. The questions were things like, "You have 60 seconds. Write down everything you can do with a pencil eraser." The following afternoon, my coach looked at me with what I took to be an expression of awe. "Bill, we've been giving this test for twenty years. No one has ever scored as high as you scored on this test." We had already learned that leaders were supposed to be modest—maybe he was checking me on that—so I said something like "Aw, shucks. I must have been having a good day."

"No," he said, "I don't think you're understanding me. We've been tracking *outcomes* for twenty years. No one who's scored this high on this test has ever succeeded as a CEO or lead manager in an enterprise. When we are advising companies, we tell them that a score like this is completely disqualifying."

That was interesting. I didn't always believe these people—they weren't the smartest people in the room—but they were professionals, and they had *data*. His story was this: To lead a large organization, creativity up to a point was a plus. But any greater, and it began to interfere with a leader's ability to deliver clear, consistent messaging. It thwarted any deliberate strategic planning process. "One week, you'll be telling your people to do one thing. Then, the next week, you'll have an even better idea. You'll tell everyone to stop doing what they were doing, and to do the new, better thing. Am I right about you, Bill?"

He was right. He had some advice: If I ever found myself in a CEO position (Lab Director at Los Alamos, just for example), I should pick a very down-to-earth operations person as my deputy. I should then always use them as a filter between me and the rest of the organization. I could be as creative as I wanted in private with my deputy, but the responsibility for distilling and communicating consistent, and only slowly-changing, policy and direction would be theirs, not mine. "Otherwise, you'll fail," he said simply. "No one will be able to, or want to, follow your leadership."

Maybe his analysis was not 100% right. Maybe only 90%. But it struck a chord. It explained a lot of my past experiences. I took it to heart. In that one week, I lost all remaining ambition to be a CEO, or a director of a national lab, or a university president. Those were important jobs for which, I now knew, I was ill-suited. I wasn't heartbroken. As deputy director, I was already discovering things about being senior management that I didn't like. I didn't like simplifying my thoughts for the stupidest person in the room. I didn't like suppressing all irony and sarcasm, because there was always someone who would misinterpret it. I understood—but didn't like—the imperative of letting no light show between my views and the organization's, even in supposed private—people would always talk and magnify the difference. I was learning a lot in my job, and having fun most of the time, but I now understood with clarity that I would be doing it only so long. How long, I couldn't yet say, but it would be finite.

57. Wen Ho Lee

Wen Ho Lee was the first of the several crises that ultimately got John Browne fired and most of his Senior Executive Team replaced. Far from starting on Browne's watch, WHL was a saga that had been unfolding for years. Soon after John was named to replace Sig as director, he was visited by two Albuquerque FBI agents. There was a Chinese spy in X-Division, they told him. X-Division was the Lab's most protected holy-of-holies, the place where bombs were designed, and where the computer codes and data files embodying fifty years of design experience were sequestered. "How long has he been there?" John asked. More than ten years, they said. In fact, it was more than twenty. Lee had been hired as a technical staff member at LANL in 1978. The more interesting questions were how long had the FBI known or suspected him; and why was Lee still fully cleared and coming to work in X-Division every day.

The answer to the second question was the hardest for John to swallow. The FBI was building a case to prosecute Lee, codenamed Kindred Spirit, and they insisted that he be left in place—even stealing nuclear secrets—until they were ready to unseal an indictment. A year before, in April, 1997, they had reluctantly acceded to director Sig Hecker's pleas that Lee be moved from the X-1 group to a less sensitive part of the division; but Lee still had access to all the division's classified computers and data. Only Sig, John, and LANL's head of security were cleared to know about the case. "If you tell anyone else, or in any way interfere, you are liable to prosecution for obstruction of justice," the agents threatened.

With hindsight, it is easy to argue that John should have been more forceful in response. It took a year for him to get permission to brief the six of us on his SET. That happened to be the week that I arrived full-time, in December, 1998. We were also sworn to secrecy on penalty of prosecution for obstruction of justice. We on the SET only slowly came to understand the incentives within FBI's criminal division: Convictions were their only measure of success. It was not a matter of concern to agents from the Albuquerque Field Office that LANL might be hemorrhaging nuclear secrets, but only whether they could *convict* someone for it. We also came to understand that this obstruction-of-justice threat was something that FBI field agents threw around very casually, the way they liked showing their badges and guns. With hindsight one can speculate on what Browne ought to have done. The situation was a conflict on national security between two federal agencies. Negotiations should not have been at the level of Los Alamos and Albuquerque; they should have been between the Secretary of Energy (at the time, New Mexico's Bill Richardson) and the FBI Director or Attorney General. "We'll each inform our chains of command, and I'll see you in Washington," is what Browne should have said to the Albuquerque Special Agent in Charge.

But, back to the first question above. For how long had the FBI suspected Wen Ho Lee? This is where things get complicated. In the early 1980s, FBI was on the trail of an earlier alleged Chinese nuclear spy, this one at Livermore Lab. That case was codenamed Tiger Trap; the suspect was Gwo-Bao Min, an aeronautical engineer, a naturalized American born in Taiwan. Min had visited mainland China and received a warm welcome there. Testing the boundaries of what was classified, the Chinese asked Min some general questions about nuclear weapons. This was possibly a pre-recruitment intelligence technique aimed at compromising him; or possibly he was already in their employ.

Weapons design was not Min's field. However, he subsequently agreed to find out more detailed answers to five questions, one of which clearly strayed into classified territory—though it remains unclear whether Min knew this or not. In 1981, when Min booked a trip to China through Hong Kong, the FBI panicked, thinking that this would be his one-way escape.

The bar for proving espionage is a high one, requiring proof of intent to harm the United States. U.S.-China relations were in a warming period, so Min's defense would surely be that he was trying to help, not harm, relations between the two countries. Even a lesser charge of mishandling classified information would require proving that he, an aeronautical engineer, understood the intricate classification rules around nuclear design. The only way to stop Min from escaping, the FBI concluded, was to arrest him on the obscure charge of exporting munitions information without an export license.

What happened at San Francisco airport on Min's day of travel has been described elsewhere in heroic terms. The official story is that special agent Bill Cleveland set a clever trap that Min avoided only by some preternatural sense that something was amiss. JASON's FBI contact, who was not there at the time, but, himself in FBI's intelligence division, was steeped in trade gossip—described it as a giant fuck-up.

Agents wrestled people out of the way to get ahead of and behind Min in line. Somebody's brilliant idea was to have the airport public address system announce repeatedly—instead of the usual "no parking in front of the terminal"—that "it is a federal crime to knowingly export weapons information without a valid export license"—not an announcement that you hear every day at airports. The Bureau's on-site lawyers had opined that this would help the trumped-up charge stick. Who could imagine that such an announcement might alert the suspect! Min flushed some items down a restroom toilet, was searched and not arrested. He resigned his job at Livermore, stayed in the United States, and started an import-export company—presumably one that was scrupulous about its licenses.

Which brings us to Wen Ho Lee, who in 1982 telephoned Min, expressed sympathy for his travails, and wondered aloud who at Livermore had turned him in—something that Wen Ho mentioned that he might be able to find out. The FBI still had Min wiretapped, so this call made Lee a large blip on their radar. They opened an investigation, interviewed Lee and, in 1984, polygraphed him. His negative answers to the questions, "have you ever given classified information to anyone unauthorized," and, "have you ever worked for a foreign intelligence agency," were deemed truthful; but some of his other answers, e.g., about foreign contacts generally, were thought to be deceptive. Nevertheless, the FBI closed their investigation at that time.

But what *was* going on with that phone call? As fellow Taiwanese emigree scientists who worked in U.S. nuclear weapons laboratories, Lee and Min surely knew each other. We know now that during the 1970s and 1980s, Taiwan had its own secret nuclear weapons program, intended (if ultimately successful) to deter a Chinese invasion of the island. It seems possible—especially in light of later events—that Lee at this time thought that Min was spying for Taiwan, not China—and that he approved of this.

To have even a chance of understanding later events, we need to zoom out to a broader view of U.S.-China nuclear developments in the 1980s and 90s. Broadly speaking, all nuclear powers start out building heavy, round nuclear weapons, suitable for use as bombs or as single-warheads in heavy ICBM missiles. Starting in the 1970s, and taking advantage of innovations developed mostly at Livermore Laboratory, the United States learned to build weapons of a different architecture, "miniaturized" for use in small cruise missiles, or for multiple-warhead missiles. ("Miniaturized" is a relative term. Modern weapons still weigh at least several hundred pounds.) The Los Alamos-designed W-76, sometimes called the "Chevrolet of nuclear weapons" was an early such design, entering the stockpile in 1978 and manufactured in quantity during the 1980s Reagan nuclear buildup. The W-88, also a Los Alamos design and the "Cadillac of nuclear weapons," was first deployed in 1989.

China's nuclear stockpile was two orders of magnitude smaller than the U.S. stockpile at its peak. Still, in the 1980s, China was able to switch from heavy, round designs closely resembling those of the U.S.S.R., to miniaturized weapons resembling those of the U.S. Was this rapid advance enabled by Chinese espionage? Was it purely indigenous—physics dictating only certain ways that small nuclear weapons could be built? Or was it something in-between? Here we enter the corridor of mirrors.

A period of diplomatic warming followed President Nixon's "opening" of China in 1972, accelerating after the establishment of full diplomatic relations in 1979. In 1982, former Los Alamos director Harold Agnew visited China and met his Chinese counterparts, including Qian Shaojun, called "the father of Chinese nuclear weapons," and Hu Side, a rising star in weapons design. Jay Keyworth was at the time Reagan's science advisor in the White House, on leave from Los Alamos where he had been Agnew's physics division director. Sino-American rapprochement was driven in part by a shared view of the Soviet Union as a common foe, as forcefully articulated in Reagan's 1983 "Evil Empire" speech. More official visits by Los Alamos and Livermore scientists followed.

Decades afterward, I asked Harold Agnew, who was by then a close friend, about his 1982 trip. He was received very warmly, he said. Qian and Hu were proud of the Chinese effort and wanted to bring it out of the shadows. There was a kinship among nuclear designers. "Did you speak at all about nuclear design?" I asked. Harold could have said, "No, of course not." But what he said instead, somewhat frostily, was, "Everything that I discussed with the Chinese was cleared in advance with Jay Keyworth and the White House."

A few years later, Danny Stillman and Terry Hawkins, Los Alamos managers with intelligence connections that the Chinese surely knew about, were invited to visit the secretive Lop Nor nuclear test site. A decade after the trip, Hawkins described to me two incidents that stuck in his mind: Breakfast with a young Chinese nuclear designer who, taking a melon in his hands as a prop, wordlessly illustrated a basic design principle of miniaturized weapons. And, a tour of the tunnels at the Chinese test site where Terry was counting steps so as to be able to later draw an accurate underground map. "You don't have to do that," his host said. "I'll tell you how many meters it is." A plausible hypothesis is that, at least through the 1980s, the Chinese were hoping for some kind of nuclear special status with the United States—not anything like the U.S.-U.K. special relationship of sharing virtually everything, but perhaps something more like the U.S. and France. Or in any case, something appropriate for cooperation in the face of a common enemy, the U.S.S.R.

Now with the above as context: in 1986, Wen Ho Lee and his wife (a data analyst at the Lab) traveled together to Beijing for a conference sponsored by the Chinese Institute of Applied Physics and Computational Mathematics—Hu Side's institute, the Chinese equivalent of LANL. There, Hu invited him to give an additional talk, separate from the conference, to IAPCM staff. At Los Alamos, Lee was a minor player, more a computer programmer than weapons designer, although his training was in physics. His managers considered his performance and ability unremarkable. He was unlikely ever to be promoted. Later on, the question was raised: Why should Hu, a high Chinese official, lavish attention on the lowly Lee—unless it was a Chinese intelligence operation, an attempted recruitment? I think this misses the point. *Of course* Hu would lavish attention on Lee—on anyone from Los Alamos. That Lee was a Chinese speaker made him the more special. Perhaps, in the wake of Agnew's and other official contacts, Hu and others entertained the hypothesis that Lee was sent by the United States to open a back-channel of unofficial communication. A more revealing unanswered question would be, how did Lee ever hear about the conference in the first place? On his own? Through Min? Cold-called by the Chinese?

I got to know Hu Side in a formal way in the 2010s, when I traveled to Beijing several times to participate in "track two" (approved, but non-official) U.S.-China exchanges on nuclear disarmament and cybersecurity issues—and I similarly got to know formally Qian Shaojun, by then retired and addressed as General Qian. By that time, exchanges with the U.S. National Academy of Sciences had been going on for more than a decade. Hu and Qian were known figures in international circles and personal friends of my older JASON colleagues Dick Garwin and Sid Drell. No one could doubt that Hu would collaborate willingly with a Chinese intelligence operation. But he was not an intelligence professional. He was a physicist who wanted to see China's nuclear achievements recognized as peer to the United States. The kind of cooperation that he would have been looking for with Wen Ho Lee (whom he surely misjudged in one way or another) was overt, not covert.

Lee made another trip to Beijing and IAPCM in 1988. A planned trip in 1989 was cancelled when LANL (which by then had a counter-intelligence office) formally disapproved it. In 1992 Lee traveled to Hong Kong. His adult daughter, Alberta, traveled without him to Shanghai. In 1994, Hu Side and a high-level Chinese delegation visited Los Alamos. Wen Ho was not invited to the official reception, but he crashed the party to greet Hu, who embraced him, saying loudly that Lee was his good friend. Was Lee's behavior appropriate for a deep-cover spy? If Hu knew him to be a spy, would he have been so public in recognizing him? Or (corridor of mirrors) was this a planned clever ruse to deflect suspicion away from Lee, because no spy and spymaster could ever be so obvious? Or did Hu still think that Lee was part of some authorized back-channel operation?

In 1995, a "walk-in," that is, a person not previously known to the U.S. intelligence agencies, handed a U.S. embassy official in a third country a Chinese document marked "Secret" that appeared to contain design information about the U.S. W-88 warhead—the Cadillac of nuclear weapons. A year elapsed before the document was competently translated and, when it was, it was less of a smoking pistol than was at first thought. Far from giving nuclear design information, the document gave information on the warhead's external dimensions and mass, intended for someone designing a missile to carry the warhead. While secret, that information would have been routinely circulated in the United States to as many as a thousand people outside the nuclear Labs. Moreover, CIA quickly assessed the whole affair as a Chinese provocation. The walk-in was believed to be a Chinese intelligence officer. The document's revelations were what John le Carré spy novels would call "chickenfeed," intended only to…to *what*, exactly, in this case? If Chinese intelligence intended to send a message, it was far too subtle for their American counterparts.

But in Washington, the walk-in document was taken as proof of rampant successful Chinese nuclear espionage (maybe *that* was the intent). It ignited a wave of Sinophobia, and also anti-DOE sentiment, in Congress and right-wing circles. In this view, it was now abundantly clear that, as far back as the early 1980s, loose-lipped physicists, including even a former Laboratory director (Agnew), had been blabbing secrets to their Chinese counterparts. Scientists were incapable of keeping secrets. New discipline was called for.

One particularly rabid Sinophobe and Lab-hater was the memorably named Notra Trulock. Trulock had been an intelligence analyst at Los Alamos before moving to DOE where he became—to no small surprise back home—the head of all DOE intelligence. At Los Alamos, he was remembered for his insistence that the town's five Chinese restaurants could not be so numerous by chance and were thus most likely run by the Chinese government as spy nests. Trulock knew about Wen Ho Lee's visits to China and apparently chummy relationship with Hu Side, and he prodded the FBI to reopen the case—newly codenamed Kindred Spirit.

Kindred Spirit eventually showed Wen Ho Lee to be a spy after all. Well, maybe it showed that. And, maybe, he was not spying for China at all, but for…well, let's continue the story, from here something closer to a personal account. After the SET was sworn in on the case, we received weekly classified briefings. Computer forensics were also now moving more rapidly, because the FBI was allowing us to assign a lab team to the project.

The evidence uncovered was damning. Off-line computer data in these years was stored on ten-inch reels of half-inch wide magnetic tape. Individual researchers had racks of such tapes in their offices for unclassified data. Tapes with classified data were stored in special rooms supervised by classified "custodians"—file clerks. Such tapes were marked with red-striped labels. When you ran a computer job with classified data, you checked out the tape from the custodian and hand-carried it to the computer center. (Yes, this today sounds very primitive!) To store new classified data, you signed out a blank tape in the stockroom, labeled it classified, and took it to the computer center, afterwards checking it in with the classified custodian for storage.

Investigation found that, over many months, Lee had signed for blank tapes from the stockroom and written to them from classified computer jobs. But, these tapes never made it into classified storage, nor were they found in his office. So, two disturbing questions were: Where were the tapes? And, what was on them?

From computer logs, it was possible to answer the second question. Lee had copied huge numbers of classified files onto the tapes. The particular inventory was alarming. Nuclear weapons design relies on computer programs, colloquially called "bomb codes," that simulate the physics of proposed designs—in a certain sense testing the weapon within the computer. These codes are classified, requiring a Q clearance for access. Bomb codes are not purely mathematical, however. They require physics input about the properties of chemical elements at high temperatures and densities—not just ordinary elements like iron or copper, but, self-evidently, materials like uranium and plutonium. That data, so-called *cross-sections* and *opacities*, had been physically measured in large, expensive experiments done at the weapons labs over decades, and from the results of hundreds of highly instrumented nuclear tests at the Nevada Test Site in the 1960s through 80s. Those data were classified and considered even more sensitive than the bomb codes themselves.

Lee's work involved bomb codes being developed to run on the largest computers in the world, the Lab's Advanced Super Computer Initiative. The ASCI codes weren't the programs that he copied, however. Instead, he dug back into the archives and copied ten- and twenty-year-old codes that had run on what were large computers then, but now (when he copied them) would run on modest machines—even desktop PCs. But, the data he copied for cross-sections and opacities were the most up-to-date. In effect, Wen Ho Lee's purloined tapes comprised a carefully curated nuclear weapons design kit for a country (or do-it-yourself group) with only modest, commercially available, computers. Illegal, incredibly dangerous, and yet not espionage until one could prove that he had given the tapes to a foreign power. Hence, the FBI's insistence that he be kept in place.

Meanwhile, in Washington, Bill Richardson had replaced Federico Peña as President Clinton's Secretary of Energy. Richardson was unsympathetic to Notra Trulock's conspiracy theories and had him demoted within DOE. Trulock took his story to congressman Chris Cox, who chaired the Select Committee on U.S. National Security in the Republican-controlled House of Representatives. It was the perfect match. Trulock wanted retribution and vindication in some combination. Cox, a right-wing Republican, wanted to discredit what he saw as the Clinton Administration's soft-on-China policies. Behind closed doors, in classified hearings, Trulock was a star witness with (now omitting the part about Chinese restaurants in Los Alamos) a clear narrative: A spy in Los Alamos gave to the Chinese the plans to the W-88, American's Cadillac. We knew who he was. He was still there because the Administration was soft on China, and because the Lab was soft on security.

On March 6, 1999, a Saturday, the *New York Times* broke the story with a page-one lead: "China Stole Nuclear Secrets For Bombs: Espionage Case at New Mexico Lab Is Said to Be Minimized by White House." It was a spectacularly inaccurate account, based on the Trulock-Cox narrative, plus an implication that it was DOE, not FBI, who insisted that the spy be left in place. The reporters, James Risen and Jeff Gerth, never contacted anyone at Los Alamos to verify any part of the story. It was the first of a succession of inaccurate stories in the *Times* over the next year. This was a part of my new education: I grew up reading the *Times* and believing what I read. Why was it known as "the gray lady," if not for its reportorial care and circumspection? How wrong I was! Risen, Gerth, and soon David Sanger and others, concocted a narratively self-consistent storyline, and they stuck to it in article after article, selecting facts that buttressed their narrative, blind to facts and sources that didn't.

That same Saturday, Bill Richardson called John Browne and told him to fire Wen Ho Lee. Richardson had belatedly realized that he was about to become the fall guy: He was responsible both *for* the DOE labs, and *to* the President on DOE's China policy. Richardson also called the *Times* to unburden himself on-the-record about his own longstanding doubts (new to us) about the University of California's ability to run a secure laboratory. That story ran on Sunday. John Browne called UC president Dick Atkinson. Looking for solidarity, he instead got the cold shoulder. No, Atkinson said, UC had no responsibility for Lab security. All fingers now pointed at Browne. *He* was going to be the fall guy. And, by the way, Atkinson said, if you want to fire Lee, be sure to follow all UC procedures.

Our SET meeting on Monday morning was unreal. Despite the damning classified evidence, there was nothing actually in Lee's employment record that, by the Lab's and UC's processes, would allow us to fire him. Still, he had to be fired. After some useless debate, and some grandstanding by Steve Younger, who was already trying to avoid becoming the fall guy himself—he was responsible for X-Division—we agreed that Lee could be fired for, in essence, failing to put classified labels on the still-missing computer tapes. Wednesday's *Times* reported that Bill Richardson personally and proactively had fired Lee for "failure to properly notify Energy Department and lab officials about contacts with people from a sensitive country, specific instances of failing to properly safeguard classified material, and apparently attempting to deceive lab officials about security matters." Richardson was ordering a thousand Los Alamos scientists to be polygraphed, the article said.

I may as well compress the next eighteen months into a few paragraphs. Thanks to Trulock, Cox, and the *New York Times*, the FBI now had to prove not just that Lee was a spy, but that he was spying for China. Nothing else would answer the political imperatives of the case. The problem was that, in all likelihood, he wasn't spying for China. What was on Lee's tapes had little if any connection to the W-88, and, anyway, the walk-in document was discredited. There was talk of a secret bank account in Lee's name in Canada that had been funded from Taiwan. There was talk that maybe he wasn't a real spy at all, but a kind of Walter Mitty who just wanted the powerful feeling of possessing a do-it-yourself kit of nuclear secrets. There was talk that he was preparing to become a spy, and that his intent was to sell the tapes to the highest bidder—any country would do. None of this could be proved. The tapes were never found. I happened to be acting lab director the day that the FBI, in desperation, asked for Lab assistance in shoveling and searching the top layers of the Los Alamos town dump. I agreed to "full cooperation," of course. Someone else found the dozen workers for this futile, smelly, patriotic task.

In December, nine months after the story broke, Wen Ho Lee was finally arrested. He was charged only with mishandling classified documents—on fifty-nine counts. Steve Younger testified to Judge James Parker that a whispered word from Lee could reveal the location of the tapes to the Chinese. As a result, Lee was held in solitary confinement under harsh conditions for 278 days. When Judge Parker finally ruled that the government would have to reveal the nature of what was on the tapes (albeit in camera and under seal), the government plea-bargained Lee down to one count of illegal retention of national security information, and he was released for time served. Judge Parker stated that he had been misled by the FBI and prosecutors and apologized to Lee for the conditions under which he had been held. A guess is that the government's turnaround occurred when they realized that the tapes, while damning, wouldn't in any way point to China—and, if not China, what was the point?

Shortly after Lee's release, the *New York Times* ran a long (1,700 word), tortured "correction". Clearly it was written by a committee that couldn't agree. The *Times*' coverage, they admitted, "attracted criticism from competing journalists and media critics and from defenders of Dr. Lee, who contended that our reporting had stimulated a political frenzy amounting to a witch hunt." The *Times* remained proud of their work, they said, but admitted that it did not "pay enough attention to the possibility that there had been a major intelligence loss in which the Los Alamos scientist was a minor player or completely uninvolved." "The *Times* should have moved more quickly to open a second line of reporting," they admitted—meaning not just a mouthpiece for Cox et al. This "correction" continued tilting back and forth between assertion and admission, yet still managed to miss the true story.

Lost in the Lab-security bashing that continued through all my time in Los Alamos was the fact that, under Browne, Los Alamos in fact had an effective counterintelligence program, directed by Ken Schiffer, a former FBI senior special agent. Counterintelligence at LANL was not the same as security. Security was "guards, gates, and guns," embodied in a large, in-your-face LANL division directed by Stan Busboom, a retired colonel from the Air Force military police. Ken Schiffer's small counterintelligence office, invisible by contrast, had reachback into intelligence data at FBI and CIA, and it vetted all foreign visitors and foreign postdoc hires at the Lab. Occasionally, a manager at the Lab was gently discouraged from making an invitation or job offer.

Ken grew up on a ranch and liked playing the cowboy—especially pulling out his false front teeth to prove that he had been kicked in the face by a horse. His claim to fame within FBI was the apprehension and 1986 conviction of Larry Wu-Tai Chin, the Chinese mole within CIA. Chin had agreed to cooperate with investigators, but committed suicide in his Virginia cell the day he was to be sentenced. At the time, conspiracy theories abounded. Ken and I got along well. I had the right intelligence clearances, including some that LANL didn't know about, but that he did. Meeting in a special room, Schiffer would show me interesting reports, and sometimes ask me questions: Did I think that a foreign postdoc's discussing a particular scientific topic in email home crossed the line into something sensitive? Virtually always the answer was no. Half of all work done at Los Alamos was unclassified and intended for open publication. Foreign nationals were restricted to the physically separate unclassified areas where such work was done. Real spies didn't pass information openly by email, we both knew. Ken was just being careful.

Oddly, Ken was a personal favorite among the foreign postdocs, despite their knowing exactly what his job was. They liked his cowboy persona. And, although the Lab had an International Office that was supposed to deal with such things, Ken's small operation was somehow better at fixing problems that foreigners encountered: visa issues, relations with landlords, how to wire money home—or get money from home. The counterintelligence office gave always friendly help. I grasped their intent immediately, and with admiration: Ken was *recruiting* spies, or at least setting the stage for someone else, perhaps years later, to do so. Had he succeeded yet? That was not something he talked about to me.

58. Cerro Grande

Bandelier National Monument is an area of fifty square miles bordering Los Alamos National Laboratory and Los Alamos County. It preserves the ruins of pueblo structures built between 1200 and 1600 AD, primitive in technology but awesome in scale. On Thursday, May 4, 2000, National Park Service personnel started a fire—a "controlled burn"—in a remote part of the monument west of LANL. Later, when it came time for finger-pointing, it would be noted that forecasts of high and gusting winds were ignored to keep the burn on schedule. On Sunday morning, with increased wind activity, the fire went out of control and started to burn east with rapidly increasing spread and intensity. It now had a name: Cerro Grande, after the mountain peak near where it started.

I was back the night before from a meeting at the Stanford Linear Accelerator Center in California. Jeffrey, James, and the dog Timber were on their way back from a search-and-rescue conference at Philmont Scout Ranch, near Cimarron, New Mexico. We had acquired Timber as a newborn puppy soon after our move to New Mexico. He was an intentional cross-breeding of two prize-winning search-and-rescue dogs, a golden retriever (belonging to our new neighbor Elaine) and a Labrador. At Elaine's urging, Jeffrey joined the local canine search-and-rescue club—Los Alamos had an abundance of clubs and civic organizations of all kinds—and started training with the dog. In fact, at Philmont, she had just passed her half of the certification test as a K-9 team. Timber's half was still a year off.

I got lunch at the local McDonalds and on the drive home noticed the large smoke plume to the west. By the time Jeffrey got home, the news was all over town that the fire was out of control and might reach Lab property. That would make it Dick Burick's concern as operations deputy. The Lab didn't have its own fire department, but instead subsidized the county to maintain an appropriately large force that was trained for all eventualities, up to and including plutonium fires (which, luckily, had never occurred).

Raise your hand if you don't want a digression about plutonium. OK, plutonium is pyrophoric, meaning that small chips or filings exposed to oxygen will burst into flame. Plutonium smoke is deadly. A single inhaled smoke particle lodged in the lung is likely over time to trigger a lung cancer. Plutonium workers pass through a full-body scanning machine at the end of every shift. During my time in Lab management there was one positive detection—of one single plutonium particle in a worker's lung. There had been no glove-box failure, so its origin was a mystery. The individual was treated with chelating agents that dissolved the particle and flushed the plutonium from his body. Had that failed, surgical removal would have been necessary.

Bulk metallic plutonium—of course in quantities less than the fission critical mass—is by comparison not very dangerous. Luis Alvarez liked to relate that, during the Manhattan Project at Los Alamos, he kept a small piece of metallic plutonium on the shelf above his workbench—chrome plated to keep it from flaking off deadly oxide particles. It was precious beyond price, and he had no actual need for it. It was just to show off.

Mindful of such risks, Burick and the Los Alamos fire chief met regularly for contingency planning. It didn't cross my mind to worry about the new fire personally. Our house on North Mesa was on the opposite side of the Lab and town from Bandelier. I was a bit miffed, however, to get a call that night from the Lab's emergency operations center telling me, first, that the lab and schools would be closed the next day—and second, that despite my being the number two Lab official, I had no assigned emergency function and should just stay home and watch the TV news like everyone else.

The plan to hold the fire at the LANL boundary marked by state highways 4 and 501 was abandoned on Monday morning when the fire jumped that line. On TV, reporters were explaining how fire could spread by repeated "spotting," jumping as much as a mile at a time in high winds; and also, with reporters' enthusiasm, speculating on how rapidly a fire might race through the long, pine-forested canyons that ran down the whole Pajarito Plateau. By Monday afternoon, a Type One incident team of 600 "Hot Shots" was brought in and tasked with establishing a defensive line running northwest from the Lab up the Camp May road to the ski hill and beyond. This was our first inkling of what the real danger was—what the officials weren't telling the public—that the fire might rapidly burn northeast across the mountains behind the town and then race down all the canyons near-simultaneously into the built-up areas of the Lab, the townsite, and the residential neighborhoods. Our house was on the edge of one of those canyons; and our property was dense with pine trees.

My Monday calendar was, as usual, packed with individual meetings. I reached many of the people by phone and got some of them to come to my house to meet. Lab computers and email servers were all operating normally, but everything else was far from normal. All day, and late into the night, we watched flames leaping up above the ridgeline on the far side of the Lab, and heard the sound of the slurry bombers flying overhead.

Tuesday was much the same. The Lab was closed. The fire line on Camp May road was strengthened by back burns and was judged to be an effective barrier. The fire on Lab property was substantially contained without approaching any buildings. However, on Tuesday afternoon, as a precaution, the county declared a "voluntary" evacuation of the Western Area, a neighborhood up against the mountains and closest to the fire's path if it jumped the Camp May line. About a thousand people were affected. Most went to houses of friends elsewhere in town. Closer to home for us was the announcement that all residents should start thinking about what they would pack, and where they would go, in the event of a general evacuation. It seemed in accord with my Lab safety training that I should take the announcement to heart, and I had little else to do. Our Cambridge friends Ed and Laura had once house-sat for a rich friend who owned a six-bedroom vacation retreat in Santa Fe. That might be comfortable! I managed to reach the owner, the wife, but got not much beyond, "*Who* gave you my number?"

People whom we personally knew were more hospitable. Hans and Vreneli Frauenfelder immediately offered the guest room of their large, modern house in Tesuque. And Phil Smith, who had recently retired as NAS executive director but was still living in Washington, gave me instructions for getting the key to his cozy vacation place, an apartment in an historic adobe-walled compound just a block from the galleries and restaurants of Canyon Road. So, we had plan and backup plan.

The Lab remained closed on Wednesday, but the smoke plume seemed much reduced. The news on KRSN, the town's single radio station, sounded optimistic. At home, I fielded a series of phone calls from people in my management span of control. I was embarrassed by the fact that I didn't know any more than they did. Around noon, I called the emergency operations center and made one more try at pulling rank. "This is the Principal Deputy Laboratory Director," I said. "I want to talk to the incident commander." This actually worked, and I got on the phone Stan Busboom, the security division director. Stan was Dick Burick's direct report, and I knew him well. He was a large man, more like a big bear than like a big cop—or colonel in the military police, as he had been. And, he was always utterly unflappable. I started by mentioning that, on the radio, things seemed pretty calm.

"Uh, well, Bill, actually things here are pretty close to what you might call a panic situation. The fire has jumped the Camp May line and is heading for the town. The county has decided to evacuate the whole town, but is not giving any hint of this to the media until the evacuation plan is in place. If I were you, I'd get out of town now before the roads jam solid." He delivered this alarming message in his normally calm tone. "Keep in touch," he concluded, as if we had just happened to meet on the street.

Busboom's advice was good, but I couldn't follow it. That morning, Jeffrey, with James and Timber, had set off with her hiking group, the Wednesday Irregulars, for Rio en Medio, a scenic part of the Santa Fe National Forest—and, taking safety into account, forty driving miles in the opposite direction from the fire. At the time, this had seemed not imprudent. She had her cell phone with her. Reception on the hiking trails was spotty, however, and I wasn't now able to reach her. I figured that, if she didn't make it back up the hill in time, she would figure out to rendezvous at the Frauenfelders. She had the newer all-wheel-drive Subaru. I had the old Honda wagon that Sara and I had driven cross-country. I turned it around in the driveway facing out, and started packing it with our computers, computer backup tapes and CDROMs, Jeffrey's several portable search-and-rescue radios (used, modified, Yaesu FT-23 five-watt handheld transceivers which also worked on the ham bands), our Motorola 465 MHz "family radios," chargers, GPS receivers—I could have filled the whole car with electronics, but remembered in time to pack some clothes from our respective sides of the closet. Later, I would realize that I had forgotten the single most valuable item in the house, my second-printing copy of Darwin's *On the Origin of Species* published in 1860 just two months after the first printing. I had bought it the year before for $3,000. (Worth much more today, it now lives in a bank safety deposit box.)

At about 1:00 p.m. the phone rang. Not Jeffrey, but the Los Alamos Emergency System (we never knew about it before) able to call all phone numbers in town simultaneously. The message, repeated over and over in a tape loop, lacked Busboom's unflappability: "THIS IS THE COUNTY. PACK YOUR CAR AND GET OUT OF LOS ALAMOS NOW!," followed by information on evacuation routes. Somewhat after 2:00 p.m., Jeffrey was able to get through briefly on her cell phone. She was just starting back for Los Alamos, with James, Timber, and three women hikers from her group. She reminded me to grab her jewelry, her mother's sterling flatware, and her jar of extra-high SPF sunscreen, a priority that the other women in the car thought hilarious.

Jeffrey rolled down the driveway just before four. Coming up the hill, she had been bucking the continuous line of cars in the reverse direction. She had gotten through several police roadblocks by asserting vehemently, "We have *three mothers* in this car and we've got to get home to evacuate our *families*." They waved her through. About fifteen minutes after she arrived, a police car, lights flashing, came down our driveway to order us out—or else be arrested. After redistributing some of our two-way radios, our family caravan of two cars had an excellent communications infrastructure, including a family list of primary and backup frequencies that Jeffrey thought laughable. There was little traffic on our road, San Ildefonso, when we pulled out of our long driveway. By now, we were near the tail end of the evacuation. Police were at the traffic circle (a major intersection), and we expected to be routed onto Diamond Drive, and thence through town and down the Main Hill Road. Instead, we were directed to continue on San Ildefonso Road across to Barranca Mesa, a direction that we thought of as leading only to a dead end.

However, in the Manhattan Project era, San Ildefonso Road was called "Escape Route 4," because it did connect to a tortuous dirt road, long considered washed out and impassable, that snaked ten miles downhill before reaching a paved highway. Doubly impassable, because the San Ildefonso Pueblo tribal government, over whose land the road ran, had blocked it with monster heavy-duty gates to counter the too many scofflaw Jeeps with bolt cutters. Secretly, the night before, an emergency road crew had graded and widened the whole ten-mile length of dirt road, so that it was now two lanes wide and completely smooth, though still steep in parts. Governor Perry Martinez of the San Ildefonso Pueblo had graciously agreed to open the Pueblo-controlled gates for the county's evacuation. It came out later that a county bulldozer was in position at the top gate in case Governor Martinez was not so gracious. So much for Indian sovereignty.

Traffic on the resurrected Escape Route 4 was a bumper-to-bumper line of cars, but moved smoothly at about 10 mph. I was reminded of evacuation scenes in the movies—*Deep Impact*, where the asteroid hits Earth, was one that came to mind—but this version was orderly, polite, and quiet. We never heard a horn honk. We saw a few stalled cars that had pulled to the side of the road; others had stopped to give assistance. When we reached the paved road, we turned left to Santa Fe. Most traffic turned right, back up the hill halfway, to White Rock, where people planned to stay with friends. As it turned out, this was a bad choice. That night at 1 a.m., the county decided to evacuate White Rock also. Police cars went door-to-door with their sirens going, to wake ten thousand people for a nighttime evacuation to school gymnasiums and other public buildings in Española and elsewhere in the Rio Grande valley. Hispanic community leaders proved to be welcoming, warm-hearted, and accommodating. Counting both the Hill and White Rock, 18,000 people were evacuated without a single mishap.

In Tesuque, the Frauenfelders welcomed us with enormous warmth and—with some trepidation—even welcomed Timber the dog, whose existence I had just somehow neglected to mention. Vreneli made an amazing dinner. Hans announced that, since it was an emergency, we would have the best wines in his oenophile collection. We spent the evening watching TV pictures of Los Alamos houses in flames. The Albuquerque stations were by now all running continuous 24-hour coverage of the fire.

We stayed with the Frauenfelders on Wednesday and Thursday nights. The second night was suspenseful, when high winds continued to spread the fire, and the maps of burned-out areas that the TV stations were showing seemed to include our neighborhood—but it was hard to be sure. We went to bed with an oddly light-hearted sense that we might have lost everything—and yet not everything, because here we were, safe and sound, drinking wine that we could otherwise never afford. It was two more days before we learned that our house was OK, based on two lines of evidence: A park ranger friend of Jeffrey's actually hiked in to our house and *touched* it; and our answering machine started answering the phone after a two-day lapse in power.

On Friday, our family moved from the Frauenfelders' to Phil Smith's house in Santa Fe. Also on Friday, the Lab's Senior Executive Team reconstituted itself, and met at the Santa Fe home of Janet Young, John Browne's chief-of-staff. This was not so easy to arrange: most of us were evacuees in unknown locations. Although most of us had cell phones, no one had ever thought to distribute a list of cell phone numbers for the SET as a whole. Back at the office, we had always relied on our admins to find people when we needed them. Without them, we were helpless.

This problem—being unable to locate people away from both their office and home phone numbers—turned into the single biggest rate-determining factor in putting the Lab back into operation. It was easy to push the "stop" button on a large, complex organization; but only after an actual experience does one discover that there is no correspondingly easy "start" button. I was reminded of this twenty years later, in the later stages of the world's coronavirus pandemic shutdown.

Saturday morning found Jeffrey and me reading newspapers in a Santa Fe café. It was hard to walk thirty feet in Santa Fe now without meeting a fellow refugee from Los Alamos—many of whom recognized me and had questions. Now that I was in contact with the SET again, I was able to pump facts (as opposed to rumors) into the exponential word-of-mouth grapevine. Raymond LaFlamme, a young Canadian scientist in T-Division, now continuously suffered jokes about his name. At his invitation we went to a potluck dinner for Los Alamos postdoc and student refugees, held at someone's girlfriend's apartment in a building that housed an avant-garde Santa Fe art colony. (LaFlamme was later famous for his discovery of quantum error correction, a prerequisite for building quantum computers.)

Sunday was my first chance to get back to the Lab. I was supposed to pass through a roadblock staffed by LANL security forces who had a list of names to let through—now including mine. But that route was closed by the new outbreak of a spot fire, so I was diverted many miles to a different roadblock, staffed by the State Police Mounted Patrol. (Most of these now drove Ford Crown Victorias, not horses.) A uniformed trooper with wraparound sunglasses stared at me and growled "What's going on?" I launched into a song and dance about how important I was, and how important it was that he let me through. "I know who you are, Bill," he said impatiently, taking off his sunglasses and showing himself to be Tom Granich, someone I worked with every day at the office. I just didn't know that he was a state police reservist. "What I'm trying to ask you is, *what's going on?*" He had been on duty for twelve hours with no instructions or communication. I told him to let through anyone with Lab ID and a good excuse.

The damage to major Lab buildings was miraculously small. The fire jumped around and over our major facilities, including those that housed explosives and plutonium. Some did read this as a divine statement of approval for the U.S. nuclear weapons program, but a more conventional explanation was that most buildings were made of concrete or other fire-resistant materials; and the Lab had followed good practices in eliminating flammable materials within defined distances around structures. The Lab had spent more than $1million the previous year on tree-thinning on its property and reducing fuel-loading near occupied areas generally. The investment proved its worth. That said, more than a quarter of the Lab's forty-three square miles of open space burned. Vast acreages of charred, dead pine trees would now need to be logged out, since, brittle and dry, they presented an enormous continuing fire hazard. The areas would need to be reforested. Over a hundred power poles burned. An even larger number would have to be replaced because the preservatives were boiled out from the wood, leaving it susceptible to rot. Ditto for hundreds of miles of power line cables with compromised insulation.

Sunday evening, the county re-opened White Rock, but kept Los Alamos closed. We stayed in Santa Fe. In the pizza restaurant where we ate dinner, people who hadn't previously known each other jumped from table to table trading stories, as if it was a big party.

During the fire and its immediate aftermath, the emergency operation center had taken over all aspects of managing the Lab, overriding line management with what was the moral equivalent of martial law. The EOC had pre-planned operational playbooks, and they followed them with remarkably good results. Now it was time to wind down the emergency and return the Lab to, in effect, "civilian" control by the line management chain. This had to be done *safely*. We didn't want researchers and office workers investigating on their own whether their labs' wiring was melted, or their building's drinking water was contaminated. In hundreds of buildings and other facilities, there would have to be a careful dance of inspectors, building engineers, line management, and personnel to effect a safe handover. Interestingly, there had been no pre-planning on this at all. Zero.

John Browne put me in charge of the re-opening. Burick would have been more logical for the operational side, but the line scientific managers—who wanted to get their people back to work—would not have trusted him to give priority to their interests: Burick was viewed as captive to the emergency operations cabal. I put together a small team and started, on Monday, May 15, to bring in people who came to be known as "Orange Badgers," for the new color of pass issued by the EOC, allowing our people to come on site. These were a steadily increasing cohort of managers and others needed to reestablish the management and programmatic chain. By Friday, May 19, we had several hundred Orange Badgers on site, and we were preparing to begin the wholesale re-occupation of Lab buildings on Monday, May 22.

I had my own command center—take that, EOC!—where I spent twelve or fourteen hours a day at a big conference table, in effect chairing a meeting with no beginning and no end and with participants coming and going. My role was to adjudicate conflicts by making decisions on the spot. Many of these involved obvious, but nevertheless unanticipated dependencies. You couldn't reopen a building until it had water and power. There couldn't be water and power until a particular person did something. That person was unlocatable. "Tell the line manager that he is authorized to assign anyone he thinks capable." Or, there were circularities that made things deadlock. The building had a storeroom containing a piece of equipment needed to get the water going, but without water for firefighting, the fire department had cordoned it off as unsafe to enter. "OK, you can open the building 'unsafely' for one eight-hour shift." No one questioned my authority to break or make up rules. It was a lot of fun, and reminded me of those movies about World War II where the hero breaks all the bureaucratic rules, saves the mission, and gets the girl—that would be Jeffrey, of course.

Monday, May 15 was also when the county re-opened Los Alamos to residents. We stayed in Santa Fe an extra day, but by Tuesday were back home. We had been evacuees for only six days. It seemed a lot longer! The smoke smell inside our house was strong, but it hadn't seeped into the walls, and it was gone within a week. There were several places on our property where lofted firebrands had landed and burned circles a foot or two across in the pine needles on the ground. It would have taken just one of these spreading fire to a dry pine tree, and our house would have been gone.

One afternoon, I spent an hour in a helicopter, flying back and forth over the burned areas of the Lab, town, and surrounding mountains. I was struck by the arbitrary variability of the damage. In the middle of untouched forest, I saw a few acres that had burned so intensely that there were not even charred trunks or logs, but just a level, uniform, vitrified grey ash. The experts told me that this required a heat of several thousand degrees. The many square miles of standing dead tree trunks were a future danger. In a few windy seasons, these would topple and produce an impassable black tangle, while the soil pulled up with their roots would be particularly susceptible to erosion.

We experienced the Cerro Grande fire as little more than an inconvenience. It was different for the four hundred families whose houses burned. Everyone in town knew many such people personally. The green, pine-forested mountains that previously framed our town, visible from most homes, were now black. The experts brought in told us that the first growth would be as a deciduous aspen forest. That would take fifty years or so. The aspens would then slowly be shaded out by growing pines. In a century (some said three) we would be back to the status quo ante. I write this now from our Los Alamos house, as it happens. It is June, twenty years after the fire. The mountains are green again, not yet forested, but blanketed with shoulder-high aspens poking up through opportunistic species of ground cover, grasses and flowers.

59. Hard Drives

At the end of May, 2000, the mood on the Senior Executive Team was high on survivor's exhilaration. There was still an enormous amount to be done—erosion and flood control were big operational concerns—but the Lab was back in its full operation of 12,000 people. Wen Ho Lee was in jail awaiting trial—the government's problem, not ours. And, thankfully, it was the government, not we, who had started the fire. Only occasionally, on slow news days, did the *New York Times* now rehash Los Alamos' bad karma. My re-start operational responsibilities were given back to Dick Burick.

Trish saw her job as filling my calendar from 8:00 a.m. to 6:00 p.m. Mentally, I sorted each day's entries into three buckets: Most were what I thought of as "keeping the prayer wheels turning," the day-to-day business of managing: meeting people I was expected to meet with, signing things I was expected to sign, and so on. My function was often no more than to encourage people to do what they wanted to do anyway, but didn't have the moxie to proceed without top cover. I was like the Wizard of Oz, meeting daily with a succession of Cowardly Lions and dispensing placebos of courage. Other times, my role was to apply a restraining force to speeding trains that were going off the rails—that being what brought them to my office in the first place. In this first bucket, anyone in my job would have done the same things—made the same decisions.

A second, smaller bucket of daily events was "tilting at windmills". Mike Anastasio, who was then my peer as a Livermore deputy laboratory director, but was later Lab Director, consecutively, at both Livermore and Los Alamos, once told me that he had a rubber stamp: "THIS WON'T HAPPEN." He used it to dispose of certain things that came across his desk. The stamp meant not that he necessarily agreed or disagreed, approved or disapproved. It was just his judgment of the mathematical probabilities—and a statement that he wouldn't waste time on the issue. I thought his approach possibly not right. Even things that WOULDN'T HAPPEN could have an influence on the system, and one could choose windmills so as to have a positive effect. Many that I tilted at were owned by the weapons program and thus by Steve Younger. Steve pushed for line-manager control of everything that fed into his program. But much was "dual use" science with both classified weapons and unclassified civilian applications.

An example was a major reorganization of supercomputing at the Lab that proposed to sweep multimillion dollar hardware acquisition, computer code development, basic computer science, and applied mathematics into a large new division—one under Younger. I battled to have it instead under Tom Meyer, whom John Browne had hired as associate laboratory director for the Lab's unclassified research. Tom had been vice chancellor for research at the University of North Carolina at Chapel Hill, like me a Lab outsider and an NAS member. This was a fight I was never going to win, but I was secretly satisfied when it resulted in more clout for basic science within the new division—even if it was under Younger.

Steve Younger, and also Don Cobb, the associate lab director for non-weapons national security programs, in turn took aim at some windmills of mine, for example trying to circumvent my new peer review rules for Laboratory Directed Research and Development. Younger and Cobb wanted to allocate LDRD funds for programmatic things that their Washington masters didn't want to pay for, often (in my view) faddish shiny objects of low scientific merit. Younger was the Jesuitical advocate for weapons. I was the Talmudic advocate for science. Cobb was neither Jesuitical nor Talmudic, but he was a salesman who never saw a research proposal that he didn't like—if it brought in funding for his own people. None of us three had much respect for John Browne's vision of SET comity.

The third bucket of events on my daily calendar, much smaller than the other two, were things that might actually advance a strategic agenda—might, when my time in the job was over, actually leave the Lab or the world a better place. Improving LDRD was one such. Another was the creation of a new biosciences division, "B-Division" at the Lab.

Going back to the days of the Atomic Energy Commission (DOE's earliest predecessor) there had always been some research in life sciences at Los Alamos. Specifically, DOE was responsible for research on the health effects of radiation. Since mutations of DNA were one such effect, all of modern genomics fell into DOE's scope. Few remember that the Human Genome Project was started within DOE—the National Institutes of Health joined only later—or that Los Alamos was an early participant in the project, doing the first sequencing of human chromosomes 16 and 5. Los Alamos was later partnered with two other DOE labs in running the Joint Genome Institute in Walnut Creek,

California—at whose board meetings I got to know Jim Watson, then in his seventies. Watson, with Francis Crick co-discoverer of the helical structure of DNA, was then just starting to undermine his scientific reputation with unsupported statements about race, some of which I heard from him directly. Jumping ahead in this sorry story, his view that darker skinned people had genetically stronger sex drives caused an uproar in 2000, and in 2007 he spoke publicly on his long-held view that people of African descent were less intelligent than whites. When, in 2019, in his nineties, he repeated these claims in an unfortunate interview for a TV documentary, the board of the Cold Spring Harbor Laboratory, his home institution, finally stripped him of his (by now emeritus) titles and cut all ties.

Unrelated to the human genome, Los Alamos researchers had invented the technology of flow cytometry, popularly called "cell sorting," an important contributor to modern massively parallel and big-data biology. In Theoretical Division, Bette Korber was developing a national database of HIV viral sequences, and would later do the same for COVID. There were other biological bits and pieces in several other divisions. I chartered a transition team headed by Jill Trewhella, a forceful Australian biophysicist who had come to LANL to use neutron scattering to map biological structures at the cellular level. I had to give this team a lot of top-cover, because "transition" was a euphemism for her stealing people—and their funding lines—from other divisions. Jill became B-Division's first director.

The first week in June, Jeffrey and I traveled to Pasadena for several days of events celebrating Kip's 60th birthday, "Kipfest". There were scientific talks by many of Kip's former students, and a day of public, popular talks including by Stephen Hawking, science-writer Timothy Ferris, and Alan Lightman—who had left Kip's group a theoretical astrophysicist, but now returned as a best-selling novelist. David Lee, Alan's graduate-school collaborator, also now returned, in his case as a billionaire, a co-founder of Global Crossing, a fiber-optic company then at its zenith. In the runup to Kip's event, Lee had given $10 million to Caltech, and was continuously fawned over by Caltech president David Baltimore.

During the Kipfest week, I got a guarded email from Trish, and then a phone call from Miles Baron, who had replaced George Kwei as my special assistant—more about Miles presently. Both communications hinted that there was some kind of turmoil at the Lab, but that it was being kept hushed-up as classified. Only when I got back did I learn the true horror of it.

Within X-Division—yes, that same one as for Wen Ho Lee—a dozen or more elite nuclear weapons designers and experimenters were members of the national Nuclear Emergency Search Team. NEST comprised personnel from the weapons labs, from the military's STRATCOM and special forces, from the intelligence agencies, and from domestic law enforcement. The team as a whole trained to respond in real time to accidents involving nuclear weapons (so-called "broken arrows"), or to terrorist threats that could involve stolen or improvised nuclear devices. The goal was assessment, emergency disablement, and (if ever required) advice to civil authorities on emergency evacuation.

At LANL, NEST members held special security clearances and went off to do days of training and exercises several times a year. Their Lab badges showed off a special designation. NEST team equipment was kept in a collection of aluminum footlockers, accessible only to NEST members and stored in special rooms. For exercises, or when the team was activated (which had occurred a few times over the years, all false alarms as far as was publicly known), there was a defined procedure for collecting the footlockers and, by van, rendezvousing at Kirtland Air Force Base in Albuquerque, where there was a special airplane able to transport the New Mexico component, which included Los Alamos and Sandia Lab people and some Kirtland military personnel, anywhere in the world.

During the Lab's evacuation for the fire, NEST members collected their equipment and moved it to a single vault room that was thought to be the most fireproof and secure. At that time, an inventory indicated two missing computer hard disk drives. This was not a particular concern, because the drives could be legitimately borrowed by team members for reference or updating. The assumption was that the drives were properly stored in someone's individual office safe within secure X-Division spaces; but there was no way to check this, because of the evacuation in progress.

By May 31, everyone was back, all NEST members had looked in their safes, and the drives were not found. John Browne and the NEST national leadership were notified. By the time I was back on June 7, a seventy-two-hour search of the whole Lab had turned up nothing, and, by then inevitably, the FBI had been called in. No fewer than sixty agents were assigned to the case. Several conference rooms and multiple offices on the third floor of the administration building (the Lab director's and my office was on the fourth floor) were taken over to establish a 24-hour law-enforcement command post. FBI was at this time coming under heavy criticism for their mishandling of the Wen Ho

Lee case, so this show of force was entirely political: Los Alamos was once again incapable of protecting nuclear secrets; the FBI would once again be the American people's protector. When I toured the FBI's commandeered spaces, I saw dozens of agents sitting around gabbing. There didn't seem to be any particular sense of urgency, or anyone actually doing anything at all.

In fact, in terms of the missing material itself, this was much worse than Wen Ho Lee's computer tapes of obsolete design codes. While the exact contents of the missing drives were never disclosed, one can guess that they contained things like precise blueprints for all known nuclear weapons—U.S. and (to the extent known from espionage or otherwise) foreign; specific procedures for disarming these weapons and overriding any anti-tamper functions; arming and disarming codes and procedures that may have been obtained by human intelligence; and, judging from the special clearances held by the team—one can only imagine what else. Today, such computer disk drives would routinely be encrypted with NSA-approved technology; but such wasn't then available.

On the other hand, there was not the slightest evidence that the drives had ever left X-Division protected spaces, that those spaces were at any time breached, or that the drives had been touched by anyone other than NEST team members, all of whom had been thoroughly vetted (by FBI, e.g.) for their high-level clearances. Most X-Division staff thought that the drives would eventually turn up and that there would prove to have been at most a minor security infraction, classified material stored within a protected classified space, but not (as procedure dictated) within a safe. That sort of infraction was regrettable, but occurred regularly. Infractions were less serious than violations or (worst) breaches. Repeated infractions were sometimes punished by a week or two forced leave without pay. But, if there was no possibility of an actual *criminal* act, then why were we hosting a floor full of FBI agents?

Predictably, it was the *New York Times* that broke the story on its front page on June 13 in an article by James Risen that also re-hashed Wen Ho Lee. Another front-page story followed the next day. The FBI made a show of conducting yet another search of X-Division, but it was mostly Potemkin. Then, they intensified their questioning of the two hundred people in X-Division. Polygraph tests were given, and of course a few people—probably people who were cheating on their spouses, or just naturally nervous—failed, a fact instantly leaked to the *New York Times*. The FBI was certain that someone *knew* where the drives were and that they just had to find that person and make him or her crack.

FBI started hauling people out of bed in the middle of the night, driving them to Albuquerque for all-night questioning, and then releasing them onto the street with no way to get home. That kind of thing drove cooperation between Lab employees and FBI (other than us senior management—we "fully cooperated") to zero. A broadside was circulated, not authorized by the Lab, explaining that, unless an actual warrant was produced, you *didn't* have to open your door to agents, you *didn't* have to come with them for questioning, they *couldn't* charge you with obstruction of justice, and—I thought interestingly—if they wanted to question you on Lab property, you could insist that a Lab manager be present. The situation was a total stalemate.

Above, I promised to say more about Miles Baron, my special assistant and among the most remarkable people I have ever known. Miles wasn't even my special assistant—not exactly. He had interviewed for the two-year rotating position, but I had hired another scientist. Miles had then gone to his manager in (yes) X-Division and requested that she assign him half-time to my office. That manager—small town life—was our neighbor Elaine, godmother to our dog, and Jeffrey's compatriot in canine search-and-rescue. "Do you want Miles?" she asked me. She would continue to cover his salary, so obviously the answer was yes.

Miles' physical appearance was jaw-dropping. He could have modeled for a nineteenth century oil portrait of a Great Plains Indian chief, or a bronze sculpture of a stereotype noble Native American on sale for twenty-five thousand dollars in an art gallery on Canyon Road. In fact, with dark skin, he *was* a bronze sculpture brought to life. He was a long-distance runner, wiry, tense, thin, who ran solo half-marathons every day over lunch hour in remote parts of the Lab's property. He was born in San Ildefonso Pueblo, but was given up at birth for adoption by a middle-class Anglo couple in Farmington, two hundred miles to the west. The family moved several times while he was growing up. Miles graduated from high school in Boca Raton, Florida, did his undergraduate work at Auburn University in Alabama, and earned a Ph.D. in physics at Texas Tech. He then moved to Los Alamos, in effect back to his birthplace.

Miles told me that he had always dreamed of someday finding his birth mother. His inquiries at San Ildefonso Pueblo were rebuffed—it was a black mark against the Pueblo that any baby had been given up for adoption. Once, when I visited San Ildefonso, its Lieutenant Governor accosted me: "Does Miles Baron work for you?" I said yes. "He's no Indian. You shouldn't let him claim to be Indian." I explained that we had no control over how people self-identified for affirmative action purposes.

Miles was outstanding at handling tasks where facts were in dispute and someone was needed to get to the bottom of things—but only if I didn't care if he broke some crockery in doing so. If finesse was not Miles' strong suit, his written memos to me were reliable summaries of *who* was doing *what* to *whom* and *why*. He was a stern judge and always disappointed when, informed by his investigations, I took less than the most extreme possible action.

Mid-morning on Friday, June 16, Miles was in my outer office wanting to see me. He had a new theory about the hard drives, he said. I was rushing off somewhere and told Trish to fit him into my afternoon's schedule. Lunchtime found me, as always, eating in the cafeteria at the geezer's table.

That deserves an elaboration: None of my colleagues on the SET ever ate lunch in public. Instead, their admins went to the cafeteria and brought back their food packaged to-go. People thought that they did this to save time; but I thought they did it to avoid being cornered by tendentious Lab employees. The geezers table in the cafeteria had existed before I arrived, the regular lunch gathering of a group of Lab old-hand physicists, some retired. My visibly eating lunch every day with this group earned me an improved reputation with Lab scientists as not just another despised and out-of-touch senior manager; and the geezers' own venerable reputation—and our loud laughter—kept at a distance lesser employees with axes to grind. Also, as Jeffrey would remind me, these geezers were actually my *friends*.

This day, Trish ran to the cafeteria and penetrated the geezer invisible shield to tell me to come with her right away. "You should have made time for Miles this morning," she said ruefully. The story was that, failing to see me, Miles had decided to put his new theory directly into practice. He had gone into X-Division's lair (for which he had the necessary clearances), told people that he was acting at my behest, and started his own search for the hard drives. "Oh no," I said, thinking of the mess that I would now have to clean up. I *hated* apologizing to Steve Younger.

"And he found them," Trish finished. I was flabbergasted. That wasn't all. Miles had then decided that his duty was to report his discovery of the drives only to his managers, that is, to our dog-neighbor Elaine and me, and not to the horde of FBI agents. So he was waiting in my office. Elaine met us at the vault door to X-Division—he had already told her. The two led me to a copying room, windowless with a large Xerox machine and a cabinet of office supplies. The door was being guarded by Elaine's admin. "The hard drives are behind the Xerox machine," Miles told me. "No one has touched them." I squeezed alongside the Xerox machine and looked without touching. There were two objects, each wrapped in bubble wrap of the sort used to protect electronics equipment, both suspended above the floor in the mess of cables and power cords behind the machine. They might have been put there by someone, or they could have been on top of the machine and fallen off the back.

I understood that every passing minute would scream cover-up, but I wanted some extra heft on my side. John Browne was out of town, so the acting Lab Director was—me. On the fourth floor, I rounded up Lab general counsel Frank Dickson, his deputy Sheila Brown, and the Director's Office chief-of-staff Janet Young. Our now-formidable delegation went down a flight to the FBI's command post and informed the special-agent-in-charge, whose name was Greg, that the drives had been located and were being guarded in situ for forensic analysis.

Foreseeably, all hell then broke loose. Greg, who was hugely obese, shouted at a dozen agents, a mob whom we guided to the copy room. Elaine offered the agents latex gloves from her search-and-rescue kit. The agents pushed her aside and all crammed into the tiny room. Greg was too fat to squeeze beside the Xerox machine, so he put his beefy arms around the machine and yanked it out from the wall. There was a clatter as the drives fell to the ground. Now able to reach, he grabbed them and ripped off the bubble wrap. So much for any follow-up forensics. "We've found them," he shouted to the other agents.

Miles himself was immediately the prime suspect. How could he have found the drives after multiple previous searches unless he himself put them there? Miles insisted that he would answer FBI questions only with a Lab manager present. Frank Dickson and I sat for the first hour and then turned the job over to Sheila Brown and others. Miles' questioning went on for days. He never wavered from exactly the same retelling of facts, calmly, over and over; but he did himself no service by including in the telling his theory of that morning. His theory was that the FBI was right: There was someone in X-Division with guilty knowledge. The drives had been stored in that perp's office or safe, perhaps a security infraction or perhaps not. Maybe the perp knew about the May 7 inventory, or maybe not. In any case, by the time the Lab was reoccupied after the fire, the perp was scared—thought he might be fired—and took steps to shield the drives from the searches, maybe a shuffle between offices, maybe a personal backpack. You could imagine ways. The perp's idea was to let them be found, eventually, in a place that didn't reveal his or her own identity. "And I woke up this morning," Miles' account always concluded, word-for-word, "with just a *feeling* that today would be the day. So, I went to look for them."

Poor Miles! You had to know him to know that he might actually have such a theory, believe and act on it. A much more plausible theory, after the fact, was that some NEST team member, in haste to gather the equipment and possibly copy some NEST-related instructions, had let the drives fall off the back of the Xerox machine. In the chaos of an evacuation in progress, the person either forgot about the drives, or thought that some other NEST member had grabbed them. The only unlikely part was that, in a month, no one except Miles thought to look behind the Xerox machine. Forensic analysis (of dust on the bubble-wrap, for example) could easily have told whether the drives had been behind the machine for six weeks undisturbed, but the FBI's elephantine agent-in-charge ruined that possibility.

I'll skip ahead. Miles, an Eagle Scout, imagined that finding the drives and telling the truth would make him a hero. It doesn't work that way. Instead, for many months, he had to pay for a high-profile criminal defense lawyer. The University of California's policy was to pay the legal expenses of an employee whose liability was incurred in the reasonable performance of their job—but only after the matter was finally settled, or not at all if the employee was convicted of a criminal charge. No one except the FBI could see a crime—as distinct from a general fuck-up—anywhere in these events. But undependable as always, UC weaseled out of paying. Miles mortgaged his house to pay for his lawyer. It was a year before the FBI grudgingly exonerated him. He had sunk into depression. He and his wife, who also worked at the Lab, were divorced. There was no home for him in X-Division, because his theory implied that an X-Division colleague was a guilty party, so no one wanted to work with him. He moved from the weapons program to intelligence programs. He died in 2019 at age 59 of metastatic prostate cancer.

But back to 2000, when John Browne was very much alive and was hanging on as Laboratory Director. John understood that Wen Ho Lee and the hard drives were two big strikes against him, in Washington and with UC. Privately, he appreciated the irony that neither incident was his fault in any personal way. But it was clear that one further big strike would be the end of his job—and, depending on circumstances, mine, too. That third strike took two more years to arrive, but it eventually did.

60. The Process

My leave of absence from Harvard was set to expire after two years, at the end of 2000. Long before then, Jeffrey and I knew that we weren't moving back to Cambridge. The Lab's existential crises, and our own family's travails—the fire evacuation especially—had unexpectedly bonded us to this peculiar small town at the end of the road. I talked to John Browne about our situation. I would stay as his deputy for as long as he wanted, but I needed something in writing specifying that, when I departed Lab management "for any reason," I would still have a job at LANL as a senior scientist. We both knew, without saying so, that my concern was really that he would, sooner or later, get fired. The Lab lawyers filled John's letter with conditions about "termination for cause" and "cannot legally but only morally bind any future director," but the intent of the letter he gave me was satisfactorily clear.

I never officially told Harvard that I wasn't coming back. Unofficially, so that they could plan, I told my two department chairs, David Nelson in physics and Ramesh Narayan in astronomy; and of course, our friends all knew. I had left behind shelves of less important professional books, and I arranged for the astronomy graduate students to pick over them, the remains going to the Bryn Mawr bookstore. We easily sold our Cambridge house to the people who had been renting it. I sent to physics laboratory director Margaret Law a list of half a dozen small Harvard research and gift accounts that I secretly controlled—in twenty years, I had done my share of academic money laundering—and requested only that the funds be put to good use.

In December, the dean's office sent me a checklist for returning from leave, which I ignored. In January, 2001, I got a letter, "Dear Professor Press, has it escaped our notice that you have returned to Cambridge? Please advise," etc. It was childish of me, but I ignored that one, too. In June, I got a letter from the Harvard general counsel reciting the university's exact rules and telling me that I was no longer a Harvard professor. So (I like to tell people), I never resigned from Harvard—I was fired. Or maybe not: Almost a year later, I got a letter from a different part of Harvard congratulating me on now having served on the faculty for twenty-five years and asking whether I wanted as my gift the engraved pewter plate or the insignia rocking chair. I chose the commemorative plate, and they duly sent it to me in Los Alamos.

In the wake of the Lab's crises John Browne made some management changes. Dick Burick retired as Deputy Laboratory Director for Operations and was replaced by Joe Salgado, who had come to the Lab a year before as the new associate lab director for business operations. I wasn't involved in John's hiring Salgado as Burick's replacement, but at the time he seemed an odd choice. Joe had been Deputy Secretary of Energy in the later Reagan years, the DOE's chief operating officer. That job was light-years above the one he took at LANL a decade later. Despite his name, the legacy of an absent father, there was nothing Hispanic about Joe. I think he once told me that his mother was Albanian. In the 1970s, he was a full-time cop in the Oakland police department. He earned night-school undergraduate and law degrees. He became a prosecuting assistant district attorney in Alameda County. He was tapped by the Reagan administration to be the associate head of enforcement for the Immigration and Naturalization Service. After that, DOE. In brief, Joe was a tough guy. That he had been under-employed in the post-Reagan years might have been a warning sign. But, almost all of us who might come to this Lab-at-the-end-of-the-road as senior management were damaged goods in one way or another, myself included. Joe might be a thug, but he was our thug.

Conveniently, Steve Younger left in mid-2001 to become director of the Defense Threat Reduction Agency in Washington. John Browne elevated Steve's position to the deputy laboratory director level and appointed John Immele. Among weaponeers, Immele was low key, almost professorial. He and I got along fine, notwithstanding that I had been subtly demoted: I was now the deputy laboratory director for science and technology, and no longer "principal" deputy. The three of us, Salgado, Immele, and I, were supposed to be co-equals. When John was out of town, it was usually Joe who was acting director.

In fact, we three did get along well. Managing "down," Salgado was a bully; but he treated Immele and me with kid gloves. On any issue that might affect my interests, Joe made a great show of consultation: "What would *you* like me to do here, Bill?" It was unmistakable that he was arrogating power to his office, but he did it by relieving me (or Immele) of burdens, not by interfering in things that either of us cared about. The same applied to the division of labor between Salgado and Browne. One increasingly heard from Joe, "I can fix that. Let's not bother John." Possibly this was the kind of principal deputy that I should have been earlier on—but neither was it in my nature, nor clearly communicated by Browne.

An interesting angle on Joe was that he was only the outward facing half of a team. The other half was Jo Ann Milam, whom he had brought with him to LANL. She had been with him from his DOE days and was the invisible brains of the outfit. Jo Ann was a "get to yes" person in the best sense, and she was *very* smart. Once I figured out their relationship (purely professional—Jo Ann had a retired husband who moved with her from job to job), I always went to Jo Ann first. She and I worked things out, and she then programmed Joe, who applied any necessary muscle. So finally, a bit too late, John Browne had the collegial Senior Executive Team that he had wanted all along.

Joe's own social situation was mysterious. He was rumored to be divorced. We invited him to dinner or parties many times, but he always sent regrets, and others had the same experience. Outside of the Lab, his life was invisible.

Meanwhile, in Washington, there were big changes. Because two of DOE's three nuclear labs were in New Mexico, Senator Pete Domenici—a major voice in Congress on defense issues generally—was all-powerful on matters that affected Los Alamos and Sandia. One of my duties as John Browne's deputy was to be frequently one of the nodding chorus of yes-persons sitting behind Domenici on the speaker's platform when he gave in-state speeches on defense issues. Previously, Domenici, a moderate Republican, had supported UC's management of Los Alamos under a contract that limited direct interference from DOE. Probably he credited UC with actually adding value; but by the time of the 1999 and 2000 crises, the university was, in any practical sense, AWOL. It was around this time that Pete Lyons, a technically capable LANL mid-level manager, joined Domenici's staff as science advisor. In his advice to the senator, Lyons was honest—and scathing—about how broken the system was.

Influenced by Lyons, Domenici swung around to the view that the Labs—and not just the Labs but the whole national nuclear weapons complex—needed firmer government supervision. DOE was too soft, too science-oriented, to do the job. A talk that he gave at the Lab came to be known as "Domenici's tough love speech," and it left little doubt about his change of heart. The senator introduced legislation to create an independent agency to take over from DOE, to be called the National Nuclear Security Administration. Domenici's proposal was opposed by the Clinton administration—or, more accurately, by Secretary of Energy and fellow New Mexican Bill Richardson. Richardson, a Democrat, imagined himself Domenici's rival. (Richardson was later New Mexico governor.) The lame compromise reached on NNSA was to create it as a sub-agency within DOE. NNSA had authority to make its own rules and regulations, but NNSA contractors—the Labs—would also be subject to all of DOE's rules and regulations, a bureaucratic nightmare.

This is worth enlarging on as an example of government gone haywire. Already in DOE, and even more so under NNSA, there existed a divide between officials responsible for funding and supervising the Lab's programs—these implementing administration policy and Congressional guidance—and those responsible for monitoring operations in the weapons complex. The latter were far more numerous; but, before the crises and the creation of NNSA, they were functionaries who checked boxes and pushed paper—and they were at least grudgingly responsive to the needs of DOE program officials whose job was to get things done. The operations side of NNSA was a vast bureaucratic web with employees based in Washington (thousands, literally), in the Albuquerque field office (hundreds) and at the Los Alamos site office (dozens). Under DOE, they had pretty much done their own thing and we at LANL (taking orders from our Washington programmatic masters, who numbered a few dozen) did ours.

LANL's operational crises and the creation of NNSA upended everything. Operations now trumped program absolutely. The Lab became newly responsible, day-to-day, to hundreds of GS-10s, 11s, and 12s—equivalent to military junior officers like lieutenants and captains. It made no difference that our intellectual bosses, the program officials, were as unhappy as we were. They were powerless.

Another internecine bureaucratic conflict within NNSA was exposed in LANL's first contract performance review under the new agency: The previous scheme for assigning performance "points" to UC was a checklist of more than three-hundred items, so-called Appendix F. Each item was the responsibility of one of those GS-11s or 12s, all of whom jealously guarded their prerogatives to award or withhold points. In our first review after Wen Ho Lee and the lost hard drives, when these points were totaled up, LANL unexpectedly did just fine! No one's assignment was to withhold points for "harboring Chinese spies" or "embarrassing high political appointees." However, Bill Richardson felt that his political future depended on his punishing Los Alamos. He ordered an immediate renegotiation of the UC contract, especially Appendix F. He wanted political appointees, not career weenies, to be able to grade us as unsatisfactory whenever the *New York Times* made it expedient.

The negotiation over Appendix F took several months. Jo Ann Milam represented the Lab. Rulon Linford, a former LANL manager now in the UC president's office, represented UC. NNSA was represented by—surprise!—its weenies. These low-level bureaucrats had no love of LANL, but they hated their politician masters even more. Subverting the charge from above was a shared priority. The new Appendix F had thirty measures—down from three hundred—and the unique contractual provision that, for each measure, the Lab must submit to NNSA its own self-assessment before every annual review.

These self-assessments were Jo Ann's brilliant invention, and it was from her that I learned a lot about how to be a tricky bureaucrat myself. The magic word was always *process*. In a rational world, each of our thirty self-assessments could have been written by the appropriate manager, no more than half a page, in an hour. But that would have carried no weight. Instead, Jo Ann and I organized a dozen separate committees, all of whom wrote lengthy reports. These fed into a Laboratory-wide Contract Performance Evaluation Board, CPEB, which I chaired under Jo Ann's invisible guidance.

Why me? That was Jo Ann's idea too. Salgado was a thug, and Immele dealt only with weapons. I had an outside scientific reputation. Also, every science or engineering division at LANL already had an outside visiting committee that met every year and produced a report. Already, a part of my job was organizing that process, appointing and meeting with each of these two dozen groups when they visited. With a slight shift in guidance to these committees, these reports by distinguished outsiders could also feed into the CPEB *process*. The great bureaucratic secret was this: The more *process* there was behind something, the less likely that any single government official, even a political appointee, would stick his or her neck out to overrule it—this despite the fact that our self-assessment process was transparently self-serving. Process was vastly more important than substance or objectivity.

Inspired by Jo Ann, I also invented another process. At my request, John Browne somewhat reluctantly delegated to me oversight of the Lab's cybersecurity. I created an InfoSec Policy Board. I was less cynical about this IPB than about the CPEB. I classed the IPB as strategic, not prayer wheel. At this time, cybersecurity was hardly a blip on the public's radar. The Lab, as a facility doing classified work, was far ahead of most places, but the measures enforced by the security professionals were often based on out-of-date checklists handed down by DOE. I knew from some of my highly classified work outside the Lab what real computer attacks by capable foreign intelligence services looked like. I couldn't use this compartmented information in any direct way, but I could use my general understanding of it to steer LANL toward effective, instead of useless, policies.

The IPB *process* was to hold weekly meetings, around a single table, of high-level managers from Lab constituencies that normally didn't communicate—or did so only by fiat directives. Security director Stan Busboom and his three top computer security people attended. So did Ken Schiffer from counterintelligence. Eyebrows were raised by my including several ordinary researchers, but they were people whom I knew to hold high level intelligence clearances and some actual understanding of the cyber threat. Finally, and importantly, I included on a rotating basis division directors and key group leaders. These were scientist-managers with previously no interest in cybersecurity other than thinking it a bureaucratic impediment in their people's work. I was trying to apply principles of safety culture to cyber. I wanted to plant the seed of the idea that, just as for safety, managers should understand cyber hazards and take responsibility for mitigating them, with the cybersecurity professionals furnishing proper tools to do so. I also wanted the security professionals—whom I liked and got along with—to have some understanding of what was workable vs. unworkable, as seen from below.

Internally in the Lab, the IPB was a great success and lasted for three years. New division leaders pressed to be invited. And, increasingly, the security people brought to the table issues that had nothing to do with cyber or computers. It was a kind of bootlegging: The IPB was the only forum at the Lab where they could meet scientific managers, hash out issues, and get to reasonable solutions.

I never got much credit from John Browne or the rest of the SET for this, because—although the IPB had a meticulous process, with written meeting minutes, draft proposals, and formal votes—our activity amounted to a way for middle management to cut their bosses (my SET colleagues) out of the loop. The SET knew that the IPB existed, and they assumed that it dealt only with arcane matters of cyber. We let them think that.

In the end, the InfoSec Policy Board was done in by NNSA. A team of auditor-inspector types was routinely sent from Washington to assess our cyber practices and routinely punish us—as inspectors always did. Stan Busboom should have known better, but he pointed with pride to progress made via the IPB. *What? Who at headquarters authorized that?* The IPB was soon dissolved. It turned out, after all, to be not a strategic advance, but just another prayer wheel, and after three years, its bearings wore out. The Lab went back to obsolete cyber checklists—now just many more of them.

In September, 2001, I was in Washington for a meeting of an NAS activity with the ungainly title Commission on Physical Sciences and Mathematical Applications. I had recently become its co-chair. CPSMA was my first scientific committee with about equal numbers of men and women. At this time, especially in the physical sciences, this was unheard of. I delicately asked my other co-chair, a male, about the anomaly during a coffee break. "Look," he said. "This isn't really such a big deal. At this level in the science pyramid there are plenty of qualified people, including plenty of qualified women. A couple of years ago, we had an unusual number of members rotating off, so we replaced them all with women." Some might debate the merits of this approach, but I took it as a lesson: If you want to make progress in diversity, *just do it*.

Years later, when I was on the NAS committee that nominated candidates for election to the NAS Council, the dynamic had shifted: If a woman ran against a man, she always won. NAS members were still dominantly white males, but seemingly they were voting with some degree of guilt or perhaps enlightenment. Towards the end of our process, a male on the nominating committee objected to our final slate: "At this rate, the Council will soon be dominated by women." The fact that the Council had been all male for its first more than one hundred years didn't seem to bother him as much. A few years after that, when I was on Council myself, two-thirds of the councilors were women, but, in paired races, the man now sometimes actually won. This was progress of a kind.

Back now in 2001, the second day of our CPSMA meeting was on September 11. Ten minutes before nine, an Academy staffer ran in and switched on the TV in our conference room. In New York, the North Tower of the World Trade Center had just been hit by an airplane. The tower was standing, and it was being reported as an accident; but at 9:02, when another airplane hit the South Tower, we knew that it wasn't. Half an hour later, a third airplane crashed into the Pentagon, just a couple of miles from where we were meeting. Then, in the next hour, the two towers in New York collapsed, killing three thousand people.

Sara, my daughter, who was living in a warehouse loft in Harlem, managed to get through to me by cell phone a couple of hours after the towers collapsed. She was thinking of bicycling ten miles downtown to have a look at things. I convinced her that this was not a good idea. No one then knew whether the airplanes were only the first sally, with the worst still to come. Nuclear terrorism was even a possibility. I told Sara that if she saw a blinding flash in the direction of downtown, she would have almost a minute to find an interior shelter. The shock wave at that distance would blow in windows, but probably leave her brick warehouse building standing. She should then walk north to the GW Bridge and across it to New Jersey. "You won't get lost," I said. "Just follow the other million people."

I learned a few days later that a friend, a computer scientist who had defected to the financial industry, was in his office in one of the towers, on a floor below the airplane strike. As a routine safety practice, his group had practiced fire drills, including walking all the way down stairwells, many floors, to street level. When the alarms went off, they didn't know anything about it, but they dutifully all walked down the stairs and out of the building before it collapsed. All were saved. Firefighters were rushing in the other direction, up the stairwell. All died. My friend walked home ten miles, most of the length of Manhattan.

Everyone then alive has their own 9/11 stories, and I'll be brief with mine. I was scheduled to fly home, but airspace over the whole United States was shut down indefinitely, so I went to stay with my parents in their Watergate apartment. We could see from their balcony the plume of black smoke from the Pentagon. There were about a hundred and fifty LANL employees in D.C. that day, a typical number. The Lab travel office knew who we were, whether we had rented cars, and what were our cell phone numbers; and they emailed a spreadsheet with the information to everyone. Cars were the limiting factor, since any unreserved rental cars were immediately taken.

Sallie Keller, a group leader whom I knew slightly, dialed the LANL list until she found someone with a car that they were not planning to use. She then called me—a courtesy to a "senior manager". The two of us drove back to Los Alamos, thirty-two driving hours over two days. The Interstates were dense with a rainbow of assorted state license plates. The rental car companies all waived one-way drop-off fees, not that fees would have deterred anyone: Driving was the only way to get anywhere at any price. At every gas stop, strangers stood around trading stories. Sallie and I became as war buddies, a connection that came to influence both of our careers: She was later a member of JASON, dean of engineering at Rice University, and president of the American Statistical Association, elected to the National Academy of Engineering in 2020.

61. The Fall

Mayor Richard Lucero's office in Española contacted Trish to invite me to spend a day walking in the mayor's shoes—about the hundredth time I had heard that expression in New Mexico. I met him at 7:30 a.m. at the Española feed store, Country Farm Supply, that his family owned. He had been mayor on and off for twenty years, and was a *patrón* among *patrones*. I had met him a few times when we both sat onstage at Pete Domenici events. Lucero had the magical charisma of a great leader or successful politician. If he had been born in Texas or Florida, he could have become a governor or senator.

In the feed store, Lucero sat on a pile of forty-pound sacks and held court. Except for the rural staging, it was exactly the famous scene in *The Godfather* movie. People with problems or disputes were ushered into the mayor's presence, and he dispensed solutions or justice. An assistant, a young man, kept notes. One petitioner was behind in his rent and facing eviction. Lucero agreed to fix it with the landlord—but beware if the man didn't catch up on rent by the next month. Another wanted the town to pave the dirt alley behind his house. Lucero would look into it, he said, but suggested that there was not much hope. A mother brought in a new baby for the mayor to, in effect, bless.

Lucero drove the two of us around town in his Cadillac. He pointed with pride to the Richard Lucero Public Library and the Richard Lucero Recreation Center, which boasted an Olympic-sized swimming pool. A current project was the construction of a soon-to-be-beautiful Plaza de Española on a dusty square of town land. Already built was a half-scale re-creation of Alhambra Gardens in Grenada (not yet with water) and, near to it, a replica Spanish *misión y convento*, a church basically. Funding for the project was a gift from Senator Pete Domenici, via an earmark on a federal appropriation bill. But this was a source of frustration to Lucero: Since the Plaza was to be officially a national historical site, the National Park Service would not allow the church to be consecrated as such. Designated as a museum, it could not be reserved for baptisms, confirmations, weddings, or funerals. Could the Lab do anything to help on the issue? he asked me. I said that I doubted it.

Still walking in the mayor's shoes, I went with him to a party in an Española family's small home. The occasion was a young woman's graduating from high school with honors and, the main cause for celebration, her getting a job as an entry-level admin at the Lab. The mayor made sure that I understood what a big deal this was. She would have a reliable income, promotions, a retirement plan, and health insurance. She would be more desirable in marriage, because her husband and children would be covered on her health insurance. The girl was in a frilly party dress and was beaming. The mayor gave her a big hug and implied that she had gotten the job by his influence and good relationship with senior administrators at the Lab—now including me, he said pointedly.

Only toward the end of my visit, cruising in his Cadillac with the air conditioning going full blast, did Lucero get to the point about what he really wanted from me. LANL long had some college scholarship funds to be awarded, based on need, to children of Lab employees. I had succeeded, after much bureaucratic to-and-fro, in getting permission from UC and NNSA to use the money for scholarships for other kids, not necessarily Lab families, in the poorer counties bordering the Lab. New Mexico politics was involved, because, if we found kids for the scholarships, much of the money would flow to the New Mexico universities, UNM in Albuquerque, NM Tech in Socorro, and NM State in Las Cruces. Mayor Lucero was well informed about this, and he didn't like it. "I don't want you to give college scholarships to our children," he told me. All over America, he insisted, small towns were dying because young people were leaving for college and not coming back. His town had a history over four centuries. He had spent the day with me to show me that it was alive and vibrant. It wasn't just the girl who was celebrating; it was her whole extended family, knowing now that she wouldn't be forced economically to move away, and that she would also anchor some lucky boy. "Don't take our children away from us," Lucero said, in some combination of plea and command. He had an alternative in mind: The Lab's money should be used to establish a barber college in Española. I gathered that this was something he had already asked Domenici and been turned down.

He wanted an answer from me. *Patrones* could make deals in parked cars and honor them. "Mr. Mayor," I said. "I've learned a lot today. I'm grateful to you. But you have your job, and I have mine. A part of my job is to get smart kids to go to college. I have no choice about this." He gave me a disappointed look and dropped me off at my car. His Cadillac versus my Subaru said it all. Lucero died in 2024 at the age of 88.

In a curious and indirect way, my experience with Mayor Lucero is related to John Browne's final fall from grace. But I need to connect a few more dots.

If someone at the Lab broke the law, you might think that they would be arrested, prosecuted, convicted, and sentenced. Not so simple! Physical security at LANL was provided by a contractor, Protective Technologies of Los Alamos. PTLA was the civilian descendant of the military force that guarded the Lab during World War II and beyond through the 1950s. The Lab was PTLA's only client. Their people dressed in battle fatigues, carried handguns and automatic weapons, and trained with heavier equipment—armored personnel carriers with machine guns—against the day that the Lab might be attacked by terrorists seeking nuclear materials like plutonium. After 9/11 this seemed not quite as fanciful as before. But PTLA was not a police force. They had no power of arrest. The Los Alamos police department had no jurisdiction on Lab property, and they, and the county council, wanted to keep it that way. The closest people with power to actually arrest someone on Lab property were FBI agents in Albuquerque.

This arrangement had sometimes awkward consequences. Every August 6, the anniversary of Hiroshima, the Lab hosted a small nuclear protest—for several years led by the actor Martin Sheen. "Hosted" is the right word. The day before, we sent a memo to all employees, warning that any harassment of peaceful protesters would subject them to disciplinary action. We cleared out the main parking lot in front of the administration building for the protesters' use. We supplied drinking water so that no one would collapse in the heat. PTLA always made an impressive show of force with its line of impassive security officers in battle gear. And, importantly, they were all checked and double checked to be sure that they had *no ammunition* in their weapons or on their person.

One year, Martin Sheen's agent contacted the Lab to say that Mr. Sheen wanted to be arrested, and that others would surely want to join him. Could we handle that? Of course! Stan Busboom was in charge on our side. He caused a bright yellow line to be painted across the parking lot. Any protester who crossed that line would be arrested for trespass. News photographers took pictures of Sheen's crossing the line, falling dramatically to his knees, and being handcuffed with plastic ties by PTLA. No matter that the photographers were themselves freely crossing the line; this was theater. About thirty more people lined up obediently to cross and get handcuffed. They were led to a bus—a borrowed Los Alamos school bus—and driven into town where, to their surprise and distress, they were released. We had tried to get the FBI or the U.S. Marshal to come up from Albuquerque and arrest at least Sheen, as his agent demanded. Wisely, these actual cops wouldn't play.

So, for petty crimes committed by Lab employees on Lab property, Lab management was in practice judge and jury, with punishments "up to and including the possibility of dismissal." This last was a big deal. Mayor Lucero's lessons for me had included the demonstration of the significance to his people of LANL employment.

In a two-billion-dollar-a-year facility situated in a poor rural area, some petty crime was inevitable. Tools or office supplies were sometimes stolen. Money was sometimes missing from office petty cash boxes. Lab purchase (credit) cards, used for small items, were sometimes abused. I once estimated, based on cases I had heard about, that the Lab lost maybe a hundred thousand dollars a year on things like this. That might sound like a lot, but, out of two billion dollars, it amounted to less than one-hundredth of one percent, and it was kept pretty stable by our catching and firing the occasional bad actor. The idea of arresting and actually prosecuting such people was ridiculous and would have generated ill will in communities such as Lucero's, far beyond any utility.

Then Joe Salgado decided, for reasons unclear to me, that we needed to find and dismiss more bad employees. Early in 2002, he created under Busboom and the latter's deputy Gene Tucker, a new investigative unit. If Busboom was a large, unflappable bear, Tucker was a compact, fighting bantam cock. I liked the bear, but didn't much like the rooster. The new investigative unit was staffed by hiring two former cops—in fact, former police chiefs—Glen Walp, who had been commissioner of the Pennsylvania state police, and Steve Doran, who had been chief in tiny Idaho City, Idaho, a distant suburb of Boise. That this job was far below Walp's previous one might indicate something about his past. Walp proved to be neither a bear, nor a rooster, but an open five-gallon can of gasoline.

Up to this point, I had some direct knowledge of events. After this point, I learned things only much later, not long before the rest of the world did. Less understandable is that John Browne was also kept in the dark by Joe Salgado. But that was consistent with, "I can fix that. Let's not bother John," which I was familiar with.

Walp and Doran found petty crime and corruption pretty much everywhere they looked. Some of it was real; some was invented, because potential whistleblowers thought that they would be rewarded, or they just came forward out of spite. In a place as complicated as New Mexico, where everyone hated everyone, there was no shortage of informants. When Walp and Doran took their smoking-gun cases to Gene Tucker, he told them not to pursue them. Gene understood the realities of doing business in Northern New Mexico, but it was not in his nature, as it might have been in Busboom's, to give patient explanations to the two investigators. He just kept throwing them out of his office. Walp and Doran concluded that Tucker, Busboom, and Salgado were all covering up the secret truth—that LANL was rotten to the core. The gasoline was now poured on the floor, needing only a match.

That match was lit by Pete Bussolini, a buildings manager with a Lab credit card and a love of the great outdoors. Over a period of months, Bussolini acquired with Lab funds, and hid in an empty bunker in Technical Area 33, two hundred and fifty-one hunting knives, eighteen pairs of binoculars, sleeping bags, Arctic jackets, hiking boots, Coleman lanterns, battery operated coolers, and multiple GPS units. It made sense for him to use TA-33 for his stash. It was the most secret area of the Lab, miles distant from other areas, where some of the most hidden projects were conducted. I had toured some of these activities, authorized not by John Browne, but by my friends in three-letter agencies back in Washington. When I now and then visited TA-33, I wasn't supposed to tell even Trish. Bussolini was undone when he sent a disgruntled underling to TA-33 to collect the stash and drive it to Bussolini's home garage.

This larceny amounted to more than $50,000, by anyone's definition not petty. When Walp and Doran (who seemed to be joined at the hip) took the case to Tucker and Busboom, the detectives cannily knew that this one couldn't be covered up. But, surprisingly, Busboom took them off the case and ordered them to hand over their investigative records. That was the moment when the match was struck and the Lab went up in flames—although unlike the Cerro Grande fire, this time only figuratively.

What might have happened in a more benign universe was this: The Lab, through general counsel Frank Dickson, would take the evidence to the U.S. Attorney's Office for the District of New Mexico and ask them to prosecute, informing UC and NNSA along the way. The assistant U.S. attorney assigned to the case would then work with the Albuquerque FBI office as needed. It would be out of Lab hands.

What actually happened was this: Walp and Doran went directly to FBI, offered to help them with the investigation on a continuing basis, and, incidentally, accused Tucker, Busboom, and Salgado of a criminal coverup. Albuquerque FBI was immediately chomping at the bit to, once again, occupy LANL in force. For a few days, Salgado and Dickson were able to hold them off. That became impossible when Walp and Doran surfaced an unrelated case where a Lab employee attempted to purchase a souped-up $30,000 Ford Mustang from a dealership in Phoenix with a Lab purchase card. The charge was instantly declined, so the transaction never happened. (The employee said it was a mistake.)

UC now got into the act by hiring an outside auditing firm to look at all of LANL's procurement—not actually a bad thing. But it was Joe Salgado who brought matters to an unfortunate head. He summoned Walp and Doran to his office and, Dickson advising and Busboom loyally at his side, fired the two investigators. The newspapers instantly got the whole story—it is not hard to guess from whom. The Mustang and the two hundred and fifty-one hunting knives made for a great read.

Christmas in 2002 fell on a Wednesday. LANL was closed for the whole week—the annual "winter energy conservation closure" that coincidentally always fell on Christmas week. UC's holiday vacation was already in progress. In this hiatus, UC president Dick Atkinson and UC Vice President for Laboratory Operations John McTague, together worked themselves into a state of hysteria. McTague was a UCLA chemist who had done a stint in Reagan's Office of Science and Technology Policy, then twelve years as a senior research manager at Ford Motor Company. His position in UC's Office of the President was a new one, established after Wen Ho Lee. McTague liked to talk tough—I think he learned that at Ford—but he probably knew no more about operations than I did; and almost surely less than I did about operations in northern New Mexico.

Buried in UC's management contract for the Los Alamos and Livermore Labs was a standard clause in all federal contracts: The government could cancel the contract at any time, immediately, "for convenience." Historically, the federal government cancelled contracts for convenience almost never. At Los Alamos, doing so would leave 12,000 workers immediately unemployed, not to mention plutonium and tritium facilities (among others) unattended. It just couldn't happen. If NNSA actually wanted to yank the contract, it would take months of negotiations for an orderly transition. Nevertheless, Atkinson and McTague convinced each other that a termination-for-convenience was imminent. Somehow, they got the idea that Secretary of Energy Spencer Abraham would make the announcement on January 2, the first working day of the New Year. Vacation week notwithstanding, they felt a need to act before then.

Two days before Christmas, Atkinson called Browne and fired him—later reported as Browne calling Atkinson to say that he "intended to resign." Former directors of national laboratories normally move to senior scientist positions, with opportunities to advise and mentor. Here, the deal offered (rumor, as I later heard it) was ugly: Browne was told to literally get himself and his family out of town within two months. If he did so, UC would pay him a reduced salary for some period in residence on a UC campus. Implied was that, if he resisted, UC would terminate his employment immediately, for cause. John agreed to pack up and leave. The day after Christmas, he called us SET members individually with the news. Secretary Abraham had been informed. Browne's resignation was announced in an all-hands memo on January 2. Salgado was fired by McTague—there was some later dispute about exactly when. Busboom and Tucker couldn't be summarily fired. We on the SET were at-will employees, but people below the associate director level could be fired only after a lengthy procedure. They negotiated individual settlements to leave the Lab. Browne eventually cut all ties with UC and, as an independent consultant, later served on a series of high-level advisory boards.

Jo Ann Milam escaped anyone's notice. She arranged to transfer herself to some far corner of the Lab. She and I kept in contact. I asked her how Joe could have been so stupid as to fire Walp and Doran, and why she didn't stop him. "It was a pissing match among three ex-cops," she said. "There was nothing I could do."

I was left untouched by these events, except that suddenly I was working for the new Acting Laboratory Director. He was Vice-Admiral Peter Nanos, USN (ret.).

62. Nanos

Who was this Vice Admiral (ret.) George P. ("Pete") Nanos whom I now worked for? Conveniently for U.C.'s Atkinson and McTague in their rush to fire John Browne, Nanos was already at Los Alamos and on the University of California payroll—and he was a retired Navy admiral with a reputation for toughness. John Browne had hired Pete to direct the Lab's global security division six months before, just after Pete left the Navy. Division director was a job well below the status of an admiral; it was generally thought that Pete would in due course succeed Don Cobb as, more suitably, an associate laboratory director. Nanos was a U.S. Naval Academy graduate, Class of 1967. After a tour at sea, he had qualified for the Navy's highly selective Burke Program, which sent young officers to get graduate degrees at elite universities. Pete moved to Princeton and, in the two years that the program allowed, got a Ph.D. in physics. My JASON colleague Curt Callan remembered him from that time thirty years earlier. Rising through the ranks, Nanos checked the required boxes for duty at sea, but his career was mainly on shore, in acquisition and procurement. He did a tour as the head of acquisition of submarine ballistic missiles. His final command, which came with promotion to three-star admiral, was leading Naval Sea Systems. NAVSEA, so-called, had 20,000 employees and a procurement budget of $23 billion.

This all seemed promising. Nanos appointed Rich Marquez to fill the deputy director position vacated by Salgado. Marquez, a skillful bureaucrat, had worked twenty years in DOE, then come to LANL as associate director for administration. Although he reported to Salgado, he had somehow remained untouched by the latter's downfall. The rest of the Senior Executive Team was as it had been. Pete assured us that we, the survivors, were not part of the problem that he had been brought in to fix. Indeed, he met more often with the full SET and associate laboratory directors than John Browne had done, another good sign.

And yet, the meetings with Pete were strange affairs. We soon learned that freeform discussion was not tolerated. We spoke when spoken to. We sat in our seats. Nanos typically jumped out of his seat and strode around the table, pointing at people and giving orders, some of which made sense, others not. He appointed Judith Kaye as his "executive" chief of staff, a title that had not before existed. Judith had previously been some kind of HR representative to the director's office, but she was now Pete's executive officer on all matters, including technical issues far beyond her capability. She wrote down everything Pete said if it sounded like an order, and then followed up to ensure our individual compliance. It was hard to know when to comply and when to ignore. If, in response to my mention of some longstanding activity, Pete said, "I don't like that," did it mean that I was supposed to terminate it forthwith? Judith often thought so. If I ignored something that Pete actually wanted—and remembered—I sometimes got an email from him ordering me to produce weekly reports on the subject until advised otherwise. I would do so for six or seven weeks and then stop, which always went unnoticed. Since most issues that I worked had timescales of months, or even years, most of my weekly reports were, "status same as last week," although I often dressed this up in bureaucratese.

Then there were excursions beyond the strange to the completely bizarre. Once, striding around the table, Nanos stopped, pointed his finger at Don Cobb, and shouted, "*You shithead!* If I don't see an immediate improvement, then *you're fucking out of here*." We all, including Don, sat dumbfounded. The preceding discussion had not even been about Don or his programs. It took us some further moments to realize that this wasn't about Don at all. Pete's outburst was supposed to be, as it were, in quotes. He was *demonstrating* for us the kind of forceful leadership that, he thought, was lacking at LANL.

As Director, Pete had ample opportunities to make such demonstrations (perhaps minus the obscenities) in many forums at the Lab and in town—and he did. Within a year, he had become the most hated man in Los Alamos County. At public events in town, where John Browne had always been surrounded by well-wishers and petitioners, Pete was shunned. I took note of an intermission at one of the town's well-subscribed classical music concerts. Half of the lobby was crowded with people chatting; in the other half were Pete, his wife, and his aging mother (whom he doted on), and a vast empty space. Vandals planted For Sale signs in his front yard, hoping that he would take the hint. It was reported that, fearing for his mother's welfare, he had a safe room installed in his house. The contrast with John Browne could not have been starker. When I had traveled with John on his regular TWA flights to Washington, ticket agents and flight attendants greeted him by name and with big hugs. The Lab received confidential quarterly reports from its pharmacy benefits manager listing in aggregate the drugs most frequently prescribed for Lab employees. Allergy medicines were always near the top of the list. Under Nanos, there was a sharp increase in anti-depressants.

In my work in Washington, I had interacted with, worked with, many active-duty and retired three- and four-star flag officers. They were an impressive group, thoughtful people who radiated a calm quality of leadership. Pete Nanos was completely different. But for two ball bearings, he might have been Captain Queeg in the Caine Mutiny novel. I was still on the board of the Institute for Defense Analyses and asked its new president, recently retired Admiral Dennis Blair if he knew Nanos. He said he did. I described some of Pete's behavior. "Why does Pete seem to know nothing about leadership?" I asked.

Denny said, "Well, Nanos, he's not really an admiral." I wasn't sure I understood him. "Oh, you mean he's not a four-star." Denny had led the U.S. Pacific Command, a four-star position. Nanos had retired with three stars. Denny said, "No. That's not what I mean. He came up through procurement. He never had a real at-sea command. In my book, he's not an admiral."

One-on-one meetings with Pete were also strange. He heard exactly what he wanted to hear. Everything else seemed to pass over or through him and be lost. It was just like my ex-wife Margaret's description of her friend, the incarcerated serial killer Tom Maimoni. OK, so I have to explain that Pete Nanos was not a serial killer (as far as I knew), and that Maimoni was not exactly Margaret's friend. She had gotten to know him while writing a true crime book about his case, and he called her once a year to sit beside him and testify at his parole board hearing. "Should this man be released?" she would be asked. "Absolutely not," she would say. "He's a hyper-intelligent psychopath who, I think, will likely kill again." After the hearing, Maimoni would thank Margaret for her praise. Of what she said, the only thing that registered was that she thought him hyper-intelligent. The next year, he would ask her again to testify.

The admins in the director's office all hated both Pete *and* Judith Kaye, so, unless those two kept a matter strictly between them, it was soon known to all the other admins and thereafter their bosses. In April, 2004, Trish told me that Judith was working for Pete on a secret project, a complete shuffling of the SET-level org chart. In their secret current draft, there was no Deputy Director for Science and Technology, but only a Chief Science Officer with very limited responsibilities. That would be me, obviously. It was by now obvious to me that Pete would himself eventually be fired. It was only a matter of time. No director could survive the hatred and disloyalty that he had brought upon himself. And, I couldn't imagine that a third director, after Browne and Nanos, would want to keep me as a deputy after I had been so close to these two imploding bosses. I had stuck with Nanos for more than a year. But now it was time for me to get out.

I prepared carefully. I told Trish the plan. I located John Browne's letter, specifying that the Lab would support my continuing as a senior scientist. On a pretext (since director's office calendars were visible to Judith Kaye and others), Trish scheduled Sallie Keller, my 9/11 war buddy and Statistics Group leader, to come to my office. "I'm stepping down," I told Sallie. "No one knows except Trish, Jeffrey, and now you. I need to burrow far, far down in the Lab. I need a safe-house. Can I join your group?"

Pete Miller had once explained to me that LANL's groups were its indestructible atoms. Higher management might create or destroy divisions and shuffle groups around within them, but it was next to impossible for them to much affect a group's inner workings. This was a fact of management not unique to the LANL: First-line managers everywhere tend to be more loyal to their own people than to the management chain above them. The Statistics Group was a particularly indestructible atom. Existing since the Manhattan Project, it had bounced around among divisions and was currently lodged in an applied division that did mostly reactor safety simulations. Sallie easily met payroll for her people by providing their services to projects all over the Lab, with the result that she was responsible pretty much to no one. In fact, her growing budget was funding a hiring spree. "We do have office space for you," Sallie said. "But, can you bring some director's office money with you for office furniture?" We consulted Trish, who said that she could arrange this without leaving any fingerprints—hers, mine, or anyone else's.

Once my new furniture was safely ordered, I went in to see Pete. I showed him Browne's letter. "I'm really a scientist at heart," I told him, trying to sound meek and academic. I reminded him that I had been a manager now for five years. Working for him was a privilege, I said, but I wanted to get back to research. I felt that I had more to contribute to LANL as a scientist. Pete read the letter and stood up from his desk. He pointed his finger at me as he had done Don Cobb.

"*You're a rat,*" he shouted. "*You're a rat leaving a sinking ship!* I won't let you get away with that. I'm going to make sure that you drown." I had not previously credited Pete with this degree of self-awareness about his own leadership, let alone easy command of nautical metaphor. I responded that, no, this had nothing to do with any external events. It was just time for me to be a scientist again. "*You're fired!*" Pete screamed. "I'm going to leave you with no salary, effective today. I'm going to call security and have you escorted off the premises!" Now this was scary, except that Pete didn't actually pick up the phone or do anything. Backing out of his office like a courtier in the presence of an angry emperor, I expressed the obsequious hope that he would reconsider before taking any such action. The encounter had not gone as I had expected.

Back at my office, I called John Birely on his cell phone. John was a sensible LANL manager who had moved to UC Office of the President as McTague's deputy. Through JASON interactions, I had known him for years. "Your director here has gone berserk," I said, as calmly as I could, but noticing that my voice was quavering.

My message to Birely, as my friend, was that, unlike in the case of John Browne, I was not going to go quietly. I mentioned the reporters at *Science* and *Nature* whom I talked to regularly, and that Sandy Blakeslee, the *New York Times* science reporter who lived in Santa Fe, was a good friend of Jeffrey's and mine. I said that my goal wasn't to smear Pete. It was unnecessary. He would collapse under his own incompetence. No, I was going to smear the University of California for their treatment of me as a senior scientist, Harvard professor (former), and NAS member. I suggested that Birely take my message directly to UC president Dick Atkinson, not to McTague. Atkinson knew me pretty well, and he would fold, whereas McTague, playing the tough guy, might have some crazy idea of backing Pete. Birely agreed with this assessment.

At LANL, I went to see Rich Marquez, the only member of the SET who ever seemed able to influence Nanos. I couldn't make the comparison at the time, but I later came to see Marquez as a kind of New Mexico version of Thomas Cromwell. Cromwell served Henry VIII effectively and with, invisibly, much greater freedom of action than he ever let on. He could make things happen without leaving traces. Marquez was the same. By now my voice was steadier, and I related my conversation with Birely. "This can all be made to go away by Pete's simply honoring John Browne's letter," I said to Rich. "I'll see what I can do," he said without emotion.

For a few nervous days, I heard nothing. Then, Birely sent me a draft letter for comment. I would be appointed a Senior Fellow, the highest rank of scientist, but my manager-level salary would be immediately cut by twenty-five percent. I would get three years of research support, but I would promise to leave the Lab after that time. "Pete won't go any farther than this," Birely said. I didn't think it was actually Pete. The letter had Marquez's non-fingerprints all over it; this was as far as he could go for me without angering his boss. Except for its trashing of John Browne's promises, it was not too bad an outcome. The promise to leave the Lab was meaningless, because I was sure that Pete would not be there in three years. I accepted the terms. Within days, I moved to my new senior-scientist corner office—with new office furniture out of nowhere—in distant building 1402, where Sallie's group was housed. Everyone around me had Ph.D.s in statistics. I was suddenly classified as a statistician, another unlikely career shift. *(The role of statistician will be performed henceforth by William H. Press.)*

A few weeks after I was there, when I was starting to get to know my new colleagues, a young scientific staff member came into my office. "May I shut the door?" she said. Then, privately, "I want to ask about your experience in management. I know what Sallie does as group leader. And I *think* I understand some of what Sallie's boss, the D-Division leader does. But you were three levels above that. What do you people on the fourth floor of the Admin Building actually *do*?" She wasn't being sarcastic. She was genuinely curious.

I thought about my previous five years of ten-hour days filled with back-to-back appointments, the crises, the prayer wheels and windmills and rare strategic accomplishments. As much as I wanted to answer my now-colleague, it struck me that we had no common vocabulary on the issue. Nothing done on the fourth floor could actually sound significant to an ambitious early-career research scientist. "It's exactly what you think," I said eventually. "We all get big salaries for just pushing paper around. We don't really do anything." She left, satisfied with the truth.

A month after I left the sinking USS Nanos, the Lab had a bad week. A student's retina was scarred by a laser beam—her supervisor inexcusably didn't enforce the wearing of safety goggles—and, in a different part of the Lab, an inventory of classified computer media indicated that two floppy disks were missing. Both incidents could have been dealt with by normal processes. The student's supervisor would be deservedly fired—and was in fact. The floppy disks would be found to not actually exist—a clerical error in tracking pre-numbered labels. But instead of normal process, Pete shut down the whole Lab and summoned all scientists to a meeting in the auditorium at which he ranted, Queeg-like, about "cowboys and buttheads" who thought they were above the rules. In an all-employees memo, he wrote, "I don't care how many people I have to fire to make it stop. If you think the rules are silly, if you think compliance is a joke, please resign now and save me the trouble." In a press conference he talked about a "culture of arrogance" and the "suicidal denial" loose at the Lab.

Scientists (including me) were ordered to do a punitive long week of ineffective video safety training. Crafts workers (electricians, plumbers, etc.), who were mostly independent contractors and mostly Hispanic, were told to report every day to the parking lot of Los Alamos High School and—if they wanted to get paid—stand around in the hot sun for eight hours. There were no restrooms, and no water was provided. Some Lab activities were shut down for months. I knew from my DuPont-style safety training under John Browne: This was exactly how *not* to produce individual responsibility for an effective safety or security culture. The U.S. Government Accountability Office later estimated that Pete's actions cost taxpayers $370 million.

Birely later told me that, after these events, Atkinson and McTague wanted Pete gone. But, this time to avoid *looking* so panicked, they gave him time, and they used their influence to find him a position elsewhere. It took almost a year to find a place willing to take him. In May, 2005, Pete abruptly resigned as Los Alamos director to become one of several associate directors at the Defense Threat Reduction Agency in Washington.

I want to loop back for a moment with further color on Denny Blair, whose only connection to my Nanos saga was his expressed belief that Pete's three stars didn't make him a real admiral. Denny was a man consumed with a sense of honor—of which he had considerable self-awareness. He once told me, by way of explanation, that he was a sixth-generation naval officer. His great-great-great grandfather was at sea aboard a U.S. Navy ship when a passing ship gave them the news that Fort Sumter had been attacked, the Civil War begun. The captain called his officers together. "If any of you cannot support the Union, I will let you ashore at a neutral port. Everyone else must now take an oath to support the United States Constitution." "I cannot take such an oath," Denny's ancestor is supposed to have said, "because I took that oath when I became an officer, and no gentleman should ever be asked to take the same oath twice." He departed the ship and left the Union Navy. Denny felt a strong kinship to this ancestor of his.

Blair was eased out as IDA president in 2007, after he got into an extended semi-public pissing match with Senator John McCain (another retired admiral) on, yes, a point of honor. McCain opposed continuing the troubled F-35 fighter acquisition, and criticized an IDA report for recommending otherwise—implying incorrectly that Blair had a financial conflict of interest. That besmirched Denny's honor, and he fired back at McCain personally. John White, a member of the IDA Board and former Deputy Secretary of Defense, called every board member individually. "Denny has every right to defend his honor," White said to each of us, "but not as CEO of a not-for-profit DoD analysis center. He has to choose which is more important to him." White worked patiently until he had unanimous agreement of the board.

Given a choice between honor (as he saw it) and a practical duty to IDA (as the board saw it), Denny chose to resign. This was the first and only time I ever witnessed a corporate board acting independently and at odds with management for the actual good of the company. I thought White a hero. Blair became Director of National Intelligence under President Obama in 2009. When he resigned the position in 2010, and it was reported that he had gotten into a pissing match with Obama officials John Brennan and Leon Panetta, I wasn't surprised.

As for Nanos, a decade after I had left Los Alamos for Austin (as will be told), I encountered Pete at a JASON cocktail reception. He greeted me warmly. He told me that he had been sorry when I decided to step down from management and go back to science. He was glad that my research at the Lab, with his support, had worked out. It was once again spookily like Margaret's serial killer Maimoni. Pete assimilated and retained only his own rosy version of events.

63. Innocents Abroad

CISAC, the Committee on International Security and Arms Control of the National Academy of Sciences, was formed in 1980 as a vehicle for so-called Track Two exchanges with Soviet scientist-policymakers. Track One exchanges are defined as government-to-government. Track Two exchanges involve private individuals, but with the light cognizance of their respective governments. Unsurprisingly, CISAC's roster included many of the older generation of JASONs (Pief Panofsky, Murph Goldberger, Dick Garwin, Josh Lederberg, Charlie Townes) along with others in this same generation of Cold-War arms control technocrats (Paul Doty, Al Flax, Spurgeon Keeny, Jack Ruina). Most CISAC members were NAS members, but others were drafted from the larger policy world.

That CISAC achieved some traction and success, generally behind the scenes, was partly due to a quirk of the way that Soviet science was organized. Their academy of sciences, Akademii Nauk SSSR, was vastly more powerful than our NAS. Most funding for science, and much for applied technology, was channeled through ANSSSR, making it the equivalent of our NAS, NSF, DOE, DARPA, and the military service labs, all combined. ANSSSR was the technology bedrock of the Soviet military-industrial establishment. Academicians who sat on the Soviet side of the table in exchanges with CISAC were powerful people with direct lines into the Kremlin. Some were Supreme Soviet members themselves. It was a convenient pretense to consider them non-governmental and able to engage in Track Two exchanges with our national academy.

That all changed with the end of the Soviet Union, and especially with the rise to power of Putin, who had little respect (or use) for the entrenched ex-Soviet scientific establishment. ANSSSR was rechartered as the Russian Academy of Sciences (RAS), with much-diminished portfolio and prestige. CISAC continued to hold exchanges with RAS academicians. Raymond Jeanloz, a materials scientist and my contemporary in JASON and NAS, became CISAC chair; but the older generation of members still dominated the meetings. By the mid-2000s, my view was that CISAC had become an obsolete forum where out-of-influence, aging arms controllers, Russian and American, could talk nostalgically about the good (bad) old days. I didn't keep this opinion to myself and casually bad-mouthed CISAC whenever the subject came up.

The perhaps foreseeable result was that Bill Colglazier, NAS's Executive Officer (essentially its COO), asked me to join an upcoming CISAC delegation to Moscow in January, 2005. This particular meeting was problematic on several counts. The American delegates, led by Rose Gottemoeller and including Dan Poneman, Orde Kittrie, and Bill Burns Sr., were career professionals in nuclear weapons non-proliferation. However, the Russian Ministry of Foreign Affairs had suddenly declared weapons non-proliferation to be off-limits for this forum. Instead, discussion would be limited to nuclear power reactor safety, an engineering discipline with few policy implications—and in which the U.S. delegation had no expertise.

A second problem was that no NAS member of CISAC was willing to travel to Moscow in January on short notice. This put the Academy in the awkward position of sponsoring a formal delegation that had no NAS members and over which it had almost no influence. Rose in particular was thought to be an entirely well-intentioned, but unguided, missile: Her actual employer (as she waited out the Republican administration) was the Carnegie Endowment for International Peace, a think tank that maintained an office in Moscow. She spoke Russian fluently—had once lived in Moscow for two years as a Russian-studies post-doc. Rose's loyalty to the United States was not in question; it was her loyalty to the National Academy of Sciences that Colglazier doubted. I was to go to Moscow, represent NAS interests, keep an eye on Rose (and the others), and report back to Colglazier whether I still thought that CISAC was useless. In short, I was a kind of Academy spy.

I had last been in Moscow more than twenty-five years before—an historical eternity. Russia's GDP plummeted after the fall of the U.S.S.R., bottoming in 2000 at a third of its pre-dissolution value. By 2005, under Putin, it had recovered to more than the pre-dissolution value and was growing rapidly. Signs of poverty were still everywhere, but literal *signs* of wealth—that is, billboards advertising luxury goods—were everywhere. When I had last seen the central department store GUM, it was a dirty warren of threadbare shops. Now it had a Cartier, a Bulgari, a Blackgama, a Gucci, an Hermes, Louis Vuitton, Piaget, Prada, Rolex, and on and on, luxury goods for the wealthiest tenth or hundredth of a percent. I wondered how this could be a stable situation, how Putin, despite his taming of the oligarchs, could possibly stay in power?

Our meetings with the Russians were as bad as I had imagined. The Russian delegation included a couple of Academicians with credentials in nuclear safety and a minder from the Ministry of Foreign Affairs, Vladimir Rybachenkov, whose job was to be sure that we accomplished nothing. There was indeed a lot of Cold-War nostalgia. When the name Sagdeev came up in our discussions, he was referred to by the Russians (dryly) as "the American physicist Sagdeev": Since I had known Sagdeev twenty-five years earlier as the director of the Soviet space sciences institute IKI, he had defected to the U.S., married Susan Eisenhower (granddaughter of the president), and was now living in Maryland.

Much more interesting than our formal sessions was accompanying Rose around Moscow when she called on her many professional acquaintances—some going back a generation to when she was that cute young Russian-studies scholar. As she had advanced in government, so had her old Russian friends. Some were high in the Ministry of Foreign Affairs, literally and figuratively. The MFA's huge edifice, one of the seven Stalinist skyscrapers, could not but impress. None of these courtesy calls could be described as substantive; but if diplomacy had the aim of keeping open channels of communication, then Rose, suffering through many lingering bearhugs, was a fine diplomat. I reported as much back to Bill Colglazier, along with my assessment that CISAC, as such, seemed useless. He responded by having me appointed to a six-year term on it.

Separately from the Russia exchange, CISAC maintained a dialog with China that was informally termed Track One-And-A-Half. Its promotion from Track Two was in recognition of the fact that anyone worth talking to in China *was* a government or Party official, effectively for life. The Chinese nuclear weapons elder statesmen valued this exchange enough to set up a dummy organization, the Chinese Scientists Group on Arms Control (CSGAC), that, as an arm of the Party, rather than the government, was technically non-governmental. Even "arms control" was a fiction, because China was not a party to any significant arms control treaty, nor was likely ever to be.

If CISAC's Russia exchange was a meeting of mostly irrelevant arms controllers given over to nostalgia, its China exchange—while not free of nostalgia—was a meeting of still-somewhat-relevant nuclear physicist-warriors, led on the Chinese side by General Qian Shaojun, "the father of Chinese nuclear weapons," and Hu Side, he who had so visibly (and ambiguously) made a show of embracing Wen Ho Lee on his visit to Los Alamos. Over the years Hu had become good friends with Dick Garwin, Harold Agnew, and others in U.S. weapons circles.

In fall, 2008, I traveled to Beijing with the CISAC delegation, my first trip to China. It was CISAC staff director Micah Lowenthal's idea that I give a talk on cybersecurity, implicitly to push the idea that the China dialog should be broadened beyond nuclear. A similar effort was being made in bio-security. The Chinese nuclear warriors knew nothing about cyber and cared less. My talk, um, bombed. (Cratered?) My American colleagues explained: You inject an idea into the Chinese establishment. It seems to disappear; but in reality it is rattling around inside somewhere in the bureaucracy. Later, some transformed version of the idea pops back out.

In this case, it took five years. In 2011, CSGAC suggested, on their own, that a cybersecurity dialog be established; even then, it took two more years before an actual meeting, in Beijing, took place. Given Qian's and Hu's stated ambivalence about cyber, we didn't want to drag onto fourteen-hour flights across the Pacific the real Washington U.S. experts—retired military and NSA officials—for what might turn out to be no more than a Chinese propaganda exercise. Our CISAC delegation was thus visibly anemic: MIT computer scientist Dave Clark, NAS staffer Herb Lin, and I were the claimed cyber experts, augmented by Dick Garwin (as self-declared expert on everything) and Iain Johnston, a Harvard government professor, fluent in Mandarin, whose field of academic study was the Chinese People's Liberation Army (PLA).

Had we known, we would have done better. Qian and Hu had used their prestige and influence to reach across bureaucratic lines in the Party and PLA, rounding up a set of people whom we judged to be high-level managers in China's equivalent of NSA. That was Iain's assessment of who they were, anyway. China did not admit to having any NSA equivalent, and cyber functions—offensive and defensive—were spread, unacknowledged, across several PLA organizations. Kong Tiesheng and Liu Xuekui gave their affiliations as "expert of Defense Ministry." Iain, mining the Chinese-language internet, found that Kong was a senior engineer in the PLA General Staff's 3rd Department (SIGINT and INFOSEC), Liu in the 4th Department (electronic countermeasures and signals processing). Zhao Gang and Zhu Xiaohui, whose affiliation was "Beijing Institute of System Engineering," seemed involved in classified research. Col. Xu Weidi and Dr. Su Jinshu were academics from the Chinese National Defense University. Their open use of titles only emphasized the fact that the others' military ranks or titles were obscured. General Qian presided formally over the meetings. He remained uninterested in cyber and occasionally napped in his seat. But his presence provided, by his rank and Party connections, the top cover necessary for the others in the room to be talking to us at all.

Before my time, the CISAC-CSGAC exchanges had alternated between Beijing and Washington. As U.S.-China relations deteriorated (Wen Ho Lee's prosecution one example), and when the U.S. started requiring fingerprints on entry, China forbade its high military officials from traveling to the United States. Hence, all the meetings on my watch were in China, all but one in Beijing.

Colonel Xu, an affable man with a perfectly idiomatic command of English, attached himself to me at the banquet dinners. His early career postings, before CNDU, were unmistakably those of an intelligence case officer. Iain told me that Xu was a fixture in international exchanges and was not taken very seriously by other Chinese. I couldn't tell whether Xu was assigned to me because the Chinese had assessed me to be a spook, or whether he had decided that he and I were the two wildcards in the group. Xu poked fun at my handling chopsticks like a four-year-old (holding them too close to the tips), but gave me his sophisticated analysis of China's position re DPRK (North Korea). Briefly, China was little concerned about DPRK nuclear weapons, because the Kim dynasty's survival was entirely dependent on Chinese largesse (the flow of oil, for example). What China did fear was the regime collapsing into anarchy, with an unstoppable flood of millions of Korean refugees across its border. China was therefore prepared to prop up the DPRK regime indefinitely—especially if it tightly controlled internal population movement. (And, he didn't add, if it continued to be a burr under the U.S. saddle.)

Edward Snowden's Wikileaks revelations about the U.S.'s ability to vacuum up communications worldwide happened just days before this 2013 meeting, and the Chinese had not yet fully assimilated their import. That was not the case in our subsequent, roughly annual, meetings in Beijing. In later years, the U.S. delegation was upgraded by the addition of Linton Brooks (a former ambassador to Russian arms control talks and for five years the U.S. nuclear weapons czar in DOE), retired admiral Richard Mies (who had commanded all U.S. strategic forces), and Chris Inglis, recently retired deputy director of NSA. (More on Inglis later.)

* * *

One of the Academy's more controversial international exchanges was with Iran. Despite the fact that the United States had, since 1980, no diplomatic relations with Iran, and despite a 1995 trade embargo, NAS staffer Glenn Schweitzer had managed to establish an annual exchange between Iranian and American scientists that was now, in 2009, in its tenth year. Meetings had been in Teheran (because the United States would not issue visas to Iranian scientists) and were rendered grudgingly legal by an export license from the U.S. State Department—just *talking* to Iranian scientists was considered an export that, without a license, would violate the embargo. The annual choice of subject alternated between the countries, and it was always non-controversial—"arid agriculture" and "earthquake-resistant construction" were typical. A year before the 2009 meeting, however, the Iranians surprised Schweitzer with their choice of topic, "the political misuses of science." The mandate for this came from high in the conservative Ahmadinejad administration, he was told. It was a transparent attempt to embarrass the United States: A 2004 report by the Union of Concerned Scientists that detailed the Bush administration's censorship and misuse of science in areas such as climate and reproductive health (abortion) had circulated widely in Iran.

If this was their intended purpose, the Iranians' timing turned out to be not so good. In the intervening year, Barack Obama was elected president. One of his first executive orders (thanks to science advisor John Holdren) set in motion a process for new standards of scientific integrity in government. Meanwhile, Iran had been caught out for failing to declare new, secret nuclear facilities to the International Atomic Energy Agency. And, not exactly a misuse of science but in the same vein, Iran's 2009 national elections had been visibly rigged to guarantee the reelection of Ahmadinejad over the more liberal Mousavi—this resulting in an aftermath of protests and arrests of students and prominent intellectuals in the more liberal urban areas. In this political climate, Schweitzer, visiting Teheran to firm up details, was himself detained and questioned by elements of the Islamic Revolutionary Guard. The upcoming meeting was quickly moved from Iran to France, to be hosted by the French Academy of Sciences at a private estate in Provence. Glenn described this all to me, and asked me on short notice to chair the American delegation.

The estate in Tourtour of the Fondation des Treilles (a charitable spinoff of the Schlumberger fortune) was about an hour's drive from Nice airport, in an area of olive groves and lavender fields. There, a main building hosted meeting and dining facilities. Participants were housed in smaller, rustic outbuildings, each with two or three bedrooms off a common area. Our delegation included Cy Goodman (international affairs and cyber policy), Kimberly Gray (energy and environment),

Norm Neureiter (then at AAAS, the figurative dean of international science exchanges), biologists Tim Stearns and Harvey Rubin (Tim, my JASON colleague), Schweitzer, and me.

The Iranian delegation comprised three academics and two clerics. The academics were delegation head Reza Roostaazad, vice president for research (later chancellor) of Sharif University—the Iranian MIT. Sharif was known to the U.S. intelligence community as a hub for the covert Iranian nuclear weapons program. Perhaps for that reason, our export license forbade us to discuss nuclear issues—a boundary that we may have stretched. Mehdi Bahadorinejad was a mechanical engineer at Sharif. Bagher Larijani was a physician and highly placed administrator at Tehran University of Medical Sciences. Larijani's greater claim to fame was as one of five brothers, all at times in government, who were sometimes called "the Kennedy brothers of Iran." The family was known as influential, secularly conservative but not populist or fundamentalist, and adaptive to the changing winds.

The clerics, who dressed in their full mullah outfits, were Mosen Javadi from Qum University (a conservative institution in the holy center of Shi'ite fundamentalism) and Hamid Shahriari, whose stated position was with the national textbook authority. Javadi was a kind of philosopher of science. Shahriari, it turned out, was a computer scientist; he was responsible for the governance and censorship of the Iranian internet. All the Iranians spoke excellent English. A sixth delegation member, Ahmed Fazaeli, would have represented the Secretariat of the Supreme Cultural Revolution Council, but he was denied a visa by the French government as a terrorist. The French Academy, on short notice, had put together a delegation that included medical researcher Patrice Binder, neuroscientist Henri Korn, and physicist Guy Laval.

At the welcoming dinner, and contrary to exhortations that we should all mix, the Iranians all sat at one table. I sat down with them and indicated that Tim and Glenn should join me. Bottles of French wine were on the sideboard. I retrieved one and poured for us three Americans. "You cannot sit at this table if you are going to drink wine," Shahriari said. He was the more militant of the two clerics. "I'm not aware of such a rule," I said brightly. "Yes," he said, "It is like a no smoking zone, but for drinking." "You are absolutely right," I said, "and every table is divided into two zones, drinking and non-drinking. Please let me know if anyone attempts to introduce alcohol into your zone at this or any other table, and I will help you to enforce the rule." My colleagues waited for the escalation of a religious war that would scuttle the whole conference, but what actually happened was that Shahriari nodded and we were thenceforth best friends, with a long dinner conversation about Iranian politics; why did Ahmadinejad seem so bellicose (answer: his rural populist constituency); why Iran was (in Shahriari's view) such a happy, peaceful country; why Shia was a religion that valued reason and science, while Sunnis (in Iraq and Saudi Arabia) were all ignorant barbarians; and so forth.

I don't think the Iranians had assembled in advance to assign talking points, but during the week's meetings, formal and informal, they seemed to converge on a small number of them. Shia versus Sunni was a major one for the clerics, whose sophistication the Americans repeatedly underestimated. At a formal session, Harvey Rubin went off-script to goad them: "Do you agree that truth is found in the results of replicable scientific experiments?" Yes, of course, they said, we are scientists. "But you believe that the Qur'an is the literal word of God?" Yes, of course, they said. "So," Harvey continued, thinking that he had them cornered, "if a physics experiment gave results that directly contradicted the Qur'an, which would you believe?"

The two mullahs looked at each other and smiled, then turned to Rubin. "Why the physics experiment of course!" Harvey was dumbfounded. Javadi looked to be the cat that ate the canary—it turned out that he had published a paper on exactly this in a Western philosophy journal. He launched into a patient explanation: God gave man three gifts: faith (Islam), reason (including experimental science), and revelation (the Qur'an). For a man of faith, any apparent contradiction between science and the Qur'an was a no more than a challenge for him to use reason to seek a more refined interpretation of the Qur'an. It was just that simple. "Of course, the Sunnis don't believe in reason," he said darkly, adding, "especially the Salafis."

The Iranian academicians steered clear of religious dogma, but they had a secular version of the same talking point: What a tragedy (in their view) that in the Iran-Iraq war of the 1980s the United States had implicitly sided with Sunni Arabs (the then-ruling Iraqis) instead of with Iran, a vastly more developed nation and heir to a great Persian civilization. (Theirs was an argument that conveniently ignored the Iran hostage crisis of 1979-81.) And why did we now side with Saudi Bedouins who were even less civilized than the Iraqis were?

Another interaction with the mullahs was their explaining to Henri Korn and me, while Larijani listened aloofly, the eight precise steps of religiosity, number eight being approximately sainthood. "These people have a great future as Talmudic scholars," Korn whispered to me during the sermon. I agreed: They reminded me of rabbis, seemingly tolerant but pedantically insistent. Korn was a Polish Jew who, as a child, escaped the Holocaust by being sent to live with a Catholic French family. The Iranians seemed to have no awareness of the fact that about half of the French and American delegations were Jewish. "In a perfect Islamic state, what step of religiosity do you think an Iranian scientist needs to be at?" I asked. Larijani interrupted—he had a squeaky high-pitched voice that reminded me of Ed Witten—and said pointedly "Step *one* is more than sufficient." "But not step *zero*," Shahriari insisted. Larijani shot him a dirty look, and it was the end of the conversation.

A second repeated Iranian talking point was on the legitimacy of a theocratic system of government, where absolute power vested in a religious Supreme Leader. One version of this was: Your government derives its legitimacy from a piece of paper signed by Thomas Jefferson. Our government derives its legitimacy from God. How can you say that yours is better? Larijani's subtler approach was a statistics-filled talk showing that Iran's medical research and medical education system was modern, emphasized medical ethics and patient-centered care, and, despite its being "culturally appropriate" (he never mentioned the word "Islamic") ended up within accepted norms of the West—even on issues like abortion, end-of-life, consent-to-treatment. He had to tiptoe around the latter issue because women were subject to permission from fathers or husbands. Still, his point was that we should not equate theocracy with backwardness—except in the case of the Saudis of course (please see talking point number one).

Shahriari was less convincing on this point in talking about internet censorship. "You already censor child pornography," he reminded us. "You put people in jail for having it on their computers. We're no different—we just have stricter moral standards."

The phrase "rule-of-law" was thrown around a lot. We claimed it as a Western value. They insisted that it was no less (and earlier in history) a value of Islam. We then said (so often as to become boring), "And what about the Revolutionary Guards?" the powerful paramilitary force that reported directly to Ayatollah Khomenei and honored no apparent rule of law. That is an exception, they said, because it is national security—and because the Supreme Leader is supreme. The Guards, as uncultured hoodlums, were clearly an embarrassment to all of them, the clerics included. All understood that Khomenei used the Guards to keep secularizing influences in check (the five Larijani brothers being a visible example).

The Iranians completely misjudged the relationship between the Americans and the French. In the formal sessions, on many issues, we and the French would go at it hammer and tongs, shouting at each other and insisting on small (Talmudic?) distinctions. The Iranians saw this as showing light between us, when it was actually just that all Frenchmen and many Americans (especially Jews) liked to argue. Most of the French delegation had done postdocs, or been professors, in the United States. The Iranian misapprehension became clear after an (unofficial) session on Iranian nuclear ambitions. Roosta lectured us on his party line: Iran had no nuclear weapons program—never did have. (This was a provably false statement then and now.) Why should the West be sabotaging Iran's legitimate development of peaceful nuclear power? We countered, "Why did you declare nuclear facilities only after they were discovered by the West?" We were magnifying an inconsequential detail, he insisted: Iran was still within its legal six-month declaration period when the facilities were discovered.

When the Iranians retired to their quarters for afternoon prayer (insisted on by the mullahs), the French and Americans gathered in the lounge for early cocktails. That was when we all compared notes on sidebar discussions. On this day, Henri gleefully reported that, after the nuclear lecture, he had taken Roosta aside. "So how do you really feel about nuclear weapons?" Now speaking in confidence to a sympathetic Frenchman, not an American, Roosta said, "Well of course we need nuclear weapons, because of the threat from Israel in the inevitable coming religious war between Israel and Islam." He added that the French should surely understand this, since Iran's need was no different from that of de Gaulle's *force de frappe*.

An unexpected third talking point that we heard repeatedly was this: "We could easily solve our differences with the U.S., but never with the U.K.!" This was history that few Americans ever learned, but that was seared into the brain of every Iranian schoolchild. From the time of the Anglo-Persian War in 1856 and through the 19th century, Britain had extracted economic concessions from Persia. By the turn of the 20th century, it was in all but name an occupying power. In 1925, it created out of whole cloth the illegitimate Pahlavi dynasty that ruled until the 1979 Islamic Revolution. By comparison the CIA's involvement in a coup strengthening Pahlavi rule in 1954 was a peccadillo. We, the Americans and French, listened politely.

From my Cold War participation in U.S.-U.S.S.R. exchanges, I had learned that every such event, however fraught, was supposed to produce a communique, jointly agreed to. My American colleagues thought this was silly; we agreed with the Iranians on virtually nothing. The Iranians, however, thought it a great idea. Overnight, I wrote a draft affirming the right of free scientific inquiry and denying any right of a government or religious establishment to interfere. These rights, along with fairness, transparency of process, and openness of results should apply to natural scientists, social scientists, policy makers, philosophers, humanists, ethicists, and religious thinkers, I wrote. I wasn't exactly trying to poke a stick into Islamic eyes, but I did want to smoke out what were the limits that their government would allow them to agree to.

The next day the Iranians, including the mullahs, happily endorsed my draft as written. The only arguments came from the French. It only then dawned on me that the overriding goal of the Iranians was favorable publicity in the West, selling the vision of an Iran existing within international norms. In service of this goal, it didn't matter what they signed. This wasn't like negotiating with Cold-War Soviets who would be debriefed by the KGB. No one was going to hold these Iranians accountable for anything they signed. At the closing banquet, Henri Korn, representing the French, said, "This has been a remarkable and friendly exchange, but we should not deceive ourselves that we have made any progress on the big issues." I concluded my remarks with, "We Americans have truly learned a lot this week about a country that has always seemed quite opaque to most of us. I mean France, of course."

64. Transitions

I've gotten ahead of myself and need to return to mid-2004 after I resigned—or Pete Nanos fired me—as Los Alamos deputy Lab Director. I had a safe home in Sallie Keller's statistical science group and three years' salary support. But, for the year that Pete was still Lab director, my goal of flying invisibly below the radar turned out to be impossible. The circulating, only slightly garbled version of the story was that I had stood up to Pete and been fired for it. I was a symbol of the resistance. Strangers came up to me in the supermarket to offer moral support. Thankfully, after Bob Kuckuck became the calm Lab director in 2005, I was able to sink into a desired obscurity.

Outside committee work and amateur science-diplomacy about half-filled my calendar. I traveled from Los Alamos for meetings of the IDA board and executive committee, and for the program advisory committees at its NSA-sponsored centers; for the Packard Fellows selection committee; for S&T advisory committees of NRO and DTRA; and for several NAS committees, including the nominating committee that chose climate scientist Ralph Cicerone to be the new NAS president (succeeding Bruce Alberts). I was again active in JASON and rejoined its steering committee. Texas billionaire Peter O'Donnell, a friend of my father's, kept me in his orbit by putting me on the visiting committee of the well-endowed (by him) Institute for Computational Science and Engineering (ICES) at UT Austin.

These part-time activities were hardly the new career that I needed to invent. For that, I spent a year absorbing the culture of my new colleagues in statistics—going to their seminars, lunches, and knocking on office doors with questions. I started writing short mathematical notes on statistics questions that interested me. These notes weren't publishable—they were either too quirky or (the opposite) not original enough—but I thought them professional enough to post on my public web page. I thought of them as markers that I was doing *something* about a career, not just serving on committees.

Interesting to me was the contrast between statistics as used by smart astronomers (including, previously, me), and statistics as the discipline of the LANL group's Ph.D. card-carrying professionals. I thought the astronomers more creative. They had unique data sets that could be the product of years of observation at a cost of, especially for space observations, millions of dollars. If a new, bespoke statistical technique could squeeze more out of a data set, then inventing and testing that technique was a small part of the overall effort. The published paper in an astronomy journal would report the astronomical observations and the scientific conclusions, and then also (often in an appendix) the derivation and validation of any new statistical methods used. Such methods might or might not ever be used again by anyone—that didn't matter.

Ph.D. statisticians, by contrast, functioned most often as consultants. Knowledgeable about a whole set of standard statistical methods—a graduate education's worth at least—the statistician parachuted into an ongoing project, learned just enough about it to select appropriate methods, applied them (or taught others to do so), and then went on to the next parachute drop. The process by which a really new method could become standard—taught in graduate schools, for example—was a long one. Proposed by someone in a paper in a theoretical statistics journal, a new idea might gradually gain acceptance as later papers by different authors proved broader and tighter theorems. Jackknife methods (so-called) were proposed by John Tukey in the 1950s and extended to bootstrap methods (so-called) by Brad Efron in the 1970s. But it was not until the 2000s that these were fully accepted in the statistics canon. If astronomers were sometimes more creative, statisticians were slower and more rigorous. I learned a lot from my statistician co-workers, but, at heart, I was in this respect always an astronomer.

But it had been eight years since I actually put pencil to paper, or fingers to keyboard, for an astrophysics calculation. The field had moved on without me. My historical knowledge was there, but my current knowledge was at about the level of a second year, pre-thesis, graduate student. At age fifty-six, I needed (at least in spirit) to go back to graduate school. For me, that meant Princeton, where John Bahcall at IAS presided over a superb, rotating group of postdocs and visitors. Across town at the university, my former student David Spergel now led a large group of graduate students and postdocs. David had recently succeeded Scott Tremaine as department chair. I couldn't move to Princeton full-time, but I started visiting IAS for a week every month or two and immersing myself again in astrophysics.

"You need to talk to Arnie Levine and meet his postdocs," Bahcall said to me suddenly on one of my trips, more an order than a suggestion. Levine was a famous biologist and ex-president of Rockefeller University. He inhabited that elite stratum of scientists who could reasonably wonder, every October, whether this was the year they would win a Nobel—in Arnie's case for the discovery of the gene *p53*, whose mutation was one of the most frequent causes of human cancer. Arnie had been appointed as an IAS professor in 2004. His new group at IAS was the Institute's third attempt at a foray into life sciences. The previous two attempts had failed.

That merits some backstory: The Institute for Advanced Study's director and board had neither the inclination, nor the facilities, to expand into experimental, laboratory biology. Therefore, any biology program at IAS would have to be purely theoretical. But *theory* was a dirty word to biologists—this going all the way back to Darwin's exhortations that living things could be studied only by meticulous observation and controlled experiment. Biologists often specifically cited Darwin in distancing themselves from the hubris of theoretical physicists, who thought that there could be a single equation for a unified theory to explain everything. Indeed, IAS was heavily invested in such physicists, notably in Ed Witten's string-theory group.

The isolated subfield *population genetics* was an exception to biology's contempt for theory, perhaps because it had been founded in the 1930s mostly by statisticians, rather than biologists. Ronald Fisher, the greatest statistician of the time (or perhaps ever) was a major figure. In the early 1980s, then-Princeton professor Bob May (the Australian-born mathematical biologist who later became Baron May of Oxford) supported a group of postdocs in population genetics at IAS. That effort folded when he decamped in 1988 back to the U.K. May's protégé Martin Nowak moved to IAS and re-founded a population genetics group in 1998, but he too left (with all his group), for Harvard in 2003, in part because (as I heard it) the mathematicians at IAS didn't consider him rigorous enough and blocked his appointment as a permanent professor.

Levine's appointment at IAS was an attempt not to repeat past mistakes. No one could question Arnie's credentials in biology, and he was *not* a population geneticist. He made no overt claim to being mathematical, but he was a math groupie who loved talking to math and string theory faculty at the daily IAS afternoon teas. (In truth, Arnie loved talking to everybody.) His vision for biology at IAS was that he would hire smart physics Ph.D.s as postdocs and train them in all forms of biology that could be done without a lab (and without ever calling them "theory"): computational biology, systems biology, bioinformatics, comparative genomics, and so forth. Arnie maintained a real lab at Rutgers; his "dry" people at IAS met weekly with his "wet" people at

Rutgers, and they often collaborated on separate parts of the same effort.

Arnie introduced me to his postdoc, Harlan Robins. Harlan, with a Ph.D. in string theory, had gone to the Weizmann Institute in Israel as a particle physicist and emerged two years later as an incipient biologist. Two years in Arnie's group were completing his transition. Recent converts are always the most observant, and Harlan was an enthusiastic proselytizer. He was also a fan of *Numerical Recipes*. Why was I wasting time trying to reclaim my lost reputation in astrophysics? he sermonized. Computational biology was a wide-open field. He would get me started—we could collaborate. Harlan became my tutor when I visited, and remotely by email. He told me to work my way through Bruce Alberts' massive *Molecular Biology of the Cell*, and it became my bedtime reading for almost a year. And, as backup, we both had Arnie, another enthusiastic proselytizer, for consultations.

Harlan's pitch resonated. Going to both the astronomy and the biology seminars at IAS, I noticed that I was more current, and asked better questions, at the latter. There were two effects: At LANL, I had been midwife to (and paid for) Jill Trewhella's new B (biosciences) Division. As Jill brought in new Ph.D.s and some more senior people as possible hires, I usually went to their talks, and I was usually their last appointment of the day, to sell them on the Lab. It now seemed that some of my immersion in bio had stuck. At least superficially, I knew more about what was exciting in biology than I did about astrophysics. The second effect was a more general phenomenon: As a computer person, if you move to a new field, you automatically bring with you a dowry: There are algorithms and programs standard in your old field that your new field hasn't assimilated or doesn't even know about. With a small amount of guidance, there is low-hanging fruit to be found.

I didn't at this time know the history of IAS's previous missteps in biology. I once went running into Arnie's office with a book on population genetics that I had found in the IAS library, undoubtedly acquired there during the May or Nowak eras. "Finally," I crowed, "a biology book that speaks to me! Look: differential equations, diffusion processes, probability distributions!" Arnie took the book and looked at it. "Yeah, population genetics. They made us take a course in it in graduate school. Don't *ever* make the mistake of thinking that it is real biology." I took this to heart as an early lesson. Arnie pushed me toward core biology, as opposed to things that might inform life sciences, but were dominated by techniques from other disciplines: biophysics, biochemistry, applied mathematics, and so on.

I learned also from Arnie that there was a "biological intuition" that the best biologists all had. I already knew what "physical intuition" meant for physicists. Physical intuition meant knowing how a calculation was going to come out before you did it. It was an ability somehow to *visualize* the behavior of physical systems, even unfamiliar ones. Feynman was famous for this. My good physical intuition was a main strength of my career. Biologists, too, had intuition, but it was different. Several times, I went to Arnie with a computational genomics idea and told him that I was thinking of spending a couple of weeks or more writing computer code to investigate it. "You're wasting your time," he said. "You won't find what you are looking for." "Well, how do you know that?" I asked. "Life just doesn't work that way." He was generally right.

Harlan's and my first paper together, published in *PNAS*, was about microRNAs, small fragments of RNA that are encoded in a certain special way in the human genome, and that become functional in post-transcriptional gene regulation. These had only recently been discovered, and it was a hot topic. We used existing published data and databases to show that certain types of genes tended to be in microRNAs' regulatory pathways, other types of genes not. It was a small contribution, but it was *biology*, so for me it was a big deal. Harlan and I wrote one other paper together. On my own, I then got interested in some mathematical aspects of medical MRI imaging and, back in Los Alamos, worked on that for a few months, yielding another publication in *PNAS*. On this thin record of publications, I began to think of myself as a biologist. My biologist friends (in JASON, for example) thought this laughable, but they were a tolerant bunch. David Haussler was an especially useful tutor during summer JASON sessions, as was his later replacement in JASON, Sean Eddy.

John Bahcall's order sending me to Arnie Levine was not, I think, a random event. John then knew, but hid from his friends and colleagues, that he was suffering from a rare, fatal blood cancer. He encouraged my work with Harlan and kept up-to-date on it, hearing about it from Harlan when I was not at IAS. In my last email from John, he looks forward to my next visit, at the end of August, 2005. He died on August 17 of that year, age 70. He had been my professional mentor and my friend for more than thirty years. In 2009 the astronauts on the Hubble Space Telescope repair mission carried John and Neta's wedding rings into orbit and back. The Hubble repair mission was a success.

Towards the end of 2005, I had a phone call from Mary Ann Rankin, the Dean of Natural Sciences at UT Austin. We had never met. I was recommended to her by Roy Schwitters because he knew that, at LANL, I had become, at least organizationally, a statistician. Mary Ann knew also that I was in Peter O'Donnell's orbit and on the visiting committee of his ICES institute in a university unit that was not under her direct control.

The reason for her call was convoluted: In decades past, there had been a functioning statistics department at UT. But, in the 1970s and 80s, statistics, as a separate academic field, went into eclipse. A lot of places, including Texas, merged their statistics departments with their math departments. Then, in the 2000s, statistics, especially computational statistics with its connections to computer science and artificial intelligence, was once again a hot field. Rankin wanted to re-establish a statistics department at UT. She envisioned an outward-looking department that could be home not only to Ph.D. statisticians, but—via joint appointments—to people in application areas. "You're an astronomer doing biology in a statistics department," she said happily. She was exaggerating my accomplishments, but not my ideals. "I want to move you to UT, she said."

I was amenable to being courted. Even though Nanos was no longer LANL director, it was time for me to move on. It was a good bet that Harvard wouldn't take me back, and, in any case, Jeffrey and I had promised each other that we would never move back to Northeastern winters. Austin, which was just beginning to become a high-tech hub, seemed like a great place to be.

It was an odd recruitment. Academic hirings are normally done by departments, but here I was dealing directly with the dean. And, Mary Ann was not a typical dean of that time (or now). At our first meeting, instead of lunch at the nearby faculty club, she led me half a mile to a crowded hamburger place full of undergraduates. For my second trip, she put Jeffrey and me up at the Four Seasons Hotel in a suite. Our flight was delayed. When we finally got to the hotel near midnight, she was waiting in the lobby to welcome us—had been waiting for several hours. Later, after we were established in Austin, I once mentioned to Mary Ann that Jeffrey and I were going to a fancy-dress ball, and that Jeffrey had not yet found something to wear. That afternoon, an assistant from the dean's office appeared at my office with an armload of expensive gowns from Mary Ann's closet. The two of them were the same size. Jeffrey should borrow whatever gowns she wanted and return them when she was done, her assistant told me.

But if Mary Ann Rankin was not the typical dean, neither was I the typical candidate for a professorship. My former job at Los Alamos was equivalent to an academic provost or vice president. I sent Mary Ann a spreadsheet calculating the relative value of UT's and LANL's benefits, and what it would take in salary to give me a modest five percent increase over Los Alamos in total compensation. I don't think she actually looked at it. She instead went to Peter O'Donnell, who provided $2 million to endow a chair, plus $1 million for my startup and research costs—but only on condition that I made my home in his ICES institute, where he would also pay for remodeling an office suite to my specifications.

Towards the end of these negotiations Mary Ann phoned to say that I should probably pay a visit to the physics department, since they would be voting on my tenure. (I couldn't occupy a position in a new statistics department, because it didn't exist yet.) Since she was supplying physics with a new faculty "line," the vote was a formality. "But I don't want to be in physics," I said. "I want a joint appointment in biology and computer science." "Oh!" she managed to say. Then, "I'll see what I can do."

It took a week. Integrative Biology (IB), which happened to be her home department, was willing to vote tenure on her say-so —despite my having only two published papers in life sciences. Computer Science insisted that I actually come and give a job talk. I did, and they then voted me in. I liked the Texan way of doing things, open to wild, crazy ideas, and able to brush aside bureaucratic obstacles. If I had been still at Harvard and proposed to my biologist friends (many of whom I knew in social settings) that they vote me tenure just because I *wanted* to be biologist, they would have thought it a joke—and not a funny one.

The biologists at Texas were welcoming. I knew Edward Marcotte from when he was a Packard Fellow, and his lab was particularly interesting to me. Soon after I arrived, I developed collaborations with his and several other labs and attended their weekly lab meetings. In departmental matters, though, I gravitated to the computer science department. Neither fish nor fowl, I was slightly more the computer scientist than biologist.

Notwithstanding, the biologists were a lot more fun to socialize with. The Integrative Biology department at UT had an interesting history. There had been a single biology department from time immemorial until the 1970s. Then, as at many places, that department fissioned into a forward-looking Department of Cellular and Molecular

Biology (CMB) and a traditional Department of Integrative Biology (IB)—the latter a dumping ground, really, for the botanists and zoologists who couldn't or wouldn't embrace the new molecular revolution—DNA, Watson and Crick, and so forth.

But over some years, IB began making unconventional appointments in fields like ecology, evolution theory, molecular systematics, bioinformatics, and the more mathematical subfields of biology generally (yes, including population genetics). Unexpectedly, the botanists and zoologists proved willing to vote professorships for people quite different from themselves. By the time I arrived in Texas, IB was ranked one of the top departments in the country (its ecology and evolution graduate program well into the top ten). CMB's national rank had stabilized in the thirties. My other department, Computer Science, was generally ranked nationally around twelve, but top-ten in certain subareas.

Ironically, the original reason I was recruited—to help nucleate a new statistics department—gradually evaporated. There did come to exist a statistics department in which I was an affiliated faculty member. But that department evolved not into the kind of broad-based, outward-looking department that Mary Ann had envisioned, but, despite my being there and involved in the discussions, into a respectable, scholarly department in theoretical statistics that hired almost exclusively people with statistics Ph.D.s—people who published in statistics journals. I was for a time listed on the department's web site, but it was a polite fiction.

Jeffrey and I decided to keep our house in Los Alamos for vacations and summers. Re-creating our Massachusetts arrangement of each owning one house solely, I sold my share of our Los Alamos house to her, with the plan that I would own our Texas house when we got one. The transaction necessitated formally transferring the jointly held title into Jeffrey's name at the title company. Small town life: Within hours, word spread that we were getting divorced. That was the only reason that joint property was ever transferred to the wife's name. People came up to Jeffrey in the supermarket to express their sympathy.

Roy Schwitters arranged for Jeffrey and me to be elected to Austin's Tuesday Club, which met monthly for dinner and a semi-intellectual talk. The club had been founded by Steve and Louise Weinberg when they moved to Texas, and it had about two hundred members. By charter, half were from the university, half from Austin public life (lawyers, business people, politicians, etc., almost all wealthy). Jeffrey learned a lesson in Texas deadpan humor—or Texans' disdain for nosy

Yankees—when she grilled a succession of non-university members about their fortunes. "Oil and gas, ma'am," some said, which could mean anything from that they once worked at a gas station to that they were retired Exxon-Mobil executives. The other standard answer was, "land and cattle, ma'am" which often meant careers in the Dallas corporate world—and a small family ranch for the weekends.

It was also at Tuesday Club, sometime later, that Jeffrey and I found ourselves at a table (with, by chance, no other university people) discussing U.S. income inequality and the socially indefensible wealth fraction of "the one percent." While wealthy, these were all progressive Austinites (almost a prerequisite for being invited to join Tuesday Club). I volunteered that, while Jeffrey and I usually made it into the one percent, the threshold had risen with stock market gains that particular year, and we had dropped off. We were in the *two* percent. The table fell silent. I could see them all thinking, "You poor bastard!" Soon after, I began to have a recurring dream: I am in a tumbrel crowded with wealthy one-percenters being carted to the guillotine. With both hands, I hoist myself over the edge, just enough to see the jeering proletarian crowds throwing rocks at us. "Let me out!" I shout. "It's all a mistake! I'm only in the *two percent!*"

65. PCAST

Before his January, 2009, inauguration, new president Barack Obama had already named John Holdren as his full-time science advisor and announced that his President's Council of Advisors on Science and Technology would be co-chaired by a triumvirate of Holdren, Harold Varmus, and Eric Lander. I noted the fact that, while I knew all three slightly or by reputation, I knew none well. Maybe I wasn't as important as I thought I was! Or maybe Obama wasn't appointing to PCAST the right people, that was to say, my best friends.

I was surprised, then, to get, in March, a phone call from Lander; and even more surprised to be asked to join PCAST. There would be twenty members. We would meet in Washington for two days every two months and hopefully meet with the president on each occasion. We would deal with big issues: climate, environment, healthcare, energy, national security, federal science funding, international technological competitiveness—whatever else we came up with.

I accepted on the spot. It turned out that all twenty of us accepted on the spot, unprecedented in any of the co-chairs' previous experience. Obama had campaigned on a message of *hope,* and we all now wanted to be part of it.

Not just the co-chairs, but this whole PCAST seemed drawn from circles entirely disjoint from mine. This was no casual observation: I had actual data. Before the inauguration, NAS president Ralph Cicerone had solicited from NAS members their suggestions for PCAST membership. I sent in a list of forty names. My list had *zero* matches to the twenty eventually chosen. Of the twenty new PCAST members, I knew only one well, Ernie Moniz, as MIT physics department chair in the early 1990s, and then (both of us in new roles) when he was Under Secretary of Energy during my first two years at Los Alamos. Ernie had a raging case of Potomac Fever. The surprise now was only that he didn't get a higher position in the administration, an error corrected four years later when he became Secretary of Energy, succeeding Steve Chu.

Five more members I knew just barely: Chris Chyba was an astrophysicist before he transitioned into public policy at Princeton's Woodrow Wilson School. Shirley Ann Jackson, a physicist, had been chair of the Nuclear Regulatory Commission under Bill Clinton, afterwards becoming president of Rensselaer Polytechnic Institute. Her multi-million-dollar annual compensation made headlines, the stories dutifully noting her pathbreaking role as an African-American woman. Shirley was on John Browne's director's advisory committee at Los Alamos, where her contributions were sensible, imperious, and never outside the box.

Barbara Schaal was a biologist at Washington University in St. Louis, a protégé of Peter Raven. Peter and I both participated on the nominating committee that chose her as NAS Vice President, but I didn't otherwise know her. Ahmed Zewail was Caltech's famous Nobelist in laser chemistry; I had met him just a couple of times. Maxine Savitz, a retired engineer from Honeywell, was Vice President of the National Academy of Engineering (NAE), and we had briefly overlapped on the NAS-NAE thirty-member Report Review Committee.

I didn't feel bad for not knowing the three PCAST members who were billionaires: Eric Schmidt was the CEO of Google, then running that giant enterprise on behalf of its laid-back founders Page and Brin. Craig Mundie had founded a computer company that was absorbed by Microsoft in the 1980s. Mundie acquired early stock options and became Microsoft's head of research and long-term strategy. David Shaw was a computer scientist who, in the 1980s, founded D.E. Shaw and Co., one of the two most successful "quant" hedge fund companies in the world. In recent years, Shaw had withdrawn from active participation in his company and now headed an institute of computational biology at Columbia—built and funded by him personally. I also knew only by reputation Rick Levin, the long-serving Yale president; Mario Molina, the Nobelist atmospheric chemist, and Chad Mirkin, a famously entrepreneurial and self-promoting materials chemist at Northwestern.

There were four members I had never even heard of: environmental scientist Rosina Bierbaum, physician Christine Cassel, biotechnology entrepreneur Ed Penhoet, and Maryland theoretical physicist Sylvester "Jim" Gates. I smile as I write this, because, over the next eight years, Chris Cassel and Jim Gates would become lifetime personal friends, even beyond the close professional bonds that I would develop with most of the others, Lander, Schaal, Mundie, Levin, and Savitz especially.

How I came to be included on the PCAST roster was something that I eventually found out. Geography helped: I could be branded as a Texan yet secretly be thought of as a Harvard professor. Interdisciplinarity helped: I could be touted as someone with experience in both physical and life sciences. JASON and Los Alamos helped: Apart from longtime expert Holdren, Chris Chyba and I were the only PCAST members with any experience at all in national security issues; and his was mostly in arms control and nonproliferation. I was the house nuclear warrior. Lander later described to me the afternoon in his Broad Institute office when he, Holdren, and Varmus wrote names on his whiteboard and moved them in different colors from column to column according to their attributes, ending up with the PCAST list.

There had been a PCAST under President George W. Bush, chaired by his science advisor Jack Marburger and co-chaired by Floyd Kvamme, a right-wing venture capitalist. Its roster included just one working scientist. Eight years earlier, in fall, 2001, I had happened to be at a meeting at Stony Brook with Jack, who was then the director of Brookhaven National Laboratory. He was called out of the room and returned, somewhat pale, to say that he had just been asked, with no warning, to become Bush's science advisor. Over the next eight years, his loyal, outspoken defense of the Bush administration's anti-science policies cost him his reputation in the scientific community. Once, sitting next to him at a private IDA Trustees dinner, I tortured him by trying to get him to admit that it was in fact a tough time for science in the White House—to get him to break character in private and off-the-record. He wouldn't do it. He was loyal to the end. He died two years after leaving office.

Obama's PCAST list had, depending on how you counted us, about ten working scientists. The membership was announced by President Obama in a speech at the NAS annual meeting in April. We twenty were seated in two reserved front rows before the president arrived. You could overhear the speculation in the packed NAS auditorium: We were mostly recognizable as NAS members, but why the reserved seats? Smarter NAS members figured it out before Obama got to that part of his talk. Immediately after the speech, we had our picture taken with the president in the NAS boardroom, under the oil painting of President Lincoln handing the Academy's founders the Academy charter.

Soon after I got back to Austin, I had a communication from a functionary in the university president's office informing me that Texas law forbade "dual positions of honor, trust or profit" in state or federal government. My professorship was one position; PCAST, even though unpaid, was an illegal second. Luckily, there was a waiver procedure. The Regents of the University of Texas in due course voted, by a docket action with hundreds of such items, to make an honest man of me. A few months after that, the White House sent me my "commission," a parchment document like a diploma, written in archaic language, the blanks filled-in in a calligraphic hand. The President of the United States reposed special trust in the "Integrity and Ability" (filled-in) of "William Henry Press of Texas" (ditto) and appointed him to PCAST (of course spelled out) "during the pleasure of the President of the United States for the time being" (ditto). Military commissions, I learned later, filled-in the first blank with "Valor and Fidelity" rather than "Integrity and Ability." My friends who had been in and out of government deflated me by mentioning their whole collections of commissions, because they got a new one every time their exact title, or the administration, changed. Still, it was a thrill for me.

Meanwhile, PCAST had started meeting, first in what was grandly called the White House Conference Center—but was actually a maze of small rooms in the connected nineteenth century townhouses fronting on Lafayette Square—and, soon after, in the more imposing and comfortable facilities of the NAS. For the prestige of hosting PCAST, the Academy had even waived its usual gold-plated room charges.

Every two-day PCAST meeting had to have, by law, a public session. These usually featured speakers of Cabinet, sub-Cabinet, or agency-head level. They were people promoting something—a program, piece of legislation, or their own careers—and happy to have a public forum with White House branding. Larry Summers (National Economic Council), Carol Browner (White House climate office), John Brennan (counter-terrorism czar), Raj Shah (USAID), Ash Carter (by then Under Secretary of Defense for Acquisition and still rising), and Regina Dugan (DARPA) were examples. Summers' joy at being rehabilitated into government was radiant. He had been in the private sector wilderness (at D.E. Shaw and Company, actually) after being forced out as president of Harvard; he must have spent those three years wondering whether his unperson status would be permanent.

Dugan was someone I knew well from a decade earlier, when she had been a DARPA program director and an enthusiastic JASON sponsor. We JASONs stayed in touch with her while she was out of government. As a new DARPA director she was the keynote speaker at the JASON 50th Anniversary dinner that was held at the Army-Navy Club. A kind of trademark, Regina always wore flashy shoes of high-fashion or high-grunge, always in bright colors. "Regina, I love your shoes!," was the way I always greeted her. We had once discussed our shared opinion that women in senior positions should present themselves not as male clones, but any different way they chose to.

The PCAST co-chairs divided us into five working groups, each of which was supposed to meet several times (extra meetings in Washington as necessary), consult non-PCAST experts, and come up with one or more studies that could produce reports and recommendations for the president. Chris Chyba and I were designated co-leads of the national security working group. There was no one else with the background or willingness to join our group, so Holdren and Lander assigned themselves to it for form's sake—they never actually met with us. The other working groups were healthcare reform, climate and carbon management, advanced manufacturing, and STEM (science, technology, engineering, and mathematics) education.

I thought that there were plenty of good national security topics for a group like PCAST to study and weigh in on: the need for new technologies and infrastructures for cyberdefense; a nuclear weapons refurbishment strategy, horrendously expensive, that was based on selling questionable science to generals; an intelligence community that was missing the strategic utility of unmanned air vehicles ("drones") and ceding their operation to the military for tactical use only; a look at "deterrence by science," the idea that adversaries took notice of the significant scientific talent at the weapons labs—and would notice if it were allowed to decay; the inchoate mass of second-rate research that embodied the portfolio of the Department of Homeland Security. Chyba didn't particularly like any of my topics. His own were things like options for space arms control, technology for non-proliferation monitoring in Iran, and the like. He was an arms-controller first and foremost.

Over most of a year, Chyba and I elaborated our list of topics, and made use of PCAST's notable convening power. PCAST had a Ph.D.-level staff of four, initially led by Debby Stine, and later by Marjory Blumenthal, whom I had long known through NAS and CSTB. The PCAST staff were not shy about cold calling people with, "Hello, I'm calling from the White House." Chyba and I could easily summon Deputy Undersecretaries and Assistant Secretaries of Defense to meet with us.

In turn, Chyba and I presented our suggestions to PCAST in the group's private sessions, each topic fleshed out with a few paragraphs of text. To most PCAST members, we might as well have been speaking in tongues. Mixing the metaphor, national security was, to them, insider baseball. Craig Mundie expressed interest in topics with a cyber dimension, but he was the exception. With a mostly mute membership, it fell to John Holdren to pass judgment on whether we should proceed to organize a full-scale study. That judgment was always no. "I'll consult with the NSC on that," he would say, diplomatically. Or, "Let's let the interagency process operate." Or, "This is too sensitive right now." Or, my favorite, "Let's for now adopt a watchful stance." Chyba's topics did hardly any better than mine.

Eventually I got the message. Our working group's existence notwithstanding, Holdren's view of PCAST was that it should stay away from national security, other than in the very broad sense of economic security (advanced manufacturing), environmental security (climate), and the like. Apologizing to Chyba, I abandoned him and formally switched my membership to the healthcare reform task force, led by Christine Cassel. Chyba renamed his working group as "international security," but I don't recall it actually giving birth to any significant report.

An important national concern at this time was the glacial pace at which electronic health records (EHRs) were being adopted by U.S. hospitals and medical practices. At the beginning of the Obama administration, fewer than 20% of clinical organizations had electronic record-keeping for anything other than billing. Paper, and rooms with shelves full of files—these often misfiled—were the norm. Following the 2008 financial crisis, Congress enacted the American Recovery and Reinvestment Act, an $800 billion economic stimulus package. Rushed into law, ARRA agglomerated the wish-lists of even low-level administration staffers. Happily, S&T figured in many of those lists. Pasted into the thousand page ARRA, its whole never completely understood by any legislator who voted for it, was the HITECH Act or, if you insist, the Health Information Technology for Economic and Clinical Health Act. This appropriated $2 billion to establish a new agency, the obtusely named Office of the National Coordinator. ONC was empowered to set technical standards for EHRs and the electronic exchange of medical data among clinicians. Then, elsewhere in HITECH, the Centers for Medicare and Medicaid Services, the agency responsible for servicing Medicare payments, was authorized to dole out $20 billion to physicians who made "meaningful use" of ONC's standards.

The problem was that these two agencies, ONC and CMS, could or would barely talk to each other, much less work together. Harvard Medical School's David Blumenthal, as the first director of ONC, recruited experts from the medical informatics and computer science communities, and organized nimble standards groups populated by first-rate people. CMS, by contrast, was a behemoth. Its job was to disburse $500 billion a year to doctors and hospitals. Its batch computer mainframes, with architectures dating back to the 1970s, could barely keep up. CMS viewed as its customers the medical practitioners who wanted faster, never-challenged payments—not the American public who wanted better healthcare. Implementing HITECH, the agency moved toward enacting a ridiculously low bar for "meaningful use"—in essence giving away $20 billion to clinicians for doing almost nothing.

PCAST, especially Cassel, directly engaged. Some people, including unfortunately Eric Lander, misinterpreted Chris' always-calm speaking tone—the physician's reassuring bedside manner—as ambivalence, or saw her as a lightweight. I soon thought otherwise. Chris always had in mind a specific outcome, and she usually achieved it. Eric, in private, spoke of Cassel as "our pediatrician." Her specialty was actually gerontology, and she was at the time president and CEO of the American Board of Internal Medicine. In later years she was the founding dean of the Kaiser Permanente School of Medicine in Pasadena, California.

In December, 2010, PCAST published a major report that endorsed ONC's approach and held CMS's feet to the fire to make meaningful use actually meaningful. More important than just the report was "shopping it around" to Congressional staffers, officials in the Department of Health and Human Services, professional societies, and more. I contributed to this effort in two ways. First, Craig Mundie and I worked together to specify the technical details of our proposed information infrastructure—the descriptions and recommendations for action. Second, it somehow fell to me to write much of the report, and to organize and edit the rest. It was a surprise to me then (and afterwards) that top levels of the U.S. power structure were filled with people who couldn't write well, or wouldn't write at all. *Well* didn't have to mean brilliantly or grippingly or luminously or poetically; just serviceably. I suppose this proves that writing ability has little to do with success. But, *some* people have to be able to write, and I remained grateful to Kip who, as my Ph.D. advisor, drove me to become one of those people.

With the eventual universal use of EHRs would come the possibility of doing better comparative effectiveness research—this a term of art for trying to observe statistically which drugs on the market were actually better than their competitors. HITECH appropriated money for this, but the big pharma companies proved able to neuter any such effort. Big Pharma made money from copycat drugs that were no better—and probably often worse—than the first to market. In a longstanding example of regulatory capture by an industry, FDA, in drug approvals, never asked for proof of *comparative* effectiveness, but only of effectiveness. Outside of PCAST, I got interested in the mathematics of how one might in theory slowly meter an effective (but perhaps inferior) drug into the market while accumulating comparative statistics. I found an objective scheme by which only the best drugs for a condition would gradually escape restrictions on use, poorer drugs eliminated. This would be, in effect, a radical new way of doing phase-three clinical trials much more efficiently, both in terms of cost and patient welfare. My paper on this was published in PNAS along with a "commentary" piece by Chris Cassel explaining it. But the idea never went anywhere. My UT colleague Peter Müller, whose own work was in medical statistics, told me that, while my scheme was sound, it was far, far too radical. Müller and colleagues were butting heads with FDA regulators just to get approval for the use in clinical trials of elementary Bayesian approaches, methods in use for fifty years in other fields.

By the end of the Obama administration EHR use had jumped from 20% to 80%. PCAST helped, I like to think, along with the in-government efforts of Blumenthal, his successors Chuck Friedman and Farzad Mostashari, Aneesh Chopra, and Todd Park. Aneesh was appointed by Obama as the first Chief Technology Officer of the United States, a corny title, I thought; he functioned in practice as John Holdren's technology deputy.

Todd Park succeeded Aneesh when the latter left to run for Lieutenant Governor of Virginia. I got along well with Aneesh and knew some months in advance that he was going to take this leap into elective politics. I wondered, didn't he need a whole political machine behind him for fundraising and the like. He told me that, as an Indian-American, he had one automatically. A network of rich Indian-Americans stood ready to support any reasonable candidate of their ilk. I was surprised for at most two seconds. Didn't Jews support Jews? Didn't Italian-Americans support Italian-Americans? Ethnic politics is a deeply American fact of life. However, Aneesh failed to win his Democratic primary.

In PCAST, my ability to write (and also organize the contents of reports) got me the reputation of a kind of script doctor. Several groups called on me to work on their draft reports. A notable example was a PCAST report on radio spectrum allocation, led by Mark Gorenberg. As subsequently adopted by the FCC, that report's recommendations affected, over the next decade, something like a trillion dollars of commerce—by far PCAST's most economically significant intervention.

In my adopted PCAST niche, most of my colleagues liked me well enough; I was useful, if often irreverent and occasionally outside the box. John Holdren was a puzzle, however. John was always friendly to me; but, in sessions that he chaired, where we were supposed to be called on in the order that we raised our tent cards, he seemed often to call on me last, or not at all—then noting that time was short. This was a petty grievance, but it was statistically significant. I used to keep track and update the significance calculation during meetings. Working with Eric was never easy, but there was a streak of genius in him that made me tolerant of his bullying. It was often, "bring me a rock," followed, after much work by me or others, by, "no, not that rock. Bring me a different rock." He was a perfectionist and wanted things done his way.

Co-chair Harold Varmus wrote fluently and brilliantly. Shirley Tilghman, as president of Princeton, once described Harold as the Shakespeare of committee reports. In mid-2010 Varmus resigned from PCAST to direct the National Cancer Institute. His was an act of patriotism: NCI director was a step down from NIH director, which he had been in the 1990s. That left Holdren and Lander as co-chairs—in practice just Eric, since John already had too much to do. After six months, the two decided to re-organize PCAST by elevating two members to the awkward title Co-Vice Chair. Maxine Savitz, who was vice president of NAE, was one of these. Symmetrically, the other ought to have been Barbara Schaal, then vice president of the NAS. But, instead, it was…me. I was surprised—and accepted.

For the next six years, John, Eric, Maxine, and I met by phone weekly for an hour, and met in Washington for dinner at one or another expensive restaurant—of John's choosing, and dividing the check evenly—the night before PCAST plenary meetings. John, a wine connoisseur, had expensive tastes, so I always tried to drink my share of the two or three bottles that he ordered for us. Because Maxine and I shared cabs to the same meetings so frequently, the doorman at the Palomar Hotel thought that we were married; if it was just one of us, he would ask about the health of the other. We never corrected him. When I checked-in once with Jeffrey, he gave me an odd and then knowing look.

66. Obama

During Obama's two terms, PCAST met with the president more than twenty times, about three times a year. It was only half as many meetings as had been originally promised; but, we learned, such promises are malleable and routinely ignored by the all-powerful presidential schedulers—or else honored, but then overtaken by events in real time. The schedulers blocked forty-five minutes for each of our meetings, but Obama often waved off the staffers who came to get him at the end of the time and typically gave us an hour, sometimes more. For some reason, perhaps just day of the week, the next thing after us on the president's calendar was often his meeting with Vice President Biden. After being kept waiting a few times, Biden began wandering into our meetings—at first at the end of our scheduled forty-five minutes, later from the start. Like all Vice Presidents, he didn't seem to have a lot to do.

Once, when Obama did leave at the scheduled time, and we were all standing respectfully for the presidential exit, as was the custom, Biden piped up, "Hey, guys, could you sit down again? I have a couple of things I want to ask you." We did, and his couple of things went on for more than another hour. PCAST members were looking at their watches and deciding whether the *vice* president of the United States was worth missing their flight home for—they would do so for Obama, of course. Had we known that Biden would one day be president, the choice would have been easier.

Our initial meetings with Obama were quite formal and were held in the large State Dining Room. We had so-called Green badges that got us though the White House gate. The PCAST staff had Blue badges, which let them escort us up the driveway and into the main White House front door. We arrived half an hour before the scheduled time and could circulate around the marbled entrance- and cross-halls (so-called) and the other ceremonial rooms, the Red, the Blue, and the Green. The single uniformed guard didn't seem to be bothered by our sitting on the sofas and chairs or touching things—not like the public tours. The White House was really just a *house*, grander and more significant than most, but in no way a palace like, say, Versailles or Buckingham or Schönbrunn. A dimly lit staircase up to the family quarters was next to the front door, seemingly (but probably not actually) unguarded. It was fun to see the originals of familiar presidential portraits, especially taking note of their positioning. Republican presidents were off in the corners. Nixon was nowhere; George W. Bush not yet painted. Prominent, and to me moving, was the famous Shikler portrait of JFK, arms folded, eyes downcast, face in shadow. Lincoln and LBJ got especially good locations.

In the State Dining Room, we were seated around the long table (assigned places with name tent cards) before the president arrived. Each place had a thin pad of White House note paper and a cheap White House pen, souvenirs that most of us kept after the meeting. A separate long row of chairs, behind us and facing the president, was for OSTP and White House staff. These first meetings with the president attracted only second- or third-tier White House staff. We were just another big group of outsiders on the presidential calendar.

That seemed to be how Obama initially also saw us. He entered. We stood. He went around the table and shook hands with each individual, flashing the charismatic campaign grin. He sat. We sat. John Holdren's "meeting memorandum" in front of him included biographical notes on some members (not including me), so that Obama could make polite chit-chat about Doctor Molina's Nobel Prize or Doctor Jackson's latest honorary degree. He addressed us all as Doctor, even the industrialists who weren't.

Co-chairs Holdren, Lander, and Varmus did most of the talking at the first meeting. Before the full PCAST's first meeting, an ad hoc group of a few outside experts (all close colleagues of the co-chairs) had produced a now PCAST-branded report on the 2009 H1N1 swine flu epidemic—what U.S. strategy should be for managing it. Obama listened to their presentation with interest; but by this time vaccination was widespread and cases were declining. A decade later, when the COVID-19 pandemic shut down the world economy and killed a million Americans, H1N1 would be looked back on ruefully as a missed opportunity to have made improvements in the U.S. public health system, including many that were recommended by PCAST and other groups.

Our second meeting with the president was a few months later, also in the State Dining Room and as formal as the first. By then, our several working groups had produced fleshed-out proposals for full reports. The plan was to summarize these to the president and to get his instructions on which should proceed. That, clearly, was not what he wanted or expected. He glared at us. When, for each topic, we arrived at, "Well,

Mr. President, what do you think?" he responded with some variation of a statement that *we* were supposed to tell *him* what was interesting, not the other way around.

One item that did get Obama's obvious interest was a prospectus by Jim Gates for a study on improving K-12 STEM education, especially for underrepresented minorities. Jim, with Shirley Jackson, was one of two African-Americans in our group. Under most circumstances, Jim's manner was polite—even courtly—a perhaps unworldly theoretical physicist. This was the academic string theorist that he in fact was. His unruly dreadlocks seemed in line with the personal eccentricities of mathematicians generally. Now, when he was forcefully lecturing a Black president about the need to create opportunities for Black kids, his appearance seemed to be making a different point. He was an activist. "We *must* raise the average level of science education for African-American children," he exhorted. Obama interrupted him: "Dr. Gates, I'm even more interested in ensuring opportunity for the very best kids—the talented ones capable of reaching the very top." Obama didn't say, "like you and me," but those words hung invisibly in the air.

This was a first glimpse—we would see many more—of Barack Obama's contradictory nature. He was a populist, but also an elitist. He was a community activist, but also a University of Chicago professor of constitutional law. He campaigned like the former, but governed like the latter. He could flash the jubilant Obama grin or (rarely seen in public, but often by us) deliver the corrosive Obama stare. Obama was Black by virtue of an African father; but he was raised in Hawaii by a white mother and white grandparents in Kansas. He attended the elite Punahou private school. He came to self-identify as African-American only later, at Columbia. We in PCAST seldom saw the populist or the African-American Obama.

Obama's dual nature has been much written about by journalists and historians. In the 2008 election, African-American leaders wrestled with whether to fully embrace Obama as one of their own. They eventually did. They had to. If we believe in Ockham's razor (explaining the most facts with the fewest hypotheses), then Bill Clinton was the first Black president, and Barack Obama was a Midwesterner. We saw a Chicago law professor, albeit (later) one with a certain playfulness and a wry sense of humor.

Jim Gates and I later talked about his presidential encounter. We shared the view that the history of America *was* the tragic history of its race relations. Race was America's central theme, monstrously enshrined in the very Constitution, and you couldn't understand America without facing the awful fact squarely. My white colleagues usually avoided talking about race with Black acquaintances. I wasn't shy in this way and felt that there was a lot that I might (and should) learn. Jim was a military brat who had grown up in or near army bases around the country. He first experienced a segregated school in high school, in Orlando, Florida. Talented in math, he applied and was admitted to MIT for college—after which, a successful academic career. He and I became friends and crossed paths many times later on.

PCAST's third meeting with President Obama was a different kind of flop. We were in the more intimate Old Family Dining Room (but this still one of the ceremonial, not West Wing, rooms). We had taken Obama's admonition to heart and, in our own private session earlier in the day, agreed on a list of important things happening in science—things that we thought the president should know about. Ten of us (not including me) were assigned topics on which to prepare snappy three-minute talks. A minute into the first talk, Obama said, "Yeah, I read about that in the *New York Times.* What's your next topic?" And so it went. There were perhaps three of the ten topics that he didn't already know about. We finished with twenty minutes still left. The president looked at his watch.

At this point, Mario Molina chose to compound the disaster by going completely off script: "Mr. President," he said. "Since we have some extra time, I'd like to talk to you about the urgency of taking action on climate change." He then launched into an earnest, but insultingly elementary, lecture on the subject. Obama's face hardened, but he waited for Molina to finish—which took a while. "Dr. Molina," he said sourly, "you don't have to convince me that human-caused climate change is real." He went on to give us a law-professor lecture about the political realities. His administration had barely succeeded in getting the Affordable Care Act ("Obamacare") enacted. The 2010 midterm election campaign was in full swing and looked to be tough for the Democrats. He didn't need PCAST weighing in with a report on climate that, however factually accurate, would be politically controversial. We could work around the edges of the climate issue on things like clean energy and environmental capital. We should stay away from climate directly, he told us.

History records that the 2010 elections were one of the most disastrous for an incumbent party in U.S. history. Democrats lost control of the House with a loss of sixty-three house seats, and barely retained control of the Senate, losing six seats. Climate was now even more politically untouchable, especially with new fears for Obama's reelection chances. Before the midterms, Obama's 2010 State of the Union Address ("SOTU" to insiders) called out the "overwhelming scientific evidence" of climate change. After the midterms, the 2011 SOTU didn't mention climate at all. John Holdren—who had become science advisor with climate at the top of his personal agenda—could not help but be disappointed. The president consoled Holdren that climate's day would come after the 2012 reelection.

Whatever were John's true feelings, he masked them to us on PCAST with anecdotes about Obama's enthusiasm for the role of science on other issues. John did not have Cabinet rank—that would come to a science advisor only in the Biden administration—but he attended Cabinet meetings, sitting in the Cabinet room with other senior advisors in an outer ring of seats. He learned to be at all times prepared to speak on all issues with technical content, because Obama sometimes called on him impromptu. "John, please explain that to everyone...," Obama might say, and John would have to deliver a lecture to Cabinet officers whose interest in science was mostly nil. Obama was sometimes intentionally shaming them on that.

We eventually did reach an effective working relationship with the president, although it took more work. It helped when we began to have a steady stream of reports ready for release and could run by him in each case our explicit action recommendations. We weren't necessarily looking for his point-by-point approval, but rather for his ideas on follow-up actions that could increase their impact, thoughts he freely gave. It was a curious inversion: his advising us, instead of the other way around.

A lesson we quickly learned was that we had to anticipate obvious next steps, and to take them *before* meeting with the president. This first came up in a report with recommendations addressed to several Cabinet departments. "Well, what do *they* think?" Obama asked. We were deer in the headlights. "Who?" we asked. He meant the Secretaries of X, Y, and Z. They didn't necessarily have to agree with our recommendations. But we shouldn't waste presidential meeting time when *knowing* their views was an obvious prerequisite to any action.

I came to understand the presidency as a job akin to CEO of a large global corporation, one with many freestanding subsidiaries—here, the Cabinet departments and federal agencies. A good CEO didn't micromanage actions many levels down across those lines. He (or she) set and communicated high-level policy, urged, guided, signaled favor or disfavor, gave advice, and (very occasionally) hired or fired senior people. But, in this model, a complication for presidents in the White House was the political necessity of preserving exactly the opposite illusion—that the president was a forceful leader who personally decided all government actions and could right any wrong.

It made a noticeable difference in the relationship when we began meeting the president in the West Wing's Roosevelt Room. Although nicely furnished, this was a cramped, windowless meeting room; but it was just across the hall from the Oval Office. In some subliminal way, that seemed to make us part of the *insider* team instead of (in the ceremonial rooms at the other end of the West Colonnade) *outside* experts. We got used to following John Holdren, in single file like a line of kindergarteners, into the basement west entrance, then up the narrow stairs to the windowless, claustrophobic reception vestibule with its single Marine guard and single receptionist. This cramped area was lined with about half a dozen chairs, like a dentist's waiting room. This was where people waited—there wasn't anywhere else—to see the Leader of the Free World. The Warner Bros. set for the TV show *The West Wing* replaced the vestibule with a grand marble lobby and enlarged all the other West Wing rooms by a factor of two at least. The humble reality of the actual West Wing would not have been believable to TV audiences.

Over time, we learned to interpret the president's reactions to our report presentations as a grade: If he said, "thank you, that's very interesting," the grade was a C. If he said, "here's what you should do," and rattled off a list of people whom we should talk to and actions that we should take, the grade was a B. If he said, "Let's do that," and turned to a senior advisor with, "Valerie [or Brian, etc.] please follow up on this and get back to me," then the grade was an A. That these close senior advisors were now in the room at all was an indication that we, as a group, were being taken seriously. The A grade meant that we had crystalized something that now, in Obama's mind, was ready for action. It could never be something completely new—it had to have what was called "heritage". Our report might give a round boulder its final push over the top of the hill.

The B grade was a license for us to continue pressing the issue in public or within government. Our report on radio spectrum allocation got a B grade from Obama, but became effective by virtue of Mark Gorenberg's hard follow-up work with FCC staff. When Craig Mundie and I spearheaded a cybersecurity report, some of its recommendations were taken up by White House cybersecurity czar Michael Daniel and his deputy Andy Ozment—for example, that federal agencies should get rid of the ten-years-obsolete Windows XP operating system. Craig, as a Microsoft senior executive, was in something of a delicate position. Junking Windows XP would mean the federal government's purchasing new licenses for its replacement, Windows 7, a windfall to Microsoft of hundreds of millions of dollars. But he was Microsoft's top-level liaison to NSA, with high-level security clearances. Craig knew better than anyone that Windows XP was a security disaster whose flaws were being exploited by foreign intelligence services.

A different cyber recommendation of ours was that the Securities and Exchange Commission should mandate routine public reporting of cyber security incidents under their existing authority to require publicly held firms to report investor risks. The existing SEC position was that a cyber incident had to be "material" to the company's survival—an impossibly high bar. Craig and I met with two high SEC staffers—and got a frostily patronizing lecture about the SEC's status as an independent agency that didn't take advice from anyone. Then, slowly over years, we watched the SEC's position slowly change in the direction that we (and later, with increasing weight, others) recommended.

Obama's C grade meant "you're wasting your time." These reports of ours were issued publicly—by law, all of our completed reports had to be—but never went anywhere. I put my heart into leading a report on the status of the U.S. scientific enterprise, working on it for more than a year. Its release event, with speeches by Holdren, Chuck Vest, and NSF director Subra Suresh was favorably reported in the science and higher-ed media. Still, it got a C. The problem was that few if any of its recommendations were actionable. The many that began "Universities should…" were DOA. In later years, I was often complimented on this report. People quoted it or plagiarized its prose in their own reports and white papers on that perennial subject. Notwithstanding, as presidential advice, it was a complete flop.

I was personally targeted by the Obama glare only once, in 2012, in connection with this same star-crossed report. I had three minutes on our agenda, and exactly three talking points. Ten seconds into my first point, Obama said, "I get it. Next point." I just didn't believe he had gotten it. My great punch line for this first point was just two sentences ahead, and I continued toward it. That was when I got the glare and an icy, "I said, *I get it. Please* go on to your next point." I did.

Again skipping forward, Obama got used to us, and we got used to him, and our meetings took on a much less formal character. He was very smart, and we learned not to waste his time. He learned some of our names (not mine, I think), and that he could let his hair down and tell us what he really thought about some issues. White House photographer Pete Souza often took pictures during PCAST meetings. It is informative to compare an early one, in the State Dining Room, with a late one, in the Roosevelt Room. In the former, Obama is flashing his campaign grin, and we are sitting stiffly, our faces respectful and serious. In the latter, Obama is slouched back in his chair looking unhappy, while, across the table, John Holdren is standing and wagging an outstretched finger at him. The rest of us are in various poses that range from laughter to disapproval.

Our good relations with the president earned us the perk of an all-day visit to Camp David. (An overnight stay, or a visit while the president was in residence, would have been far above our station.) For form's sake we held a brief, non-public, PCAST meeting—this in one of the well-appointed "cabins" that figured prominently in the 1978 Sadat-Begin negotiations (a plaque on the wall explained this); but mostly it was a junket, and we had free run of the facility. The helpful Navy enlisteds who staffed the place—Camp David was technically a Navy base—set each of us up with an electric golf cart for getting around. In the president's absence there were no security restrictions. By tradition, every president sponsored a signature addition to Camp David. Obama's was an indoor basketball court. Some PCAST members shot hoops for bragging rights. I raced my cart around the whole perimeter road inside the high fence—several miles circumference. There were walking paths through dense woods in which I tried to spot the camouflaged surveillance and security sensors that I was sure must be there. I didn't see any. The real highlight was the visit to the Shangri-La gift shop in Hickory Lodge, where we all bought Camp David logo T-shirts, hoodies, mugs, medallions, etc. Collectively we dropped several thousand dollars on souvenirs.

With a candor that surprised us, Vice President Biden once said to us about Obama, after the latter had left the room, "This man is a *good* president, but he could have been a *great* president." Biden elaborated on his thesis: Technology progress—this the connection to PCAST business—left groups of people behind. Manufacturing had gone overseas. Unions had lost influence. Middle-aged factory workers could no longer count on lifting their families securely into the middle class. Obama, Biden said, hadn't succeeded at providing a vision for these people—a light at the end of the tunnel. It wasn't sufficient to create job training programs that were beneath the dignity of the people who needed them. It didn't help to tell people that, in a global economy, realistically, manufacturing in the U.S. was gone forever. Obama campaigned on hope, Biden reminded us. The president wasn't delivering hope to these people who needed it. At the time, we were shocked at the vice president's seeming disloyalty to his boss—his talking so freely to outsiders like us. Later, after Donald Trump won the 2016 election, I came to understand Biden's rant to us not as disloyalty, but as a frustrated prognostication.

67. Alice and Bob

In the holiday period between Thanksgiving and New Year's, when my external academic world slowed to a crawl, I liked to investigate an offbeat, recreational topic and then write an explanatory paper for posting on my web page. One Christmas, I wrote about Albrecht Dürer's famous magic square, which provided the plot resolution for Dan Brown's pulp thriller *The Lost Symbol*. Another year, it was a subtle method for finding mathematical coincidences, formulas that looked (and were) impossible, but numerically convincing. (A simple example was $\pi = (12 / \log 889)^2 + e^{-4}$, which is false, but true to sixteen digits.) Another year, I found a simple model for U.S. income inequality that could predict with good statistical accuracy how many individuals would file tax returns showing incomes at various stratospheric levels. Yet another year, I developed a new algorithm (and computer program) for calculating the winner in a so-called Condorcet election. And yet another was a paper on distorted cartographic maps, the kind that (e.g.) makes states' map areas proportional to their populations, or any other variable—like their number of members in the National Academy of Sciences, where the U.S. became two bloated coasts with nothing between.

None of these investigations were profound, or even good enough to submit to a reputable journal. Nor did I expect anything different in 2011, when my chosen topic was the Iterated Prisoner's Dilemma game, known as IPD. Some background explanation is helpful. The game imagines that Alice and Bob are both arrested on suspicion of committing a serious crime. The detective questions them in separate rooms. Each has sworn not to betray the other, here termed *cooperate* (meaning with each other, not with the detective). "I already have enough evidence to convict you both of a misdemeanor," the detective says to each. "That will put you away for one year. But if you *defect* and rat out your partner I'll let you go scot-free. Your partner will get a felony conviction, six years in the pen."

"What if we both defect?" Alice and Bob each ask.

"Well, I can't let you both go free," the detective says. "You'll each get three years."

Alice reasons as follows: There are only two possibilities. Either Bob will *cooperate* (with her) or else he'll *defect* (to the detective). If he defects, then she'll get six years—unless she defects also, in which case she'll get just three years. So, if he rats, it is better for her to rat too. But what if Bob doesn't rat? Then she can still rat on him and be out immediately. So, either way, she is better off defecting. Bob employs the same reasoning and rats out Alice. They each get three years. They spend the time wishing that they had both kept their promises not to betray each other and escaped with misdemeanor convictions. This seeming paradox is known as the Prisoner's Dilemma, or PD.

IPD differs from PD only in that Alice and Bob play the game not once, but repeatedly. Perhaps, after both keep defecting for a time, they'll wise up and start both cooperating. That idea is grandly named "the evolution of cooperation." It is much studied by evolutionary biologists, because cooperation is observed in many species in the natural world. IPD, as a mathematical model for the evolution of cooperation, was widely studied in the 1960s and 1970s. The famous political scientist Robert Axelrod called it "the *E. coli* of social psychology," meaning everyone's favorite laboratory model. Based on computer experiments, and actual trials with college undergraduates, Axelrod conjectured that the optimal strategy for IPD was some variant of what he called *tit-for-tat*: If on the previous round your opponent cooperated, then you should cooperate on the present round; if he defected then, then you should defect now. In other words, punish a defection with a defection—but only one.

I thought it odd that, forty years later, there was in the literature still no mathematical demonstration that *tit-for-tat* (or any variant) was in fact optimal. That assertion remained only conjecture. I hoped that with vastly more computer power than was available to Axelrod (just on my laptop!), I would be able to straightforwardly calculate the game's so-called Nash equilibrium, the optimal strategy for both players. Here omitting the details, I reframed the problem as one that any computational scientist could recognize as a numerical saddle-point optimization on an eight-dimensional hypercube. I then wrote a computer code to solve exactly that problem.

In real time, I watched the progress of the optimization on my computer. The program seemed to be approaching a Nash equilibrium—until it crashed. I tried again, starting from a different point. Again, the program crashed, but at a different position in the hypercube. I automated the procedure and ran the program a thousand times—and got a thousand crashes. But, I found, they weren't at random points. All of the crashes occurred on a particular four-dimensional hyperplane. I traced their cause to a faulty assumption in my program. It assumed that when Bob *changed* his strategy, there would be some effect on his own average performance, and similarly for Alice. How could this not be true? Apparently it wasn't. The computer could find instances, but not explain them.

Freeman Dyson was then eighty-eight years old, but I knew that I needed his help. The exact complement to computer intelligence, as yin to yang, was Freeman Dyson intelligence. Over thirty years of JASON summer studies, I had gotten good at taking problems to just the right level of well-posedness to engage Freeman's mathematical genius. Within JASON, I was considered a kind of Dyson expert in this regard. Word of this had gotten out, and I contributed several quotes (some for attribution, some definitely not) to a 2009 *New York Times Magazine* profile of Freeman.

Now, I emailed Dyson a description of my puzzling results. A week later, he sent back a note with the general result all worked out. When the confusion was hacked away, it came down to a simple equation in high-school algebra. Hidden within my hypercube and overlooked in fifty years of research on IPD was the mathematical fact that Alice had the ability, by a certain strategy, to herself dictate Bob's score. No matter how Bob played, that score, on average, would be his outcome. And Bob, correspondingly, could do the same thing to Alice. Game theorists already had a name for this situation: an ultimatum game. Each player could present to the other an ultimatum: this will be your score. Game theorists had no idea that there was an ultimatum game hidden inside IPD. An even more striking variant was that either player could cause their own reward (avoided jail time, say) to be a *multiple* of the other player's—an unfairly extortionate result, in other words.

It was only a short jump to the realization that, in this circumstance, the outcome of the game was influenced by whether each player had a *theory of mind* about the other. Psychologists used the term theory of mind to mean the ability to attribute mental states—such as belief and intention—to others. Noticing that Alice's winnings are always five times his own, Bob thinks, "I am being extorted! I will temporarily sacrifice to make us both get zero—five times zero is zero. I have a theory of mind about Alice: she'll notice this and not like it. Then, we can negotiate." But Bob may be wrong about Alice. She may have put her extortion strategy on autopilot, and then disappeared. Bob' signaling can only hurt himself. His best option is to accede to the extortion and take the meager winnings that Alice allows him. "Press and I have solved the

Prisoner's Dilemma game," Dyson told people jocularly. "The winning strategy is to go to lunch."

I wrote up our paper over Christmas vacation. Josh Plotkin, whom I had first known as a Packard Fellow, and Harvard game theorist Drew Fudenberg were referees for *PNAS*. The paper was published in mid-2012 and caused a stir. People reacted with surprising emotion. Science writer Bill Poundstone suggested applications to abusive marriages, terrorism, the then-current U.S. Congressional deadlock, and income inequality. A blogger wrote, "This and similar 'quant' nonsense is precisely what has led to the too-big-to-fail banking disaster we are currently confronting. These studies are worse than useless, they are parasitic cancers on society." Elsewhere, a kinder blogger wrote, "Once again, physicists invade a field and add value." Even that seemed barbed.

As I write today, more than two hundred papers have been published on one or another aspect of our "zero-determinant strategies". Neither Freeman nor I did any further work in game theory. I was never sure what lesson could be drawn from all this. That a couple of physicists with complementary abilities—the more senior, eighty-eight years old—could invade a field and, over a holiday vacation, find an undiscovered nugget capable of attracting such attention may say something about serendipity, or about the genius of Freeman Dyson; or it may suggest that subfields of science can easily become too set in their ways, and that the scientific enterprise should seek institutional mechanisms that encourage more cross-fertilization across scientific boundaries.

A couple of years after our paper, the Institute for Advanced Study held a celebration in honor of Freeman's ninetieth birthday to which I was invited as a speaker—to recount the above story of what was thought likely to be Freeman's last mathematical paper—although he did continue to write book reviews. I accepted on condition that Jeffrey and I could stay at Marquand House, which was otherwise being reserved for the IAS trustees and big donors. Freeman was consulted, and we were given a very nice room.

After many decades, Mrs. Moriarty had retired as the Marquand housekeeper. Her replacement had the updated title of house manager. Breakfasts were offered at staggered times, so that each could be cooked to order. Our time slot found us sharing the breakfast table with billionaire Charles Simonyi and not-quite-as-rich Danny Hillis. In the 1980s, Simonyi had founded Microsoft's applications group and wrote the first version of a word processor that become Microsoft Word. He was also famous as a space tourist, paying the Russian space agency, twice, for visits to the International Space Station. Hillis, whom I had last interacted with on CSTB and at his Thinking Machines Corp. in the 1990s (and who was the brother of my UT colleague David Hillis), was volubly telling Simonyi about his Long Now Foundation and the construction of a purely mechanical clock with a 200-foot pendulum length that would keep time for ten-thousand years, powered by temperature fluctuations. When he mentioned bran, it was not the breakfast cereal on the table but Bran Ferren. "Bill" (said with awe) was not me—I was an uncredited extra in this movie—but you-know-who, that founder of Microsoft. It seemed that no last names were necessary among billionaires. Simonyi said very little, but somehow, during the meal, he took a liking to Jeffrey. He silently followed her around most of the day during the breaks between talks. When she and I took our provided box lunches to an out-of-the-way bench to eat an almost intimate meal, Simonyi sat down next to her with his own box lunch—and said nothing, not a word, the whole time she and I were eating and conversing. As I now write, he is the excellent chair of the IAS board.

I continued to send Freeman occasional math problems of interest to me. He generally responded with thanks but no further results. In his nineties he continued to attend the JASON summer studies. He didn't bill for his daily stipend as the rest of us did, or, as far as I could tell, do much. My reputation as a Dyson expert continued to spread in the media, however. I think this annoyed Freeman, since, although we had worked together now and then in JASON for thirty years, we were never really friends. I can't recall even a single conversation with him about anything remotely personal—although Jeffrey had many such conversations with Imme.

But, when Dyson died in 2020 at the age of ninety-six, several publications, taking for granted that we had been personally close, asked me to write obituaries or remembrances of some kind. (Some years earlier, I had written a warm biographical remembrance of Harold Agnew for the journal *PNAS*.) George Dyson, Freeman's son, sent me a photograph of the handwritten page of mathematics that was on top on Freeman's desk at IAS the day he died, seemingly the last thing he was working on. It related to a problem in the statistics of inheritance of polygenic traits that I had sent to him ten months earlier.

I felt that there was a debt of irony that I somehow needed to work off. Over the years I had, I was sure, casually claimed to have read each of Freeman's many popular books; but in fact I had bothered to read only one or two. Now, I bought them all from used-book websites, used an X-Acto knife to slice them into loose pages, and scanned all the pages into my computer. After I cross-indexed all the files for word search, I could read all the books in order of publication and cross-reference every later use of the same anecdotes in later books—Freeman did tend to reuse his own best material. His ur-source became clear only in 2018 with his publication of a volume of his lifelong weekly letters to his mother and sister. All the best anecdotes were there, contemporaneously recorded. During the COVID-19 lockdown, on a twelve-hour drive straight through from Austin to our house in Los Alamos (at that time a safer place to isolate), I listened to almost that many hours of Freeman's recorded interviews and oral histories.

If, before, I was only pretending to be a Dyson expert, now I felt genuinely to be one. I wrote a 1,000-word obituary for *Physics Today,* a 5,000-word piece for *Inference*, and, with Ann Finkbeiner (a science writer and friend who had written a book on JASON and also knew Freeman well), a 7,000-word biographical memoir for publication by the National Academy of Sciences. Unlike some other biographers, Ann and I did not avoid the dodgy issues of Freeman's strange pantheism, his teleological belief that the universe bent to a purpose, and his heretical stance on the reality of global climate change. We built a case that these were all related, that Freeman was at heart a mystic. With trepidation, we ran our draft past George Dyson. He objected to nothing, so, by way of thanks, I sent him my cross-indexed all-Dyson database file.

68. Elections

When I was twelve or thirteen, my father gave me a copy of Leó Szilárd's recently published book of gentle science-fiction stories, *The Voice of the Dolphins*. Frank had met Szilárd—now most remembered for drafting Einstein's letter to FDR that led to the atomic bomb—at a PSAC meeting in Washington. In substance, the stories were wry political satire, at the time far over my head. But several of them developed one of Szilárd's favorite themes: that aging scientists hold back, rather than advance, scientific progress; and that their negative influence is best mitigated by inducing them to spend their time on honorific, but meaningless, committees.

Szilárd's *Dolphins* came to my mind fifty years later in 2010 when Alan Leshner, the long-serving executive director and CEO of the American Association for the Advancement of Science, telephoned to ask if I would be willing to stand for election as president of AAAS (pronounced by all as "Triple-A-Ess"), the organization that billed itself as "the world's largest general scientific society."

It was a fluke that I was even a member of AAAS. Anyone could join for a hundred dollars a year, and membership provided the benefit of a paper subscription to the weekly *Science* magazine. Before electronic publishing and university site desktop licenses, joining the AAAS was the *only* way to subscribe to this top scientific journal, so most of AAAS's one hundred thousand active memberships existed on its rolls for that reason alone. I joined the AAAS when, at Texas, I started pretending to be a biologist: I knew that biologists all read *Science*. Three years later, the issues piled up on my desk made me eligible to run for the association's top job.

AAAS existed on three nonintersecting parallel planes. The magazine was on one. Its professional staff, many of them lifers, jealously guarded a haughty independence from the rest of the organization. A second plane was significant only "inside the D.C. Beltway," practically invisible elsewhere: AAAS was a ready host for soft-money, government-agency sponsored projects in science policy that could not find a home at a university. "Soft money" meant that entrepreneurial individuals who got such funding became AAAS professional staffers for exactly as long as they had funds to pay their own salaries—as well as paying the overhead charged for their space.

AAAS occupied a modern, architecturally impressive building in the center of downtown Washington—the gift of financier Bill Golden, who was for thirty years the association's treasurer, and whom I had known as a member of Harvard's astronomy department visiting committee when I was chair.

On a third parallel plane, AAAS was a professional scientific society with twenty-four self-governing sections in disciplines ranging from the more traditional astronomy, physics, chemistry, and so on, to agriculture, dentistry, "societal impacts," and even "general interest in science." In none of these fields was it the top professional society, but, even so, it offered opportunities for otherwise idle scientists. Szilárd would have been proud: Each section had its own officers, committees, and annual agendas of make-work. AAAS annual meetings were held in large convention centers—a different city each year—and typically hosted 120 separate sessions, with also invited plenary talks by famous people. Attendance was in the many thousands.

The fragility of AAAS governance was that the organization's standing (on two of its planes) depended on its having elected officers and board members with national and international name recognition; but its election process tended to produce boards mixed in reputation and accomplishment. Add to this the peculiarity of the so-called presidential line: Members (those who actually bothered to vote—a small minority of *Science* subscribers) elected not a president, but a president-*elect* who would serve for one year, advance to president for a second year, and become board chair in their third and final year. The other board members had four-year terms. Like me, individuals in the AAAS presidential line often had little previous involvement with the organization. They parachuted in, briefly lent their name, and departed—with an unwritten lifetime obligation to continue contributing to the annual fund.

Perhaps not so bluntly as the above, but still very clearly, Alan Leshner explained this all to me in his telephone call. He also made clear that *he*, not the presidential line or the board, actually ran the organization—so I needn't worry that I might actually have to do anything. Leshner was a well-known and well-liked presence around Washington. He had an unpretentious, folksy manner, was excellent on panels, and could always make any audience laugh. He was the Jewish uncle that you never knew you had, but immediately loved. Not unlike other uncles, he also made sure you knew certain things about him, for example that he was the second-highest-paid scientist inside the Beltway.

Only the CEO of the industry-dominated American Chemical Society made slightly more—at the time about a million dollars a year.

"And anyway," Alan's telephone pitch to me concluded, "you won't get elected." I would be running against a charismatic female candidate who was sure to win. That was good enough for me, and I accepted the nomination. But while the other candidate may have been charismatic, she turned out to be also controversial. To everyone's surprise, I won—my first election victory since student council in junior high school. I would be the 165th president of the AAAS.

Alice Huang was outgoing chair in my year as president-elect. In 1980s Cambridge, her (and David Baltimore's) daughter Teak was a school friend of Sara's, and Alice now felt free to give me her unvarnished advice on matters large and small. Small: The shortest path from the Willard-Intercontinental Hotel to the AAAS building took advantage of a diagonal block at the Inter-American Development Bank to save twenty seconds' walk. (Leshner always had the board put up at five-star hotels.) Large: A lot of things at AAAS needed fixing, and Leshner wasn't going to lead in fixing them. It would take concerted effort by all three in the presidential line, and a lot of arm-twisting of the board, to get any kinds of reforms even started. And, unfortunately, serving between Alice and me in line (that year as president) was Nina Federoff—known as someone who rarely cooperated with anyone on anything and, in fact, on leave in Saudi Arabia at King Abdulla University of Science and Technology. During the two years that Nina and I overlapped, she rarely attended board meetings.

Every new AAAS president gives a plenary address to some thousands of people at that year's annual meeting. Held in Chicago that year, Rahm Emanuel, then mayor, was on the plenary's agenda to welcome the thousands of us. He arrived with police escort at the convention center with only minutes to spare. Backstage, I introduced myself as the plenary session's chair. "What d'ya want?" he said. I didn't know what he meant. Was I supposed to ask for something? Kickback for bringing in the business? A patronage job in the department of sanitation? "What d'ya want?" he repeated impatiently. I got it! He wanted me to write his speech for him. I reeled off: "Welcome, Chicago, industry, technology, basic science, educating the next generation, great University of Chicago, Enrico Fermi then to Fermilab now, keep doing all your good works …" He stopped me and went onstage to deliver a polished ten-minute welcome, elaborating on all of my points and coming across as meticulously prepared.

Phil Sharp, Gerry Fink, and I (the presidential line in my final year) did succeed in getting Alan Leshner to initiate some modest efforts at change—for example, chartering an internal committee that could possibly, at some indefinite future time, actually produce a strategic plan for the organization. He announced this to the full board with great fanfare and, at the same meeting, took Phil, Gerry, and me aside to tell us that he intended to retire in a year, but that we were not authorized to tell a single other soul. It was his way of telling us to get off his back—that nothing was going to happen of any consequence in his (or our) remaining time. Alan did retire, to praise from all quarters for his long service to AAAS, the nation, and science.

Did I accomplish nothing at AAAS? Not quite. It was on my watch that Bruce Alberts decided to step down as editor-in-chief of *Science*, a job he had taken after finishing his term as NAS president. Bruce did the work part-time from California, at most a few hours a week, leaving almost all decisions to his full-time managing editor in Washington. His extraordinarily generous salary was set by Leshner. I chaired the search committee to find his replacement. *Science* had been coasting on past reputation and was now far behind its international competitor, *Nature*. I insisted that we look for someone who would do the job full time and be aggressive about making changes—even if they lacked the prestige of a former NAS president. Alan, on the committee, was skeptical that we could find any good scientist willing to make such an irreversible career shift. Mostly he was right. Candidates whom we approached were uniformly enthusiastic—that is, about doing the job part time, staying in their existing labs, and getting paid like Bruce.

Marcia McNutt was an exception. As director of the U.S. Geological Survey, she was an Obama political appointee who would, in any case, be out of a job when Obama's second term ended in less than three years. With originally a Ph.D. from Scripps Institution of Oceanography, she had risen to full professor in Frank's department at MIT. I knew Marcia from that time, but better from her job after that, as director of the Packard-supported Monterey Bay Aquarium Research Institute, where she was often at the annual Packard Fellows meetings.

Marcia's reputation in science was more the do-er than the theorizer. While in graduate school, she qualified in a Navy SEALS training program in underwater demolition and explosives handling. She competed in rodeos—barrel racing. At MIT she ran the joint oceanographic program with Woods Hole, doing fifteen expeditions, chief scientist on half of them. She was in the process of reorganizing USGS with an energy not there seen in fifty years. She had been the government's senior on-the-ground official in Houston for many weeks during the Deepwater Horizon oil spill in the Gulf of Mexico.

I was not surprised when the search committee's reaction to Marcia was mixed. The problem, never openly articulated by either the men or the women on the committee, was her appearance. You could describe Marcia as glamorous, with straw-blonde hair, piercing blue eyes and a perfect smile. Or, you could see her as—well—inappropriately dressed in tight clothes with revealing necklines, too-short skirts, stiletto heels or (maybe worse) decorated cowboy boots. Marcia's appearance would not have been out of place as a hostess in a Las Vegas cocktail lounge—an upscale one to be sure. She was a rarity (at least then), a woman scientist who dressed exactly the way she wanted to and didn't give a fuck about what other people thought. The eccentricity of Marcia's appearance had no effect on her manner, which was always businesslike and direct. She was a killer change-agent in a tight pencil skirt.

It was hard to bring the search committee around, especially when none were willing to say what all were thinking, that she didn't *look* like the editor-in-chief of a preeminent scientific journal. Detractors on the committee were sure that no one who looked like that could write well. I had to phone Marcia to ask apologetically for a writing sample. "Well, I do a weekly letter to our eight thousand USGS employees. Would that do?" She submitted a dozen of these, which proved to be well-written essays on all manner of scientific and administrative topics. I was relieved when the committee agreed to offer her the job, the more so when she accepted it. My involvement with AAAS ended soon after that, as did, I thought, my professional entanglement with McNutt's career.

But, as it turned out, it was only a hiatus. In 2015, I was asked to serve on NAS's nominations committee for councilors, officers, and—this happened less often than once a decade—a new NAS president. Then-president Ralph Cicerone had quietly asked the NAS Council to shorten his second six-year term by a year. In fact, Ralph looked terrible, but he affected surprise and indignation when anyone asked about his health. He was fine, he always said. In the event, he died less than six months after stepping down, at age seventy-three, with no cause of death ever announced. I was a known quantity for the NAS nominations committee. I had been on it three times before: first, eleven years earlier when Cicerone was nominated; then a year later as chair (with only council members, not officers, elected in that round); and again, a few years after that, when Barbara Schaal was elected to her second term as academy Vice President. Barbara was now to be chair of this new nominations committee. Together on PCAST, she had become one of my best friends, and she had been on the AAAS nominating committee that proposed me for that association's president. I always found the degree of inbreeding at this level of scientific leadership staggering, even while benefiting from it.

Because every NAS scientific section was allocated a member, it was a huge committee—twenty-five in all. At the first meeting, before any possible names for president were even mentioned, Barbara took me aside. "We think it's about time to have a woman as NAS president." Guessing that the eight women on the committee were telepathically linked, I didn't have to ask who "we" were. "I'm with you," I said. I don't know who else she lined up in so direct a fashion. Nothing was said in the presence of the full committee, but I think Schaal soon secretly had the necessary thirteen fully committed votes. After that it would be just process—lengthy, but foreordained.

We did still have to *find* the woman, however. Luckily, by a quirk of the NAS election process, we only needed one. While all offices other than president were decided in contested elections with two candidates, for NAS president, the nominations committee's single candidate was submitted to a yes/no vote of the membership. The president position was a demanding full-time job in Washington, for as long as twelve years; finding two candidates to so commit, one of whom would then very publicly lose, had been long ago rejected as impractical.

At home, Jeffrey told me that the choice ought to be Marcia. But what about her clothes? "She just always looks great," Jeffrey said dismissively. But, when I mentioned Marcia privately to Barbara (who dressed conservatively and was by now a dean at her university), she looked doubtful. It might be cool to have a fashion-minded Las Vegas cocktail hostess as the public face of the NAS, but the world might not be ready for that. There were other highly qualified women candidates. But, in the end, in the final round, none were willing to be considered for the job. There was no shortage of willing male candidates, but none were standouts. Over a lengthy six-month process with multiple interviews, McNutt emerged on the merits as the clear front runner.

Then, unexpectedly, the process stalled. Three committee members, all male, all respected scientists active in academy affairs, announced *en bloc* that they would not only vote against McNutt, they would actively oppose her—appealing to the NAS Council if necessary to block her nomination. It was extraordinary. Theirs were merely three votes out of twenty-five, but they spoke as if they had special standing as secret representatives of a ruling patriarchy that I, somehow, had not been invited to join—or even know about. The remaining twenty-two of us were aghast. None of us doubted that unspoken gender discrimination existed, but no one expected it to be so crudely expressed. Ultimately, Barbara forced the vote, and Marcia won decisively. The attempt at a three-male appeal was decisively rebuffed by the NAS Council, and McNutt became the 22nd president of the National Academy of Sciences, its first female president, in July, 2016.

With all this high drama, the lesser business of the committee—nominating two candidates for a four-year term as NAS treasurer, a part-time job—should have been an anticlimax. Jerry Ostriker, the outgoing treasurer, had done very little. NAS and its operating arm, the fiscally much larger NRC, had a full-time Chief Financial Officer who had a staff of a hundred. Other than his or her being an officer on Council, the elected treasurer's job was largely ceremonial. We already had a willing and well-qualified candidate, a respected medical school dean who had spent some of his career in big pharma and was thought to be personally wealthy—his research was the basis for a best-selling drug.

Still, statutorily, we had to find an additional candidate to run against him and had, thus far, failed. I called several candidates trying to drum up interest, but to no avail. It was getting late in the day, and the committee was tired. I stood up and announced that I was going to step out of the room—and that if this gave the committee any ideas about who might be the sacrificial candidate to run against the pharma dean, then that was OK with me. When I came back ten minutes later, I had been chosen to run.

Word must have spread fast through the NAS building. In less than twenty minutes, we heard that Ralph Cicerone wanted to address the committee on the necessary qualifications for treasurer and, after another ten minutes, he was there. It was very important, he said, looking directly at me, that the new NAS treasurer have industry experience. I couldn't believe it! I had just been nominated, and the president of the National Academy of Sciences was already campaigning against me! It reminded me most of when Scoutmaster Thomasson had campaigned against me in my (rigged) election as patrol leader—and probably for many of the same reasons then as now. I think that Ralph, a cautious and possibly unwell man, couldn't stomach the idea of a cocktail hostess as NAS president *and* a smart-aleck dilletante as NAS treasurer, both at the same time. Jerry Ostriker's slack performance as treasurer—he, another astrophysicist—might also have been a factor. Ralph clearly wanted me to withdraw as a candidate, to let the committee to find someone else. But we had tried; there wasn't anyone else.

Winning the election was a surprise, but less so than AAAS had been. Academy members in Class II (life sciences) didn't know me, but saw that I was listed as a professor of biology—and I was *not* from big pharma. That was enough to split their votes. Members in Class I (physical sciences) knew that I was secretly one of their own and nearly all probably voted for me. Some nonagenarian members probably thought they were voting for my father. I became NAS treasurer the same day that Marcia became president. Three out of five Council officers were now women, plus eight out of twelve Council members at large. It seemed to be the beginning of a new era. And then, four months later, Donald Trump was elected U.S. president.

69. Trumped

On U.S. election day, November, 2016, I was sitting in a meeting room in Beijing's Crowne Plaza Zhongguancun as a participant in CISAC's latest bilateral China discussion. Our Chinese counterparts were by this time familiar faces. We were all seeming friends, while also adversaries. The Chinese participants expressed a moderate interest in the American election, mainly to make a point of their own system's superiority, as they saw it. The permanence of the Chinese Communist Party guaranteed them immunity from any disruptive change. These Chinese, mostly People's Liberation Army affiliated, didn't much differentiate candidates Hillary Clinton and Donald Trump. The implied criticism was more generic: You Americans go crazy every four years.

At least this time we could counter the accusation: The new Clinton administration would be very much a continuation of Obama's. Hillary's policy advisors were from the same universities and think-tanks as Obama's now-serving officials. Many of these would stay in government. That Hillary would win the election was simply assumed. During a tea break, I explained to a small Chinese audience that Trump had won his party's nomination by mobilizing a minority of the extreme right who were not typical of the electorate at large who would reject his extreme views. The polls all comfortably favored Clinton.

I went on to explain the so-called Blue Wall—the union-heavy, solidly Democratic states, Pennsylvania, Michigan, and Wisconsin. My Chinese colleagues' eyes glazed over. They had less interest in the demographics of Pennsylvania, Michigan, and Wisconsin than we would have had in Hebei, Hubei, and Henan. To them, I was an Obama political appointee—a kind of low-level, bureaucrat-mandarin—unless I was also an intelligence agent under academic cover, to them a possibility. My formal presentation to the current session could support both hypotheses: I gave a historical lecture on the intricacies of the legal interface between U.S. military and U.S. civilian cyber, both offense and defense. The Chinese interest in Title 10 versus Title 50, and in *posse comitatus,* was even less than their interest in Wisconsin dairy politics.

The rock star of these 2016 cyber talks was Chris Inglis, our team's recently retired deputy director of the National Security Agency, NSA. Likely in part the result of the Snowden reveals, the PLA had embarked on a multi-year plan to consolidate and modernize its cyber forces, both offensive (which publicly didn't exist) and defensive. At breaks, the Chinese peppered Chris with questions about exactly how NSA and U.S. Cyber Command were organized and how the two interacted. They were visibly surprised at Inglis' willingness to answer. His answers had a military directness—no *posse comitatus* obfuscation from him. This wasn't loose lips: U.S. doctrine was that mutual transparency in military structure and order-of-battle was strategically stabilizing. NSA leadership wanted nothing better than to have a Chinese NSA with exactly its same structure, so that they would know exactly who their opposite numbers were, and could communicate with them in time of crisis to avoid misunderstandings and escalations. Thanks to Track One "mil-mil" exchanges, contact information was already exchanged in some military realms other than cyber. Our Track Two exchange was trying to move the needle in the cyber realm.

Because of the twelve-hour time zone difference, late evening election returns were for us late morning. While pretending to listen to Chinese presentations on expected norms of cyber military operations—aimed purely at the U.S., because they didn't admit to having such operations—we watched our mobile phones with increasing horror. Although the Chinese "Great Firewall" blocked direct internet connections to most U.S. news sites, China Telecom tunneled the internet packets of foreign mobile phones directly to the foreign carrier—also maybe keeping a copy for themselves. I alternated between the CNN and New York Times sites. Around 11 a.m., Florida and North Carolina were called for Trump—contrary to pre-election polling. The Blue Wall fell, starting with Pennsylvania, during our elaborate lunch banquet in the hotel restaurant. At the afternoon session the Americans were ashen-faced. The Chinese were polite and triumphant. This was full circle from our 2013 meeting, at which they, the Chinese, were rendered mute by the Snowden revelations of U.S. information dominance.

Back home, the weekly PCAST leadership telephone meetings provided glimpses of the Obama-Trump transition through OSTP eyes. With their staffs, Holdren's OSTP associate directors (environment, national security, science, technology) prepared hand-off memos and filled thick loose-leaved briefing books with tutorial white papers, government org charts, and the names, job descriptions, and phone contacts of the career civil servants who would span the two administrations. Then, they waited for visits from their opposite numbers on the Trump team. Those contacts never came. The experience of the National Security Council staff was much the same, as was that at OMB. In the sense of any modern presidential campaign, there simply *was no* policy infrastructure on the Trump side. It was painfully clear that the Trump people had never really expected to win and had made no plans for the eventuality. By contrast, Clinton's campaign had dozens of policy committees covering all subjects, each ready to fission into formal transition teams (who would vet individuals for jobs) and the ambitious subset looking to be appointed. The number of "Schedule C" presidential appointments throughout the government—these below the level requiring Senate confirmation—was more than a thousand.

Apart from the small number of dog-whistle issues that Trump had campaigned on—anti-immigrant, anti-abortion, anti-regulation, and so on—there were no Trump policy advisors other than the Trump children (Donald Jr., Ivanka, and Eric) and, figuratively, the more thumb-worn cards in the Fox News rolodex. The Schedule C vacuum was so big, and decision-making in the Trump White House so dysfunctional, that it took the first half of Trump's term for the vacuum to fill—and when it did, it did so with opportunists, refugees from Fox News, Federalist Society zealots, and many pure loonies, including, sadly, a couple of my former colleagues. I always thought myself well-informed on American history. I could think of no parallel to this since Andrew Jackson's invasion of the White House. Jackson ended up being treated more favorably by history than he perhaps deserved. I couldn't imagine any such redemption for Donald Trump.

PCAST had its final meeting with President Obama in January, a couple of weeks before Inauguration Day. Some of us were crying. Our eight years of offering advice to the president now came down to his advising (that is, consoling) us. He must have had dozens of similar meetings with other groups, and had settled on a few comforting aphorisms. "Progress doesn't travel in a straight line. It zigs and zags in fits and starts," he reassured us. It sounded like a quote, and, when I Googled, I found that he had used this line in a speech four years earlier, when Democrats had failed to re-take the House. He was noncommittal to us about his future plans, nor did he say anything about maintaining any level of contact. Behind the aphorisms, he seemed *done*. And, reportedly, he was disgusted with the Clinton campaign's failure to have taken advice from either him or Bill Clinton—two of the greatest retail campaigners in recent history.

A few days after this meeting, we received individual emails from the White House Personnel Office telling us how and where to send in our resignation letters, which were due a week before the inauguration. We floated among ourselves the idea of not resigning. Legally, PCAST was chartered under the Federal Advisory Committee Act. That law was clear: FACA committee appointments didn't automatically expire at the end of an administration. We knew that lower-level committees at NSF and NASA were continuing. What, then, if we didn't resign and continued meeting as before? Would anyone in the Trump White House even notice? But, some spoilsport leaked the nascent (and not entirely serious) plan to OSTP, and we soon received a new email from the personnel office clarifying that, if we didn't resign, we would on January 19 be formally fired for cause—and thereby be ineligible for any future federal employment. Obama had decreed that there were to be no dirty tricks, by us or anyone.

After January 20, OSTP continued its statutory functions with a skeleton number of career civil servants. Ted Wackler, who had been Holdren's Chief-of-Staff, became acting director. Several self-organized groups of prominent scientists tried to find a contact point in the new administration to make the case for science—minimally that a science advisor to the president should be appointed. Someone knew someone who knew someone who thought they could get an audience with Jared Kushner, the president's son-in-law, but that didn't pan out. The only Trump advisor with any technical chops at all was Peter Thiel, a Silicon Valley billionaire (co-founder of PayPal) who supported libertarian political candidates and gave scholarships for bright kids *not* to attend college, but to drop out and try their luck as entrepreneurs. Thiel indeed suggested two candidates for science advisor, both extreme climate doubters: Will Happer (my JASON colleague, then approaching age eighty) and computer scientist David Gelernter, whose public statements included his view that women should stay home to care for their children. Luckily, neither got the job. For FDA administrator, Thiel's favorite candidate was on record saying that drug regulation did more harm than good.

Twenty months in, an OSTP director was finally appointed, Oklahoma Secretary of Science and Technology (who knew!) Kelvin Droegemeier, who was named at the instigation of Oklahoma senator Jim Inhofe. The science community reacted with relief, because Droegemeier was an actual, and mostly apolitical, scientist—and it could have been so much worse. But Droegemeier was never given the parallel appointment of science advisor to the president. Trump never had a science advisor. An anemic PCAST was reconstituted thirty-three months into the forty-eight-month term, mostly of business executives. Among the few scientists willing (or invited) to serve was my former

Harvard astronomy colleague Avi Loeb. Avi was so widely disapproved of anyway that this made hardly any difference to his reputation. Will Happer eventually took a short-lived position on the NSC staff. Even he came away disappointed. One night, Jeffrey asked me: Taking into account that, from the inside, one can work to mitigate harm, were there *any* circumstances under which I would serve in the Trump administration if asked. "None," I said. "Good," she said. "That's the right answer."

As the anti-science (indeed, anti-reason) agendas of the Trump administration came into focus, an increasing number of National Academy members wanted the NAS to denounce the administration's actions. A year into Trump's term, a thousand NAS members (about half the membership, and including me) signed a statement deploring U.S. withdrawal from the Paris climate accords and the denigration of science and harassment of scientists generally. But this was a statement by individuals, not an official NAS position.

Marcia McNutt was in a tight situation. NAS's charter, going back to President Lincoln's wartime need, was to advise the federal government—not just certain administrations that we liked better than others. In heated council meetings, it became clear that a majority of the NAS council wanted the NAS to be a kind of science government-in-exile, exposing the administration's hypocrisy in fiery public denunciation. Better to be a de Gaulle in London, the argument went, than to be a Pétain in Vichy. I was emotionally sympathetic to this group, but understood that the collateral damage to U.S. science of a war between Trump and the NAS would be "huge" (a favorite word in Trump's rhetoric). This was essentially also Marcia's position. When the five officers all backed Marcia, enough of the other twelve council members stepped back from the brink to avert a showdown.

In fact, NAS/NRC business was, under Trump, booming. Agencies were chartering NRC studies at a record clip, and many new studies were being mandated by Congress. It was another indication that the thin political layer of generally incompetent Trump appointees had little control over their own bureaucracies. Civil servants a few levels down from the top understood that strong NRC recommendations on issues would put their political masters in the position of looking bad if they routinely countermanded them. On the rare occasions that I watched Fox News (usually in hotel rooms), I felt satisfaction at their commentators' rants against the Deep State and its obstruction of the supposed will of the American people. I thanked God for the Deep State.

Two years into his term, Trump's appointees first sufficiently penetrated the Department of Defense that JASON became visible on their radar. The JASON Program Office at MITRE (JASON's administrative home) received an immediate stop-work order from a low-level Pentagon contracting officer. The letter said that DoD was considering more economical alternatives for getting programmatic advice, and that no further charges to the JASON contract would be honored. Obvious care had been taken to leave no fingerprints of the individual who actually ordered this action, and it took several days of phoning our friends in the Deep State to trace it to Under Secretary of Defense for Research and Engineering Michael Griffin, who had been confirmed in office two months earlier. Although Griffin had served in previous Republican administrations (as an unremarkable NASA head under George W. Bush, for example), he was not particularly doctrinaire, or indeed forceful in any way. Why, then, take sudden aim at JASON? It took more calls to trace the action further, to Griffin's deputy, Lisa Porter. Porter had now worked with Griffin as his deputy for nearly two decades, at NASA, at CIA's research arm In-Q-Tel, and now in DoD. Porter was the power behind the scenes, people told us. Griffin was the front man. It now all made sense, because Porter was a long-time JASON-hater. We never knew why—it must have been something in the distant past—but her enmity ran very deep. Now, when she saw a chance to kill us, she took it.

Except not successfully. Even this late in their term, Trump appointees had not managed to gain control of the quasi-independent National Nuclear Security Agency within the Department of Energy. Much of NNSA's professional staff came from Los Alamos and Livermore, and they held JASON in high regard. Lisa Gordon-Haggarty was the Trump administration appointee heading the agency. Her strong nuclear credentials and service in the Bush administration may have misled them. She was no Trumpite. She was Deep State—Livermore chapter—to the core. Breaking all records for the speed of government contracting, JASON was soon again alive and well under NNSA sponsorship. We died by the hand of one bureaucrat named Lisa, were resurrected by the hand of another.

So, there were a few small battles won. But, overall, the effect that Trumpism and resurgent, intolerant populism worldwide had on me and my cohort was spiritually devastating. In Cambridge for my Harvard undergraduate class's fiftieth reunion, I was put on a panel with the magisterial title, The State of Our Democracy. When my turn came, I grandly intoned that we should all consider the possibility that the world had come to the end of its four hundred years of the Enlightenment. What if fact-based decision-making was not the normal human condition? What if the natural state of government was authoritarian? The natural state of economies, monopolist kleptocracies? The natural state within and between nations, tribal conflict? What if the moral universe *didn't* bend toward justice? It was not the fall of the Roman Empire that we should be thinking about, I said, but the centuries-earlier collapse of the Roman Republic—the rise of permanent autocracy. That seemed to be where American civilization was headed. Sherry Turkle, a classmate also on our panel, in turn helped drive the audience to despair by talking about the commoditization of the individual by big-data high-tech. At the reception following, the bar ran out of hard liquor.

70. COVID and Beyond

I was in Washington on February 21, 2020, for an NAS-hosted Wikipedia "edit-a-thon," an organized group of volunteers who met to improve Wikipedia science content. I wasn't a regular member of the group, but I was already going to be in New York for a Simons Foundation board meeting the day before and could easily take the Acela train to DC. And, under several noms de plume, I had for years been a regular Wikipedia contributor.

I had gotten to know Jim Simons, the hedge fund billionaire, after becoming NAS treasurer. Simons had been elected to NAS membership in the mathematics section a few years before, deservedly for Chern-Simons theory and not for inventing the hedge fund; but his being a billionaire hadn't hurt him any. He had agreed to serve on its investment committee, which, two years later, I came to chair as treasurer.

The Simons Foundation, in New York City, had existed for twenty years as the vehicle for Jim and Marilyn Simons' personal philanthropy, supporting an ambitious program in autism research—their adult daughter Audrey was autistic—and also a mix of scientific and institutional awards. In the 2010s, the foundation began a conscious shift toward more permanent professionalism. Marilyn Simons' experience on a number of philanthropic boards had made her a proponent of what were thought of as best practices. Jim had less interest in best practices (other than his own), but tolerated Marilyn's with an affectionate deference. He did take an interest in formalities that would nail down the foundation's mission of supporting basic scientific research beyond his own lifetime. I was one of several scientists newly appointed to the Simons Foundation board, along with Shirley Tilghman and Peter Littlewood. At our first board meeting, in October, 2017, we learned that Jim and Marilyn had amended their binding donor letter to *require* that a majority of the board be scientists.

Getting from NYC to DC was, on this particular February occasion, unpleasant. On a Thursday, at rush hour, the Penn Station waiting room was packed. Freezing outside, the train station was overheated, and a sweaty crowd was standing shoulder to shoulder. The next day's Wiki session was less interesting than I had hoped, but at least I was there.

The next Monday, back home in Austin, I was in bed, sick with what seemed to be a bad cold, but which laid me out flat for the next three days—for me unusually severe. I was supposed to travel again that week, but I cancelled. COVID didn't seem like a possibility, because it was then only in the Seattle area (it was thought). And, I didn't have the symptoms that were being talked about—difficulty breathing, loss of sense of smell (anosmia), etc. The first confirmed case of COVID in New York City came a week later, on Leap Year Day, February 29. I got out of the Big Apple just in time, it seemed. Or did I?

The day the word "hoarding" first appeared in a newspaper article, we rushed to Costco and scored a several-month supply of toilet paper. UT students left for a spring break that was initially extended for a few days, then made permanent for the whole semester—spring classes would be continued only remotely using then-newfangled Zoom. By late March, the whole country—the whole world—was on lockdown. When the *New York Times* made its raw data of COVID cases, by county, available on Github, I plotted a weekly graph of the counties of interest to us: Travis in Texas, and also Dallas and Harris (Houston) for comparison; Los Alamos in New Mexico, but also Santa Fe and Bernalillo (Albuquerque), because, in NM, we might end up getting hospitalized in one of those.

Isolated at home, I tracked in the news and on less-reputable web sites the ups and downs of claimed COVID prophylactics and treatments, sorting them into three categories: (1) certainly wrong and probably harmful; (2) probably wrong but not harmful; and (3) possibly real but unproved. President Trump's public suggestion that people inject bleach solution into their veins—later walked back as "only kidding"—was the poster child in category one. But I did enthusiastically stock our household with every nostrum in categories two and three that I could get ahold of before supplies were exhausted—by crazy people like me. I classified high-dose vitamin D and alpha lipoic acid food supplements in category two. Pepcid (famotidine) acid reducer was in category three: Jim Simons was enough of a believer to fund an ongoing clinical trial in a Long Island hospital, which failed for lack of sufficient enrollment, not on its merits. I managed to score 50 grams of non-human-grade hydroxychloroquine at an outrageous price from a fly-by-night chemical supplier, but decided to treat it as category one—dangerous—the more so since I had no way to verify what was actually in the bottle. If I had fallen seriously sick in this early period, before vaccines or effective antivirals, I might have tried the Pepcid; I hope not the others, but people were increasingly desperate.

By June, 2020, my chart showed five times as many people (per capita) dying in Austin as in Los Alamos or Santa Fe. Austin hospitals were overflowing; the city was making plans to convert the Austin Convention Center into what was being politely called a field hospital—but really it would be no more than a large quarantine facility with beds and little else. There were still no effective medical treatments. If you got sick, your chance of dying was one or two percent. If you needed hospitalization, it was ten or twenty percent. If you were intubated on a respirator (only later understood as inappropriate treatment), it was fifty percent. Jeffrey and I packed our two cars and drove in caravan to New Mexico, stopping only for gas, about a twelve-hour drive.

We sheltered continuously in our house in Los Alamos until late fall. I largely worked on my autobiography, and participated remotely that summer in JASON's minimal, all-Zoom (and thus all unclassified) summer study. We drove back to Austin, again nonstop in caravan, when the mortality statistics had reversed in its favor. Our relief at Biden's narrow victory over Trump in November was indescribable.

By this time, it was clear that my February "cold" might have been COVID after all. It was now known that the virus was circulating in NYC during most of February, and crowded Penn Station would have been a perfect superspreader locus. Long COVID had become a recognized syndrome, and, over the summer, my sense of smell had gone seriously off kilter—not anosmia (loss of the sense), but *parosmia*, where I perceived many ordinarily good smells as, instead, a strong odor of sweat-socks, or as things not so unpleasant but just as weirdly wrong. There could be other causes of this than long COVID, but, pre-COVID, they were all very rare conditions. A serum antibody test for COVID was negative, but, after so many months, the test was not considered reliable. An ENT specialist found nothing to explain the parosmia and had little interest.

Normally, we would have traveled back to Los Alamos for the 2020-2021 winter holiday—Jeffrey driving via Lubbock with Lewis the dog, me flying. But my graphs of case statistics showed northern New Mexico now no safer than Travis County. Vaccine was being promised within weeks. We worried that as Texans (per our driver's licenses), we might not be able to get, in New Mexico, vaccine doses allocated to that state. So, we stayed in Austin through the winter. In early 2021, we were among the first to get vaccinated.

I was able to get the Pfizer mRNA vaccine through UT, both doses. Although distribution within Texas was chaotic, UT was able to secure a supply. Jeffrey had a harder time. At the dog park a block from our house, she learned of a "vaccine angel" in our neighborhood, a volunteer young person, isolated in lockdown, who trolled continuously the web sites of local clinics and pharmacies looking for available vaccine. There existed a self-organized network of such angels who shared information on the who-what-when-where of vaccine shipments, often by getting tips from other young people working at drug stores and distribution sites. There existed no more formal way to find a shot: Coordination at the county, state, and federal levels was completely broken. Our local angel eventually got Jeffrey an appointment at a Walmart ten miles away. I had zero reaction to the two vaccine doses, or, later, to the booster shot—not even a sore arm—another indication that I might already have had the disease as one of the first—or maybe not. I'll never know.

For the one-year-plus of the pandemic's complete disruption of life, my human contacts populated small squares on Zoom video screens. These were sometimes a source of global intelligence. Early in the lockdown, as the ineffectiveness of the then-Trump administration's response became obvious, John Holdren organized a subset of us former Obama PCAST members for an unofficial PCAST-like role: Chris Cassel, Chris Chyba, Susan Graham, Harold Varmus, Eric Lander, Rick Levin, Maxine Savitz, Ed Penhoet, me. On individual topics—the national PPE stockpile, the role of contact tracing, obstacles to public health data flow, the state of computer modeling, and so forth—we interviewed (by Zoom) national and international experts, came to consensus recommendations, and wrote white papers.

I was amazed at how everyone we contacted wanted to talk to us. It wasn't that we had any actual authority—there were at least a dozen self-organized groups like ours with comparable claims to legitimacy. The phenomenon was that *everyone* we contacted wanted to contribute *somehow*, and, given the vacuums in Washington (the White House) and Atlanta (CDC), no one knew exactly where the fulcrum points were, or where the loci of effectiveness might turn out to be. So, everyone wanted to talk to everyone. It was chaos trying to self-organize—encouraging, but not efficient.

Our group put up a web site. (Of course I was the webmaster.) We sent our papers individually to Holdren's private email list of powerful (or once-powerful) people—House and Senate members, present and former federal officials, media editors and columnists, state government leaders, academicians, and so forth. In various subsets of two or three, we published op-ed pieces. Chris Cassel, Sue Graham, and I had an editorial on "data" in *Modern Healthcare*, a widely read trade paper; Rick

Levin and I had one on epidemiological modeling in *Science*. All of us signed onto a policy paper drafted by Chris Chyba on the coming need for a formal COVID commission; this was published in *Science* after a year's delay. Perhaps our efforts had a beneficial effect. It is hard to know. The effort had a beneficial effect on *us*, at any rate. We old PCAST friends enjoyed getting together on Zoom every week. A sense of camaraderie helped us individually through the pandemic's darkest time. Interestingly, none of us actually caught COVID during this whole period, though we all knew people who had, and knew or knew *of* people who had died. That anecdotal fact was roughly consistent, numerically, with the pandemic's killing about one in three hundred Americans.

A completely different Zoom group, though with a similar sociology and mode of operation, was organized by Bob Conn, then retiring as president of Kavli Foundation. Conn's targets, much more specifically than OPCAST, were the various science bills being introduced in Congress. Multiple Congressional offices—the most important being majority leader Senator Chuck Schumer's—had concluded that a pandemic was the right time to garner public support for a big increase in federal science funding—and not just health related. Beyond that idea, there was little consensus, however. Should the National Science Foundation be steered into more applied research, less basic? Should NIH have a new DARPA-like organization for less-conventional health research? Should we invest in semiconductors and materials, the better to compete with China? Or build quantum computers? Or create new regional research centers—anywhere except California and Massachusetts—and in what areas of technology?

Conn assembled a high-profile group that could advertise itself as honest brokers on Capitol Hill: David Baltimore, Mary Sue Coleman, Tom Rosenbaum, Shirley Tilghman, Harold Varmus. Most were current or ex- university presidents. David Spergel and I got added when Conn learned that Jim Simons was close to Schumer and spoke to him regularly. This so-called Conn Salon met weekly by Zoom, and had Zoom calls with staffers from one or more Congressional offices, and with carefully matched Democratic and Republican staff from House and Senate committees. The group's work product, not in writing, was the patient, careful explanation to non-scientist staff of the implications, good and bad, of their proposed bills' provisions. I was often reminded of the old adage that legislation was a very blunt instrument and hard to get right. The Conn Salon had less camaraderie than OPCAST, but I enjoyed seeing Harold Varmus (in both groups) and Shirley Tilghman, whom I liked a lot, regularly.

By mid-2021, it seemed that things were inching back to a new definition of normal. JASON held an in-person summer study in La Jolla—ordinary-seeming, except that vaccinations were required, and we wore masks and did antigen self-tests at least twice a week. I started traveling again to NAS and Simons activities. There was one Simons event that Jeffrey and I would not have missed, even with the continuing COVID risk: David Spergel was formally installed as the president of the Simons Foundation. Jim and Marilyn remained on the board, but were stepping back from day-to-day foundation affairs. David was going to run the show. Shirley and I had known for a year that this was in the works—known from David in confidence. David had been rumored to be a candidate to succeed Shirley as president of Princeton, but didn't get that job. Simons Foundation president was, in a way, a consolation prize, less visible and prestigious, but a lot more fun.

Marilyn and Jim viewed David with not just with deserved professional respect; they seemed to love him as a kind of late-adopted son. Except for Shirley and me, the rest of the Simons Foundation board was taken by surprise when Jim announced the succession. A long-serving board member, who may have had his own ambitions for the position, was brave enough to ask, "Shouldn't we engage a search firm and have an international search for so important a position before taking a vote?" "I don't think so," Jim said. The vote in favor was unanimous.

* * *

Soon after the U.S. presidential election, Eric Lander called me with the news that he was going to be President Biden's science advisor, a fact to be kept secret. Eric invited me to send him a list of suggested PCAST co-chairs and members. I submitted a list of twenty excellent women—and no men. I wanted to make the point, as unsubtly as possible, that Eric's personal reputation put him on shaky ground. The influential #MeToo movement was calling out men for misogynistic behaviors. I tended to think of Eric as an Equal Opportunity bully, not biased toward women. But his *ad hominem* attacks on Jennifer Doudna in their institutions' CRISPR patent dispute were noticed, and not just by scientists. There were said to be other skeletons in his closet. Deserved or not, the liabilities were there. I included Doudna on my proposed PCAST co-chair list, with a prickly comment, "Nobelist, whether you approve or not." Including Doudna, I added, might be seen by the scientific community as an indication of his willingness to put scientific and personal differences aside in his new position. "You are more controversial than you might think, and this would help," I added. He wrote back, thanking me for my comments.

In January, President Biden did announce the nomination of Eric Lander as his science advisor. Also announced was the choice of two PCAST co-chairs, Nobel Laureate Frances Arnold and MIT vice president Maria Zuber, both near the top of my list.

Frances, by now a friend of some thirty years, called me a few weeks after that to invite me to be a PCAST member as (apart from Eric) the one renomination of an Obama PCAST member. A mark of how overloaded the Biden White House was with the pandemic and with cleaning up after Trump, it was September before the public announcement was made. In the press release, I was just another rank-and-file member, a professor in Texas somewhere, alphabetically under P. Still, my appointment was comforting evidence that I hadn't yet become entirely irrelevant. In the wake of COVID's isolation, that had some significance.

Lander resigned, said to have been fired by Biden, the following February. His bullying and mistreatment of subordinates, especially women, was stated as a reason. I was not too surprised. Mentally, I listed the ways that I was similar to and different from Eric. Both of us were (in the contemporary idiom) male, white, and privileged. He was a genius. I wasn't. But perhaps I was quicker to recognize an overdue societal reckoning. In that case, I thought I might still have a future.

Index

www.ingramcontent.com/pod-product-compliance
Lightning Source LLC
LaVergne TN
LVHW010339220926
848484LV00004BB/931
* 9 7 8 9 8 1 9 8 0 1 7 6 3 *